T0311969

# Measuring Capacity to Care Using Nursing Data

# Measuring Capacity to Care Using Nursing Data

Evelyn J.S. Hovenga RN PhD
FAIDH FACS FACN FIAHSI

*CEO and Director, eHealth Education Pty Ltd, East Melbourne, VIC, Australia, Registered Training Provider No32279*
*Managing Director, the Open eHealth Collaborative Trust, trading as the Global eHealth Collaborative (GeHCo), East Melbourne, VIC, Australia*
*Retired Professor, Health Informatics, East Melbourne, VIC, Australia*

Cherrie Lowe RN RM Dip Teaching (Nursing), PG Dip Hospital Admin, AFACHM MACN

*CEO and Director, Trend Care Systems Pty Ltd, Brisbane, QLD, Australia*
*CEO and Director, Trend Care Systems UK Ltd, Manchester, United Kingdom*
*Founder and designer of the TrendCare product*

**ACADEMIC PRESS**
An imprint of Elsevier

ELSEVIER

Academic Press is an imprint of Elsevier
125 London Wall, London EC2Y 5AS, United Kingdom
525 B Street, Suite 1650, San Diego, CA 92101, United States
50 Hampshire Street, 5th Floor, Cambridge, MA 02139, United States
The Boulevard, Langford Lane, Kidlington, Oxford OX5 1GB, United Kingdom

**Notices**
Knowledge and best practice in this field are constantly changing. As new research and experience broaden our understanding, changes in research methods, professional practices, or medical treatment may become necessary.

Practitioners and researchers must always rely on their own experience and knowledge in evaluating and using any information, methods, compounds, or experiments described herein. In using such information or methods they should be mindful of their own safety and the safety of others, including parties for whom they have a professional responsibility.

To the fullest extent of the law, neither the Publisher nor the authors, contributors, or editors, assume any liability for any injury and/or damage to persons or property as a matter of products liability, negligence or otherwise, or from any use or operation of any methods, products, instructions, or ideas contained in the material herein.

**Library of Congress Cataloging-in-Publication Data**
A catalog record for this book is available from the Library of Congress

**British Library Cataloguing-in-Publication Data**
A catalogue record for this book is available from the British Library

ISBN 978-0-12-816977-3

For information on all Academic Press publications
visit our website at https://www.elsevier.com/books-and-journals

*Publisher:* Stacy Masucci
*Acquisitions Editor:* Rafael Teixeira
*Editorial Project Manager:* Sam Young
*Production Project Manager:* Kiruthika Govindaraju
*Cover Designer:* Mark Rogers

Typeset by SPi Global, India

# Contents

# About the authors

**Evelyn J.S. Hovenga**

Evelyn J.S. Hovenga brings global experience, strong scientific and professional leadership in Health and Nursing Informatics, she is widely published. Her experience and expertise covers many factors of health and nursing informatics, especially standards development pertaining to health terminology and electronic health records, including knowledge management, ontology and semantic interoperability. Evelyn undertook all her graduate studies whilst working full-time as a divorced mother of two daughters, and was awarded a doctorate in Health Administration from the University of New South Wales.

Her healthcare industry background began as a Nurse in paediatrics, obstetrics, general medical and surgical, organ imaging and ended as an operating room suite unit manager. This was followed by a new appointment as a Health Service Management (Workstudy) Consultant for the Victorian Government Health Commission, where her career progressed to her taking on the role of senior nursing advisor and researcher for a Ministerial enquiry into nursing in that State. Evelyn developed a nursing acuity system in response to a need to address major nursing industrial unrest over workload, nursing career structures and pay. This acuity system was in use by over 100 hospitals in three Australian States for close to fifteen years. It was outlawed in Victoria by the then Minister of Health, as the information made available to nurses was used by them to manage their workloads. He considered this to be in conflict with his new policy to reduce surgical waiting lists. Evelyn then worked privately as a consultant and undertook numerous additional patient acuity studies and nursing productivity reviews in Victoria and other Australian States.

Evelyn met up with her co-author, Cherrie Lowe when undertaking major research for the Private Hospitals Association in Victoria and Queensland, where Cherrie was a Director of Nursing of a private hospital. They recognised their shared vision and entrepreneurial mindsets while working together on this project. The research project's aim was to establish a nursing career structure.

Evelyn accepted an invitation from the Director of Nursing of the London Hospital, Maureen Scholes, who had established a Nursing Informatics working group for the International Medical Informatics Association, to represent Australian nurses internationally as a member of that working group. This provided her with an international network of other nursing informatics researchers, with opportunities to collaborate with them. She was elected to Chair this group following her successful hosting of the International Nursing Informatics conference in Melbourne (NI'91). Profits made were used to establish the Health Informatics Society of Australia, of which Evelyn is a founding and lifelong member. She is a founding Fellow and life member of the Australasian College of Health Informatics, and the International Academy of Health Sciences Informatics. She was awarded fellowships by the Australian College and Nursing, the Australian Computer Society and the Australian College of Health Executives.

Evelyn again diverted from her career path by accepting a University appointment to develop post graduate programs in health informatics and administration, and to establish a research centre. Evelyn became a founding member of a newly established Health Informatics Standards Development Committee (IT/14) by Standards Australia. She has participated in many standards development activities as a volunteer for Standards Australia, ISO TC215 and HL7 Australia, and continues to work with the openEHR foundation at the University College of London. One of her Post Doctoral fellows, Dr Sebastian Garde established the first ontology based online clinical knowledge repository under her guidance. Evelyn was an external advisor for the EU funded NIGHTINGALE, TELENursing and ICNP development projects during the 1990s and has witnessed many new technical advances during her career. She retired as a full professor in 2007 and is continuing her work in her current positions.

## Cherrie Lowe

Cherrie Lowe is a registered nurse, midwife, an innovator and business manager, who brings local, national and international health service executive management, research, software development and system implementation experiences. Her health industry experience includes past roles as a Nurse Educator, Quality Manager, Director of Nursing, Director of Clinical Services, hospital accreditation surveyor and medico-legal expert witness.

Her executive level industry experience began as a Director of Nursing for Mercy Health and Aged Care where she maintained an efficient nursing service and improved the hospital's profit margin by making use of her patient acuity system. Cherrie initiated the development of a hospital promotion campaign, which included a television video that significantly increased the hospital's bed occupancy. The success of this campaign achieved the Australian Council on Healthcare Standards (ACHS) National Quality Award for large hospitals.

As Director of Clinical Services for Ramsay Health Care Cherrie played a major role in managing the transition of a large Commonwealth funded veteran hospital to Australia's largest private teaching hospital. Here she developed a strong, efficient and dynamic nursing service and allied health team and assisted in the expansion of clinical services, including: Cardiac, Gynaecology and Neurosurgery. She again achieved the ACHS National Quality Award for large hospitals and the hospital was also awarded the Employer of the Year Award for large organizations in Brisbane. Cherrie was again responsible for generating a significant profit margin for that organisation by maintaining a high level of efficiency in clinical services, an achievement made possible through the use of her patient acuity system.

During her years as a nurse executive, Cherrie managed her family, undertook her post graduate studies as an external student, was a surveyor for the Australian Health Care Council, developed, tested and made use of a patient acuity system, and undertook various consultancies. She partnered in business with a software developer and her system was fully computerized taking advantage of ongoing technical developments. Cherrie shared her research findings with other Directors of Nursing who then worked with her by facilitating ongoing research and development activities in their facilities. This research was presented at a world informatics conference in San Antonio in 1994. During the mid 1990s both Cherrie's and Evelyn's patient acuity systems were used by numerous Queensland hospitals. The Queensland Government funded a validation study enabling a comparison to be made between these two systems using the same patient populations which validated both systems, as the use of their systems provided comparable results.

The success of Cherrie's automated and highly interoperable TrendCare system enabled her to assume the CEO, researcher and developer role for Trend Care Systems Pty Ltd on a full-time basis. Her primary focus has always been to take on the many ensuing challenges to benefit the nursing and midwifery professions. In 1997 Cherrie received a Nursing Excellence award from the Royal College of Nursing for her contribution to nursing in Australia.

Developing and continuously improving the reliability of an evidence based patient acuity and workload management system for nursing and midwifery has been a challenging undertaking, and during the past 25 years Cherrie has had to overcome many barriers. These include (1) convincing nursing and midwifery leaders, colleges and unions that nursing services need to collect and present their own evidence of nursing demand in order for nursing services to be adequately resourced, (2) convincing health service senior executives, including CEO's, finance managers and chief information officers of the methodologies that are best suited to measuring nursing demand, and the value of nursing demand measurements for effective budget management and accurate costings of episodes of care, (3) Convincing nurses and midwives of the importance of collecting nursing and midwifery data, so that safe staffing and fair workloads can be a reality. These barriers have been overcome in some countries but are still ongoing in others.

Developing a viable small business, while trying to provide an affordable software product to health services that are financially stretched, has tested Cherrie's business skills. Transforming a small local business to an international business with a customer footprint across six countries in the health care environment is testament to her determination, commitment and sound business strategies.

Cherrie has won the AustCham Business Award in Singapore, the Australian national and state Microsoft eHealth iAwards for innovation in IT development and the Australian national ICT exporter of the Year Award.

# Preface

*The Need for a Better Understanding of Contributions Made by the Nursing and Midwifery Professions to the Health of any Nation.*

This book seeks to meet the need for a better understanding of the many and varied very significant contributions made by the nurses and midwives within any country's health system. The nursing and midwifery workforce is three times as large as that of any other health professional group. It is viewed by many as the largest health cost burden rather than for being responsible for the most significant cost savings to the health industry by enabling many individuals to continue to contribute productively to the nation's overall economic status.

We are passionate advocates for the nursing, midwifery and informatics professions and have spent a lifetime endeavoring to make our profession more visible to decision makers. We have fought many battles and encountered many barriers. Despite the many solutions developed over the years, we continue to witness a very poor understanding of the nursing and midwifery human resources' needs relative to meeting health service demands. Nurses and midwives deserve to be provided with the capacity to care for our fellow human beings during their most intimate and often traumatic life events. That represents a highly valued investment for the population at large. Nurses and midwives, most of whom are women, continue to be compelled to fight and engage in industrial action to get a fair deal. This needs to change.

This book is addressed to a wide readership, including policy makers, health service executives, managers in healthcare settings, nurses and midwives, health IT, health information and ICT professional, researchers and educators. It is intended to contribute knowledge and provide an argument for why nursing and midwifery data need to be a focal point for identifying evidence of practice and facilitating health care service planning and resource management at all levels within any organization, region, nation and the global health ecosystem.

The book's content is based on extensive literary scoping reviews plus the authors' real life past and current local, national and international work and research experiences. The many complexities associated with nursing and midwifery work and their human resource management are decomposed. Many tried and tested solutions to address identified issues are provided together with detailed descriptions of the underpinning research and development activities we have undertaken. This includes some case studies from public and private sector, acute, sub-acute and long term care healthcare organizations. Our findings have an unequalled large evidence base.

We provide detailed insights into the complexities associated with any health system to improve future decision making by using digital transformation. It is our desire to influence our readers so that they will be in a better position to negotiate, innovate and find suitable solutions to issues they encounter in the healthcare environment. We have identified numerous impacts of past high-level decisions made and associated lessons learned, so that those in highly influential positions will be able to improve their future decision making.

Attaining the capacity to care requires the inclusion of nursing data and the effective use of routine operational data collected at any point of care. Readers are exposed to evidence provided from numerous systematic reviews and real-world case studies from multiple countries to provide them with new knowledge. Our analysis of the many issues identified has shown a need for improvements in clinical data management, digital transformation implementation strategies, greater collaboration and a stronger focus on working effectively using multidisciplinary teams, and to make use of small teams for direct care delivery in the acute sector.

We have provided evidence based solutions for the adoption of safe nurse staffing principles and optimum use of operational data enabling the implementation of health service delivery strategies that will result in improved patient and organizational outcomes. Our recommended solutions provide organizational transparency enabling every health worker to view information that enables them to effectively contribute as a team member to achieve optimal patient and organizational outcomes. Readers can learn to make better use of informatics to collect, share, link and process data collected at the source for the purpose of providing real time information to decision makers at every level of the organization, enabling optimum use of available human and other resources required to meet health service demands at any point in time and place.

## Organization of the book

The first two chapters provide context and a big picture view by describing current dynamic health care environments and explaining the concept of the capacity to care. This is presented within the context of any national health system's framework and its primary building blocks that determine goals and outcomes. Our focus is on achieving efficient operational processes in order to optimize our capacity to care. This requires us to examine and decompose input, process and output factors together with the metadata needs for measuring operational efficiency and effectiveness.

The next four chapters provide an overview of current data use to determine nursing workloads and analyses four different nurse staffing methodologies. This reveals significant variations and limitations leading to the need to identify the multitude of variables known to influence service demand and to develop an agreed standard nursing metadata set that is well suited to maximizing desired outcomes from any digital transformation. This is followed by an analysis of nursing work measurement methods and an assessment of their construct validity and use for benchmarking. As nurses and midwives constitute the largest group of health professionals delivering care it is important to consider how best to match not only numbers but also skills, knowledge and competencies within this workforce with service demands. This analysis identifies the many difficulties encountered in our attempts to make this happen.

Whilst acknowledging the global shortages of nurses and midwives, we provide suggestions for the re-engineering of clinical service delivery methods by implementing the appropriate use of non-nursing support staff. This section ends with an examination of nursing/midwifery professional models of care, including care plans and organizational models that focus on work allocation and distribution. Possible linkages between nursing documentation, electronic care plans and electronic patient records are explored to identify how best to improve data collection and data use efficiencies whilst supporting the effectiveness of services delivered.

After achieving a better understanding of how to measure service demand, there is a need to explore how to manage human resources to meet these demands in a manner that enables the provision of quality care, facilitates the best possible use of the available skill mix at any time in any location and be cost effective. Rostering staff to cover a 24 hour service is challenging. Detailed instructions for roster development and re-engineering are provided. This is followed by an examination of workforce planning to ensure the supply continues to meet service demand needs over time. We again found many shortcomings in current practice. We provide suggestions for meeting future service demands.

Throughout the book we identify numerous data collection, data use, information and communication flow limitations. These are best resolved by making better use of available digital technologies. Despite an extensive body of work associated with health (medical, clinical and nursing) informatics, we were able to identify only relatively small pockets of success. We understood the need for system connectivity many years ago and have been working as participants in multidisciplinary teams to ensure that interoperability schema adopted were able to accurately meet functional needs. Chapter 9 explains the need to understand the relationships between data characteristics and attributes, and how these are technically processed whilst retaining meaning and accuracy.

Health care is global. Clinical trials and medical advances are shared globally. There is a need to consider how healthcare providers are best able to share, link, aggregate and compare data, to attain reliable and accurate big data sources and facilitate data analytics to benefit global health. Doing so ensures that local needs are met in a timely and sustainable manner. Key technical features required to optimize the results of local and national digital transformation strategies, including system connectivity, interoperability, health data exchanges and secondary data use are provided.

The chapter that follows focuses on how best to manage local digital transformations in a manner that ensures the nursing data are incorporated and able to be used to suit multiple purposes. Considerable detail is provided about how to overcome change management barriers and how to go about designing new work processes. This includes a generic example of a nursing acuity system implementation process. This chapter ends with guidelines for a system implementation evaluation and the provision of an evaluation framework. This leads in to methods of measuring the quality of services provided and the need for adopting meaningful measurements, trend analysis, continuous monitoring and evidence-based practice. The benefits of including nursing data is explained and outcomes research methods currently in use are explored.

Chapter 12 applies all topics discussed in previous chapters to the residential long-term care and community care sectors to identify significant differences. This is of paramount importance given the growing ageing population and an expected growing increase in service demand. Patient acuity is also valid in this sector and needs to incorporate lifestyle support needs. The primary differences between acute and residential long-term care are the funding mechanisms which in turn determine data collection and reporting requirements, without a focus of supporting the capacity to provide holistic day to day care.

The final chapter summarizes findings and recommendations by looking to the future highlighting the unique contributions made by nurses and midwives and initiatives underway to provide greater public visibility. A detailed list of benefits to be obtained from using nursing data is provided along with descriptions of cutting-edge digital transformation activities as examples for others to follow. Two case studies detailing the authors patient acuity research studies are provided as appendices.

**Evelyn J.S. Hovenga**

**Cherrie Lowe**

# *Acknowledgments*

This book is dedicated to the thousands of nurses and midwives who have contributed by their data collection efforts for the many research studies we have independently undertaken in numerous hospitals, and aged care facilities located in six countries. These studies would not have been possible without the support of the many senior health service executives who believed in us. We are indebted to our co-workers, staff and our external national and international collegial networks who have contributed to the knowledge gained, challenged us throughout our journeys, and accommodated and rewarded us for our relenting persistence towards realizing our vision.

A special acknowledgement to the international health and nursing informatics communities who have supported us, and the many technical standards development experts and researchers from multiple disciplines participating in the many stimulating cutting-edge technical discussions aimed at developing solutions, have collectively contributed to new knowledge gained by sharing practical experiences.

We also thank our family and friends who have cared for us, supported and encouraged us when times were tough.

**Evelyn J.S. Hovenga**
**Cherrie Lowe**

# CHAPTER 1

# Dynamic health care environments

## How is capacity to care defined?

The term "capacity" refers to one's ability to successfully undertake any type of activity to achieve a desired objective or outcome, including the ability to apply a degree of competence associated with any physical or cognitive activity. Capacity also refers to a quantity of things that can produce or deliver required objects or services. Within the healthcare service industry there is an emphasis on building workforce capacity in terms of numbers and skill mix. Capacity building may be defined as:

> promoting an environment that increases the potential of individuals, organisations and communities to receive and possess knowledge and skills as well as to become qualified in planning, developing, implementing and sustaining health related activities according to changing or emerging needs [1].

Crisp et al. [2] identified four capacity building strategies, a bottom-up organizational approach, a top-down organizational approach, the use of partnerships, and a community organizing approach. In essence this is about building social capital. We're interested in measuring the nursing and midwifery capacity to care, where caring is the desired outcome measure and in building capacity among those who need to plan, develop, implement and sustain nursing and midwifery service delivery.

Caring processes may be referred to as personalization, participation and responsiveness as applied when meeting a person's health and care needs while making them feel "cared for" [3]. These three concepts were defined following extensive research by Strachan as follows:

- *Personalization is the degree to which the healthcare team gets to know the person. This includes those interpersonal behaviors that demonstrate: connecting, knowing and empathizing.*
- *Participation is the degree to which the healthcare team respects the involvement of the person, and those close to them, in their healthcare. This includes those interpersonal behaviors that demonstrate: involving, goal setting and sharing decisions.*
- *Responsiveness is the degree to which the healthcare team monitors and responds to the person's health & care needs. This includes those interpersonal behaviors that demonstrate: being attentive, anticipating and reciprocity.*

Measuring Capacity To Care Using Nursing Data. https://doi.org/10.1016/B978-0-12-816977-3.00001-0

A capacity to care requires sufficient human resources with the appropriate knowledge and skills to achieve these desired outcomes when and wherever health services are provided. Nurses and midwives are at the center. The data and information collected and used by nurses and midwives are fundamental to our ability to measure our collective capacity to care. This group of health professionals apply their scientific knowledge and skills, as members of multidisciplinary teams of health professionals, supported by lesser qualified staff within a large variety of health care environments.

Balancing the many factors contributing to any nation's health system's capacity to care is very challenging, given the continuing significant changes in the world around them such as workforce availability, technology changes and increasing service demands. Health systems and organizations need to be adaptive. Within the international medical informatics community it is an accepted fact that sustainable health systems require successful implementations of future proof digital technologies within every healthcare organization delivering services [4]. This is also required to enable us to measure our capacity to care. Sustainable information systems need to be semantically interoperable to realize operational effectiveness and efficiencies through the retention of meaning (context) despite electronic data transfers and processing. This requires the linking of data elements to standard terminologies and associated ontologies as a foundation and capability of machine processing. Semantic interoperability, and its significance in terms of resulting system functionality, potential return on investments, data integrity for decision support, ability to aggregate valid data for public health use and practice evaluation, appear not to be well understood by key decision makers and many software vendors.

## Healthcare environments

Healthcare environments can be described from any one of many different perspectives, such as financial, organizational, industrial, behavioral, philosophical, physical (healthcare building design) or population health trends. Health systems overall are influenced not only by service demand but also by external factors such as Government policies and legislation. Recipients of health services tend to view their healthcare environment from the perspective of access to services, service effectiveness and their experiences related to caring and welcoming aspects. Health service providers are likely to view their healthcare environment in terms of location, organizational facilities and culture, type and amount of service demand, resource availability, available support services, equipment, supplies and technologies, research opportunities, or environmental factors influencing health outcomes. From a workforce perspective, the adoption of environmental standards pertaining to the supply of workers, labeling, work protocols and procedures can contribute to improved patient safety.

The literature on healthcare environments primarily considers physical environments in terms of interior design, color schemes, acoustics, lighting, space usage, room and unit configuration, ventilation, environmental hygiene or links between internal and external environments [5].

The design of healthcare facilities is ideally influenced by a desire to design "healing environments." Healthcare facility design must consider the needs and cultural preferences of the patient, family and staff. In addition designs need to meet various safety requirements to reduce opportunities for infection transmission, visitor, patient and staff injuries, enhance workflow patterns and processes, and minimize cleaning, building maintenance and heating/cooling costs.

In recognition of available evidence regarding the critical role of nurses in patient safety, the Institute of Medicine was asked to undertake a study some years ago to identify the key aspects of the work environment for nurses likely to have an impact on patient safety [6]. This study found evidence indicating that organizational management practices, workforce deployment practices, work design and organizational culture, that collectively make up the nurses' work environment, all contributed to many serious threats to patient safety. Nursing and midwifery working environments are covered in some depth in Chapters 6 and 11.

## What influences the capacity to care?

This book's view of the health care environment is from a workforce capacity to care perspective. Such capacity is influenced not only by the knowledge, skills and numbers of staff that make up the available workforce, but also by all the nuances identified above, that collectively make up their working environment. Many of these factors are constantly changing based on service demands at any point in time. Changes occurring at other levels within the health industry may also have an impact at the point of care and/or influence service demand. Collectively these factors create a dynamic work environment for those directly or indirectly engaged in meeting health service demands. It is imperative that individuals making up the health service workforce have sound contextual knowledge of their dynamic working environment.

In situations where any of these supporting environments have deficits, health professionals tend to innovate and problem solve in an effort to minimize the impact on patient care. Their actions and behaviors directly address the actual survival of their patients/customers. This makes getting it right the first time imperative. Health service delivery is very much dependent upon collaborative teamwork. Individual team members depend on effective communication and information flows.

Fundamental to the delivery of health services is access to the right information and ease of access to this information in a timely manner. Information guides actions and assists in decision making. The absence of the right information is likely to cause delays in workflows, which in turn can result in extended periods of patient discomfort or adverse patient outcomes. The overall efficiency and effectiveness of health services actually delivered (productivity) is very much dependent upon service co-ordination, communication and information transfer strategies adopted.

Society and healthcare delivery systems have, and continue to experience major changes resulting from our ability to generate more data, and to transfer more information faster to more people at any one time than ever before. As a consequence, people's expectations are changing and the health industry, being information and knowledge intensive, is well suited to maximize the benefits of the new digital world. There is an urgent need for the health workforce as a whole to think differently about our communication methods, data and information flows, data collections and the way our key data assets are managed and made available for use.

Technically it is possible to almost simultaneously collect, aggregate and process lots of data about all possible confounding variables associated with any specific health issue and gain new insight regarding the best possible treatment or care options in a very short space of time. It's about our ability to collect practice based evidence. We need to be able to collect every bit of data once at the point of encounter, and use it many times to suit multiple purposes. Health system performance can be measured in a variety of ways. How should accountability boundaries be described? Murray and Frenk [7] argue that *"it is unfair to hold health systems accountable for things that are not completely under their control; and health systems can achieve greatest impact through influencing non-health system determinants of health."* The latter is what contributes significantly to health service demand that in turn influences the capacity to care. They have developed a framework for health system performance measurement on the basis of:

1. The levels to which health system goals have been attained irrespective of the reasons that explain the results.
2. A country's control of the level of non-health system determinants through effective intersectoral action such as tobacco smoking, safety requirements such as helmets for motorbike riders, and road safety measures.
3. A narrower scope of accountability that refers to sub-systems and institutions within an overall national health system architecture.

According to the World Health Organisation (WHO) [8], health system performance may be viewed according to Tanahashi's model of successive cascading levels, each dependent on the previous level. From the target population perspective each of these levels need to perform optimally. The highest best performing level that influences all other levels is accountability and coverage, this is followed by the supply level, required to meet demands, then quality. The least well performing level globally is financial coverage. Each of these cascading levels has great potential for performance improvements through the use of digital health interventions. The latter is addressed in Chapter 9.

The Australian Institute of Health and Welfare has developed a conceptual framework against which to understand and evaluate the health of Australians and the Australian health system. It has 14 health dimensions grouped under three domains; health status, determinants of health and health system performance [9]. The latter has six categories that collectively indicate effectiveness in terms of relevance to client needs, accessibility in terms of universal access, continuity of care, responsiveness, safety, efficiency and sustainability in terms of achieving the desired results with the most effective use of resources and health system capacity.

The WHO has developed a Health System framework [10] consisting of the following six building blocks that can be adopted to measure the overall performance of national health systems:

- Leadership and governance (accountability and coverage)
- Healthcare financing (financial coverage)
- Health workforce (supply)
- Medical products, devices and technologies (supply)
- Health service delivery and (meeting demand)
- Information and research (quality).

## Leadership and governance

Any healthcare workforce consists of many who are among the most highly educated within a service industry. Relations between health professionals directly responsible for the provision of clinical services are built on trust and collaboration. These complex interrelationships are further influenced by those occupying positions of organizational power. Leadership is provided in various ways by different people within any one organization. This may be provided by those with formal positional authority or by those who are recognized for their specific area of expertize or professional standing or personal attributes.

Governance is about ensuring compliance with legal, regulatory, professional, ethical, policy and procedural requirements. Organizational behaviors, applied leadership and governance strategies at any level within a national health system or healthcare facility, influence healthcare delivery environments, patient safety and the ability to optimize the capacity to care.

## Healthcare financing

The funding of health services is a fundamental requirement. Financial capacity determines which services can or cannot be accessed or provided by location or by type of healthcare facility or service. Any nation's health budget is a large component of its Gross Domestic Product (GDP). This ranges considerably between nations averaging at around 10%. There are also significant variations regarding the distribution of health funds between primary care and hospital services. In many instances the provision of hospital services represents the most costly component of any nation's health budget. Healthcare funding is subject to any number of national and local policy initiatives. Healthcare financing influences the demand for service and determines who needs to pay for the service, their capacity to pay and the urgency of the healthcare need.

Nursing budgets tend to be the largest component of any hospital budget. As a consequence, financial managers often consider this budget to be an easy target to use to prop up other departmental budgets. In our experience nursing directors are not always provided with the most accurate information about their budgets by their co-executives, they tend to have to "make do" with their budget allocation. The number of staff, skill mix and salaries paid, plus on costs including leave arrangements and penalty rates for over time or working unsociable hours, relative to service demands, ultimately determine to a large extent any healthcare facility's capacity to care. It is therefore imperative to manage these resources in a manner that optimizes the match between service demand and resource allocation. A number of chapters in this publication focus on this complexity by exploring the many variables that influence staff resource allocations and utilization.

## Health workforce

The health workforce is large and diverse covering many occupations, ranging from support staff to highly qualified professionals. The availability of health workers with the necessary knowledge and skill mix in appropriate numbers where and when needed is fundamental to any nation's health outcomes. Costs associated with contractual employment agreements or regulatory working conditions that apply to specified groups of employees or occupations within the healthcare sector, is influenced by industrial and political activities undertaken by various professional organizations and unions who represent various categories of staff or contractors. This can provide limitations or constraints for managers regarding the allocation of human resources to individual workplaces. This is of particular concern regarding the nursing, midwifery, and to a lesser extent, other clinical professions, as these groups need to be required to provide their services 24/7 for 365 days per annum. Thus service scheduling and rostering is an important component of managing costs while maximizing the capacity to care.

Effective workforce planning together with sound educational opportunities are required to ensure the continuing availability of staff and contractors with the required knowledge and skills. In addition, it's important to optimize workforce participation by those suitably qualified. Dysfunctional work environments lead to high staff turnover, a reduction in hours actually worked by individuals and reduced participation rates. This impacts on any health facility's capacity to care. Workforce planning is covered in more detail in Chapter 8.

## Medical products, devices and technologies

The availability of and access to various medical products, devices and technologies is influencing the work methods and processes employed for the delivery of various health care services. This includes the use of telehealth and information systems. Their use changes workflows, communication patterns and support service needs. Their use may enhance or impede any facility's capacity to care. Their use is referenced and explored further where relevant in many of the following chapters.

### Health service delivery

The delivery of health services covers all clinical and caring services provided in any location or type of health service and is labor intensive. These services rely heavily on any number of support services all of which need to be well co-ordinated to avoid delays and the inefficient use of available resources. The challenge is to maintain a continuity of care throughout any patient journey. Good planning plus a sound supportive infrastructure contribute to a health service's capacity to care. The following chapters detail how the capacity to care can be achieved.

### Information and research

Research adds knowledge, a fundamental requirement of critical importance to the health industry. Existing and new knowledge must be made available to all relevant health care workers in any health care delivery location. Ideally data and information captured at the point of care, and entered into an information system, can be processed immediately and reported on in a timely manner. Real time operational systems providing routine and standard information as well as facilitating the processing of ad hoc requests, are instrumental in supporting the capacity to care at all times in any location. A digital health care ecosystem is highly recommended as it is very time consuming to do this manually. A number of chapters to follow provide the rationale and ideal system requirements.

## What are the desired health system outcomes?

The WHO health framework [4] includes four high level goals or outcomes. These are:

- Improved health — both level and equity
- Responsiveness
- Financial risk protection
- Improved efficiency

### Improved health, efficiency, responsiveness and caring

Health systems consist of many health care providers, both organizations and individuals whose primary interest is to promote, restore or maintain health. This is achieved by delivering and coordinating direct health services in a timely, safe and effective manner making use of available support services and resources with minimum resource wastage to those who need them when and where needed. Health systems which perform at a high level of quality and productivity generally utilize a well connected digital infrastructure supporting maximum automation and using real time operational data. This type of automation ensures the production, analysis, dissemination and use of reliable and timely information on health determinants, service demand, resource distribution and use, and the ability to provide practice based evidence of service delivery effectiveness and patient outcomes.

A well performing and effective health workforce is responsive to meeting service needs in a competent, fair, efficient and effective manner to optimize health outcomes within the available location, resource and circumstantial constraints. The concept of nursing is best described by Henderson's definition adopted by the International Council of Nursing (ICN):

> *The unique function of nurses in caring for individuals, sick or well, is to assess their responses to their health status and to assist them in the performance of those activities contributing to health or recovery or to dignified death that they would perform unaided if they had the necessary strength, will, or knowledge and to do this in such a way as to help them gain full or partial independence as rapidly as possible [11].*

The ICN also notes that:

> *within the total health care environment, nurses share with other health professionals and those in other sectors of public service the functions of planning, implementation, and evaluation to ensure the adequacy of the health system for promoting health, preventing illness, and caring for ill and disabled people.*

Health and caring services are not only provided by nurses or midwives, however it is clear from this definition that nurses contribute to, must have access to and make use of, all data and information that collectively describes both an individual's health status, required diagnostic, treatment and caring services and supporting infrastructures. There is an increasing desire to be able to demonstrate nursing's contributions to improving the quality of services delivered and cost reductions as well as evaluate and improve an understanding of care inputs, processes and outcomes [12].

## *Nursing data at the center*

We argue that nursing and other data used by nurses regarding all aspects associated with health care service delivery, need to be at the center of all decision making. From a workforce perspective nurses and midwives are the most prevalent in terms of numbers and they are responsible for the bulk of all direct health care delivery services globally. They provide a crucial support service to all specialized clinical services, are available on a 24/7 basis and represent a critical link between the recipients of care, their families and communities as well as between the many providers of care. Nurses and midwives fill the gaps especially after hours when other health professionals are not on duty. They work with and contribute to real time operational data. The continuity of care provided by nurses and midwives enhances data accuracy and provides important information to the entire health care team enabling them to make informed and timely decisions, impacting on their capacity to care.

Nursing and midwifery contributions to overall health outcomes are not well known and tend not to be officially measured. Given that nurses and midwives undertake many preventative and risk management activities, their absence may be identified by increases in the number of reported adverse events and poor patient outcomes including early deaths. With an increasing use of information systems, it is now becoming more feasible to identify and collect relevant information to demonstrate and quantify the value of nursing and midwifery services.

This book is about contributing knowledge to enable its readers to undertake data analytics relevant to the costing of nursing services, such as Activity Based Diagnosis Related Group (DRG) costing and funding, nursing service demand, workforce planning, best practice resource management and identifying increasing trends of acuity relative to an aging population and a higher incidence of co-morbidities. This requires in depth coverage of those factors known to influence the capacity to care, including the use of new technologies, medical advances and changes to health care delivery patterns and methods.

Fig. 1 has positioned patient care experiences at the center. These are influenced by the number of and type of patients who need to receive care, the available staff numbers and mix and how service delivery is organized to produce overall performance outcomes. Workforce availability

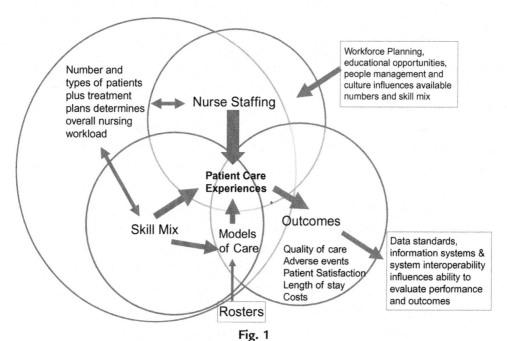

**Fig. 1**
Nursing workload influencing patient care experiences.

is influenced by workforce planning, educational opportunities, people management and culture. The ability to provide evidence of overall performance is subject to the adoption of data standards, information systems and system linkages/connectivity.

Content provided in this publication is based on real world problems and case studies encountered in both the public and private sectors of healthcare service organizations. Details regarding effective solutions developed, tested and implemented are provided in the chapters to follow. This includes the provision of examples of detailed involvement of many stakeholders and explanations regarding their many different perspectives and local health care delivery environments. This includes data/information and technology governance, automated secondary data use, the need for effective data interpretation, costly adverse events, skill mix relative to service demand, work measurement methods, models of care relative to nursing resource use, patient outcomes, rostering, scheduling, human resource management, workforce planning, system connectivity, data sharing and linking, small and big data, change management and a future vision. Stakeholders need to understand these concepts to enable them to be part of solution development. Proposed solutions provided are based on real world evidence obtained from numerous research studies as reported in the literature and undertaken by the authors across multiple countries.

Data generated and used by nurses and midwives have the potential to provide them, as well as other stakeholders, to acquire detailed insights of any health system's complexity. Better use of routine operational data collected at the source enables the demonstration of the link between these data and its use for efficient and effective healthcare service management. This highlights the need to work effectively in multidisciplinary teams by adopting sound collaborative approaches. This in turn enables future occupants of highly influential positions to improve their negotiations in order to contribute to discussions regarding health system transformation needs, referring to the many limitations currently experienced and their impact on health service delivery, to find suitable solutions and improve their decision making to generate benefits for many.

Organizational transparency enables every health worker to view information that enables them to effectively contribute as a team member. The adoption of safe staffing principles and optimum use of operational data are key factors toward improving patient and health care organizational outcomes. This must be supported by the better use of informatics to collect, share, link and process data collected operationally for the purpose of providing real time information to decision makers at every level of the organization. This enables collaboration and optimum use of available human and other resources to meet health service demands at any point in time. Such rich data also enables the collection of practice-based evidence and supports the undertaking of continuous research to keep systems up to date and relevant.

The following chapters collectively explore the many facets policy developers and health service managers need to understand to enable them to best manage available resources while meeting health service demands. Health care operational inefficiencies and costly events are explored in the next chapter.

# References

[1] Markaki A, Lionis C. Capacity building within primary healthcare nursing: a current European challenge. Qual Prim Care 2008;16(3):141–3.

[2] Crisp BR, Swerissen H, Duckett SJ. Four approaches to capacity building in health: consequences for measurement and accountability. Health Promot Int 2000;15(2):99–107.

[3] Strachan H. Person-centred caring: its conceptualisation and measurement through three instruments (personalisation, participation and responsiveness) [Doctoral thesis]. Glasgow: Glasgow Caledonian University; 2016.

[4] Coiera E, Hovenga EJ. Building a sustainable health system. Yearb Med Inform 2007;11–8.

[5] Anåker A, Heylighen A, Nordin S, Elf M. Design quality in the context of healthcare environments: a scoping review. HERD 2017;10(4):136–50.

[6] IOM. Keeping patients safe: transforming the work environment of nurses. Washington, DC: Institute of Medicine, National Academies Press; 2004.

[7] Murray JLC, Frenk J. A framework for assessing the performance of health systems. Bull World Health Organ 2000;78(6):717–31.

[8] WHO. Guideline: recommendations on digital interventions for health system strengthening. Geneva: World Health Organisation; 2019. [cited 18 April 2019]. Licence: CCBY-NC-SA3.0IGO. Available from: https://www.who.int/reproductivehealth/publications/digital-interventions-health-system-strengthening/en/.

[9] AIHW. The national health performance framework, Australian Institute of Health and Welfare; 2009. Available from: https://www.aihw.gov.au/getmedia/0473c334-bb4d-4eca-8fd7-29f15a2ac94f/national-health-performance-framework-figure-31Aug17.pdf.aspx.

[10] WHO. The WHO health systems framework. Available from: http://www.wpro.who.int/health_services/health_systems_framework/en/.

[11] Henderson V. The concept of nursing. J Adv Nurs 2006;53(1):21–31.

[12] Westra BL, Clancy TR, Sensmeier J, Warren JJ, Weaver C, Delaney CW. Nursing knowledge: big data science—implications for nurse leaders. Nurs Adm Q 2015;39(4):304–10.

# CHAPTER 2

# Health care operational inefficiencies: Costly events

## Workforce management

A poor understanding and/or the mismanagement of dynamic health service demands has a high probability of resulting in healthcare operational inefficiencies and adverse events. The results of inefficiencies are identifiable by health service acquired morbidity or mortality, increases in lengths of stay and an increase in incurred costs. Inefficiencies are the result of poor service coordination, delays, unsafe staffing levels, a poor match between knowledge and skills available and the type of care required, processes adopted such as poor communication, incomplete information, non-adherence to evidence based practice guidelines or the non-availability of required equipment or supplies at the time they are needed. The complexity associated with the delivery of health services means that operations can be streamlined by adopting various evidence based standard practice processes. Operational efficiency is all about the ability to meet service demand with all the "right" inputs and processes to provide safe patient care and achieve the desired clinical and financial outcomes.

Those concerned with health service planning and management have identified issues associated with unequal human resource distribution within healthcare facilities at all levels. The World Health Organisation (WHO) published an approach to adjust staffing levels to effect a fair and optimal distribution in 1998 in an effort to balance the workforce within and between health care facilities. Its use by many healthcare facilities around the globe was documented and evaluated over 10 years or so and lessons learned resulted in a new manual detailing the Workload Indicators of Staffing Needs (WISN) method [1, 2]. The overall aim for human resource management is to have

- The right number of people
- With the right skills
- In the right place
- At the right time
- With the right attitude

Measuring Capacity To Care Using Nursing Data. https://doi.org/10.1016/B978-0-12-816977-3.00002-2

- Doing the right work
- At the right cost
- With the right work output

The WISN method is based on health worker's workload, with activity (time) standards applied for each workload component. The aim for any workload measurement method is that it should be:

- Simple to operate using data collected for operational management purposes
- Simple to use by providing the right information in the right format for ease of making staffing decisions in real-time to accommodate unexpected workload changes, and for planning and budgeting purposes at all relevant decision-making levels
- Technically acceptable to all users
- Comprehensible to non-clinical managers, finance directors, policy makers and researchers
- Realistic with a high degree of accuracy

The chapters that follow reflect in-depth evaluations of nursing and midwifery workload management methodologies and systems in use, their development and their application for nurse staffing strategies based on extensive scoping literature reviews over many years and many practical research studies undertaken by the authors.

## Nursing workloads and nurse staffing methods

Nurse staffing and nursing workloads continue to be of great importance to many stakeholders, but how are they measured or evaluated? How do nursing workloads relate to operational efficiencies and costs? Perceived heavy nursing workloads have resulted in numerous instances of industrial action by nurses to address a mismatch between available nurses and service demand. This leads to a desire to measure and monitor workloads relative to the number and type of nurses and nursing care support workers made available to meet service demands. Reaching agreement regarding what constitutes a reasonable workload for any one nurse is frequently controversial and difficult to achieve. Ensuring the maintenance of reasonable workloads for all nurses and nursing care support workers in the workforce is highly desirable as this is most cost effective and less likely to result in operational inefficiencies or costly events. Nursing workloads influence the potential for conflict between management and nursing staff, staff turnover rates, sick leave, patient satisfaction, patient safety, the quality of care provided, length of stay, the number of hospital acquired adverse events, staff wellbeing and overall organizational performance. There is general consensus from all concerned, the nursing and midwifery professions, and administrators, that we are all trying to achieve better care at lower cost so that we'll have sustainable health systems.

A systematic review undertaken to examine the impact of nursing workload and staffing on creating and maintaining healthy work environments defined a healthy work environment as "a practice setting that maximises the health and wellbeing of nurses, quality patient outcomes and organisational performance" [3]. The collective evidence identified during this review suggests strong correlations between patient characteristics and work environments, workload and staffing and the quality of outcomes for patients, nurses and the system/organization. These authors found that the greater the proportion of highly qualified nursing staff was associated with improved outcomes as measured by the Functional Independence Measure score, the Short Form Health Survey (SF-36) vitality score, patient satisfaction with nursing care, patient adverse events including atelectasis, decubitus ulcers, falls, pneumonia, postsurgical and treatment infection and urinary infections. Such adverse outcomes influence the length of stay, overall costs and the occurrences of failure to rescue, resulting in premature mortality.

Another literature review of the nursing workload concept identified five defining attributes [4]. These are:

1. Amount of "nursing time" that is spent to perform all nursing care.
2. Level of knowledge, skills and behavior (Nursing Competency) that nurses are expected to exhibit in order to meet the physical, psychological, social and spiritual needs of the patient.
3. Weight of nursing intensity (Direct-patient Care) that is carried out directly to the patient, excluding the non-patient related work.
4. All the physical exertion, mental process and emotional effort performed by nurses, including but not limited to bending, lifting, pushing, moving, carrying, caring, thinking, planning, problem-solving and decision-making.
5. Care complexity — ability of the nurse to change the plan during the shift between different patients with different acuity levels, attending unexpected patient complications and sudden changes in severity of illness, changing the nursing procedure and missing supplies.

A concluding definition of nursing workload was *the amount of time and care that a nurse can devote (directly and indirectly) toward patients, the workplace, and professional development.* A panel roundtable discussion of nurse leaders explored the issue of using patient acuity systems. They noted that these systems have been around for many years, had been shown to improve patient outcomes but required extra work by nurses to undertake the necessary data collection and processing. They presented research and real-life examples illustrating that the use of acuity systems can optimize patient outcomes and help balance nursing workloads [5]. One valuable reported component associated with successful acuity development, implementation and use, was the adoption of a shared governance model. This model gave nurses a voice, promoted innovation, allowed nurses to manage and control their practice with permission to influence beyond their unit walls and share decision making processes while keeping patient and families as the central focus.

A significant body of research is available regarding the use of patient acuity/nurse dependency, but in almost every case these studies are relatively small, confined to one ward or hospital or period of time or group of patient types. Methodologies used to develop acuity systems are many and varied preventing validation, other than face validation studies, or meaningful comparisons. This has limited the building of our body of knowledge in meaningful ways to allow for continuous improvements. Every acuity system in use has unique features.

Despite the many studies undertaken healthcare provider organizations, administrators and nurse managers continue to encounter difficulties when attempting to best match service demand with the right mix and numbers of staff to provide quality care. Effective management of nursing and midwifery resources is imperative, not only to provide quality care but also to meet desired financial outcomes. Paulsen [6] has provided an excellent overview of the many and varied nurse staffing methods in use in the United States. She concluded that to conduct the types of studies needed to advance nurse staffing research and allow for its effective use by nurse managers at the unit level, researchers need access to unit-level staffing and patient outcomes data and that this should now be possible given the wide use of hospital information systems. We have undertaken such studies.

The following chapters provide the results of our detailed analysis and decomposition of the complexities associated with the measurement of nursing workloads and our capacity to care and the results of our research and development activities undertaken at the unit level, in numerous hospitals located in several countries over many years. Our concept analysis highlights the need to consider the many nuances that in totality make up the type and amount of service demand, nursing working conditions and work environments, the importance of continuing professional development (knowledge and skills), fair workload distribution as well as teamwork, communication and collaborative strategies.

## *Measuring operational activity and efficiency*

Measuring nursing workload represents one very large and significant operational issue that needs to be well understood by all concerned, especially by those who make resource allocation decisions. This will be presented in some detail in the next chapter. This chapter will first explore many other related operational activity and efficiency measures.

Governments, Healthcare service executives and funders have a keen interest in health care service operational activity and efficiency measures. These indicators are used to decide how to allocate and distribute available resources or how best to optimize available resource usage. This includes the need to estimate current and future service demands. Although healthcare service demand is not totally predictable, patterns of activity generally do emerge for each service. Such patterns are discernible from various data collections with the use of any one of many work study and operations research methods. The key success factor is how to identify critical indicators and collect the right data in a timely manner to strategically fix fundamental issues and facilitate productive operational planning.

There are many key healthcare optimization issues, including service planning, resource scheduling logistics, medical therapeutics, disease diagnosis and preventative care. All of these relate to a lesser or greater extent to nursing service delivery and/or a need to make use of the same data and information collected and/or used by nurses. The productivity of any healthcare delivery operation consists of a specified target group of care recipients, a set of resource inputs, any number of activity processes and desired outcomes. It is imperative that each of these components can be clearly identified and described for the purpose of effectiveness evaluation, performance monitoring and capacity planning.

## Care recipient characteristics

There are multiple ways healthcare recipients may be described, by number, age, provisional or confirmed diagnosis, diagnosis related group (DRG), injury type, population type, socio-economic status, disability status, location, residential status etc. How are these characteristics quantified and measured? The number of patients accessing a health care service is fairly simple or is it? This needs a qualifier such as a number of patients attending a service per day, or treated/cared for per day or reviewed per service per day. Then we need to define the service type and patient day. Is someone who has spent 6 hr in the day surgery unit or an acute hospital bed classed as an inpatient day? Patient day is a commonly used denominator for many statistics, thus it is important to ensure consistency in interpretation and data use. To overcome these issues and ensure consistency in interpretation one needs to agree on metadata.

The Australian Institute of Health and Welfare (AIHW) has established a repository for national metadata standards for health, housing and community services statistics and information known as METeOR. Other countries have similar arrangements. The AIHW repository is based on the ISO/IEC 11179 Metadata registries standard [7]. Metadata is information about how data are defined, structured and represented. Once endorsed these metadata are referred to as data standards. These standards improve the quality, relevance, consistency and availability of health data. This provides meaning and context to data. It also describes how data is captured and the business rules for collecting data. Individual data elements may be used as indicators of relevance to agreed National Minimum Data Sets used to calculate and present national operational statistics.

Individual patients may be diagnosed with a condition which then becomes the reason for them to access the health system and receive a service. Diagnosis is a commonly used term. It is the foundation for the identification of health trends and statistics globally. It refers to any one of the diagnosis as listed in the International Classification of Diseases (ICD) [8]. Such classifications do change over time. They are published according to version number. Any classification system requires the application of rules and the use of several data elements to enable coders to make a judgment to arrive at a valid code. The interpretation and application of such rules need to be reliable to enable meaningful comparisons of findings to be made. These ICD codes plus other data may be used to form the basis for the application of a grouping algorithm to reflect estimated average resources used. The resulting groups are known by various names such as Diagnosis Related Groups (DRGs) or Health Resource Groups (HRGs).

In summary, metadata describing data used in accordance with minimum data set collections and used for statistical reporting are known as data standards that need to be governed. This same principle regarding data standards and governance applies to clinical specialties interested in collecting data about, for example, specific disease indicators, or, organizations wishing to collect any type of data describing care recipient characteristics.

### Types of resource input

Inputs consist of the number and type of staff, their knowledge and skills, scheduling, work practices and processes adopted to complete work, the availability and correct use of the right equipment, supplies and support services. Resource input factors are many and varied. From a financial perspective it is important to differentiate between one off capital costs and recurrent costs. Capital costs cover buildings, building service infrastructures such a plumbing, wiring, and equipment. Each of these items require recurrent spending on maintenance and each item has its own predicted lifespan of use and replacement timeline. Recurrent expenditures cover all other resources that collectively enable a health service to function. This includes resources such as staff, medical and surgical supplies, and medications. Their source location, availability and distribution logistics need to be considered from an operational perspective relative to volume of use. There is a direct relationship between care recipient characteristics and the use of such supplies.

The health workforce makes up the largest component of all resource inputs; they collectively consume the largest amounts of recurrent expenditures. Support services and processes such as service scheduling, staff allocation and rostering logistics are activities that largely determine any health service's degree of effectiveness and efficiency relative to service demand. Overall performance effectiveness and efficiency is determined by how well matched the available resources are to service demand and the level of achievement of pre-set patient and organizational outcome goals. Operational workforce performance is dependent on availability in terms of numbers, type, knowledge, skill and the accuracy of decisions made, in combination with the use of other material resources,. This publication focuses primarily on the allocation and use of nursing and midwifery resources as these represent most resources used for the delivery of health care services.

### Healthcare activity processes

Any work activity undertaken by any member of the workforce is part of delivering a service. Each work activity consists of preparation, doing and putting away activities. Each of these three steps can be further analyzed as a means of evaluating methods used for each sub-activity process to determine overall efficiency and performance in terms of competency, timeliness, degree of accuracy or compliance with protocols. This means there is a hierarchy of activity consisting of several levels of detail one is able to employ to measure healthcare activity processes. Each level will have its own set of terms that could be used to undertake and document the results of such measurements. Actual health services provided are commonly defined by type of service.

Clinical services are most commonly defined by diagnosis, therapeutic procedures or interventions undertaken. Clinical intervention refers to information about the surgical and non-surgical interventions including invasive and non-invasive procedures and cognitive interventions. Clinical activities directly relate to care recipients and are integrated with the prepare, do and put away functions associated with direct patient care. Various coding systems exist to describe each of these clinical services. These data elements provide the basis for analysis of health service usage, especially in relation to the use of specialized resources, for example operating theaters, equipment, medical and surgical supplies and human resources.

Clinicians, including nurses and midwives, need to be able to make informed decisions regarding how best to respond to an individual's health care needs. In doing so they need to draw on various sources of current knowledge and patient information while considering patient preferences. This includes the need to consider new practice based research results where available. One solution is to provide clinicians at any point of care with access to clinical decision support systems that are frequently updated in accordance with practice based evidence.

This publication explores nursing service activity processes in some detail, relative to care recipient characteristics and associated health service demands. It's about the gathering of data that describe care recipients' characteristics known to indicate a need for nursing care, plus data known to quantify nursing intensity, and the type of knowledge and skills required to provide the service safely and efficiently so as to result in the desired patient outcomes. There are several terminologies that describe nursing practice such as the International Classification of Nursing Practice (ICNP®) [9], and the Clinical Care Classification (CCC™) [10]. Another standard reference terminology that complements this use is the Systematised Nomenclature of Medicine Clinical Terms-SNOMED-CT.

## Measuring health outcomes

Health service outcome measures are essentially the result of any person's health assessment relative to the results of a previous health assessment such as on admission to hospital or 4 weeks following the discharge health status. These include the identification of adverse outcomes such as unplanned readmissions within 28 days of separation, or the rate of hospital-acquired bacteraemia or the rate of wound infection following clean and contaminated surgery.

Such outcomes can be linked to medical treatment methods, and processes adopted to provide care. Australia has identified some clinical indicators relative to specific clinical services. Such indicators are agreed to by Clinicians; they are part of the Australian Council on Healthcare Standards Clinical Indicator program used to review trends and variations to suggest areas where there is the greatest scope to improve clinical practice.

Indicators are flags that signal whether or not desired health outcomes have been achieved. Because they are indirect measures, they do not in themselves demonstrate quality or explain variation in outcomes, they can only indicate the presence or absence of quality processes and outcomes. In addition to process and outcomes indicators, indicators are further broken down into sentinel event indicators and aggregate data indicators. Sentinel event indicators identify specific (usually undesirable) events that are always investigated. Aggregate data indicators can be discrete variable indicators (rate-based) or continuous variable indicators. The type of aggregate data indicator most commonly referred to in the literature is rate-based. Rate-based indicators refer to a proportion, or rate, of expected events. That is, a certain level of a specific event is expected to occur, and if the rate of occurrence changes to above the expected rate, it signals the area should be investigated for possible improvement.

The USA 2015 Precision Medicine Initiative [11] has recognized the need for greater collaboration to address global health issues by improving translational research processes to get more treatments to more patients more quickly. It has emphasized that the biological, environmental and behavioral influences on disease and individual variability in genes, environment and lifestyle for each person plays a significant part [12]. Similarly the United States based Patient-Centred Outcomes Research Institute (PCORI) [13] focuses on funding projects designed to improve care and outcomes for patients living with high-burden health conditions. It's about helping patients and those who care for them make better informed decisions about healthcare choices. It's critical for the nursing profession to contribute and gain new knowledge from these initiatives by ensuring we have the capacity to contribute nursing data [14]. The Nursing profession needs to ensure that it is able to collect such data and make their contributions to human health visible. These transformational activities have major implications for nurse leaders [15].

A variety of specific health outcomes measures are available across the health care industry. It is common practice to simply identify the number of people treated or discharged or transferred to a nursing home or a rehabilitation center or community care. Other examples are: nursing outcome measures included in the ICNP®, the Rand Health 36 item short form survey (SF-36), which consists of a set of generic, coherent and easily administered quality-of-life measures [16], the Barthel Index [17] or the Functional Independence Measure (FIM®) [18] or the use of patient satisfaction surveys. Although patient satisfaction surveys are popular, the satisfaction concept is often ill defined due to the need for subjective judgements about various aspects of the health care experience. The degree of satisfaction has a strong relationship with expectations. Another measure of patient satisfaction is the number and type of complaints received. These may be received locally at the point of care and/or the organization level.

Countries frequently develop and use their own set of outcome measures for performance measurement and reporting purposes. For example, the Canadian Health Outcomes for Better Information and Care (C-HOBIC) project that makes use of a systematic, structured language to admission and discharge assessments of patients receiving acute care, complex continuing care, long-term care or home care. This facilitates the abstraction into provincial databases or Electronic Health Records (EHRs) [19]. This project is of particular interest as it is one of the few that includes Nursing-Related Patient Outcomes in EHRs using both the ICNP® and SNOMED CT terminologies. The outcomes have a concept definition, a valid and reliable measure and empirical evidence linking them to nursing inputs or interventions. These are:

- Functional status — using the inter-RAI instrument [20]
- Therapeutic self-care (readiness for discharge) — using a Canadian developed instrument [21]
- Symptom management (pain, nausea, fatigue, dyspnoea)
- Safety (falls, pressure ulcers)

## *Learning health systems*

There is an increasing desire to collect, assemble, analyze and interpret operational data for reporting and other purposes. Data collected and recorded in any system has many uses. In particular data are able to provide new information and capture new knowledge that, when used appropriately, has the potential to improve care for patients, reduce costs and improve health outcomes. It's about capturing trends, identifying delays, communication hiccups, efficient service scheduling, examining cause and impact of actions taken or services provided. Of great importance to the health industry is to learn which treatment or care plans have the best possible outcomes. This requires learning from individual and groups of patients' care experiences as documented in a variety of information systems including electronic health records and patient satisfaction surveys. Such learning then needs to be used to make the necessary changes in both real time and by the adoption of a change management process. This can apply to individual departments, clinical disciplines, healthcare facilities, healthcare delivery networks or used to make changes to health policy and funding arrangements.

A learning healthcare system was defined as a system in which

> *science, informatics, incentives, and culture are aligned for continuous improvement and innovation—with best practices seamlessly embedded in the care process and new knowledge captured as an integral by-product of the care experience [22].*

Data are only useful if the meaning is clear and if it is collected in standard formats. This requires the adoption of data and information governance principles. Health data and information are primary assets of the health industry. Such assets need to be used optimally to ensure we are able to meet the health and health care needs of the population in a timely, responsive and sustainable manner. That requires the ability to appropriately collect, consistently define, accurately aggregate, link, relate to knowledge and machine process health data accurately. Health language is extensive and complex, includes a lot of jargon and abbreviations. Computers require consistency for accurate, comparable data and information processing, making the management of health data and information in this new and continually expanding digital environment a major challenge. Yet the situation encountered by Florence Nightingale continues to this day in many places.

> *In attempting to arrive at the truth, I have applied everywhere for information but in scarcely an instance have I been able to obtain hospital records fit for any purpose of comparison. If they could be obtained, they would enable us to decide many other questions besides the one alluded to. They would show subscribers how their money was being spent, what amount of good was really being done with it or whether the money was not doing mischief rather than good.*
> **Florence Nightingale (1820–1910), Founder of modern nursing**

The World Health Organisation (WHO) compiles health related data for its 194 member states although only 34 of its members are able to provide reliable quality health data. The accuracy and amount of information collected and processed varies considerably between nations. WHO makes use of a number of databases, some of which are maintained by a range of other organizations including the United Nations International Telecommunication Union (ITU), the United Nations Department of Economics and Social Affairs (UNDESA), the United Nations Educational, Scientific and Cultural Organisation (UNESCO), the United Nations Children's Fund (UNICEF) and the World Bank. WHO's Global Health Observatory (GHO) provides access to data and analysis for its monitoring of the global health situation [23]. Little if any of these data collections tell a story about nursing service demand or provide an indication of our capacity to care in a manner that is useful for new policy decision making that will benefit both the nursing profession and those cared for. Hospital capacity is measured via episode statistics in terms of patient numbers by ICD (International Classification of Diseases) or DRG (Diagnosis Related Groups) or equivalent casemix codes. Patient recorded outcomes tend to be measured relative to specific medical conditions treated, not nursing services received.

Historically data sets and collections were largely fiscal or statistical and collected administratively. Such data collections had varying levels of governance. The emergence of the electronic health record and clinical decision support systems require a consistent structure of the data as well as consistent representation of each individual thing in the record (each concept). These structures and codes must be meaningful at the time of collection in the clinical environment but also retain their original meaning in computer systems over time and support extraction and aggregation of data to meet the longer standing reporting requirements of fiscal, planning and statistical collections.

Current directions are toward transitioning from concentrating on these volume based statistics and start to focus on value based healthcare. This requires the adoption of new data standards to reflect outcomes of care. By doing so we are in a better position to learn based on real world evidence of practice. The World Economic Forum launched such a project in July 2016. It's goals are to [24–26]:

- Develop a comprehensive understanding of the key components of value-based health systems
- Draw general lessons about the effective implementation of value-based healthcare by codifying best practice at leading healthcare institutions around the world
- Identify the potential obstacles preventing health systems from delivering better outcomes that matter to patients, and at lower cost
- Define priorities for industry stakeholders to accelerate the adoption of value-based models for delivering care

This has set the scene toward a stronger focus on patient centered care with a focus on evidence based care, health outcomes and sustainability. Four enablers were identified as key to accelerating the adoption of value-based care, health informatics, benchmarking, value based payments and innovations in organizing care delivery. Achieving this requires clinical decisions to be supported by accurate, timely, and up-to-date clinical information, to reflect the best available evidence [27].

### Making better use of data and information

Data reflect raw facts and figures. When such data are processed new information is created to make the data more meaningful and valuable to users. This then becomes knowledge. New data or information may then be added to existing knowledge to create new knowledge. Information and knowledge are then used by many different decision makers at all levels within any nation's healthcare system. Ease of use is greatly facilitated where health information systems are able to connect at all levels in a manner that facilitates data linkage. Unfortunately there are few systems with this capacity in use, most health data continue to be located in numerous individual databases, each with their own unique structures making data linkages a cumbersome and time consuming, if not impossible task. These health data "silos" exist within many healthcare organizations, some of which are known to have more than 300 separate databases (silos).

This situation came about due to the perceived need to create a new data base to suit specific data collection and reporting requirements. It is further exacerbated by a customary desire to evaluate the effectiveness of individual health program initiatives. This usually results in the need to collect very specific data sets. There is a known overlap of various data elements in any number of data sets requiring collection; this represents a costly additional administrative burden. New technologies enable such data collections to be streamlined but this requires effective health data governance at all levels. Choices made are influenced by what the decision maker values most.

Governance is about steering or directing any concept or entity in the desired direction by an entity (governing body) who has the authority to do so. In addition to this general governance concept, it is important when transferring data between information systems within the healthcare industry, to also manage trust and reliability so that clinicians can confidently make use of information received. Healthcare organizations tend to make use of many data collection paper forms. Each form has numerous headings and special sections that are used as prompts to collect and record all the required information — a group of data elements, referred to as a data set, to suit each purpose. Some of the data collected via forms, such as patient demographics or a record number, is used to enable filing. The way such data are recorded may vary, it may be spelled out in full or abbreviated or it may be described in different ways. However once such data is collected and entered into a computer it needs to be in a format that computers can meaningfully use. This is where a standard way of entering each data element's content within a set of data (purpose) needs to be agreed upon. Computers are then able to consistently record and access such data, display or present it in different forms, such as screens or reports, in a manner that is consistent to the meaning originally intended. This principle is often called record once, use often.

Each data element within any dataset needs to be defined, as do the codes used to represent different meanings within that data element, to ensure all users apply the same meaning, much like any dictionary, also known as metadata. Metadata includes the specification of the data element and the codes used to represent that data element. When used to suit computer data entry such metadata needs to not only have an agreed definition it also needs to describe any other features, such as use, characteristics and rules that need to be applied, such as the entry of this information is required or this field may be left blank. Such data specifications assists system developers and users to clearly understand what data needs to be collected, the level of detail, the way it is to be collected and how it may be used. Metadata has a standard specified structure and needs to be governed to ensure consistent use and interpretation of all data collected and used.

Any clinical data governance process, irrespective of whether national, jurisdictional, professional, or organizational, begins with the development of a data set consisting of data elements that collectively meets the needs of a specific purpose: such as to support a research methodology, or to be able to answer specific questions to assist policy development, or to enable effective population health status monitoring, or to support clinical research. Such work needs to be undertaken by those experts who have a clear understanding of why the data are collected, the need for and uses of that data. In addition there needs to be consideration about the diverse information systems/databases containing the data elements required for collection, so that an effort can be made to develop common standards for concepts, classifications, terminologies, data values, data types etc. for ease of data collection. Data governance is about ensuring that everyone engaged in facilitating any type of data collection makes use of the data standards as agreed by all relevant stakeholders.

A number of national and other jurisdictions have recognized the need for governing the meaningful usefulness of health information. As numerous health information systems are now connecting with each other and sharing or using the same data this need is becoming more urgent. In the past the focus was on coordinating the development, collection and dissemination of nationally reported health information about the full range of health services, and a range of population health status parameters largely for fiscal or statistical analytical purposes. Such information continues to be used to form the basis of funding arrangements and for policy development. Unfortunately there are still systems that build data collection and reporting requirements to justify funding. These data collections have no relevance or utility at the workplace and therefore the data are collected separately from actual practice which puts the quality of that data into question. This practice results in costly data collection systems with minimal quality and utility. An increasing desire to link all clinical data collected, and to compile and share electronic health records, has created a critical need for governance to ensure that such data is accurate and can be linked in a meaningful manner. Fig. 1 shows two governance pillars that need to be adopted to achieve safe and efficient healthcare.

Health data, information and knowledge governance has been shown to be pre-requisites to providing accurate, reliable, meaningful and timely new information about the many factors

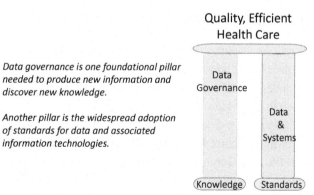

**Fig. 1**
Governance pillars underpinning quality healthcare [28].

from simple care to national policy making. Data influences clinical decision making, resource allocation, procurement, workforce planning, infrastructure investments, and health outcomes within that context.

## *Operational research*

Our ability to undertake useful operational research studies is very much dependent on the availability and quality of current and historical data. This in turn means that healthcare organizations and clinical disciplines need to consider which data elements need to be standardized and governed to facilitate reliable data capture, and make the use of various analytical processes possible. The previous section explored the many possibilities for each of the input, process and outcome factors. Input influences process which in turn influences outcomes. Operations research is about exploring these relationships to arrive at an optimum combination to meet known or estimated service demand.

Every healthcare service has its own issues of high volume or high cost or poor performance outcome areas that would benefit from an optimization process. Each scenario has its own unique data set requirement. There are many similarities and common occurrences between healthcare services. For such instances we can make use of well tested methodologies and operations research mathematical modeling approaches such as queuing or the use of probability theory. These can be contextualized as required to best address the desired objectives. We can calculate the probability of experiencing demand that exceeds our capacity, such as the number of available respirators, or labor wards or resuscitation teams or a clinical specialist at any point in time as long we have the right current and historical data.

This publication is focused on capacity planning and management by providing numerous examples based on real world experiences using data collected to support nursing service delivery as the operational center. Our experience demonstrates that the use of these data creates more useful and timely information to improve input resource management.

Chapter 3 explores how the resources required for the provision of nursing services is commonly measured and the analyses the four different nurse staffing methodologies in use, Nursing Hours Per Patient Day, nurse staffing ratios, patient/client types (acuity) and patient classification. This requires the need to identify and examine the many confounding variables known to influence nursing service demand. Significant variations and limitations in terms of staffing methodology reliability, degree of accuracy and data/information collection methods are highly likely. We then identify a need for the adoption of data standards and report on the development of a beginning structured metadata set to represent this complex knowledge domain in preparation of the digital transformation now in progress across the healthcare sector. All of these factors are part of operational management. Operational inefficiencies are the result of gaps and/or deficiencies in any of these workload defining factors.

# References

[1] WHO. Workload indicators of staffing need (WISN): selected country implementation experiences. Geneva: World Health Organisation; 2016. [cited 17 December 2018]. Available from: http://apps.who.int/iris/bitstream/handle/10665/205943/9789241510059_eng.pdf?sequence=1.

[2] WHO. Workload indicators of staffing need tool and software. Geneva: World Health Organisation; 2010. [cited 18 December 2018]. Available from: https://www.who.int/workforcealliance/knowledge/toolkit/17/en/.

[3] Pearson A, O'Brien Pallas L, Thomson D, Doucette E, Tucker D, Wiechula R, et al. Systematic review of evidence on the impact of nursing workload and staffing on establishing healthy work environments. Int J Evid Based Healthc 2006;4(4):337–84.

[4] Alghamdi MG. Nursing workload: a concept analysis. J Nurs Manag 2016;24:449–57.

[5] O'Keefe M. Acuity-adjusted staffing: a proven strategy to optimise patient care. Am Nurse Today 2016;11(3):28–34.

[6] Paulsen RA. Taking nurse staffing research to the unit level. Nurs Manage 2018;49(7):42–8.

[7] AIHW. Metadata standards. Australian Institute for Health and Welfare; 2018. Available from: http://meteor.aihw.gov.au/content/index.phtml/itemId/181162.

[8] WHO. International classification of diseases (ICD). World Health Organisation; 2018. [cited 18 August 2018]. Available from: http://www.who.int/classifications/icd/en/.

[9] ICN. International Classification for Nursing Practice (ICNP®) International Council of Nurses; 2018.

[10] Sabacare. Clinical Care Classification (CCC™) System. Available from: https://www.sabacare.com/; 2018.

[11] President Obama. The precision medicine initiative. [cited 8 September 2018]. Available from: https://obamawhitehouse.archives.gov/precision-medicine; 2015.

[12] Delaney CW, Westra BL. Big data: data science in nursing. West J Nurs Res 2017;39(1):3–4.

[13] PCORI. Patient-Centred Outcomes Research Institute (PCORI). [cited 8 September 2018]. Available from: https://www.pcori.org/about-us; 2018.

[14] Brennan PF, Bakken S. Nursing needs big data and big data needs nursing. J Nurs Scholarsh 2015;47(5):477–84.

[15] Westra BL, Clancy TR, Sensmeier J, Warren JJ, Weaver C, Delaney CW. Nursing knowledge: big data science—implications for nurse leaders. Nurs Adm Q 2015;39(4):304–10.

[16] RAND-Health. 36-Item short form survey (SF-36). [cited 21 August 2018]. Available from: https://www.rand.org/health/surveys_tools/mos/36-item-short-form.html; 2018.

[17] Mahoney FI, Barthel DW. Functional evaluation: the Barthel index. Md State Med J 1965;14:61–5.

[18] Uniform, Data, System, for, Medical, Rehabilitation. The FIM® Instrument: its background, structure, and usefulness. 1–31 pp. [cited 21 August 2018]. Available from: https://www.udsmr.org/Documents/The_FIM_Instrument_Background_Structure_and_Usefulness.pdf; 2012.

[19] C-HOBIC. Canadian Health Outcomes for Better Information and Care project Canada. [cited 22 August 2018]. Available from: https://c-hobic.cna-aiic.ca/about/default_e.aspx; 2018.

[20] interRAI. interRAI—a comprehensive assessment system. [cited 22 August 2018]. Available from: http://www.interrai.org/; 2018.

[21] Doran DI, Sidani S, Kearings M, Doidge D. An empirical test of the nursing role effectiveness model. J Adv Nurs 2002;38(1):29–39.

[22] Grossmann C, Powers B, Sanders J. Digital data improvement priorities for continuous learning in health and health care, roundtable on value and science-driven health care. Washington, DC: Institute of Medicine; 2012.

[23] WHO. Global Health Observatory (GHO) data—world health statistics. [cited 31 October 2018]. Available from: http://www.who.int/gho/publications/world_health_statistics/en/; 2018.

[24] World-Economic-Forum. Shaping the future implications of digital media for society valuing personal data and rebuilding trust: end-user perspectives on digital media survey: summary report—white paper. Geneva; 2017.

[25] World-Economic-Forum. Digital transformation of industries healthcare industry—white paper. Geneva; 2017.

[26] World-Economic-Forum. Value in healthcare laying the Foundation for health system transformation. Geneva; 2017.

[27] IOM. Digital data improvement priorities for continuous learning in health and health care: workshop summary. Washington, DC: Institute of Medicine; 2013.

[28] Hovenga EJS. National healthcare systems and the need for health information governance. In: Hovenga EJS, Grain H, editors. Health information governance in a digital Environment. Amsterdam: IOS Press; 2013.

# CHAPTER 3

# Digital transformation needs to measure nursing and midwifery care demands and workloads

## What determines nursing workloads?

The first two chapters provided an overview of Healthcare systems, their management, known operational inefficiencies and costs. Nursing services were identified as the largest service representing a major cost of health care delivery in the majority of health care settings, particularly in hospitals. This chapter aims to identify and define the many factors influencing nurse staffing budgets and allocations and to analyze the variety of arbitrarily defined criteria used as foundations for the measurement of nursing care demand and other activities that make up nursing workloads. Hospitals and residential aged care facilities exhibit wide variations both in nurse to bed ratios and in the mix of nursing staff relative to their in-patient populations. Nursing care demand will vary with the clinical composition of the patient population requiring such a service.

Trends in the demand for nursing services need to be identified and measured to assist with the estimation of future service demands as well as with workforce planning. It is highly desirable to be able to quantify the many variables known to influence nursing resource usage and costs so as to achieve equity in the system, manage nursing service provision more efficiently and to ensure there is the capacity to care to achieve the best possible outcomes for all patients and residents in need of nursing care. One systematic review and meta-analysis examined the association between nurse staffing levels and nurse-sensitive patient outcomes in acute specialist units. It found that nurse-to-patient ratios influence many patient outcomes, including nurse-sensitive outcomes, but most markedly was the impact on in-hospital mortality. A number of significant limitations were reported from this review of heterogeneous studies, including a diverse mix of patient types, and the way that nurse to patient ratios were calculated [1].

There are many factors that influence the demand for nursing services and nursing workloads. Actual nursing hours provided and the staff skill mix have a major impact on patient outcomes and on the cost of nursing services. Rostering practices relative to employment conditions also have a major cost impact. All of these factors must be considered when measuring nursing care demands and the organizational capacity to care. Ideally one needs to be able to group patients/residents in meaningful ways by representing homogeneity of nursing resource usage using data analytics. This requires the adoption of data standards.

This publication aims to unpack the many variable factors known to influence nursing care demand and nursing workloads, and will refer to extensive research studies which provide the evidence that demonstrates their significance. Today's information systems are able to produce data as a by-product of patient care systems. Data collection need not be a separate impost on those providing care. Information is today's most powerful healthcare resource. We propose the adoption of greatly improved datasets to optimize information use. These new datasets enable empirical research to be undertaken to determine associations between staffing methodologies in use, balanced workloads and the health service's capacity to care. Such capacity needs to be demonstrated by a low incidence of adverse events, positive patient outcomes as well as patient and nurse satisfaction.

This evidence-based approach is supported by the Nursing Profession at large. The International Council of Nurses (ICN) has released a position statement on evidence-based safe nurse staffing [2]. This document indicates that:

> *Safe nurse staffing is a critical issue for patient safety and the quality of care in hospitals, community and all settings in which care is provided. Inadequate or insufficient nurse staffing levels increase the risk of care being compromised, adverse events for patients, inferior clinical outcomes, in-patient death in hospitals and poorer patient experience of care.*

The previous chapter referred to a systematic review that examined the impact of nursing workload and staffing [3] and another study that identified five attributes that defined nursing workload [4]. This chapter focuses on the actual data used to measure nursing care demand. One research study [5] of the state of this science, set out to identify key outcome variables as at 2005. These authors concluded that "much creative use has been made of existing data sources" and noted that: "data in large-scale datasets on a wider range of clinically relevant process and outcomes is needed to answer pressing questions regarding the management of nursing services in health systems worldwide."

A more recent systematic review of the literature [6] evaluated available research on nurse staffing methods from the perspective of links with patient outcomes and equity of nursing workloads. Previous findings were supported as they concluded that "there is a need to develop an evidence-based nurse-sensitive outcomes measure upon which staffing levels for safety, quality and workplace equity could be established." These authors identified the need for an instrument that is valid and reliably projects nurse staffing requirements in a variety of clinical settings. They noted that Nurse-sensitive indicators reflect elements of patient care that are directly affected by nursing practice.

## Nurse staffing methods in use or recommended

The American Nurses Association (ANA) founded a national database of nursing quality indicators (NDNQI) to develop national data which identifies relationships between nurse staffing and patient outcomes and to provide participant hospitals with national comparative unit level data for use in quality improvement activities. The indicators developed reflect structure, process and outcomes of nursing care. Indicators for structure include Nursing Hours Per Patient Day, nursing turnover rate, RN education/Certification and staff mix.

A number of indicators are for both process and outcome such as inpatient fall rates, length of stay in hospital and pressure area injury rates. Process indicators are about measuring methods of patient assessment and nursing interventions undertaken. Nosocomial infection rates, the incident rate of deep vein thrombosis post-surgery, and readmission rates are outcome indicators and were some of the indicators included in the Joint Commission's core measures. This database was acquired by Press Ganey and commercialized in 2014 [7]. This is used by 2000 hospitals nationwide and tracks up to 19 nursing-sensitive quality measures based on unit level data [8]. It is the richest database of nursing performance in the United States.

The ANA collaborated with health care insight leaders Avalere, a panel of top nurse researchers, nurse leaders and managers of nursing services, to explore the clinical case for the effect of nurse staffing models [9]. This has resulted in a white paper that concluded that "staffing levels in a value-based health care system should not be fixed, as day-to-day hospital requirements are constantly in flux." Their report recommends that staff levels depend on the following factors:

- Patient complexity, acuity or stability
- Number of admissions, discharges, and transfers
- Professional nursing and other staff skill level and expertise
- Physical space and layout of the nursing unit
- Availability of technical support and other resources

The US Veteran Affairs have a long history of reviewing and testing staffing methodologies [10]. As from the early 1990s they have made use of an Expert Panel Nurse Staffing and Resource Management Method to determine staffing levels for over 170 VHA hospitals. Ten years later this was viewed as being overly complex. A new staffing methodology consisting of a multi-step process designed to lead to projections of full-time equivalent (FTE) employees required for safe and effective care across all in-patient units, was introduced in 2011. Lessons learned following an evaluation of its implementation indicate the importance of leadership buy-in commitment at all levels of the facility. Learning to interpret and rely on data requires a considerable shift in thinking for many. Staff need to see positive change, if not then apathy and cynicism sets in [11]. This staffing methodology was updated and became the new directive in 2017 [12]. It is described as:

> *a budgeting and forecasting process to determine resource requirements based on an analysis of multiple variables to include Veteran care needs, environmental factors, organisational supports, trends in performance metrics and professional judgment to provide safe, effective, quality care at various points of care.*

The latest Grattan report regarding Australia's health system, indicates that safer care saves money [13]. Their recommendations require hospitals to provide very detailed comparative data to see opportunities to make unit by unit improvements, track rates of complications, transparent measurements and reporting and hold hospitals to account. The reduction of complication rates was estimated to lead to national savings of $1.5 billion each year. Nurses are well placed to make very significant contributions to realize such improvements.

To examine these factors in more detail we first examined commonly used nursing care demand measurement methods in use, followed by an analysis of the validity of data used. This is followed by an examination of known influencing con-founding variables, information flows and patient journeys as a means to also identify relationships between these variables. The complexity and number of concepts one needs to consider when measuring nursing workloads, makes it clear that there is a need to develop an agreed metadata set, identify existing data standards and data gaps to facilitate a digital transformation. This is especially important for the production of an evidence based staffing methodology [14, 15].

## Methods in use to measure nursing care demand

A desire to measure hospital nursing care demand is historical [10]. The literature has detailed numerous studies each adopting various methodologies over the years since the 1950s. A major literature review undertaken during the early 1970s concluded that "Techniques are loosely described, samples are small or ill defined, instrument testing is deficient, rationales are weak, and in most instances care quality is assumed, based on judgments of the nurses involved [10]" (p. 42). Many studies have been reported since, using a variety of methods, where each study examined one of many possible variables, most based on small sample sizes, such as one unit, or one hospital, or one particular clinical specialty. Others have lost their acceptance as being reliable or valid over time.

Two systematic literature reviews were undertaken more recently, one looked for evidence related to mandated nurse staffing ratios in acute hospitals, another looked for evidence associated with nurse staffing relationships with outcomes [6, 16]. Another scoping literature review confirmed that Nurse staffing levels have a direct relationship with nursing working conditions, including workloads, and patient outcomes [17]. A low level of registered nurse staffing was found to be associated with omissions of essential care in another literature review [18]. These latter authors concluded that adequate staffing levels are essential to avoid missed care.

In 2010 the Royal College of Nursing published a very comprehensive guide on safe nurse staffing levels in the United Kingdom that details a number of workload measurement approaches in use plus their strengths and weaknesses. This publication includes recommended minimum staffing levels for a number of patient types [19]. Other staffing methods in use are Professional judgements or timed-task activity methods or patient classification systems also referred to as patient dependency or acuity systems. Min and Scott [20] identified the following six common nurse staffing measures:

1. nurse-to-patient ratios
2. full-time equivalents nursing staff per patient day
3. total Nursing Hours Per Patient Day (NHPPD) and/or RN hours per patient day
4. nursing skill mix
5. nurse-perceived staffing adequacy
6. nurse-reported number of assigned patients

Many studies have focused on finding evidence relative to specific patient outcomes. A Cochrane review [21] explored the effect of hospital nurse staffing models on patient and staff related outcomes. It identified 6202 studies that were potentially relevant to their review but only 15 studies were included. These authors found that the quality of the evidence overall was very limited and that the evidence in relation to the impact of replacing Registered Nurses with unqualified nursing assistants on patient outcomes, is very limited. However, it is suggested that specialist support staff, such as dietary assistants, may have an important impact on patient outcomes. One nurse staffing evaluation study found that the greatest differences in staffing measurements arise when unit level data are compared with hospital level aggregated data as reported in large administrative databases. These authors found that differences between databases may account for differences in research findings [22].

How best to determine nurse staffing requirements is a key issue for many. Our research and practical experiences have revealed many complexities associated with this activity. A recent study reviewed the approaches used to date, to investigate the idea of safe staffing, in an attempt to elicit knowledge to gain a deeper understanding of the inter-relational nature of this problem [23]. These researchers found that this topic is not limited to the nursing literature, so their search was expanded accordingly with a focus on acute hospital staffing. They found that the literature revealed a wide range of approaches used, these ranged from descriptive studies to operational mathematical models. It was concluded that the evidence in the nursing literature appeared to offer no firm guidance on staffing models or absolute solutions thus demonstrating the complexity of the problem.

An innovative nursing informatics method for measuring nursing services is the adoption of an electronic method referred to as the Workload Action Measures Method (WAMM) which makes use of the Clinical Care Classification (CCC) information model [24]. The CCC model is incorporated in the electronic health record system in use. A coded patient care plan can thus be documented. This plan is tracked for a specific inpatient episode of care. All relevant documented and coded nursing intervention actions are aggregated on discharge. These data are then converted into:

1. Care values consisting of nursing actions by action type and service time are calculated to become Relative Value Units (RVU).
2. Acuity Values consisting of nursing service requirements in four Healthcare patterns based on nursing diagnosis become Base Value Units (BVUs).

This method has demonstrated that the use of coded nursing data advances calculations of nursing services, in terms of workload and time, based on pre-determined RVUs reflecting the value of nursing time.

What has become apparent from these literature reviews is that there is no common or agreed framework enabling comparative evaluations of these various methodologies to be made. There is no consistency regarding the variables considered, their definitions and/or measures used for any of these methods, even where the nursing care demand measuring methods appear to be the same. At best one can group the methodologies employed according to some common features.

Inadequate nurse staffing levels tend to be the result of poor planning and allocations of nurse staffing establishments and budgets and/or faulty methods used to determine nursing care demands and/or poor staff distribution and rostering practices. The most recent, widely used and popular methodologies are the use of Nursing Hours Per Patient Day (NHPPD) or Hours Per Patient Day (HPPD) (using a bed census figure as a denominator) and the Nurse Staffing Ratio approach. Are these methods reliable and how fair and reasonable are these measures of nursing workload? This needs to be evaluated from a data consistency and degree of accuracy perspective. Few reported studies permit close scrutiny to facilitate an informed assessment of the validity of the final workload measure adopted. Of the studies providing actual Nursing Hours Per Patient Day reviewed, the validity and reliability were seen in most instances to be questionable.

### Nursing Hours Per Patient Day

The use of Nursing Hours Per Patient Day (NHPPD) was found to be the most popular nurse staffing variable among international experts [5]. One comparative study of nurse workload approaches noted that the use of Nursing Hours Per Patient Day (NHPPD) as applied to the American hospital system, makes use of the number of patients occupying a bed at midnight. This number is then multiplied with an agreed average of Nursing Hours Per Patient Day to arrive at the number of nursing hours required [18]. The use of bed occupancy at midnight as the indicator of a patient day, does not account for variable staffing needs over any 24 hr period that occurs when patients are cared for during the day and discharged prior to midnight [25]. Table 1 shows an average bed utilization of 6.67 patients over three shifts, compared with a midnight census of 3 patients reflecting an increasingly common occurrence of single day episodes.

**Table 1 Bed midnight census vs actual bed occupancy across three shifts in an acute hospital**

| | Day | Evening | Night | 2400 MN |
|---|---|---|---|---|
| Bed 1 | Discharge 1 @09:00<br>Admit 1 @11:00<br>Transfer out to ortho | Admit 1 @14.00 | Death 1 @23:00 | 0 |
| Bed 2 | Discharge 1 @10:00<br>Admit 1 @12:00 | 1 | 1 | 1 |
| Bed 3 | 1 | Discharge 1 @16:00<br>Admit 1 @ 18:00<br>Transfer out to ICU<br>Admit 1 @22:00 | 1 | 1 |
| Bed 4 | 1 | Discharge 1 @16:30 | – | 0 |
| Bed 5 | 1 | 1 | 1 | 1 |
| Bed 6 | Discharge @09.00 | – | Admit 1 @02:00 | 0 |
| Total bed occupancy | 8 | 7 | 5 | 3 |

*From the TrendCare® General Acuity PowerPoint Presentation.*

Nurse administrators allocate staff to care for every patient as they arrive, while Chief Financial Officers allocate nurse staffing budgets sufficient only for those who generate room and board charges at midnight. This causes numerous budget discrepancies [18]. These authors noted that although NHPPD is commonly used to examine nurse staffing levels, the reliability and validity of this measure have rarely been examined. Studies reviewed by them had noted that there was no clear operational definition of the measure, there was no single reliable and valid database and the databases used for various studies provided different levels of data (hospital-level and unit-level). Thus, they concluded that these diverse data collection processes and aggregation methods used to identify nurse staffing, meant that accurate comparisons across studies could not be made or be meaningful.

Both NHPPD and skill mix are usually derived from administrative data [26–29]. One needs to consider the accuracy of such data by questioning the definition of a patient day as well as how the hours used are obtained and calculated. One needs to ask: do the number of nursing hours used to derive these measures, represent actual hours worked (productive hours) or scheduled hours, or hours paid which include various types of leave provisions? This issue was addressed by the American National Quality Forum that has endorsed a standard definition for the Nursing Hours Per Patient Day (NHPPD) Measure [30]. This NHPPD is defined as:

*The number of productive hours worked by nursing staff (RN, LPN/LVN, and UAP) with direct patient care responsibilities per patient day for each in-patient unit in a calendar month.*

The numerator is defined as the:

*Total number of productive hours worked by nursing staff with direct patient care responsibilities for each hospital in-patient unit during the calendar month.*

The denominator's definition is:

*the total number of patient days for each in-patient unit during the calendar month. Patient days must be from the same unit in which nursing care hours are reported.*

Exclusions for the denominator are:

*Patient days from some non-reporting unit types, such as Emergency Department, peri-operative unit, and obstetrics.*

For these definitions to be meaningful and consistently interpreted, there would also need to be definitions for "direct patient care responsibilities," as well as for "patient day," "in-patient unit," and "productive hours." The US Veteran Affairs staffing methodology includes the use of NHPPD which is defined as "the total number of nursing care hours available divided by the number of patients in a 24 hour period" [12].

This NHPPD definition is a good start but is this measure in common use? Do these metadata comply with any metadata representation standard such as ISO/IEC 11179 to ensure consistent interpretation? How are the productive hours worked meant to be calculated? Actual NHPPD thus arrived at are simply based on the number of staff allocated and the number of patients occupying a bed at midnight or at the beginning of each shift? This measure can be grossly inaccurate, non-comparable between healthcare facilities and does not indicate real nursing service demand. NHPPD in this context is a retrospective measure of nursing hours worked which is reflective of the supply of nursing hours and does not reflect the actual required nursing hours/demand for nursing hours. These retrospective measures when used to plan staffing establishments can perpetuate the current under or over staffing in nursing services and not address the actual demand for nursing hours.

### Nurse staffing ratios

Mandating nurse staffing ratios is a popular industrial requirement for many nursing unions. This requires an agreed ratio of registered nurses (RNs) for different types of in-patient care units. In some instances, these requirements are included in legislation making this a legal requirement. Unions view this methodology as the best practical method to ensure patient safety as, subject to resource availability, it is viewed as fairly easy to implement. The use of this method is not subject to a separate data collection method as staff resource data as well as patient day data are readily available from administrative systems. The staffing numbers' degree of accuracy is unknown.

Nurse patient ratios are generally measured and applied at the beginning of a period/shift with no adjustments being made if the number of patients decrease or increase during the period/shift. The application of nurse patient ratios does not accommodate variances in acuity between patients, unpredicted changes in patient acuity or fluctuations in patient numbers throughout the period/shift.

Virginia Plummer completed a 4-year comprehensive review of the TrendCare patient dependency/acuity system and compared its use to the nurse/patient ratio staffing methodology used in Victoria, Australia. In her conclusion she stated:

> *TrendCare can provide fairer and more equitable workloads and at a lower cost than the Victorian mandated nurse/patient ratios [31].*

A 2004 systematic review that investigated the effects of nurse staffing in the acute care setting on patient, nurse employee and hospital outcomes found no evidence to support nurse-patient ratios [32]. Other variables such as patient acuity, skill mix, nurse competence, nursing process variables, technological sophistication and institutional support of nursing were recommended alternative variables to be considered when arranging staffing requirements.

In 2015, The Safe Staffing and Healthy Workplace Unit of the New Zealand Ministry of Health commissioned a significant review of their safe staffing methodology which included a review of the TrendCare acuity tool used to provide the acuity data for their care capacity management workforce calculations. The TrendCare acuity tool was described in the final report as "currently the only validated patient acuity tool available" [33].

The 2015 updated Victorian legislation makes use of a greater number of nursing service types for which nurse to patient ratios are listed [34]. This includes a list of many definitions for terms referred to in the Act. Ratios represent minimum requirements and apply to the actual number of patients in each ward or unit of service. Ratios differ for night shifts and for afternoon shifts by hospital type ranging from level 1 to 4 as shown in Table 2. There is also a provision regarding skill mix. Amendments were introduced in February 2019.

Table 2 Victorian legislated nurse to patient ratios by patient type

| Safe patient care (nurse to patient and midwife to patient ratios: Act 2015 | Victorian legislated nurse to patient ratios | | |
|---|---|---|---|
| | Morning shift | Afternoon shift | Night shift |
| General medical or surgical – Level 1 hospital – plus nurse in charge | 1:4 | 1:4 | 1:8 |
| General medical or surgical – Level 2 hospital – plus nurse in charge | 1:4 | 1:5 | 1:8 |
| General medical or surgical – Level 3 hospital – plus nurse in charge | 1:5 | 1:6 | 1:10 |
| General medical or surgical – Level 4 hospital – plus nurse in charge | 1:6 | 1:7 | 1:10 |
| Aged high care residential – plus nurse in charge | 1:7 | 1:8 | 1:15 |
| Emergency department p-plus 1 nurse in charge and 1 triage nurse AM, 2 triage PM, 1 night[a] | 1:3 | 1:3 | 1:3 |
| Coronary care units – plus 1 nurse in charge | 1:2 | 1:2 | 1:3 |
| High dependency unit[a] | 1:2 | 1:2 | 1:2 |
| Palliative care inpatient units – plus 1 nurse in charge | 1:4 | 1:5 | 1:8 |
| Rehabilitation and geriatric evaluation management – plus 1 nurse in charge[a] | 1:5 | 1:5 | 1:10 |
| Operating theaters – instrument, circulating and anesthetic nurse for each theater[a] | 1:3 | 1:3 | 1:3 |
| Post-anesthetic recovery room – per each unconscious patient | 1:1 | 1:1 | 1:1 |
| Special care nurseries[a] | 1:4/1:3 | 1:4/1:3 | 1:4/1:3 |
| Neonatal intensive care units – plus 1 nurse in charge | 1:2 | 1:2 | 1:2 |
| Antenatal and postnatal wards plus 1 midwife in charge – patient does not include newborn | 1:4 | 1:4 | 1:6 |
| Delivery suites[a] | 2:3 | 2:3 | 2:3 |

[a]Additional instructions apply.

Hospitals need to be able to provide documentary evidence demonstrating compliance. There need to be rules regarding rounding up or down as the number of patients in any one ward can rarely be divided into groups of 3, 4, or 5. Similarly the use of the midnight census to count patient numbers varies from equating one patient day with a 24 hr occupied bed. The Victorian legislation has many additional interpretive clauses to handle such situations. It's not clear how healthcare facilities need to handle their inability to meet these requirements when they are unable to recruit sufficient numbers of nurses due to workforce shortages including staff absences due to illness. The evidence listed as supporting these Victorian ratios were from studies undertaken by Aiken et al. published in 2002 [26] and one other based on pediatric patients published in 2013 [35, 36]. It's apparent from the mandated Californian and Victorian ratios that there is no standard.

California adopted this approach in 2004 by mandating minimum ratios "at all times," this includes during meal breaks [29]. This latter requirement adds to the implementation costs. The California legislation identifies different ratios for different types of patients as shown in Table 3.

Table 3 Californian mandated patient/nurse ratios

| Unit type | Patient/nurse workload mandated by California legislation |
| --- | --- |
| Medical-surgical | 5:1 |
| Pediatric | 4:1 |
| Intensive care units | 2:1 |
| Telemetry | 5:1 |
| Oncology | 5:1 |
| Psychiatric | 6:1 |
| Labor/delivery | 3:1 |

By January 2013, 14 States and the District of Columbia had enacted some type of hospital nurse staffing regulation ranging from nurse-to patient ratios, staffing committees and public disclosure of staffing. An overview of the types of staffing regulations in effect across the United States found that associated enforcement provisions were weak or absent [37]. These authors argue that compliance data should be viewed as a critical piece of data in research on outcomes. A national campaign by the National Nurses United (NNU) to achieve nurse staffing regulation for other American States continues [38]. NNU are of the view that safe RN ratios have been proven to improve the quality of care and nurse recruitment and retention in Californian hospitals as a consequence. These findings are not supported in the literature [6, 21, 39].

Associations were reported from a study by the Agency for Healthcare Research and Quality (AHRQ) following a review of observational studies. It was noted that these associations are not necessarily causal. They concluded that estimates of the size of the nursing effect must be tempered by provider characteristics including hospital commitment to high quality care, which is not considered in most of the studies. Meta-analysis was used to test the consistency of the association between nurse staffing and patient outcomes. Classes of patients and hospital characteristics were analyzed separately [40]. Their result indicated that:

* Higher registered nurse staffing was associated with less hospital-related mortality, failure to rescue, cardiac arrest, hospital acquired pneumonia and other adverse events.
* Increased registered nurse staffing had a strong and consistent effect on patient safety in intensive care units and for surgical patients.
* A greater number of registered nurse hours spent on direct patient care was associated with a decreased risk of hospital related death and shorter lengths of stay.
* More overtime hours were associated with an increase in hospital related mortality, nosocomial infections, shock and bloodstream infections.
* No studies directly examined the factors that influence nurse staffing policy
* Few studies addressed the role of agency staff.
* No studies evaluated the role of internationally educated nurse staffing policies.

The purpose of the studies reviewed has been to identify links between any type of patient and/ or performance outcome with nurse staffing levels expressed as patient to nurse workloads

[5, 27–29, 36]. Research methods commonly used were retrospective observational and/or cross-sectional studies linking nurse surveys, inpatient administrative data and information from other relevant sources. This was necessary as more accurate reliable and relevant nursing data have not been readily available from current information systems in use. Essentially these studies do demonstrate that appropriate nurse staffing levels contributes favorably to a healthcare facilities' capacity to care, cost effective care and to improving patient safety and patient outcomes.

The adoption of easy to obtain administrative data and their use as indicators of nursing service demand or workloads such as NHPD or nurse/patient ratios is easy to understand given the complexities and relationships between the number of confounding variables identified. The initiative to regulate nurse staffing may well meet nursing union requirements and ensure that nursing service funding needs are given high priority. One needs to ask how were these ratios arrived at? How well do these represent actual nursing service demands? These studies are not about measuring actual nursing service demands or workloads. The adoption of nurse-patient ratio or Nursing Hours Per Patient Day methodologies is costly and inefficient. The adoption of these nursing staffing methodologies will result in periods of both over and under staffing relative to service demand needs. Although nurse patient ratios relate to specific groups of patients, these groups are very broad. Each group includes a wide variety of patient types with significant variances in nursing intensity required for care for each type. Given the highly variable service demands in acute care settings, the many different clinical specialties and associated treatment and diagnostic services provided, and the many associated special clinical knowledge and skills required, these nurse staffing methods do not ensure the best possible use of available resources relative to individual patient care needs.

### Patient/client types

Every patient/client is an individual with unique health care needs. Identifying possible types of patients for the purpose of establishing or measuring nursing service demand and/or actual workloads, may be achieved using a variety of criteria. Nurses provide services to anyone with a need including families, caregivers and the population at large. They undertake a variety of health monitoring and advisory preventative and ongoing health maintenance services within the community or any healthcare facility. There are enormous variations in nursing service needs between patients allocated to wards that may or may not be named general medical/ surgical; many wards in large teaching hospitals are named according to the clinical specialty such as orthopedics or cardio-vascular representing the types of clinical cases treated and cared for. Our focus is primarily on nursing workloads in acute care although these principles apply equally to any sub-acute or long term residential facility.

The nurse to patient ratio staffing methodologies adopted by acute care in-patient services have tended to classify patients according to clinical specialty or service type in very general terms. Patient classification systems have been around for a very long time. They refer to the ordering of phenomena into sets of descriptors or characteristics which determine a person's status as a patient at any point in time. These may be used as indicators of dependence on nursing services. Patient classification systems represent an attempt to standardize important components of patient phenomena. Each set of descriptors, class or category is given an appropriate label and needs to be accurately defined. Such classification systems are also known as "acuity" systems. Such systems could be viewed as "nursing casemix systems." They may be used to form the basis for activity-based funding.

## Patient classification

Two types of patient classification systems exist, prototype and factorial type. A prototype classification system uses an overall description of a typical patient in each category, such as acutely ill, moderately ill, or mildly ill or medical/surgical, intensive care etc. For a midwifery example, low risk labor ward patients may be described as being in labor with a full term live single pregnancy, vertex presentation. Patients are graded or classified to represent a patient's greater or lesser nursing dependency.

A factorial type system uses clearly defined indicators (factors) of dependency, rather than clinical specialty, by which to classify individual patients. It relies on the identification of readily observed patient characteristics and/or specific elements of care. Such a classification method will have instructions regarding how such characteristics and/or care elements are combined in order to assign a patient to an exclusive class. Factors may be mobility assistance needs, hygiene assistance needs, drug administration, treatments, teaching and emotional support needs, and other known significant indicators of nursing workload. Factorial type systems tend to be more reliable in terms of inter-rater reliability than prototype systems. These are also readily audited as all indicators identified as being applicable to a particular patient are discernible from patient documentation. With appropriate electronic health record (EHR) system design such indicators should not need to be collected separately. Both the TrendCare and the PAIS acuity systems are factorial type systems used to identify a patient's level of nursing dependency.

Once discrete classes are described using the prototype or factorial method, work measurement techniques may be applied to quantify the average nursing resource usage per patient type. Such work measurement studies were undertaken as a component of the PAIS methodology and work measurement studies continue to be undertaken for the TrendCare system. Nursing work and work measurement methods are described in some detail in Chapter 4. As a result of these very extensive data sources the TrendCare system has been able to carefully differentiate between patients on the basis of actual nursing resource usage, to the point where it has now defined over 200 distinct patient types with associated standard time values (care hours required per patient per day HPPD-24 hours) that can be used as benchmarks. Table 4 shows some established TrendCare benchmarks for Patient types using bed utilization as a denominator and the calculated HPPD for nurse patient ratios of morning (1:4), evening (1:5), night (1:8).

**Table 4 Established TrendCare benchmarks by patient type**

| Patient types | Acuity HPPD | HPPD by ratio (1:4) (1:5) (1:8) |
| --- | --- | --- |
| Medical — stroke/CVA | 5.1 | 4.6 |
| Medical respiratory | 4.5 | 4.6 |
| Medical palliative care end stage | 6.5 | 4.6 |
| Medical cardiology | 4.3 | 4.6 |
| Surgical colorectal | 5.2 | 4.6 |
| Surgical urology | 4.4 | 4.6 |
| Surgical orthopedic | 4.8 | 4.6 |
| Surgical gynecology/urology | 4.2 | 4.6 |

From the TrendCare® Recommended Benchmarking Booklet.

It is important to note that HPPD for different patient types cannot be added and averaged to obtain an average HPPD for a ward/service, as the % of patients in each patient type may vary greatly.

The use of these time values for each patient type can be used to establish the most likely future nursing demand and is far more accurate and efficient than the use of ratios or NHHPD calculated from nursing hours worked. These data are being produced as a by-product of day-to-day patient care documentation as opposed to the need to extract ratio data and nursing hours worked from administrative systems. Work measurement methods can only be applied when the work to be measured is accurately described, but before we do that there is a need to identify the many confounding variables that influence nursing work. Once these are known it will be possible to establish various relationships and use this knowledge to take action with the intent to minimize their impact on nursing workloads.

Another example is the RAFAELA® Patient Classification system originally developed at the Vasa Central Hospital in Finland during 1995–2000. It was tested and standardized in a national multi-centered study between 2000 and 2002 and has been used for systematic benchmarking nationally since. Its use to predict hospital mortality was validated in 2016 [41]. This system is now used by most Finnish hospitals and is being rolled out in Europe and Asia [42]. RAFAELA® makes use of the Oulu Patient Classification (OPCq) instrument measuring five levels of nursing intensity, and the Professional Assessment of Optimal Nursing Care Intensity Level (PAONCIL) classification method developed as an alternative to time studies [43]. It makes use of six areas of nursing as follows:

1. Planning and co-ordination of nursing care
2. Status of health
3. Medication and nutrition
4. Hygiene and excretion
5. Activity/movement, sleep, and rest
6. Teaching, guidance, care follow up and emotional support

One of four requirement levels (minimum 1 point, maximum 4 points) is then selected for each of the above areas of nursing care for each patient based on care actually received. These are then totalled to represent an overall nursing intensity (NI) ranging from 6 to 24 points for each patient. These totals are then used to classify patients into one of five patient categories [42, 44]. The OPCq has now been cross mapped with the Finnish Care Classification (FinCC) nursing terminology to provide standard nursing codes used by the structured Finnish documentation model in use and incorporated within electronic health records [45]. This has made it possible to evaluate the relationship between coded nursing data (FinCC) and NI data collected using the RAFAELA system. A clear statistical relation was found between the number of nursing diagnoses used and documented interventions from the FinCC system, and the NI categories measured by the OPCq. This has demonstrated the potential to automate and re-use coded patient documentation for nurse staffing and workload analysis [46]. Electronic nursing documentation is detailed in Chapter 6.

In 2016 The Irish Health Services Executive (HSE) funded a major nursing research project to evaluate the established nurse staffing framework for all HSE acute hospitals in Ireland. Professor Johnathon Drennan from the Cork University led the research team and worked with an advisory panel of leading researchers in the field of nurse staffing methodologies. Professor Christine Duffield from Australia was included on the advisory panel. The TrendCare Patient acuity and workload management system was selected and used to collect the data for this research project. This choice was based on TrendCare's ability to integrate with organizational level patient information management systems and its capacity to electronically capture all components of the recommendations included in the nurse staffing framework including:

- patient acuity measures
- skill mix measures
- workload management and patient allocation
- calculation of NHPPD (required, actual, and variance)
- agency staff use, one to-one specialling
- overtime and absenteeism

Electronic data collection and system integration were particularly important to enable the future development of nursing intensity weight based costing relative to patient Diagnostic Related Groups [47].

## How do nursing service demand measurement methods compare?

An analysis of such events exposes the many political, industrial, management and personal agendas as well as power and governance issues at play. It demonstrates the need for all stakeholders to gain a better understanding and appreciation of the contributions made by the nursing profession to patient outcomes and the health system as a whole. As a consequence of these types of events the TrendCare system is now able to calculate NHPD as well as nurse-patient ratios and compare these methods to the use of an acuity measurement approach. Table 5 provides a list of general medical/surgical patients, representing a typical patient cohort on any general medical/surgical ward, with their associated level of acuity.

**Table 5** A nurse staffing comparison between patient acuity, NHHPD and nurse/patient ratios for one group of general medical/surgical patients in the first 12 beds of a 30 bed ward

| Patient type | Acuity hrs per day | Day | Evening | Night |
|---|---|---|---|---|
| Bed 1 | 9:35 | Discharge 1 @09:00 Admit 1 @11:00 Transfer out to ortho | Admit 1 @15.00 | Death 1 @23:00 |
| Bed 2 | 7:35 | Discharge 1 @10:00 Admit 1 @12:00 | 1 | 1 |
| Bed 3 | 11:45 | 1 | Discharge 1 @16:00 Admit 1 @18:00 Trans out to ICU @19:30 Admit 1 @22:00 | 1 |
| Bed 4 | 2:50 | 1 | Discharge 1 @16:30 | – |
| Bed 5 | 5:20 | 1 | 1 | 1 |
| Bed 6 | 3:40 | Discharge 1@09.00 | – | Admit 1 @ 02:00 |
| Bed 7 | 4:40 | 1 | 1 | 1 |
| Bed 8 | | 1 | 1 | 1 |
| Bed 9 | | Trans in 1 @10:00 | 1 | 1 |
| Bed 10 | 12:40 | Death 1 @12:00 Admit 1 @14:00 | 1 | Trans out 1 to ICU @23:30 Admit 1 @05:00 |
| Bed 11 | 3:40 | 1 | 1 | 1 |
| Bed 12 | 4:30 | 1 | 1 | 1 |
| Total acuity hrs | 66:45 | 36 | 23:45 | 8 |
| Bed utilization | 13.33 | 5 | 13 | 12 |
| Bed occupancy by census | 9.67 | 11 | 10 | 8 |
| **EFT acuity** | **8** | **4.5** | **3** | **1.5** |
| **Min staffing** | **6** | **2** | **2** | **2** |
| **EFT ratio** | **6** | **3** | **2** | **1** |

*Copyright Trendcare.*

The bottom line in this table provides a comparative analysis in terms of staff allocations per shift of the use of these three methodologies for that list of patients. This includes the consideration of likely patient discharges and new admissions based on an average length of stay of 3–4 days. We know that any bed vacancies in acute hospitals tend to be filled within 2 hours. This means an average daily turnover for a ward of 28 patients is 7 to 8 patients. These small per shift staffing number variations have a significant impact on any organisational nurse staffing establishment and costs. An example of patient turnover and the associated complexity of calculating nursing service demand using the NHPPD or the ratio method can be appreciated from Table 1. The types of patients occupying this fictious ward are described using the TrendCare patient type description.

It's important to note that this example only applies to this mix of patients and their associated turnover activity. This varies greatly between clinical service and patient types. In every instance it will be possible to establish activity patterns which are useful for planning purposes. In every instance the EFT Acuity approach to measuring nursing workload is the most accurate reflection of staffing requirements. This instance demonstrates why nurses are dissatisfied and become burned out due to excessive work demands not well met when minimum staffing or EFT ratio staffing methods are adopted. For other groups of patient types the reverse is true, reflecting unnecessary costs due to excessive nurse staffing levels relative to nursing service needs. Another factor to be considered is the minimum staff required to ensure patient safety, such as for many types of patients two nurses are required to work together to reposition patients in bed or to facilitate checks for dangerous medication administration including blood transfusions.

### Patient type and treatment protocol patterns by clinical speciality

The patient/nurse ratio nursing workload measurement method relies on a small selection of commonly occurring patterns associated with patient types such as surgical, medical, intensive care etc. Such methods are useful for safe minimal nurse staffing estimation purposes in non-digital environments. In a digital age we are in a much better position to refine these further. The ability to process greater detail, combined with increasing clinical complexity makes it imperative to better match knowledge and skills with care and treatment demands. The identification of patient types and associated care and treatment protocols within every clinical specialty enables the development of commonly occurring patterns within each patient co-hort. Such patterns can be used as benchmarks by making use of them as a standard and assessing any deviation from that for individual patients. Many clinical guidelines have been developed but these are not necessarily made available in a format that enables a computer to interpret and make use of them [48–50]. The use of such protocols and guidelines enables practice-based evidence to be collected. The development of Patient types based on nursing resource usage is explained in Chapter 4.

## Variables influencing nursing service demands, workloads, and costs

To evaluate the efficacy of any of the above methods in use one needs to have a sound understanding of the many variables that may need to be considered, their definitions, metadata and the many relationships between metadata items. These are those variables that influence and/or generate nursing service demand, describe nurse allocation and rostering practices, nursing working conditions, workplace inefficiencies and nursing services that influence patient outcomes. Numerous reported studies undertaken over many years have identified any one or several of these variables and evaluated their significance [5, 51, 52]. These have been further evaluated in extensive research undertaken by the authors in many hospitals. Known significant influencing factors and variables are summarized in Table 6.

**Table 6 Variables known to influence resource usage and in-patient costs**

| Patient/client admission | 1. Community morbidity patterns<br>2. Choice of health services available<br>3. Consumer expectations<br>4. Health status/degree of risk for poor health outcomes<br>5. Organization's role/purpose<br>6. Suitability of mix of services provided<br>7. Service capacity<br>   a. Bed availability<br>   b. Nursing staff availability<br>   c. No of patients waiting for service<br>   d. Service utilization policies<br>8. Admission policies | Variables indirectly influencing patient/client admission |
| --- | --- | --- |
| Length of stay or period of service provided | 9. Availability of service mix required<br>10. Actual treatment/care services provided<br>11. Philosophies, treatment/care appropriateness, safety and quality of care, adverse events, complications<br>12. Service provision efficiency and effectiveness — service scheduling<br>13. Organizational work and information flows, resource allocation efficiency<br>14. Discharge policies — may be based on casemix, funding rules | Variables indirectly influencing length of stay or period of service and costs |
| Nursing service demands and workloads | 15. Patient characteristics<br>16. Number of patients, their severity of illness and complexity of care requirements for which any nurse is responsible at any one time, such as:<br>   a. Activities of daily living support requirements<br>   b. Mobility assistance<br>   c. Nutrition support<br>   d. Continence care<br>   e. Medication management<br>   f. Technical activities, e.g., medical emergency response<br>   g. Wound care<br>   h. Observations and assessments<br>   i. Patient education/instructions/reassurance<br>   j. Admission and discharge activities<br>17. Availability of supporting resources, meal delivery, patient transfers, lifting assistance, specimen collection, equipment, linen, surgical supplies<br>18. Documentation requirements, information flow methods and processes<br>19. Nursing staff distribution, skill mix and availability<br>20. Use of casual/agency/supplemental staff<br>21. Rostering practices<br>22. Models of care, team vs patient allocation vs task allocation<br>23. Ability to adhere to professional standards, organizational policies and procedures<br>24. Organizational safety policies and practices<br>25. Research participation, practice evaluation opportunities<br>26. Required administrative contributions, meetings<br>27. Paid continuing professional development<br>28. Nursing work environments, multidisciplinary staff relationships | Variables directly influencing nursing workloads, patient outcomes and costs |

The chapters that follow describe many of the most significant and manageable variables, complete with their relationships, in considerable detail based on the authors' research results and change management experiences. A sound knowledge regarding all nursing data and information needs, plus appropriate data collection strategies, data processing and the use of analytics enables organizational operational performance to be improved and achieve significant cost saving while maintaining patient safety and improving the delivery of quality care.

The ICN's position statement [2] lists the following elements for consideration to achieve evidence based safe nurse staffing:

- Real-time patient needs assessment.
- Local assessment of nurse staffing requirements to provide a service.
- Nursing and interdisciplinary care delivery models that enable nurses to work to their optimal scope of practice.
- Effective human resource practices to recruit and retain nurses.
- Healthy work environments and occupational health and safety policies and services that support high quality professional practice.
- Workforce planning systems to ensure that the supply of staff meets patient needs.
- Tools to support workload measurement and its management.
- Rostering to ensure scheduling meets anticipated fluctuations in workload.
- Metric to assess the impact of nurse staffing on patient care and policies that guide and support best practice across all of these.

## Information flows and patient/client journeys

Patient/client health journeys tend to be unique for every individual from the beginning of any episode requiring a health service. It consists of an end to end sequence consisting of many steps required to provide a complete clinical service. Each journey consists of interactions at various (one to many) points of care with any number of members of the health workforce and healthcare facilities over any period of time. Along this journey one needs to consider the physical transport aspects between point of care locations as well as demographic, clinical information and communication flows to ensure care continuity. Thus, patient transport support services are critical; in their absence Nurses frequently fulfill this role. In other instances, nurses need to accompany patients being transported from one department to another. In such situations they may be required to provide a "handover" to another health professional regarding continuing required care.

Admission to any health facility is dependent upon any of the factors identified in section 1 in Table 6. When required services are not available at the first point of care, then the patient is transferred to another service. This may be via any mode of transport or alternatively, where available, additional services may be provided via digital and/or telehealth arrangements. One example is a trauma case in a rural or remote area, where images or direct physiological monitoring data may be transmitted for external specialist advice. Nurses and first responders make use of such facilities where indicated and possible. Other examples relate to the management of any number of chronic diseases or for the collaborative multidisciplinary management of lengthy treatment plans requiring inpatient and out patient services at various times. For those with chronic long-term conditions or conditions requiring adherence to a complex treatment regime, nurses may provide a case management or patient advocate role.

An example associated with bed availability may be the result of management's poor use of historical data to establish a norm as a percentage of emergency attendances that are likely to be admitted. Such admissions are prioritized over elective surgical cases on waiting lists. When a conflict occurs, elective surgical cases are canceled. Such events impact on projected nursing workloads in ward areas.

Relevant data and information collected along the way needs to be accessible at each point of care. This information may be transported by the individuals themselves or their carers and/ or be communicated via any suitable or available communication method. Access may be supported by frontline clerical staff; in their absence nurses take on such roles. In a digital health ecosystem such information is stored in a digital format and accessible from any device by those who need to make use of it and who have authorized access. Nurses are frequently required to access this data to develop nursing care strategies that facilitates continuity of care and minimize the risk of delayed care.

Once a patient is admitted to a health service, nurses make use of the data that were used to determine that admission, to form the basis for the development of the patients' nursing care plan. Data and information are added to the record from others providing a clinical service, this information is also used by nurses for their decision making. Clinical care documentation consists of five distinct entry types [53]

1. Observation-results of observations made, anything measured by a nurse, other clinicians, a laboratory, or reported by the patient as a symptom, event or concern.
2. Instructions-regarding proposed interventions including medication and treatment orders.
3. Action-documentation regarding actions taken (e.g. wound care), medication administration, completed treatment interventions, etc.
4. Evaluation-opinions based on evaluations made such as risk potential, goals to be achieved, recommendations for follow up care.
5. Admin-entry-administrative information including, appointments, consent etc.

Nurses make use of and contribute to this information as they are the ones who may need to prepare patients for proposed interventions and/or actually complete the required interventions. They add their own observations, instructions and opinions for other clinicians to use, to ensure care continuity. Nurses are usually responsible for coordinating patient centric activities, including patient transport as required, as well as allocating additional resources or redistributing available resources to meet ad hoc unplanned new service demands at any point in time. Clinical care is very much a team effort requiring good collaboration and communication practices. Nurses often take on the responsibility of ensuring that continuing and timely clinical decision making is not compromised. Delays concerning information flow are costly and have the potential to lead to adverse clinical events due to missing information.

An analysis and mapping of data and information flows tends to identify areas of shortcomings where changes can make great improvements to reduce costs and improve operational efficiencies. This process enables the identifications of summary data required at various points of decision making throughout a patient's health journey. This knowledge forms a useful basis for designing information system dashboards where data from multiple sources are presented on one screen to assist decision making at any level within the healthcare facility.

## Digital transformation enabling nursing data inclusion

Health data represents the most valuable asset within the health industry. All patient information is primary data collected at the source where it is important to ensure these data accurately reflect the concepts that need to be conveyed. Data sets derived from these data are referred to as secondary data. It is important to make use of data standards as the means of ensuring that data collected in one facility means the same as in another to allow for comparisons to be made between these facilities. Standardized health data are reflected in health care terminologies which may be referred to as "artefacts that provide standardised meaning of human language expressions used in oral or written communication within a given domain." Collectively terminologies represent a terminology ecosystem. There need to be mechanisms to facilitate co-operation including common areas of governance across national terminology or equivalent centers [54, 55].

Given the current digital transformation we are now in a far better position to more accurately calculate nurse staffing requirements by measuring actual service demands. The projection of staffing requirements for nursing services requires the use of an information system with the capability of measuring all elements of nursing work, calculating the variances between supply and demand of nursing hours and enabling real time adjustments to be made at short notice in response to unplanned events. It is also necessary to analyze the factors influencing nursing work in a manner that enables the automation of such calculations based on real time operational data. This requires the adoption of coded data and data standards.

## Nursing minimum data sets

Minimum data sets consist of an agreed core set of well-defined standard data elements used for mandatory collection and reporting purposes. The need for a Nursing Minimum Data Set (NMDS) was first identified during the 1970s and published in 1988 [56]. Since then similar work has been undertaken in other countries. One evidence-based Nursing Minimum Data Set (NMDS) has been used by all Belgium hospitals and reported nationally since 1988. This has resulted in an extensive databank. This system was updated in 2002 focusing on care of the elderly, intensive care, chronic care, pediatrics, cardiology and oncology. This covered day clinics, short stay, in-patient and intensive care stays [57]. This minimum set of items of information has always enjoyed uniform definitions and categories pertaining to patient demographics, 23 nursing interventions and nurse staffing data. Data were collected for 5 out of 15 days every three months between 1988 and 2007 covering 19 million nursing records. In 2007 valid, reliable and usable nursing weights per NMDS intervention and cost weights for DRGs were developed [52]. That study concluded that it is possible to weight nursing care based on an appropriate staffing level instead of actual staffing levels. Its second version was integrated and linked with hospital discharge datasets, it included 78 nursing interventions and the nurse staffing data now includes qualification levels. This was validated in 66 hospitals.

To date a number of NMDS' exist, including data sets used for a variety of patient classification systems and for other purposes, but data standards adopted to make up each NMDS are inconsistent and non-comparable [58]. There continues to be no consistent, standardized, concise method for nurses to record information about their patients and clients that is conducive to store, retrieve, and use in patient and client care; to improve professional self-development; and to use in collaboration with patients and clients, their families, other nurses, doctors, hospitals and health systems [59]. It was recommended that Virginia Henderson's basic principles of nursing care as published by the International Council of Nursing in 1960 continue to be used. These foundational nursing principles have been translated into more than 30 languages, used in authoritative textbooks linking human needs to professional nursing. This is considered to be a valid method for nurses to record their clinical impressions for retrieval, analysis and application to patient care.

Attempts have been made to establish an International Nursing Minimum Data Set (i-NMDS) consisting of an essential data set of items of information with uniform definitions and categories concerning the specific dimension of nursing internationally [60]. Such a data set is considered to be useful for a multitude of purposes to meet information needs of multiple health data users worldwide. An i-NMDS needs to:

- Refer to the International Council of Nurses (ICN) and WHO strategic goals for nursing and midwifery.
- Focus on concrete and operational objectives such as nursing workload measurement.

- Institute a prospective design to facilitate big data analytics.
- Layout a data set for collection (data variables, codes, values, consistent with international standard for data specification {ISO/IEC 11179).
- Formalize the collection method.
- Identify the requirements for collection.
- Allow local terminology to be used and mapped to the International Classification of Nursing Practice (ICNP®).
- Make use of relevant reference terminologies or project endorsed classification systems.
- Enable its use to contribute to evidence-based nursing practice and allow for global comparisons to be made.

There is a need to further develop this i-NMDS to enable the nursing profession's contribution to health service outcomes to be recognized. Such datasets do require formal governance and maintenance arrangements. The Nursing Management Minimum Data Set developed during the 1990s consists of two metadata items, Environment and Nurse Resources. These collectively contain 16 data elements providing standard administrative data definitions and codes, to measure care delivery context and enable its integration into current health systems [61]. These authors consider this functionality as the means to provide nursing leaders with evidence enabling them to measure and manage decisions that lead to better patient, nurse staffing and financial outcomes. A scoping literature review around the existence and adoption of minimum data sets for nursing practice and fundamental care concluded that the use of minimum data sets could be useful for benchmarking and comparing clinical practices [62].

### Nursing data and standards

Similarly, there is a need to identify minimum data sets that represent most, if not all other confounding variables to be used as indicators for the various related concepts that cannot be measured directly. Many of the confounding variables listed in Table 6 are in fact rules representing limitations and constraints. Others do have associated data standards that need to be identified and made use of. A number of national minimum data sets and terminology/classification systems exist and may be adopted as data standards. Examples of classification systems in use in Australia are listed at its metadata online registry (METeOR) [63] as well as the World Health Organisation's Family of International Classifications [64] including the International Classification of Diseases (ICD), the International Classification of Functioning, Disability and Health (ICF), and the International Classification of Health Interventions (ICHI). The latter includes a related classification, the International Classification of Nursing Practice (ICNP).

The need for a standard nursing terminology designed to enhance nursing visibility was first identified by Clark and Lang in 1992 [65]. This resulted in the development of the

ICNP [66]. The recognition that Nursing terminology needs to be coded to describe the "essence of care" was used to develop the Clinical Care Classification system as explained by Saba and Taylor in 2007 [67]. These and other nursing terminologies have been cross-mapped with the widely recognized Systematized Nomenclature of Medical Clinical Terms (SNOMED-CT) widely used in EHRs [68] as well as the Unified Medical Language System (UMLS) [69]. It needs to be understood that terminologies consist of a multitude of atomic level data used in accordance with the relevant structure or classification rules to arrive at the ICF or ICD code for example. Ideally information systems collect atomic level data in the first instance where possible. The most effective way to be able to account for and make use of these many confounding variables is to adopt an object modeling approach to represent concepts in computable formats. This is explained in more detail in Chapter 9.

The Clinical Care Classification System's (CCC) [70] has a framework suitable for coding concepts included in this system, refer Fig. 1. There are four levels making it possible to code concepts for each level separately so that the coded data can be aggregated upward for summary information or parsed downward to view the atomic level data elements. The four healthcare patterns are (1) physiological, (2) psychological, (3) functional, and (4) health behavioral. There are three outcome qualifiers and four action type qualifiers. The framework links the diagnosis to interventions to outcomes and to each other. This structure was designed to suit the digital age. All terminological concepts use a five alpha-numeric character code for information exchange and interoperability. Coding can be used to track patient care for the entire episode of illness and measure care over time in all healthcare settings.

The signs and symptoms indicating the need for admission and nursing care, interventions, and actions are all coded and are mapped to SNOMED CT, Logical Observations Identifiers Names and Codes (LOINC), and the ICNP reference terminology.

| CCC SYSTEM FRAMEWORK | |
|:---|:---|
| Four (4) Health Care Patterns | |
| Twenty One (21) Care Component Classes | |
| 176 Diagnosis | 804 Interventions |
| 3 Outcomes * 176 Diagnosis = 528 Outcomes | 4 Action Types* 201 Treatments |

**Fig. 1**
Clinical Care Classification (CCC) System™.

A foundational requirement enabling nursing data to be integrated into clinical data repositories for big data and science, is the implementation of standardized nursing terminologies, common data models, and standardized information structures within Electronic Health Records (EHRs) [71]. An information model is a "structured specification, expressed graphically and/or narratively, of the information requirements of a domain. An information model describes the classes of information required and the properties of those classes, including attributes, relationships, and states" [72] (www.skmt.org). Structuring the language used within the health industry in a way that readily facilitates computer processing in not an easy task. Health concepts can be represented in many different ways as shown in Fig. 2.

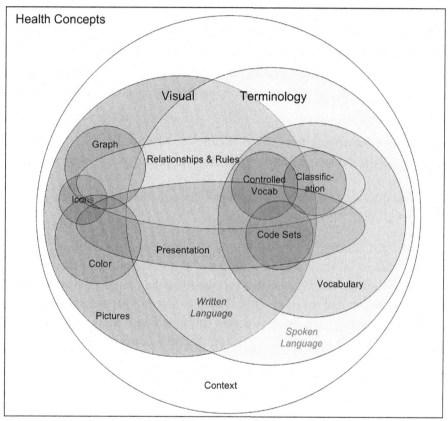

**Fig. 2**
Health concept representation methods [73, 74] p. 8.

An effort to facilitate the representation of nursing diagnosis and nursing action concepts and their relationships in a manner suitable for computer processing, led to the development of an ISO standard (ISO18104:2003) [75]. Subsequently evidence became available indicating that the Reference Terminology Model (RTM) developed can be used as a framework for analyzing nursing practice and for developing nursing content in electronic health record systems [76]. This ISO RTM is able to represent oriental nursing actions as well. It was concluded that there is a need for an integrated RTM for nursing [77]. These categorical structures were also found to be applicable to systems using natural language processing of nursing narratives [78]. This ISO reference terminology model for nursing was found to be applicable to represent the Detailed Clinical Model (DCM) structure developed for perinatal care nursing assessment. However, it was found that more qualifiers of the Judgment semantic domain are required in order to clearly and fully represent all of the entities and attributes of the DCMs used for nursing assessment [79].

This ISO standard was reviewed and updated in 2014 based on this new evidence regarding its use [80]. The categorical structures define the structure of terminological expressions for nursing diagnoses and nursing actions including the professional meaning of these constructs and their relationship to other record components. This standard's overall aim is to support interoperability in the exchange of meaningful information between information systems in respect of nursing diagnosis and nursing actions. These two categorial structures have features in common with

1.  the more general framework for clinical findings as described in ISO/TS 22789
2.  the domain-specific categorial structure for surgical procedures as described in ISO 1828:2012
3.  the World Health Organisation's International Classification of Healthcare Interventions (ICHI)

This ISO 18104:2014 standard may therefore inform development of other general and domain-specific categorial structures in healthcare. It is recommended that terminology and information system developers consider this model in their ongoing system development, evaluation, maintenance, and revisions [81]. There is a need to represent nursing data in a common format to enable comparisons across multiple institutions and facilitate Big Data science. This requires agreement regarding essential concepts and the use of standardized terminology.

## Reference terminologies

There are two widely used terminologies, the SNOMED Clinical Terms (SNOMED CT®), a reference terminology that includes specific and granular concepts necessary for capturing clinical information, and the Logical Observations Identifiers Names and Codes (LOINC®). The latter is publicly available to provide a set of universal names and codes identifying laboratory and clinical test results and it now also includes assessment data elements specifically relevant to nursing [82]. These authors made use of these terminologies plus the Clinical Care Classification (CCC)'s 21 care components [70] to develop a framework for organizing nursing assessments that support the organization, exchange, and aggregation of comparable nursing data across health systems. The CCC has also been harmonized with the International Classification of Nursing Practice (ICNP) [83]. The CCC's nursing diagnosis concepts were later translated and integrated with the ICNP version 2.0. This work demonstrated the importance of formal ontology definitions providing clear and clinically relevant descriptions to understand the ICNP concept meaning [82]. Only then is it possible to semantically map concepts between terminologies.

A reference terminology such as SNOMED CT®, representing the bio-medical knowledge domain, or the ICNP representing the nursing knowledge domain, is a terminology in which every concept designation has a formal, machine-usable definition supporting data aggregation and retrieval. It consists of atomic level designations structured to support representations of both simple and compositional concepts independent of human language (within machine) [72]. Each reference terminology's structure represents a domain ontology and may be referred to as a reference information model. Such models reflect specifications of the information requirements of a knowledge domain and may be expressed graphically and/or narratively. The root and top level concepts making up the entire domain's terminology are defined as metadata. Table 7 shows how the root and top level concepts of SNOMED CT [84] relates to top level concepts of ICNP. This structure differs greatly from the ICNP structure [85] which consists of three high level components:

1. Nursing phenomena classification
2. Nursing interventions/actions classification
3. Nursing diagnosis/outcomes classification — divided into negative and positive organizing concepts

**Table 7 SNOMED-CT and ICNP root and top-level concepts**

| Medical root and top level concepts<br>SNOMED-CT | Axis representing top level concepts<br>ICNP |
| --- | --- |
| Body structure | Location |
| Clinical finding | Judgment |
| Environment or geographical location | |
| Event | Means |
| Observable entity | Focus |
| Organism | Client |
| Pharmaceutical/biologic product | |
| Physical force | |
| Physical object | |
| Procedure | Action |
| Qualifier value | |
| Record artifact | Time |
| Situation with explicit context | |
| SNOMED CT model component | |
| Social context | |
| Special concept | |
| Specimen | |
| Staging and scales | |
| Substance | |

Each of the concepts included in Table 7 reflect the best possible match between SNOMED-CT and ICNP. Meaning is allocated through the use of defined terms contained in each of the following ICNP axis, and top level concepts listed. This demonstrates the difficulties encountered when mapping concepts between terminologies and the need to decompose concepts to the lowest atomic level to obtain equivalent meaning, making this a very time-consuming exercise to achieve the best possible match. In addition, terminologies are reviewed and updated from time to time requiring such maps to be reviewed and updated accordingly. Other nursing terminologies in use are the standardized classifications of nursing diagnoses, nursing interventions and nursing-sensitive patient outcomes known as NANDA, NIC, and NOC [86]. These have also been added to SNOMED-CT. A comparison of knowledge representation concepts in NIC and SNOMED-CT using expert human judgment revealed that some were mis-assigned or inappropriately represented [68].

Despite these differences between nursing terminologies and SNOMED CT (Interventions and Diagnosis), the equivalency tables released in 2016 by SNOMED International form part of their production package [87]. These differences are best accommodated through the use of object models where the most appropriate term from any terminology can be selected to provide meaning and represent the coded attributes at the atomic level included in each model. Objects models provide context and consistent structures for atomic level data ensuring meaning is retained [88].

An interface terminology assists data entry and display of information, this is defined as a "maintained set of unique identified terms designed to be compatible with the natural language of the user, used to mediate between a user's colloquial conceptualisations of concept descriptions and an underlying reference terminology" [72]. Again the use of object models enables the translation of all concepts thus represented into any language yet retain consistent meaning. Interface terminology is mapped to the standard within the system in use.

The availability and use of multiple health, medical or nursing terminologies makes it difficult to share, compare and link relevant nursing data between nursing information systems. One solution is to cross map existing nursing concepts. Such activities are time consuming, costly and prone to inconsistencies or error. One such research study concluded that on-going collaboration among terminology developers is required to augment the interoperability of nursing data [69].

To fully realize the potential use and benefits of these data it is important to consider the need for ensuring quality source data. It's been recognized that with the convergence of electronic health record adoption, there is an increase in the availability of electronic health data that potentially lends itself to the development and use of a variety of analytical techniques to improve, not only the delivery of healthcare services, but also resource management [89]. This requires the adoption of quality informatics that focuses on collecting accurate healthcare data, analyzing those data and applying findings to improve value and benefits. The current lack of unification, data inconsistency and the poor adoption of agreed standards are major impediments to achieving these outcomes. Overcoming this barrier requires national if not global collaborative action towards a digital transformation. This topic is explored in some depth in Chapter 9. But first we explore how best to create quality data required to better manage nursing and midwifery resources enabling a continuing capacity to care.

## Use of metadata

Caring for people requiring any health service, needs to be at the center when considering our capacity to do so. Here we have identified the many and varied data "containers" and show how the many complex data structures with a multitude of characteristics, relate to each other within and between these desired "containers" or data repositories to represent concepts such as service capacity. Digital environments require standard data definitions as these provide digital identification, facilitate legacy resource integration and system interoperability as described in Chapter 9. Collectively these metadata represent the caring domain within any health service, that is the caring domain's vocabulary ontology.

From an information science perspective, an ontology formally represents, names and defines categories, properties and relations between the concepts, data and entities that substantiate one, many or all domains. Metadata refers to structured information that describes, explains, locates, or otherwise makes it easier to retrieve, use, or manage an information source. Its use enables the content to be understood by both people and machines. Metadata is key to ensuring that resources will survive and continue to be accessible into the future [90]. Any data element described in a metadata set has a number of characteristics to describe meaning as shown in the example in Table 8 [91]. This type of metadata description can be further standardized via the application of relevant controlled terminologies to concepts described.

**Table 8 Example of metadata description**

| Example: living arrangement | |
|---|---|
| Definition | Whether a person usually resides alone or with others |
| Context | Client support needs |
| Datatype | Numeric |
| Maximum size | 1 |
| Data Domain | 1 Lives alone |
| | 2 Lives with others |
| | 9 Not stated/inadequately described |
| Guide for use | This item does not seek to describe the quality of the arrangements but merely the fact of the arrangement. It is recognized that this item may change on a number of occasions during the course of an episode of care |

Australian health metadata are freely accessible via its metadata online registry (METeOR) that is based on the International Organization for Standardization and the International

Electrotechnical Commission 11179 Metadata Registries standard (ISO/IEC 11179) [63]. The United States Health Information Knowledgebase is accessible via the Agency for Healthcare Research and Quality (AHRQ) website [92]. However there are no specific Nursing and/or Midwifery minimum data sets included in these registries although categorical structures exist as an international standard as described previously.

The nursing profession needs to ensure that the relevant nursing data are routinely collected for operational use in a manner that enables not only the identification of nursing workload demand but also the collection of practice-based evidence. This requirement is fundamental for any health service as this enables a service to publish the evidence relevant to its capacity to care. An agreed set of metadata to suit nursing data needs must comply with the principles applicable to any scientific data as described by Wilkinson et al. [93] (Box 1). This requires improvements in the infrastructure as described so that it will be possible to reliably re-use these data. The many limitations identified prevents us from realizing maximum benefits from significant investments made to date.

---

**BOX 1  The FAIR guiding principles [93]**

---

*To be findable*
F1    (Meta)data are assigned a globally unique and persistent identifier
F2    Data are described with rich metadata (defined by R1 below)
F3    Metadata clearly and explicitly include the identifier of the data it describes
F4    (Meta)data are registered or indexed in a searchable resource

*To be accessible*
A1    (Meta)data are retrievable by their identifier using a standardized communications protocol
A1.1   The protocol is open, free, and universally implementable
A1.2   The protocol allows for an authentication and authorisation procedure, where necessary
A2    Metadata are accessible, even when the data are no longer available

*To be interoperable*
I1    (Meta)data use a formal, accessible, shared, and broadly applicable language for knowledge representation
I2    (Meta)data use vocabularies that follow FAIR principles
I3    (Meta)data include qualified references to other (meta)data

*To be reusable*
R1    (Meta)data are richly described with a plurality of accurate and relevant attributes
R1.1   (Meta)data are released with a clear and accessible data usage license
R1.2   (Meta)data are associated with detailed provenance
R1.3   (Meta)data meet domain-relevant community standards

Guidelines for metadata development are detailed in a handbook published by Standards Australia [91]. This requires the development and use of a data management and stewardship plan, the adoption of an "open platform" and vendor neutral data repositories to suit each unique digital health ecosystem as described in Chapters 9 and 11.

## Nursing service demand metadata

Metadata representing variables influencing nursing and midwifery workloads reveal many highly distributed data sources that remain under the control of original data owners contained in and used by a variety of information systems and used for a variety of purposes. A proposed nursing domain metadata structure was developed representing an exploration of possibilities that need to be reviewed and agreed to by the Nursing profession. This proposed nursing metadata structure is based on the variables influencing nursing work as shown in Table 6 and Virginia Henderson's Nursing Theory to reflect efficiency productivity parameters, input, process and outcomes. An outline of the structure is shown in Fig. 3. Further detailed inclusions for each of these are provided in the following mindmaps and boxes. An agreed nursing metadata structure can be converted to nursing clinical domain models to entirely represent nursing knowledge in computable formats [88].

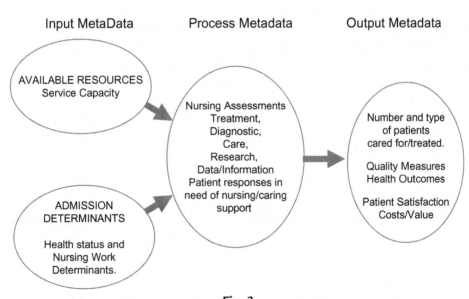

**Fig. 3**
Suggested nursing domain metadata structure.

High level metadata detail a number of factors that need to be considered as indicators or contributors to service capacity as presented in Box 2 and in Mindmap 1. Existing metadata includes available healthcare service care types, clinical specialties including nursing specific clinical specialties and nursing interventions.

---

**BOX 2  Service capacity includes**

---

o   Bed/cubicle/chair availability
o   Professional staff availability — as determined for any time period by type of health service or clinical specialty
o   Treatment/diagnostic services available
o   Type of healthcare services available within one facility
o   Waiting lists — anyone on a waiting list may require assistance during their waiting period. Elective surgical cases on a waiting list are fairly predictable in terms of resource demand needs. These can be planned for in advance to ensure resource availability

---

### Service capacity — Identifying required nursing skill mix

The Australian Institute of Health and Welfare (AIHW) lists more than 70 clinical specialties [63] plus there are numerous sub-specialties. Nurses also specialize by obtaining practical experience working as team members in any specific clinical specialty and/or they specialize in nursing specific areas of practice by obtaining post graduate qualifications or by undertaking specific professional development activities. For example Australia has established a Coalition of National Nursing and Midwifery Organisations (CONNMO) that has 54 members, most of which represent a nursing specialty [94]. One Australian metadata example is the Nurse Practitioner Workforce Planning Minimum Data Set which is included in METeOR [95]. This represents an agreed set of de-identified client level data related to services provided by Nurse Practitioners in public funded health, aged care and community settings in all States/Territories for the purpose of health workforce planning. Another example is the Belgian Nursing Minimum Data Set [5, 52]. These skill mix issues are explored in more detail in Chapter 5.

Nursing skill mix data needs to be included in the organization's human resources management system. It should be clear from the number of possible skill categories, as detailed in Chapter 5, that matching skill mix with service demand is not a simple task. The number and skill mix of nurses allocated to provide care for a group of patients influences the time taken to provide nursing services. As a rule, work expands to fill the time available. A high staffing level does not mean better care, it often means that activities are undertaken which do not necessarily contribute to the wellbeing of patients, such as social communication among staff. On the other hand, insufficient or an inappropriate skill mix of staff is highly likely to result in the omission of some activities which should be undertaken.

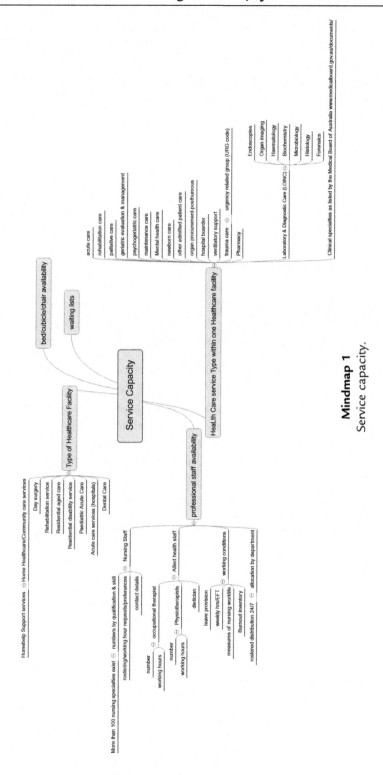

**Mindmap 1**
Service capacity.

It is not just a matter of total staff hours either as very experienced and proficient staff can complete more care activities and provide a better quality of care than staff who are less experienced and/or less competent. The aim needs to be, to match patient needs with staff who are best able to meet these needs in a proficient manner as accurately as possible. This is where rostering practices come into play as detailed in Chapter 5.

### Service capacity — Nursing working conditions

The World Health Organisation (WHO) defines a healthy environment as a place of "physical, mental, and social well-being". Nursing working environments and conditions, including physical layouts, the availability of equipment and support staff, and rostering practices, influence a department's ability to meet nursing service demands as well as patient safety and outcomes. Costs and workforce availability are influenced by various leave provisions and the number of hours per week that equate with one full time equivalent nurse. Nursing workloads and working conditions represent a "hot" industrial issue. On one side are those who fund health services with a desire to deliver a quality service at minimal cost, and on the other side is the nursing workforce who also want to be able to deliver a quality service and who don't wish to be exploited by funders wishing to save costs or to be unable to provide all required nursing services.

The nursing profession is very concerned about the need to minimize risks to the quality of care provided. Measuring nursing service demand is all about the ability to provide empirical evidence on how best to ensure that the resources provided match service demand. Major issues frequently encountered are the existence of a finite budget and/or workforce availability. These potential limitations lead to the need for management to make decisions regarding which services should not be made available at any one time to ensure that the safety of existing patients is maintained. There appears to be a dearth of explicit decision-making frameworks within which to consider rationing nursing care.

Policy makers and health service managers tend to assume that nurses will continue to provide full care, yet nursing care left undone is a direct response to overwhelming demands on nursing resources [96]. This is also about health system sustainability in the face of increasing service demands and decreasing workforce availability or workforce participation rates. We therefore need to consider the need to work smarter and make the best possible use of available information and communication technologies. Nursing working conditions and its association with the capacity to care is further explored in Chapters 6 and 11.

### Admission and continuing service determinants

Many other factors to be considered relate to admission determinants as shown in Box 3 and Mindmap 2. These were explored from a nursing perspective and identified in terms of their likelihood of influencing nursing service demand as shown in Box 4. Once an individual is

---

**BOX 3 Admission determinants refer Mindmap 2**

- o Population's disease burden as defined by the International Classification of Diseases (ICD)
- o Condition onset flag — used as a qualifier for each coded diagnosis to indicate the onset of the condition relative to the beginning of the episode of care, as represented by a code [63]. This is an indicator of the likely severity of the disease at the time of admission

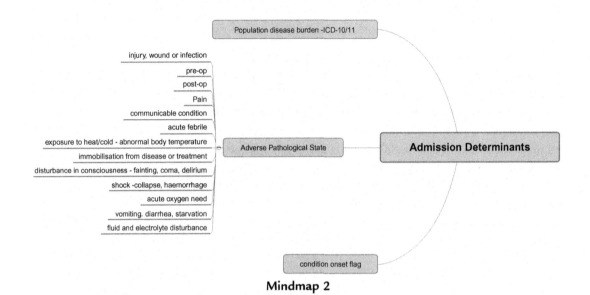

**Mindmap 2**
Admission determinants.

---

**BOX 4 Contributing admission and continuing nursing work determinants**

- o Disturbances in fluid and electrolytes
- o Vomiting
- o Diarrhea
- o Starvation
- o Acute oxygen want
- o Shock — collapse, hemorrhage
- o Disturbances in consciousness — fitting, coma, delirium
- o Exposure to heat/cold with abnormal body temperature
- o Acute febrile states — all causes
- o Local injury, wound and/or infection
- o Communicable condition
- o Immobilization from disease or prescribed treatment
- o Persistent or intractable pain — scale 0–10
- o Pre-op state
- o Post-op state

admitted to receive a health service, nursing service demand is further influenced by individual patient characteristics including age, gender, languages spoken, care preferences and specific assistance needs as described by Henderson [96a] or as identified for the Canadian C-HOBIC project that has adopted 24 data elements for collection on admission and 24 data elements at time of discharge.

From a nursing perspective any of the following individual's adverse pathological states on admission and at any time during their episode of care may represent a specific care need as listed by Henderson's theory of nursing [59] shown in Box 4.

Treatment, diagnostic procedures and care requirements are usually represented by a terminology in use for funding purposes, such as the Australian Medicines Terminology (AMT) or the Current Procedural Terminology (CPT) or the Medicare Benefits Schedule (MBS). Other terminologies not used for funding purposes such as the Nursing interventions are included in the ICNP reference terminology or the Nutrition Intervention terminology or any available Allied Health or Rehabilitation intervention terminology. These also need to be considered for use. In some research settings it is important to also consider additional nursing workload generated by specific clinical research requirements [66].

Box 5 presents the C-HOBIC data elements and includes any related data elements from Henderson's list. Box 6 presents additional patient characteristics identified by Henderson as influencing nursing practice. Content shown in Boxes 5 and 6 are represented in two different mindmaps, one contains metadata for treatment/diagnostic/care and research services provided (Mindmap 3) and the other Nursing Assessed Patient Characteristics (Mindmap 4).

**Mindmap 3**
Treatment/diagnostic/care/research services provided.

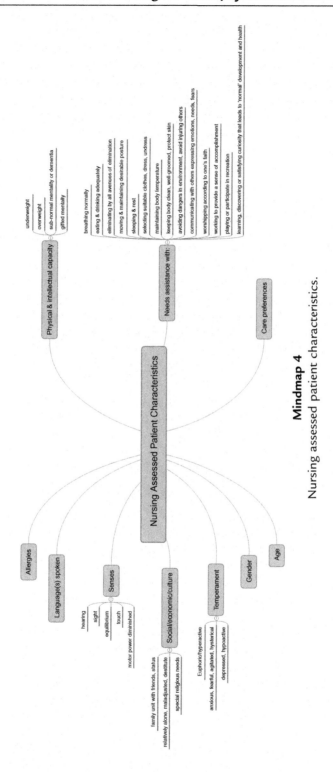

**Mindmap 4**
Nursing assessed patient characteristics.

The list in Box 5 presents both the C-HOBIC data elements followed by the equivalent Henderson's list of potential nursing assistance needs [97, 98]. C-HOBIC associated measures are those described by the interRAI assessment system [99] that codes activities according to one of eight levels indicating the degree of assistance required relative to the listed status. Further details about the interRAI assessment system are provided in Chapter 12. Measures for the therapeutic self-care category of data elements were developed specifically for the C-HOBIC project [100, 101].

---

**BOX 5  C-HOBIC data elements chosen for collection relative to Henderson's list**

---

*Functional status and continence (ADLs/instrumental activities of daily living)*

| | |
|---|---|
| Bathing | Keeping body clean, well-groomed, protect skin |
| Personal hygiene | |
| Walking | Moving and maintaining desirable posture |
| Bed mobility | |
| Toilet transfer | Eliminating by all avenues of elimination |
| Toilet use | |
| Bladder continence | |
| Dressing | Selecting suitable clothes, dress, undress |
| Eating | Eating and drinking adequately |

*Symptoms*

| | |
|---|---|
| Pain-frequency | Persistent or intractable pain — scale 0–10 |
| Pain-intensity | |
| Fatique | Immobilization from disease or prescribed as treatment |
| Dyspnea | Acute oxygen want |
| Nausea | Vomiting |

*Safety*

| | |
|---|---|
| Falls | Equilibrium |
| Pressure ulcer | Local injury, wound, and/or infection |

*Therapeutic self-care*
Knowledge of current medications
Knowledge about why you are taking current medications
Ability to take medications as prescribed
Recognition of changes in body (symptoms) related to health
Carry out treatments to manage symptoms
Ability to do everyday things like bathing, shopping
Someone to call if help is needed
Knowledge of whom to contact in case of a medical emergency

---

Clearly perspective and purpose influences the selection of data elements to be used for data collection purposes. Henderson's list was based on her theory about nursing practice and did not include any items for inclusion under the therapeutic self-care category. Henderson's list is about documenting human activities and conditions that may require nursing support or that need to be considered when providing nursing care. In addition to items listed above, individuals may need nursing/caring assistance with items shown in Box 6.

---

**BOX 6  Additional human activities and conditions that may require nursing support**

- Breathing normally
- Sleeping and rest
- Maintaining body temperature — adjust clothes, change environment
- Avoiding dangers in environment, avoid injuring others
- Communicating with others expressing emotions, needs, fears
- Worshipping according to one's faith
- Working to provide a sense of accomplishment
- Playing or participate in recreation
- Learning, discovering or satisfying curiosity that leads to "normal" development and health

Henderson also noted ever present factors that affect needs and influence nursing care

- Temperament
  - euphoric/hyperactive
  - anxious, fearful, agitated, hysterical
  - depressed, hypoactive
- Social/culture
  - family unit with friends, status
  - relatively alone, maladjusted, destitute
- Physical and intellectual capacity
  - underweight
  - overweight
  - sub-normal mentality or dementia
  - gifted mentally
- Senses
  - hearing, sight, touch, motor power diminished, equilibrium

---

Henderson's list essentially provides the basis from which nurses undertake their assessment of nursing need from which they develop their care plans. These include nursing activities associated with various treatment and/or diagnostic service requirements such as medication administration, specimen collection, taking specific observations at prescribed intervals. Many interventions undertaken by nurses are time related.

## Indicators of nursing care demand

How does one identify the many variables and select the best indicators that collectively reflect patient characteristics? Does the nursing profession have an agreed taxonomy of nursing practice? One could argue that the ICNP represents the nursing practice taxonomy as it has named and classified the concepts that collectively represent nursing practice.

Hovenga explored these questions prior to the existence of the ICNP, as part of her role to allocate hospital nurse staffing establishments for an Australian State Government department responsible for budget allocations to over 160 hospitals of various types and sizes during the 1980s. Hovenga's subsequent research [51] found that the greater the number of patient characteristics identified as dependency indicators applicable per patient, the greater the demand for nursing resources. This work resulted in the development of a Patient Assessment and Information System (PAIS) that was used to resolve numerous industrial disputes over nursing workloads, to match service demand with available nursing resources and for the conduct of numerous hospital nursing productivity reviews. This patient classification system consisted of six dependency categories with associated standard time values. It was used as a nursing workload management system by over 100 Australian hospitals for well over 10 years.

Lowe explored these same questions to develop a practical solution to support the management of nursing resources while occupying a Director of Nursing Services role in another Australian State. This was tested and refined by Lowe while working in another large hospital and later successfully commercialized as the TrendCare System [102], a workforce planning and workload management system that provides dynamic data for clinicians, department managers, hospital executives and high level healthcare planners. This system's development was the result of Lowe's use of reliable nursing data to solve many real hospital management problems.

A shared outcome of Hovenga and Lowe's independent research efforts indicated that different levels of dependency, represented significant variations in terms of average nursing resource usage per patient between dependency levels. Further research undertaken over many years by Lowe in numerous hospitals, has resulted in considerable refinement toward greater homogeneity in terms of average nursing resource usage per patient within each level of dependency. There are now around 200 well defined patient types each with a number of dependency categories with associated standard nursing time values. It follows that it is then also possible to determine the cost of these nursing "products" by tracing defined input costs (actual resource usage) relative to each of these patient group types.

The data elements or factors used to classify patients may also be used for other purposes. For example, the percentage of patients for whom certain dependency indicators are identified, provides an overview of nursing work or a patient profile. The monitoring of trends or changes over specified time periods provides valuable information regarding changes in the patient population and their care requirements throughout their inpatient journey. This in turn may be directly related to other resource uses such as disposable supplies or equipment to explain variations in material resource usage. It is also necessary to capture patient data such as admission and discharge times, to estimate staffing requirements per patient day or shift for future demand estimates. Such estimates can then be updated and modified using the same data operationally in real time. These updates reflect any staffing changes made in response to changes in patients' service requirements or changes in staff availability that resulted from unplanned absences due to sick leave.

A variety of outcome metadata is in use including data in use for statistical reporting such as ICD-10, which differs from a reference terminology as it is a classification system. Outcome measures in use are listed in Box 7 and shown in Mindmap 5.

---

**BOX 7  Outcome data in use for statistical reporting**

- Discharge diagnosis — ICD and DRG or equivalent
- Discharge status
  - death
  - transfer to another health service
  - to self-care
- Number of patients/clients serviced during any time period
- Average length of stay or episode of care
  - per number of patients/clients treated or
  - per ICD or per DRG or equivalent
- Re-admission due to
  - condition not noted as arising during the episode of admitted patient care
- Length of stay in ICU
- Complication — condition with onset during the episode of admitted patient care. These are nurses sensitive outcome measures
  - pressure ulcers
  - falls
  - acquired infection
  - failure to rescue
- Patient satisfaction

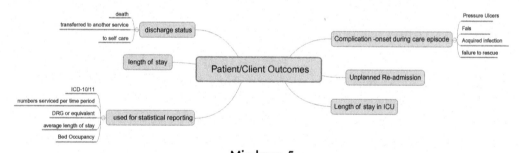

**Mindmap 5**
Patient/client outcomes.

## Metadata enabling the evaluation of nursing service contributions relative to patient outcomes

Patient outcomes essentially represent the results of a new patient assessment at a later point in time to determine change in a patient's health status. Qualifiers such as any deterioration, no change, or improvement, are used to depict expected outcomes or goals as well as actual outcomes following nursing interventions.

There are many other possible outcome indicators or performance measures in use as described in some detail in Chapter 11. For example, the use of a special service during an episode of care such as the length of stay in critical care. Health information systems may also be programmed to report performance indicators associated with accreditation requirements. There have been a number of studies that identified adverse events representing unintended and at times harmful occurrences associated with the use of medicines, equipment, or various health service practices employed, including associations with staffing levels and skill mix [5, 6, 21, 103, 104]. The National Health Service in the United Kingdom (UK) identified around 50 outcome indicators pertaining to five areas of quality [105, 106]. These are:

1. Preventing people dying prematurely
2. Enhancing quality of life for those with long term conditions
3. Helping people to recover from episodes of injury or ill health
4. Ensuring people have a positive experience of care
5. Treating and caring for people in a safe environment and protecting them from avoidable harm

These were designed to be of use to the Secretary of State for Health and Social Care to hold the NHS to account. ICHOM, the International Consortium for Health Outcomes Measurement Inc. (United States) [107], is developing a new paradigm focused on health outcomes in a form that matters most to patients. This represents a future view of value-based health care. Their published standard sets are standardized outcomes, measurement tools, time points and risk adjustment factors for given medical conditions based on patient priorities.

The Canadian nurses have managed to address nursing outcomes via their C-HOBIC project [98]. Other stated requirements of this system were that these measures must also consider patient satisfaction, workload and staffing, clinical risks and other measures of quality and safety of care as well as nurses' work satisfaction. A more detailed examination of outcome measurements is provided in Chapter 11 as this relates to the quality of services provided.

From a nursing perspective it is important to be able to identify those outcomes that are known to or are likely to have been impacted by nursing service performance. These are reported as nurse sensitive outcomes and will vary in accordance with the type(s) of health services provided. Such measures are in fact indicators based on demonstrated significant research findings regarding relationships between practices adopted and any number of specified patient outcomes. Not all relevant concepts or entities can be measured or described using specific data element standards, nor is this an essential requirement. In most instances it is sufficient to recognize significant indicators although these also need to be standardized. It is apparent that a number of terminologies and coding systems relevant to nursing workload metadata exist with, many being country specific.

## Nursing workload management metadata need

Driscoll et al. [1] concluded that "there needs to be greater homogeneity in the nurse-sensitive end points measured and the calculation of the nurse to patient ratios" and that "such metrics should not be used in isolation." An agreed use of data standards pertaining to nurse-sensitive end points can contribute to a "triangulated" approach to the decision-making process about safe and sustainable nurse staffing levels. We have argued that the measurement of nursing care demand in a consistent and comparable manner requires an agreed Nursing Workload Management Meta-dataset. The need to enable data processing in a timely manner has been highlighted. Adherence to common metadata standards enables organizations to efficiently share data and search metadata records. This needs to start by identifying metadata needs at each point of care. From there secondary data use (re-use) and required data aggregation needs to be considered. An agreed nursing metadata specification standard needs to be based on a nursing domain information model that defines its structure and a value domain standard (code system) that defines both the data content and context, and enables unique identification.

Many if not all of these nursing work determinants identified in this chapter may or should be derived from electronic health records (EHRs). These are simply a start. There is a need for the Nursing Profession to develop a Nursing Workload Metadata standard to enable the ICN's nurse staffing position statement [2] requirements to be fully implemented and monitored.

International EHR metadata standards such as the ISO 13606 Part 1 and Part 2 [108] and HL7 CDA2 [109] or the HL7 Fast Health Information Resources (FHIR) need to be considered when developing this nursing workload management metadata standard. The ISO 13606 Part 2 standard complies with the openEHR specifications that includes an electronic record reference information model [110]. The Clinical Models (Archetypes) provide both structure and a link to standard terminologies where available via their attributes [111]. These standards provide syntax rules for how to construct values, describe content, and address what needs to be included, they provide a way to format and exchange metadata and improve interoperability [112]. Once an agreed nursing workload metadata set exists, there is a need for these data standards to be governed and maintained [113–115].

## Optimizing workplace efficiencies

Before any measurement of nursing care demand takes place, it is important to optimize workplace efficiencies. The efficient delivery of nursing services requires the provision of best value from a patient's perspective, the best possible use of available people talents and the use of the fewest resources to avoid waste. The application of "lean" methodology in healthcare requires consideration of the following areas where organizational waste is very likely to occur. These are: unused human potential, waiting, inventory, transportation, defects, motion, overproduction and processing [116]. Reducing wastage in any healthcare requires leadership and teamwork.

It starts by adopting a questioning and problem-solving approach regarding each of these potential areas of waste. By doing so you may identify poor staff allocation methods and/or rostering resulting in times or areas of heavy workloads and other times or areas of having more nurses available than needed. Waiting can refer to waiting for a porter to assist with lifting or patient transfers, or for a piece of equipment to become available. Inventory refers to the amount of materials, surgical supplies or medications to be on hand and in stock elsewhere. Where there is an excess store of sterile supplies or drugs they may go out of date resulting in wastage. Insufficient supplies may lead to delays.

Transportation is very much concerned with the movement or discharge of patients. Defects may apply to hospital beds, a malfunctioning calling system, a broken pan flusher or malfunctioning lifting machine resulting in having to make compromises. Motion relates to the amount of walking nurses need to do to obtain the necessary supplies or equipment needed to provide nursing services and/or to initiate communication and information flows. Over production examples include undertaking unnecessary vital signs observations, excessive reporting and/or duplication of documentation. Processing refers to the actual methods or models of care employed to deliver nursing care.

The questioning approach is about asking if this activity is necessary? Does it have to be done at this time or as frequently? Who or what staff category is best placed to take responsibility for this activity? Are the most appropriate supplies and supporting services being employed for this activity? Can the desired objectives be achieved using a different method? Is this activity having a negative impact elsewhere? In all cases one needs to ensure patient safety remains a primary consideration. It's all about identifying and assessing real world experiences.

The problem-solving approach is about identifying the reasons for any further action to be undertaken. One needs to document what is happening now and the intended outcome using a new approach. This may result in the identification of a service gap or excessive use of time to complete the desired activity. Such findings can then be analyzed in terms of possible solutions. These solutions may need to be trialed prior to standardization and full adoption to ensure sustainability. These concepts are explained in greater depth in Chapter 10.

### Political, professional, managerial, and industrial influencers

The provision of service capacity, including nurse staffing, has always been and continues to be a political and managerial issue. This conflict is exacerbated by three factors:

1. Miscalculations of nurse staffing requirements and budgets as explained in Chapter 7.
2. Medical dominance [117–119], along with a poor appreciation of the nursing scientific discipline and how this complements the provision of medical services.

3.  The fact that nurses are mostly female, many with personalities to please, accommodate and find solutions to care for their patients in the best possible way, no matter the resource shortcomings.

On the other hand, health service funders and those responsible for resource distribution tend to be male with a strong desire to meet budget limitations. Unlike nurses, many do not appreciate the impact their decisions have at the point of care. From a financial perspective Nursing budgets are frequently viewed as having the capacity to absorb cuts and subsidize other departments as desired by management due to the size of the nursing budget. Such underpinning values greatly influence nursing workloads and patient outcomes. There comes a time though when such situations can no longer be tolerated.

One large European retrospective observational study interpreted their findings as showing that Nurses' staffing cuts to save money might adversely affect patient outcomes. They concluded that an increased emphasis on bachelor's education for nurses could reduce preventable hospital deaths [28]. Mandating nurse staffing ratios, or any other regulatory initiative, is one way for nurses to have access to information that empowers them. Ideally both parties work collaboratively to achieve a common goal and make use of reliable data. The ICN's position statement of evidence-based safe nurse staffing [2] lists the following element any healthcare organization needs to adopt to provide the necessary evidence:

o   real-time patient needs assessment,
o   local assessment of nurse staffing requirements to provide a service,
o   nursing and interdisciplinary care delivery models that enable nurses to work to their optimal scope of practice,
o   good human resource practices to recruit and retain nurses,
o   healthy work environments and occupational health and safety policies and services that support high quality professional practice,
o   workforce planning systems to ensure that the supply of staff meets patient needs,
o   tools to support workload measurement and its management,
o   rostering to ensure scheduling meets anticipated fluctuations in workload,
o   metrics to assess the impact of nurse staffing on patient care and policies that guide and support best practice across all of these.

An international agreement about metadata standards to be adopted for this purpose is highly desirable to enable consistent and meaningful evidence of nursing's contribution to health care to be documented.

The PAIS case study details a story about very real political influence, conflicting views and who should take on the responsibility of nurse staffing (refer Appendix 1). It demonstrates the many contentious and conflicting issues associated with the management of nursing workloads. Similar stories emerge around the globe, such as the rationale for why California ended up legislating the nurse staffing ratios. Years later again in Victoria, the use of Lowe's TrendCare system was subject to similar conflicting industrial forces. The nursing union wanted to adopt nurse patient ratios in preference to the adoption of Lowe's more cost-effective patient acuity system. A similar scenario has recently played out in New Zealand (NZ) where management didn't respond to measured nurse staffing shortcomings by arguing the funds weren't available. In NZ the continuing use of the TrendCare acuity system has prevailed as a solution as detailed in the Safe Staffing Accord reached between district health boards and the New Zealand Nurses Organisation in July 2018 [120]. Around the globe Nursing staff continue to be expected to compromise their provision of services regardless of their ability to provide safe quality care.

## Conclusion

This chapter has covered many influencing factors and confounding variables as well as many interrelationships that need to be considered in order to establish nurse staffing needs relative to service demand. It demonstrates that the data and information of relevance to managing a nursing service are fundamental and central to all health data required for decision making by many stakeholders at every level within a national health system. Health service delivery complexity is further complicated by related governance structures, various conflicting areas of responsibility, organizational behaviors and the random, variable and ad hoc nature of health service demand.

Apart from impacts of high-level political influencing factors and changes in funding arrangements, health service managers need to be able to appreciate issues such as the impact of operating schedules on inpatient nursing service demands. They need to be able to appreciate many and varied relationships between nurse rostering practices, nurse staffing allocation, skill mix management, service demand and patient outcomes. Service managers also need to be able to identify the number of beds that need to be available at various points of time to accommodate high probabilities of trauma and acute disease episodes needing admission from the accident and emergency department. Nurses need to be able to respond at anytime to any change in service demand. Such changes are always documented in a variety of ways for operational use.

Information used by nurses, including nursing specific data provide real time operational data that improves decision making and enables efficiency improvements to be made from current and historical facts. It's clear that from this analysis of available metadata and associated terminologies in use that this mix alone does not facilitate optimal digital transformation. The next phase along this digital transformation journey requires a greater use of appropriate standards plus the use of clinical modeling, to ensure health record structures and associated information systems have sufficient flexibility to meet the information needs of all stakeholders at all levels within a health systems' hierarchy. Medical and technical advances continue at a great pace. The nursing profession is a major stakeholder. The chapters that follow aim to provide case study examples regarding the use of these data to achieve and manage the capacity to care.

Chapter 4 describes and explores a number of nursing work measurement study methods, their construct validity and the use of work measurement results. It does so within the context of input, process and output factors as the relationships between them need to be part of any work measurement study. We make use of experiences gained and the work measurement data collections of our own very extensive and unequaled research and development activities. These studies were undertaken over many years, our results have remained in the gray literature and continue to be in use or have been in use by numerous hospitals across several countries as explained in the two case studies provided as appendices.

## References

[1] Driscoll A, Grant MJ, Carroll D, Dalton S, Deaton C, Jones I, et al. The effect of nurse-to-patient ratios on nurse-sensitive patient outcomes in acute specialist units: a systematic review and meta-analysis. Eur J Cardiovasc Nurs 2017;17(1):6–22.

[2] ICN. Evidence-based safe nurse staffing. Position statement. Geneva: International Council of Nurses; 2018. p.1–7.

[3] Pearson A, O'Brien Pallas L, Thomson D, Doucette E, Tucker D, Wiechula R, et al. Systematic review of evidence on the impact of nursing workload and staffing on establishing healthy work environments. Int J Evid Based Healthc 2006;4(4):337–84.

[4] Alghamdi MG. Nursing workload: a concept analysis. J Nurs Manag 2016;24:449–57.

[5] Van den Heede K, Clarke SP, Sermeus W, Vleugels A, Aiken LH. International experts' perspectives on the state of the nurse staffing and patient outcomes literature. J Nurs Scholarsh 2007;39(4):290–7.

[6] Olley R, Edwards I, Avery M, Cooper H. Systematic review of the evidence related to mandated nurse staffing ratios in acute hospitals. Aust Health Rev 2019;43(3). https://doi.org/10.1071/AH16252 [Epub ahead of print].

[7] Press Ganey. Nursing quality (NDNQI). [cited 12 September 2018]. Available from: http://www.pressganey.com/solutions/clinical-excellence/nursing-quality; 2018.

[8] Press Ganey. Acquires national database of nursing quality indicators (NDNQI®) [Press release]. June 10; 2014.

[9] ANA. Nurse staffing. [cited 12 September 2018]. Available from: https://www.nursingworld.org/practice-policy/work-environment/nurse-staffing/; 2017.

[10] Aydelotte M. Nurse staffing methodology: a review and critique of selected literature. National Institutes of Health; 1973 Contract No.: USDHEW Pub. No. (NIH) 73-433.

[11] Taylor B, Yankey N, Robinson C, Annis A, Haddock KS, Alt-White A, et al. Evaluating the Veterans Health Administration's staffing methodology model: a reliable approach. Nurs Econ 2015;33(1):36–40. 66.

[12] Clancy CM. Staffing methodology for VHA nursing personnel. Washington, DC: Department of Veterans Affairs, Veterans Health Administration; 2017. [cited 27 May 2019]. Available from: https://www.va.gov/vhapublications/ViewPublication.asp?pub_ID=5712.

[13] Duckett S, Jorm C, Moran G, Parsonage H. Safer care saves money: how to improve patient care and save public money at the same time. Melbourne: Grattan Institute; 2018.

[14] Harper EM. Staffing based on evidence: can health information technology make it possible? Nurs Econ 2012;30(5):262–7. 81.

[15] Lavin MA, Harper E, Barr N. Health information technology, patient safety, and professional nursing care documentation in acute care settings. Online J Issues Nurs 2015;20(2):6.

[16] Shin S, Park JH, Bae SH. Nurse staffing and nurse outcomes: a systematic review and meta-analysis. Nurs Outlook 2018;66(3):273–82.

[17] Barrientos S, Vega-Vázquez L, Diego-Cordero R, Badanta-Romero B, Porcel-Gálvez AM. Interventions to improve working conditions of nursing staff in acute care hospitals: scoping review. J Nurs Manag 2017;26 (2):94–107.

[18] Beswick S, Hill PD, Anderson MA. Comparison of nurse workload approaches. J Nurs Manag 2010;18:592–8.

[19] Ball J. Guidance on safe nurse staffing levels in the UK. London: Royal College of Nursing; 2010.

[20] Min A, Scott LD. Evaluating nursing hours per patient day as a nurse staffing measure. J Nurs Manag 2015; 24(4):439–48.

[21] Butler M, Collins R, Drennan J, Halligan P, O'Mathúna DP, Schultz TJ, et al. Hospital nurse staffing models and patient and staff-related outcomes. Cochrane Database Syst Rev 2011;(7). https://doi.org/ 10.1002/14651858.CD007019.pub2.

[22] Spetz J, Donaldson N, Aydin C, Brown DS. How many nurses per patient? Measurements of nurse staffing in health services research. Health Serv Res 2008;43(5 Pt 1):1674–92.

[23] Leary A, Punshon G. Determining acute nurse staffing: a hermeneutic review of an evolving science. BMJ Open 2019;9(3):e025654.

[24] Saba V, Whittenburg L. Electronic method for measuring nursing: workload action measures method (WAMM). Stud Health Technol Inform 2018;250:196.

[25] Welton JM. Measuring patient acuity: implications for nurse staffing and assignment. J Nurs Adm 2017; 47(10):471.

[26] Aiken LH, Clarke SP, Sloane DM, Sochalski J, Silber JH. Hospital nurse staffing and patient mortality, nurse burnout, and job dissatisfaction. JAMA 2002;288(16):1987–93.

[27] Aiken LH, Sloane DM, Ball J, Bruyneel L, Rafferty AM, Griffiths P. Patient satisfaction with hospital care and nurses in England: an observational study. BMJ Open 2018;8(1):1–8. https://doi.org/10.1136/bmjopen-2017-019189.

[28] Aiken LH, Sloane DM, Bruyneel L, Van den Heede K, Griffiths P, Busse R, et al. Nurse staffing and education and hospital mortality in nine European countries: a retrospective observational study. Lancet 2014;383 (9931):1824–30.

[29] Aiken LH, Sloane DM, Cimiotti JP, Clarke SP, Flynn L, Seago JA, et al. Implications of the California nurse staffing mandate for other states. Health Serv Res 2010;45(4):904–21.

[30] NQF. Nursing hours per patient day. [cited 23 August 2018]. Available from: http://www.qualityforum.org/ WorkArea/linkit.aspx?LinkIdentifier=id&ItemID=70962; 2012.

[31] Plummer V. An analysis of patient dependency data utilizing the TrendCare system. Melbourne: Monash; 2005.

[32] Lang TA, Hodge M, Olson V, Romano PS, Kravitz RL. Nurse-patient ratios: a systematic review on the effects of nurse staffing on patient, nurse employee, and hospital outcomes. J Nurs Adm 2004;34(7–8):326–37.

[33] Hendry C, Aileone L, Kyle M. An evaluation of the implementation, outcomes and opportunities of the care capacity demand management (CCDM) programme-final report. Christchurch, NZ: NZ Institute of Community Health Care; 2015.

[34] Safe Patient Care. (Nurse to patient and midwife to patient ratios) Act 2015 no.51. Melbourne [cited 30 November 2018]. Available from: https://www2.health.vic.gov.au/health-workforce/nursing-and-midwifery/safe-patient-care-act; 2015.

[35] ANMF VB. Nurse/midwife: patient ratios—it's a matter of saving lives; 2015, Australian Nurses and Midwives Federation, Victorian Branch; 2015

[36] Tubbs-Cooley HL, Cimiotti JP, Silber JH, Sloane DM, Aiken LH. An observational study of nurse staffing ratios and hospital readmission among children admitted for common conditions. BMJ Qual Saf 2013; 22(9):735–42.

[37] Serratt T, Meyer S, Chapman SA. Enforcement of hospital nurse staffing regulations across the United States: progress or stalemate? Policy Polit Nurs Pract 2014;15(1–2):21–9.

[38] NNU. National campaign for safe RN-to-patient staffing ratios. [cited 24 August 2018]. Available from: https://www.nationalnursesunited.org/ratios; 2018.

[39] Bolton LB, Aydin CE, Donaldson N, Brown DS, Snadhu M, Fridman M, et al. Mandated nurse staffing ratios in California: a comparison of staffing and nursing-sensitive outcomes pre- and postregulation. Policy Polit Nurs Pract 2007;8(4):238–50.

[40] Kane RL, Shamliyan T, Mueller C, Duval S, Wilt T. Nursing staffing and quality of patient care. Evidence report/technology assessment no. 151 (prepared by the Minnesota Evidence-based Practice Center under contract no. 290-02-0009.). Rockville, MD. Contract no.: AHRQ publication no. 07-E005.

[41] Junttila JK, Koivu A, Fagerström L, Haatainen K, Nykänen P. Hospital mortality and optimality of nursing workload: a study on the predictive validity of the RAFAELA Nursing Intensity and Staffing system. Int J Nurs Stud 2016;60:46–53.

[42] Fagerstrom L, Lonning K, Andersen MH. The RAFAELA system: a workforce planning tool for nurse staffing and human resource management. Nurs Manag (Harrow) 2014;21(2):30–6.

[43] Fagerstrom L, Rainio AK, Rauhala A, Nojonen K. Professional assessment of optimal nursing care intensity level. A new method for resource allocation as an alternative to classical time studies. Scand J Caring Sci 2000;14(2):97–104.

[44] Andersen MH, Lonning K, Fagerstrom L. Testing reliability and validity of the Oulu Patient Classification Instrument—the first step in evaluating the RAFAELA System in Norway. Open J Nurs 2014;4:303–11.

[45] Kinnunen UM, Junttila K, Liljamo P, Sonninen AL, Harkonen M, Ensio A. FinCC and the national documentation model in EHR—user feedback and development suggestions. Stud Health Technol Inform 2014;201:196–202.

[46] Liljamo P, Kinnunen U-M, Saranto K. Assessing the relation of the coded nursing care and nursing intensity data: towards the exploitation of clinical data for administrative use and the design of nursing workload. Health Informatics J 2018. 1460458218813613.

[47] Moore A. Data doesn't lie. Focus 2018;26(9):26–7.

[48] NHMRC. Australian clinical practice guidelines: Australian Government, National Health and Medical Research Council [cited 13 September 2018]. Available from: https://www.clinicalguidelines.gov.au/about.

[49] RCH. Clinical practice guidelines: The Royal Children's Hospital Melbourne; [cited 13 September 2018]. Available from: https://www.rch.org.au/clinicalguide/.

[50] Council C. Clinical guidelines. [cited 13 September 2018]. Available from: https://www.cancer.org.au/health-professionals/clinical-guidelines/.

[51] Hovenga E. Casemix, hospital nursing resource usage and costs. [PhD thesis]. Sydney: University of New South Wales (UNSW); 1995.

[52] Sermeus W, Gillain D, Gillet P, Grietens J, Laport N, Michiels D, et al. From a Belgian Nursing minimum dataset to a nursing cost-weight per DRG. BMC Health Serv Res 2007;7(Suppl. 1):1–2. https://doi.org/10.1186/1472-6963-7-S1-A6.

[53] openEHR. openEHR ENTRY types FAQs. [cited 13 September 2018]. Available from: https://openehr.atlassian.net/wiki/spaces/resources/pages/4554768/openEHR+ENTRY+Types+FAQs.

[54] Gøeg KR, Birov S, Thiel R, Stroetmann V, Piesche K, Dewenter H, et al. Assessing SNOMED CT for Large Scale eHealth Deployments in the EU-D3.3 cost-benefit analysis and impact assessment—final report; 2016.

[55] Kalra D, Schulz S, Karlsson D, Stichele RV, Cornet R, Rosenbeck-Gøeg K, et al. Assessing SNOMED CT for Large Scale eHealth Deployments in the EU: ASSESS CT recommendations; 2016.

[56] Werley H, Lang NE. Identification of the nursing minimum data set. New York: Springer-Verlag; 1988.

[57] Sermeus W, Delesie L, Van den Heede K. Updating the Belgian Nursing Minimum Data Set: framework and methodology. In: al FHRFe, editor. eHealth in Belgium and in the Netherlands. vol. 93. Amsterdam: IOS Press; 2002. p. 89–93.

[58] Goossen WTF, Epping PJM, Feuth T, Dassen TWN, Hasman A, van den Heuvel WJA. A comparison of nursing minimal data sets. J Am Med Inform Assoc 1998;5:152–63.

[59] Halloran EJ, Halloran DC. Nurses' own recordkeeping: the nursing minimum data set revisited. Comput Inform Nurs 2015;33(11):487–94.

[60] Goossen WTF, Delaney CW, Coenen A, Saba VK, Sermeus W, Warren JJ, et al. The International Nursing Minimum Data Set (i-NMDS). In: Weaver C, Delaney CW, Weber P, Carr RL, editors. Nursing and informatics for the 21st century. Chicago: Healthcare Information and Management Systems Society (HIMSS); 2006.

[61] Pruinelli L, Delaney CW, Garcia A, Caspers B, Westra BL. Nursing management minimum data set: cost-effective tool to demonstrate the value of nurse staffing in the big data science era. Nurs Econ 2016;34(2):66.

[62] Muntlin Athlin A. Methods, metrics and research gaps around minimum data sets for nursing practice and fundamental care: a scoping literature review. J Clin Nurs 2018;27(11 – 12):2230–47.

[63] AIHW. Metadata Online Registry (METeOR). [cited 5 September 2018]. Available from: http://meteor.aihw. gov.au/content/index.phtml/itemId/181162.

[64] WHO. Family of international classifications Geneva. [cited 8 September 2018]. Available from: http://www. who.int/classifications/en/.

[65] Clark J, Lang N. Nursing's next advance: an internal classification for nursing practice. Int Nurs Rev 1992;39 (4):109–11. 28.

[66] ICN. International classification for nursing practice (ICNP®). [cited 21 August 2018]. Available from: http:// www.icn.ch/what-we-do/international-classification-for-nursing-practice-icnpr/.

[67] Saba VK, Taylor SL. Moving past theory: use of a standardized, coded nursing terminology to enhance nursing visibility. Comput Inform Nurs 2007;25(6):324–31 [quiz 32–3].

[68] Park HT, Lu DF, Konicek D, Delaney C. Nursing interventions classification in systematized nomenclature of medicine clinical terms: a cross-mapping validation. Comput Inform Nurs 2007;25(4):198–208 [quiz 9–10].

[69] Kim TY, Coenen A, Hardiker N. Semantic mappings and locality of nursing diagnostic concepts in UMLS. J Biomed Inform 2012;45(1):93–100.

[70] Sabacare. Clinical Care Classification (CCC™) System. Available from: https://www.sabacare.com/; 2018.

[71] Westra BL, Latimer GE, Matney SA, Park JI, Sensmeier J, Simpson RL, et al. A national action plan for sharable and comparable nursing data to support practice and translational research for transforming health care. J Am Med Inform Assoc 2015;22(3):600–7.

[72] JIGSH. Joint Initiative for Global Standards Harmonization (JIGSH) Health Informatics Document Registry and Glossary, Standards knowledge management tool (SKMT) [web page]. Available from: http://www. skmtglossary.org/.

[73] AS5021:2005. The language of health concept representation. [cited 3 November 2018]. Available from: http://www.e-health.standards.org.au/Home/Publications.aspx; 2005.

[74] Hovenga EJ, Grain H. Health data and data governance. Stud Health Technol Inform 2013;193:67–92.

[75] Saba V, Hovenga E, Coenen A, McCormick K, Bakken S. Nursing language-terminology models for nurses. ISO Bull 2003;16–8.

[76] Andison M, Moss J. What nurses do: use of the ISO reference terminology model for nursing action as a framework for analyzing MICU nursing practice patterns. AMIA Ann Symp Proc 2007;2007:21–5.

[77] Hwang J-I, Park H-A. Exploring the usability of the ISO reference terminology model for nursing actions in representing oriental nursing actions. Int J Med Inform 2009;78(10):656–62.

[78] Bakken S, Hyun S, Friedman C, Johnson SB. ISO reference terminology models for nursing: applicability for natural language processing of nursing narratives. Int J Med Inform 2005;74(7):615–22.

[79] Min YH, Park H-A. Applicability of the ISO reference terminology model for nursing to the detailed clinical models of perinatal care nursing assessments. Healthc Inform Res 2011;17(4):199–204.

[80] ISO18104. Health informatics—categorial structures for representation of nursing diagnoses and nursing actions in terminological systems. International Organisation of Standardisation: ISO; 2014.

[81] Moss J, Coenen A, Mills ME. Evaluation of the draft international standard for a reference terminology model for nursing actions. J Biomed Inform 2003;36(4):271–8.

[82] Matney SA, Settergren T, Carrington JM, Richesson RL, Sheide A, Westra BL. Standardizing physiologic assessment data to enable big data analytics. West J Nurs Res 2016;39(1):63–77.

[83] Jansen K, Kim TY, Coenen A, Saba VK, Hardiker N, editors. Harmonising Nursing Terminologies Uning a Conceptual Framework. Nursing Informatics 2016. Geneva: IOS Press; 2016.

[84] SNOMED-CT. 2.2.1 Root and top-level concepts. [cited 3 November 2018]. Available from: https:// confluence.ihtsdotools.org/display/DOCEG/2.2.1+Root+and+Top-level+Concepts.

[85] ICNP. ICNP browser. [cited 3 November 2018]. Available from: https://www.icn.ch/what-we-do/projects/ ehealth/icnp-browser; 2017.

[86] NANDA-International. The International Nursing Knowledge Association—knowledge base. [cited 22 May 2019]. Available from: http://nanda.org/.

[87] SNOMED-CT. ICNP to SNOMED CT (diagnosis and interventions) equivalence table release notes—July 2016 SNOMED International; [cited 3 November 2018]. Available from: https://confluence.ihtsdotools.org/display/RMT/ICNP+to+SNOMED+CT+%28Diagnoses%29+Equivalency+Table+Release+Notes+-+July+2016.

[88] Hovenga E, Garde S, Heard S. Nursing constraint models for electronic health records: a vision for domain knowledge governance. Int J Med Inform 2005;74(11–12):886–98.

[89] Coppersmith NA, Sarkar IN, Chen ES. Quality informatics: the convergence of healthcare data, analytics, and clinical excellence. Appl Clin Inform 2019;10(02):272–7.

[90] NISO. Understanding metadata. National Information Standards Organisation; 2004. [cited 5 September 2018]. Available from: https://web.archive.org/web/20141107022958/http://www.niso.org/publications/press/UnderstandingMetadata.pdf.

[91] SA. HB 291-2007 Health informatics-guide to data development in health, Sydney: Standards Australia; 2007. [cited 3 May 2019]. Available from: http://www.e-health.standards.org.au/Home/Publications.aspx.

[92] USHIK. United States Health Information Knowledgebase registry and repositry of healthcare-related metadata, specifications and standards. USA: Agency for Helathcare Research and Quality; [cited 26 September 2018]. Available from: https://ushik.ahrq.gov/mdr/portals.

[93] Wilkinson MD, Dumontier M, Aalbersberg IJ, Appleton G, Axton M, Baak A, et al. The FAIR guiding principles for scientific data management and stewardship. Sci Data 2016;3:160018.

[94] CONNMO. Coalition of National Nursing and Midwifery Organisations. [cited 9 September 2018]. Available from: http://connmo.org.au/.

[95] AIHW. Nurse practitioner workforce planning minimum data set. Canberra: AIHW; 2006. [cited 26 September 2018]. Available from: http://www.dhs.vic.gov.au/nnnet/downloads/rec5_npfinal.pdf.

[96] Scott PA, Harvey C, Felzmann H, Suhonen R, Habermann M, Halvorsen K, et al. Resource allocation and rationing in nursing care: a discussion paper. Nurs Ethics 2018. 0969733018759831.

[96a] Henderson V. The concept of nursing. J Adv Nurs 2006;53(1):21–31.

[97] CIHI. Inclusion of nursing-related patient outcomes in electronic health records, Canadian Institute for Health Information; 2018. Available from: https://www.cihi.ca/en/c-hobic-infosheet_en.pdf.

[98] C-HOBIC. Canadian Health Outcomes for Better Information and Care project Canada. [cited 22 August 2018]. Available from: https://c-hobic.cna-aiic.ca/about/default_e.aspx; 2018.

[99] interRAI. interRAI—a comprehensive assessment system. [cited 22 August 2018]. Available from: http://www.interrai.org/.

[100] Doran DI, Sidani S, Kearings M, Doidge D. An empirical test of the nursing role effectiveness model. J Adv Nurs 2002;38(1):29–39.

[101] Doran DM, Sidani S. Outcomes-focused knowledge translation: a framework for knowledge translation and patient outcomes improvement. Worldviews Evid Based Nurs 2007;4(1):3–13.

[102] Lowe C. Trendcare Systems Pty Ltd. [cited 22 August 2018]. Available from: http://www.trendcare.com.au.

[103] Kohn LT, Corrigan JM, Donaldson MS, editors. To err is human: building a safer health system. Washington, DC: Institute of Medicine, National Academy Press; 2007.

[104] Spence Laschinger HK, Leiter MP. The impact of nursing work environments on patient safety outcomes: the mediating role of burnout engagement. J Nurs Admin 2006;36(5):259–67.

[105] NHS-Digital. Information and technology for better care: Health and Social Care Information Centre Strategy 2015–2020. [cited 3 May 2019]. Available from: https://digital.nhs.uk/article/249/Our-Strategy; 2015, www.hscic.gov.uk.

[106] NHS. Digital, data, insights and statistics, [cited 31 October 2018]. Available from:. https://digital.nhs.uk/data-and-information/data-insights-and-statistics.

[107] ICHOM. International Consortium for Health Outcomes Measurement, Inc. (US). [cited 31 October 2018]. Available from: https://www.ichom.org/.

[108] ISO 13606-1:2008. Health Informatics-Electronic health record communication—part 1: reference model. Geneva: International Organisation for Standardisation.

[109] International H. HL7 version 3 Clinical Document Architecturen (CDA) release 2. [cited 26 September 2018]. Available from: http://www.hl7.org/implement/standards/product_brief.cfm?product_id=7.

[110] openEHR-Foundation. openEHR reference model (RM)—latest. openEHR Foundation; 2017.

[111] openEHR. Clinical knowledge manager. [cited 26 September 2018]. Available from: https://www.openehr.org/ckm/.

[112] Olsen D. Metadata: a primer for indexers. Key Words 2009;17(1):18–20.

[113] Hovenga EJS. National healthcare systems and the need for health information governance. In: Hovenga EJS, Grain H, editors. Health information governance in a digital environment. Amsterdam: IOS Press; 2013.

[114] Hovenga EJS. Impact of data governance on a nation's healthcare system building blocks. In: Hovenga EJS, Grain H, editors. Health information governance in a digital environment. Amsterdam: IOS Press; 2013.

[115] Hovenga EJS, Grain H. Our health language and data collections. Stud Health Technol Inform 2013;193:93–107.

[116] Kimsey DB. Lean methodology in health care. AORN J 2010;92(1):53–60.

[117] Adamson BJ, Kenny DT, Wilson-Barnett J. The impact of perceived medical dominance on the workplace satisfaction of Australian and British nurses. J Adv Nurs 1995;21(1):172–83.

[118] Kenny A. Medical dominance and power: a rural perspective. Health Sociol Rev 2004;13(2):158–65.

[119] Willis E. Medical dominance: the division of labour in Australian health care. Sydney: Allen & Unwin; 1990.

[120] McBeth R. IT solutions key to matching capacity to care with patient demand, ehealthNewsnz: Health Informatics New Zealand 2018. [cited 26 October 2018]. Available from: https://www.hinz.org.nz/news/423365/IT-solutions-key-to-matching-capacity-to-care-with-patient-demand.htm.

# CHAPTER 4

# Nursing and midwifery work measurement methods and use

Work measurement is about establishing the amount of time one or more qualified nurses, and relevant nursing support workers need to complete clearly defined work activities at a predefined level of performance to produce a stated product or outcome. It's about measuring the nurse patient contact time required to achieve safe, efficient and productive patient care. Work measurement is a process adopted to translate reality into numbers. The common unit used is fixed units of time, expressed as day, hour, minute or second, relative to defined work. Various techniques may be employed for this purpose. Work content of any job needs to be related to its most suitable outcome characteristic when work is expressed in units of time. In the health industry such outcome characteristics may be expressed as service(s) provided by service type or patients attended/treated/cared for (products produced) per day. The context as determined by confounding variables for nursing work measurement is shown in Fig. 1. Work measurement applies to process relative to input and output factors.

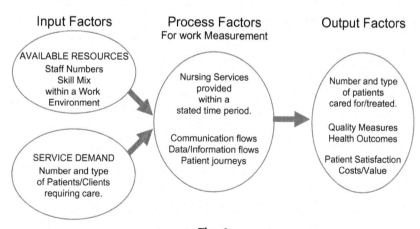

**Fig. 1**
Nursing work measurement context.

Work measurement techniques include time and motion study and activity sampling. From the Industrial Engineering perspective, work measurement should not be viewed in isolation of all facets influencing productivity, they are intricately linked. Work Study as a science draws on

Measuring Capacity To Care Using Nursing Data. https://doi.org/10.1016/B978-0-12-816977-3.00004-6

numerous concepts within disciplines such as economics, engineering, ergonomics, behavioral sciences, computer science, statistics and management science.

Evaluation of a system's validity requires an examination of the construct validity of measures of job performance that is the degree to which the method adopted measures what it claims or purports to be measuring, versus the true score obtained within a stated margin of error. The reliability of a system is determined by an examination of whether the scores obtained through the use of the system, consistently reflect the true score, assuming that the measurement process adopted is valid.

A number of factors need to be considered prior to choosing a suitable nursing work measurement technique. There is a trade off between a number of these factors, which are

- **Purpose** — what is the aim and how are the results intended to be used?
  **Example 1** — The Health Commission Victoria (HCV) had a legislative responsibility to determine the nursing staff numbers by skill mix, including associated budget allocations for all 169 public hospitals. Its Health Management Services group was given the responsibility for establishing a uniform and practical approach to meet this requirement [1].
  **Example 2** — In 2018 the New Zealand Ministry of Health identified that there was a real need for New Zealand nursing services nationally to have a reliable and validated patient nurse dependency tool capable of measuring nursing workloads in a cross section of nursing services with a broad scope of patient types [2].
  **Example 3** — In 2018 The Irish Department of Health established a taskforce to adopt an innovative approach to addressing workforce planning requirements by the development of a framework to support the determination of safe nurse staffing and skill mix (whereby nurse staffing refers to the nursing team including both the nurse and healthcare assistant roles) in a range of major specialities [3].
- **Level of detail** — What level of accuracy is required for the end results to be reliable and able to be used relative to the intended purpose?
  **For example:** easy to apply with consistency, require minimal resources to implement and use. Be sensitive to changes in demands made on nursing services over time to minimize the need for frequent in-depth reviews. Avoid further staffing anomalies.
- **Time availability** to undertake the study — the adoption of any nursing work measurement methodology requires significant time investments.
  **Example 1** — the original Patient Assessment and Information System (PAIS) study was undertaken by staff from the Health Management Service Group under this author's leadership [4]. Numerous further studies were staffed and funded by the Victorian Government's Health Department.
  **Example 2** — The work studies undertaken to develop and validate the current TrendCare® patient nurse dependency and workload management system continue to

be undertaken as a collaborative between user sites and the researcher/developer of the system. The total study period to date spans across three decades and six countries.

- **Existing data availability** and potential for use — many studies referred to in the previous chapter made use of administrative data. In such instances data accuracy, completeness and reliability needs to be considered.

    **For example** both the PAIS and TrendCare® studies applied original research methodologies and used unique data generated by their respective work studies.

- **Boundaries of work** to be measured — is the scope of the study, defining what is included and measured. What is described as the nursing work? Who is actually undertaking this work? What is universally recognized as constituting nursing work?

    **For example,** the PAIS study focused on service demand as generated by patients, which was measured irrespective of staff category doing the work. Ward clerks and other nursing support workers were included if they were part of that ward's allocated workforce.

- **Periods of time** over which the work measurement study is undertaken to account for seasonal differences in the demand for nursing services and/or staff availability in any 1 year.

    **Example 1** — the original PAIS work measurement study took place over several weeks across three Victorian hospitals. The resulting patient classification methodology was tested in another hospital over more than 1 week to generate a sufficiently large and representative data sample, and refined prior to its first implementation and use.

    **Example 2** — The original TrendCare® study took place over a period of 2 years and was tested across six public hospitals and five private hospitals in the State of Queensland, Australia.

- **Continuing use** — If an evidence based methodology is to remain valid and reliable regular work studies must be continued and the system updated as required to accommodate changing patient populations and changes in nursing practice.

    **Example 1** — The PAIS methodology was used by any Australian hospital requesting to do so for multiple purposes. The Victorian PAIS User group was active from 1983 to 1993, it's purpose was to ensure consistency in interpretation by standardizing PAIS terminology and processes in use [5]. Another PAIS interest group was active in NSW to support its users. PAIS was selected as the corporate system for Queensland Health in 1993, supported by their nursing unit. Additional work measurements studies were undertaken as validation to confirm time values in use. Its final validation took place for Queensland Health in 1995 when one patient population's nursing requirements were compared with those generated by the TrendCare® system resulting in very similar results [6].

    **Example 2** — The TrendCare® methodology became fully electronic in 2000 and the system is now extensively used in both public and private hospitals across the Australian state of Queensland, New Zealand, Singapore and Thailand and in 2018 it was implemented into eight hospitals across Ireland, and England. Work studies, system enhancements and validation testing occurs annually to maintain system integrity validity and reliability.

- **Costs** associated with any nursing work measurement study in terms of both $ and other non-tangible costs.

   **Example 1** — The PAIS study was an investment using existing staff [1].

   **Example 2** — The ongoing TrendCare® work studies make use of hospital staff from TrendCare® user sites to collect nurse/patient contact times. Data collection activities related to the work study are easily incorporated into the nurses shift routine and hence there is no additional cost to any participating hospitals. There is however ongoing research and development costs relating to data collation, analysis, and distribution which is picked up by the software developer. Ministry of Health staffing units, nurse and midwifery Unions and Colleges, Universities, specialist nurses and other significant stake holders conduct user advisory forums providing valuable feedback (in kind).

- **Use of work measurement data** — An accurate measure of nursing work across shifts of the day, days of the week and months of the year facilitates the redesign of roster patterns so that the rostered nursing hours are matched to peaks and troughs in demand, promoting safe staffing, efficiency and productivity.

   **Example 1** — The resulting patient classification system PAIS was used to manage an annual $500 M nursing/midwifery staffing budget (in1982), resolve industrial disputes over nursing workloads and maintain industrial harmony with the Nursing Profession across more than 160 hospitals. Additional investments were made by other users around Australia supporting ongoing use and benefits.

   **Example 2** — TrendCare® is now the dominant nursing work measurement system across Australasia and is used to develop and manage nursing budgets in a large number of public and private hospitals across six countries.

There are many benefits and uses of nursing work measurement studies that need to be considered. One major benefit of following the adoption of a standard framework and methodology is that results can be aggregated to achieve larger sample sizes, providing improved reliability in relation to data analysis. A larger sample size results in greater accuracy, and facilitates the ability to undertake comparisons between patient types, hospitals or nursing activities. For example to explore the impact of variances between nursing work in certain types of work environments and/or between well defined different patient populations. It's also possible to apply two work measurement methodologies at the same time to establish results validity.

Nursing work measurement studies require the following data sets:

- Nursing hours available by skill mix during the study period
- Organizational/work environment characteristics — to assist with explaining variable findings

- Patient characteristics, treatment and care requirements relative to bed occupancy
  - ○ i.e., service demand
- A nursing work taxonomy — variables included for measurement purposes
  - ○ i.e., based on decisions about what needs to be measured
- Compliance with professional practice standards, performance quality measures
  - ○ To demonstrate that service demand was met in accordance with consumer expectations and professional requirements during the study period.

The previous chapter explored available metadata and the availability of standard terminologies applicable to the many confounding variables identified. This highlighted the complexity of measuring nursing service demand and the difficulties associated with defining the multitude of variables in accordance with an agreed standard. This chapter explores these factors from a work measurement perspective.

## Describing nursing work

Before any work measurement can take place, one needs to have a sound understanding of what nursing work consists of. Nursing work could be classified as an industry which, from an industrial engineering perspective, uses a lot of applied knowledge, plant and machinery, and produces a complex product — i.e., patient outcomes. Nursing work consists of actual work done with the "tools of the trade," plus other work which includes planning, getting and preparing, plus work involving walking to get equipment, supplies or information to and from the point of care.

The actual time when the various nursing activities need to be performed is dependent upon ward routine or work organization of routine activities, requirements of various treatment and care regimes, individual patient demand, and priorities relative to the entire patient population within the nurses' care.

Some nursing work requires considerable physical exertion while other aspects of nursing work requires manual dexterity. Nursing work requires observational and intellectual ability. Many of the nursing variables identified relate to direct patient interaction, thus interpersonal and communication skills are of great importance. Although nursing consists of many tasks it is not task based. It is a behavioral science requiring various degrees of judgment, decision making and caring activities.

Clearly patient characteristics and their individual demands influence the time required to perform many nursing activities both interactively with the patient or away from the patient; in particular the preparation for and putting away of equipment and supplies required for direct patient/nurse interaction. Patients also need to be prepared by having the planned activity/intervention explained and/or patients may need to be positioned prior to any intervention.

Some nursing activities are common to all patient types, but not necessarily required by every patient, while other nursing activities are specific to certain major disease categories or specialty areas. Nurses need to prioritize their activities according to urgency and specific time schedules. Some nursing activities may be delayed from 1 day to the next, such as washing a patient's hair, cutting toenails or changing all the bed linen. Baths may also be deferred when the nurse perceives other activities as having a higher priority and need to be performed within the time available. The rate at which all activities are performed will vary between nurses, and with pressure of time. Methods used to perform nursing activities are modified to suit individual patients with special needs, the situation and time available.

Many indirect nursing activities are not determined by patient characteristics but by organizational factors. These include ward administration, attending meetings, or activities associated with the preparation for and clean up activities associated with direct patient care. Nursing time required for patient documentation, and communication between the multidisciplinary care team, family members of the patient and other relevant departments. This is influenced by the number and type of patients in the ward, their severity of illness, as distinct from the disease itself and patients' respective social support structures and cultural norms.

Nursing theories form the basis for nursing curriculum development and nursing models of practice. The assist/doing for a patient theory model is likely to result in greater patient nurse dependence over a period of time. Whereas the self-care theory model is likely to require more nursing time initially but considerably less nursing time further along a patient's health journey, as the patient progresses to provide self-care. Each organization's mission statement, organizational culture and philosophies adopted influence what is done, by whom and when. Thus nursing theories and various philosophies influence the detail of nursing practice.

Nursing work has a variety of other characteristics. It may be considered to contain short to long cycles of work, or range from rarely performed procedures to repetitive work. Some work is performed routinely in accordance with clearly defined methods, other work is performed on an ad hoc basis using a method considered appropriate to the situation.

The hospital nurse performs much of the work while standing or walking. At times nurses are required to assume unnatural postures. Rarely is 1 day's work identical to another, although much of the work is performed on a daily basis, but in varied sequences. Individual nursing procedures may be performed using a variety of different methods. Some work elements, which may be included in the total procedure, may frequently be omitted when the procedure is performed. Such shortcuts may occur due to short staffing and time constraints; this may compromise patient safety and/or lead to adverse events. Interruptions may be minimal or extensive and are usually outside the control of the nurse. Interruptions often necessitate changing methods in use and/or an adjustment of previously established priorities.

Many nursing activities are about risk management, taking action to prevent adverse health consequences. Thus nurses diagnose the potential for a health problem and intervene to prevent its occurrence, in order to optimize outcomes. The overall quality of all health services provided generally influence subsequent medical interventions and hence nursing requirements. Continual patient assessments, followed by feedback regarding outcomes, lead to adjustments in services provided. Consequently some averages will need to be applied regarding the nursing time required to meet service demand. Known industrial engineering work measurement techniques need to be modified to suit the nurses' work environment, although fundamental work measurement principles need to be adhered to.

### Boundaries or scope of nursing/midwifery practice

Boundaries of the universe of nursing work are relative to the personnel policies of the hospital and the availability of various categories of support staff within any 24/7 period. The role of the nurse is dependent upon the boundaries of nursing practice as defined by position, a nurse's qualifications and experience, legislation, hospital policies, work organization and the boundaries of other health professional roles. An in-patient population generates ward work. This work is divided among various categories of staff, including nursing, allied health and non-nursing support staff. However, nurses will perform any type of work when other staff categories are not available, based on the work required to be performed in the interest of patient safety, comfort and care. These include providing drinks and meal service, supervising exercise routines, portering, lifting, completing clerical services and relaying messages, ordering, stocking and rotating supplies, prescribing and dispensing pharmaceuticals, cleaning and tidying the environment, and others.

In many organizations nurses perform an expanded role including activities previously performed by medical practitioners. Defining boundaries of practice is difficult as boundaries of individual practitioner types tend to vary between locations, hospitals, wards, specialties, time of day and situations. For example in a ward environment nursing staff and/or laboratory staff can perform venepuncture to collect blood samples and in an emergency department advanced nurse practitioners can diagnose, prescribe medications and initiate procedures.

It is within this context that nurses establish their own methods of nursing service delivery to best suit the environment. Hence methods of nursing service delivery vary. To overcome these inconsistencies between hospitals it is recommended that work measurement studies include all staff undertaking direct and indirect patient related activities for recognizable periods of time within the study's location.

## Analyzing nursing work to be measured

Nursing Practice may be analyzed from a number of different perspectives. The perspective chosen is dependent upon the purpose of the analysis. The perspective determines which work characteristics are highlighted and noted or which need to be measured.

It provides a focus. The following points of view from which to analyze nursing work have been identified:

- Time required to perform the job (work measurement).
- Organization and methods employed, including the availability and use of labor saving devices (method study).
- Cost of providing the nursing service (costing study).
- Numbers of patients cared for (output).
- Education, skills, experience and physical attributes required to perform the job (skill mix).
- Boundaries of nursing practice and relationships with other health workers (human resource input factors).
- Philosophical basis of work performed and its relationship to objectives to be achieved (philosophy of practice).
- Working conditions relative to health care workers, industrial agreements and legislative requirements (human resource input factors).
- The value of the job to individuals, organizations and society (outcomes).
- The level of agreement between the actual job with the job description or duty statement (compliance).
- Level of job satisfaction by the incumbent (potential turnover-workforce planning).
- Opportunity for staff development and career progression.
- The quality of the product in terms of
  - (a) meeting organizational objectives/expectations
  - (b) meeting professional standards of practice
  - (c) compliance with legal and statutory regulations
  - (d) meeting consumer expectations
  - (e) meeting other health workers expectations
  - (f) meeting accreditation requirements

Defining nursing work to be measured requires one to undertake a nursing job analysis using any of the following methods:

1. Direct observation
2. Questionnaire
3. Checklist
4. Individual interview
5. Observation and interview
6. Group interview
7. Technical conference using "experts"
8. Diary or self-reporting
9. Critical incident recording

Job analysis is a precursor to work measurement. It is during this process that both the universe of the work to be measured, as well as the data elements making up this universe, are identified. Knowledge of the universe of work measured is also used when evaluating work measurement studies. When only some components of nursing practice are empirically measured, one needs to assess how the remainder of nursing work can be accounted for.

Nursing work cannot be dissected into a set of clearly defined activities where each activity has known start and finish points, as is the case in the manufacturing industry. Patient individual differences and needs may result in variances to the time required to perform an activity. Distances walked to pick up supplies or equipment for use at a point of care or to answer patient call bells vary throughout any working day or shift based on patient location, the physical layout of the facility, where items are stored, frequency of need and number of equipment items, such as the availability of frequently used blood pressure machines. Personal time, meetings and organizational administrative support functions also need to be considered.

Any process, including any nursing activity, data/information flows, communications between all relevant parties, patient journeys through the health system, requires what used to be referred to as "methods study." According to the International Labour Organisation [7] work may be quantified in terms of staff time required to perform the work. Work essentially consists of productive work and inefficiencies as listed below. Such inefficiencies impact upon healthcare's operational effectiveness (output factors). These inefficiencies are areas any Lean Six Sigma methodology needs to focus on in order to eliminate or minimize ineffective performance outcomes.

| | |
|---|---|
| Productive work | • Basic work content |
| Inefficiencies | • Work content added by defects in equipment or insufficient or difficult to access supplies<br>• Deficits in knowledge, skill or treatment/care regimes used<br>• Work content added through the use of inefficient methods<br>• Time lost due to organizational or management shortcomings<br>• Time lost as a result of inefficient time management by the worker |
| Ineffective outcome measures | • Adverse events, preventable deaths<br>• Poor health outcomes<br>• Staff and patient/client dissatisfaction<br>• Unnecessary delays, costs<br>• Non availability of data/information/equipment/services when required |

Nursing workload measurement systems essentially represent a method for determining nursing productivity relative to patients cared for.

## *Work measurement methods*

Work measurement is a process adopted to translate reality into numbers and may be defined [7] as

> *the application of techniques designed to establish the time for a qualified worker to carry out a specified job at a defined level of performance.*

Work measurement is explained in considerable detail in seminal industrial engineering (work study) textbooks [8, 9]. It is apparent from the literature that these are rarely if ever consulted by those undertaking health service work measurement research. A 2014 study of healthcare and biomedical literature, indexed according to the term "time motion study" (a MeSH term) [10] revealed a common and re-occurring misunderstanding regarding the definition and scope of time and motion studies. One study identified a long list of methodological or result reporting inconsistencies from a time and motion studies' review [11].

Any work measurement methodology needs to clearly indicate and define the staff, their performance, the work activities undertaken, and associated "product" — patients or outcomes relative to time taken as measured by the method adopted. In other words, the relationships between the time taken by whom (input), the work performed (process) and what was achieved (outcome) within a stated observation/work measurement period, must be clear.

Standard performance is defined [7] as:

> *The rate of output which qualified workers will naturally achieve without over exertion as an average over the working day or shift, provided that they know and adhere to the specified method and provided that they are motivated to apply themselves to their work.*

Time study including work sampling, can be undertaken by external observers, or be self-reported either continuously or intermittently over any period of time. Performance ratings are used to adjust times arrived at by means of time study. Another issue relevant to work measurement is that of allowances. It is recognized that no worker can be expected to work consistently without taking time out to meet personal needs, for rest and other interruptions beyond the worker's control. Standard time values need to be realistic and be applicable to the total job, thus specific allowances are added to the basic time as measured in accordance with the specific demands for the work studied. These may include allowances for physical strain, stress, posture, restrictive clothing, highly repetitive work, noise etc. Some special allowances may be included in each worker's industrial award.

Continuous observations require one external observer per nurse on duty. This method is intrusive, unpopular with staff, likely to influence the results through a "Hawthorn" effect and is expensive. Intermittent observations, at fixed intervals or at random, require one observer per group of nurses. This is more cost effective and much less intrusive. Time study of individual nursing activities using a stop watch is unlikely to measure the universe of work generated by a group of patients. In addition the range of time required to undertake any specific patient related activity is usually large. This requires a large sample collectively representing a normal distribution in order to obtain the desired degree of accuracy. However time studies that record nurse/patient interaction and associated patient related activities are able to measure differences of nursing resource use by patient type.

Self-reporting is an inexpensive alternative, and can be adopted in a variety of ways. Limitations are the risk of missing data when nurses get busy or are lacking commitment. There is a high probability of inaccuracy where data are recorded from memory much later following an activity, and there is a perceived inability to calculate the degree of accuracy.

### Nursing staff availability and performance — Input variables

Staffing "hours" made available at the point of care during the entire study period must be accounted for. This requires the time used for non-inpatient care and time spent out of the ward/ department to be monitored and to be subtracted from the hours worked. For example; time spent ordering supplies, waiting in the X-ray department, completing patient escorts or conducting any other activities not included in the scope of work being measured. Staff performance rating is extremely difficult to apply, even to the actual work done, as the rate and effort required in the performance of nursing work is influenced not only by the nurse but also by the situation.

The level of performance is ideally measured in terms of patient satisfaction and patient outcomes but may be measured in terms of adherence to defined professional and procedural standards. Unnecessary nursing actions, inefficient methods of work organization, scheduling and operational management deficits, all add to the time required to perform the work. Although all nurses are qualified to practice nursing, and as such they may be defined as "qualified workers," some are more qualified and/or experienced than others in certain nursing environments, for example in units where specialized nursing care is provided. A lack of relevant skills and competence impacts on performance.

Due to prevailing rostering practices the composition of the nursing team per shift tends to vary; paramedical and medical staff come and go throughout a patient's hospital stay. Such constantly changing staff and patient group dynamics influence the time required to care for all patients in a ward/department. Team building to achieve optimum performance is difficult to achieve in the absence of team consistency. Efficiency and effectiveness would greatly improve if the teams were kept constant. Good examples of the impact upon performance by the use of constant teams may be found in sports such as rowing or sailing or in healthcare, to perform a heart transplant operation. Such teams improve performance over time as they learn how to overcome individual weaknesses and capitalize on individual strengths as a cohesive team.

The extent to which staff can devote time to low priority care activities following peak periods, denotes just how busy the staff are. Nursing activities can be grouped according to priority rating ranging from non-productive activities to activities which must be carried out at a specific time and can neither be postponed nor neglected. The nursing time allocated to care for a group of patients is the critical factor. It may be that in some wards, nurses are rarely able to perform low priority activities and in others this would be a usual occurrence. The relative distribution of time spent between patients is also likely to be influenced by the total time made available.

Many of the variables identified previously as influencing time required for nursing practice cannot be controlled. Exceptions are the number of nurses and nursing support staff provided or allocated to work in a particular area and to some extent the physical environment within which nursing is practised. Insufficient staff is highly likely to result in shortened tea and meal breaks leading to lower productivity and an increased risk of human error rates, and/or the omission of some activities which should be undertaken in the interest of patient comfort and well being.

### *A nursing practice taxonomy — Process variables*

The choice of variables to be measured needs to be restricted to those variables known to have a significant influence on time required. Previous literature review findings provided this information for the Hovenga (PAIS) and Lowe (TrendCare®) study designs in the first instance. The results of numerous studies undertaken over several years in many hospitals from small country or regional hospitals to very large teaching hospitals, have demonstrated that the significance of individual nursing activities varies according to the type of patient population/ clinical specialty studied. For example there are few technical activities undertaken in residential aged care facilities where the time required for supporting activities of daily living is all consuming. These studies are reported separately as case studies. Their findings are referred to as examples in this chapter and other chapters where relevant.

The universe of work to be studied, and the level of detail from which it is to be studied, must first be defined irrespective of work measurement technique to be used. This requires the establishment of a nursing practice taxonomy consisting of mutually exclusive well defined data elements. Such a taxonomy is influenced by the proposed purpose of the study, research design, preferred work measurement method to be adopted and the nursing work variables that need to be considered. There is general agreement in the literature that the universe of nursing work in most areas consists of the following:

(i)   Direct patient/nurse interaction.
(ii)  Indirect but patient related work such as preparation for and putting away of equipment and supplies needed for direct care activities, plus referring or adding to patient related documentation; a number of these activities may be termed "collective patient care" as they are not easily attributed to an individual patient.
(iii) Indirect ward related work which is not dependent upon the nature of the patient population such as ward administration, in-service education, committee attendance, meal and personal breaks. These tend to vary in length based on work demands.

Each of these broad components of nursing work may be further divided into smaller components such as basic (core) activities related to daily living applicable to most patients some time during the length of stay or technical nursing work such as wound care or medication administration. The identification of elements for inclusion in the taxonomy to be used for measurement purposes needs to be guided by the study's purpose and/or chosen work measurement methodology. Some work components are readily observable and identifiable, others are not.

The PAIS taxonomy shown in Fig. 2, enabled observations to be made by an external observer relative to specific direct and indirect patient related activities such as IV therapy, drug administration to be combined [1]. The Direct activities consisted of 14 basic nursing activities and 10 technical nursing activities. A total of 28 indirect activities were included plus four non-productive. It was decided to include all personal time, tea and meal breaks as these tend to take up less time then officially allocated when nurses are busy. The selection of activities for this study also enabled one to assess workload generated for various staff categories. A total of 58 activities were included, although activities with few observations were grouped resulting in results reporting for a total of 27 activities.

The TrendCare® taxonomy shown in Fig. 3, was developed on the basis of making use of self-reporting time studies. This taxonomy enables total nurse/midwifery/patient contact timings to include a combination of direct and indirect care, and all other non-clinical activities related to a patient included in the study. The parameters of an activity time include "set up," "clean up," and "write up." The number of activity groups and specific activities defined for a study varies according to the patient types included in the study and the scope of care activities performed by the nursing/midwifery service. The selection of activity groups and activities is reviewed with each ward/department participating in a study and is adjusted, if required, to ensure that all relevant activities are included. All activities included in a study are individually coded and are recorded against each timed contact in the study. Meal breaks, tea breaks and personal time are excluded from the timings, as these are extremely variable and are not part of the actual patient demand for care which is the focus of TrendCare® timing studies. An allowance for paid breaks and personal interruptions to work is added into the resulting TrendCare® category timings together with an additional allocation of time for unpredicted work.

A recent Belgian work sampling study used the list of activity groups that is part of the nursing workload management method developed by the Dutch Hospital Institute as its foundation. A Delphi method was then used to evaluate and develop a definitive activity group list consisting of four major activity categories, direct patient care, collective patient care, general tasks and other (non-productive) tasks consisting of personal time, coffee and meal breaks. These four groups were further divided into 24 major activity groups that consisted of more specific activities. This final taxonomy was used for their work sampling study [12].

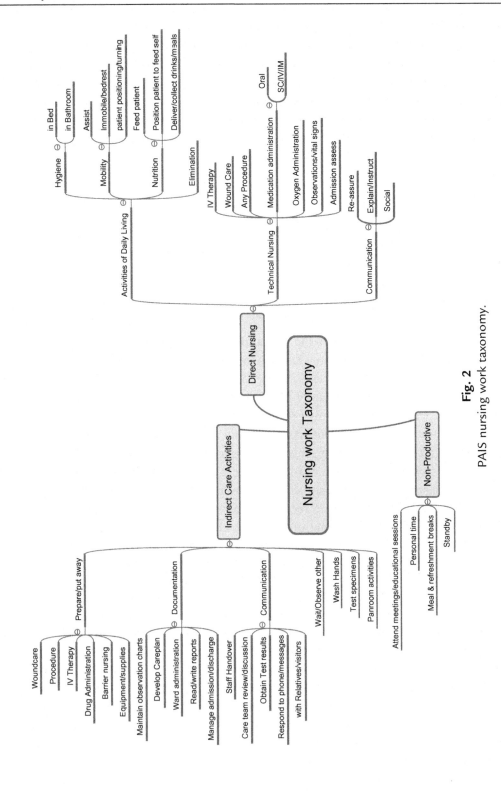

**Fig. 2**
PAIS nursing work taxonomy.

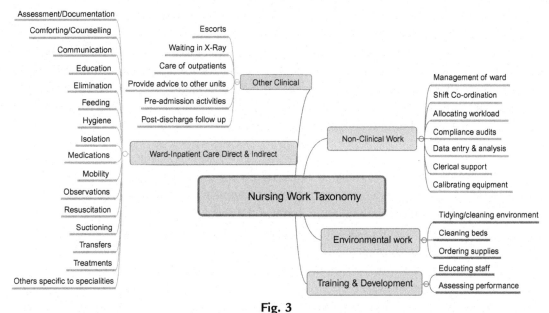

**Fig. 3**

TrendCare® nursing/midwifery taxonomy. *From the TrendCare® Taxonomy Diagram.*

## Time study methodology

Time study may be used not only to determine the time required by a qualified person to perform a specified task or set of tasks or to deliver a service, but also to form the basis for

- measuring the demand for services
- determining and controlling staffing costs
- redesigning work and staff rosters to match peaks and troughs in demand
- undertake additional planning activities
- to maintain safe, efficient and productive health care delivery services

Prior to undertaking any time study, it is desirable to undertake an evaluation of work methods in use as the existence of numerous inefficiencies is a reality at varying degrees across all health care organizations. These evaluation activities essentially represent the application of "Lean" and "Six Sigma" [13–15] methodologies, as explained further in Chapter 12, to

- minimize any work content added by defects in equipment
- identify and correct inefficient treatment/care regimes used and skills of the worker
- remove work content added through the use of inefficient methods
- identify and correct time lost due to organizational or management shortcomings
- time lost as a result of inefficient time management or worker skill deficits

An observational time study can commence once the observer has been suitably instructed about the study and familiarized with the observation chart used to record results. A decimal hour stopwatch is preferred as this provides readings directly as fractions of an hour, making these observations easier to process. An observation sheet is used to document time values relative to activities being measured. Data collection form design is an essential component of any work measurement study. Actual time study procedures can vary according to the purpose and intended data use. From an industrial engineering perspective eight steps are usually required. These are [8]

1. Identify and describe the activity to be measured.
2. Divide the activity into elements and record a complete description of the work method.
3. Observe and record the time taken to perform the activity.
4. Determine the number of cycles to be timed.
5. Rate the worker's performance.
6. Ensure a sufficient number of activity cycles are timed.
7. Determine relevant allowances.
8. Determine the time standard for the activity timed.

When adopting the time study approach, it is necessary to conduct many individual time studies of the same task performed by a large variety of qualified workers in order to ensure that the study is representative of normal conditions and that a normal distribution of occurrence was in fact measured. There are many variables which influence the number of observations required. They include the length of the work cycle, number of repetitive elements in each work cycle, skill of the worker, variations in method and the number of interruptions. Ideally observers adopt a standard agreed start and finish location of each activity timed to ensure consistency. Method and equipment used should be identified and recorded as well where relevant.

It is beneficial to break down any activity into smaller elements with identifiable beginning and end points as this enables the analyst to identify variations and determine why these occur. This in turn can assist in improving procedural methods or workflows. Activities can be timed continuously, where the finish time of each element is noted, or each element can be timed individually using a "snap-back" method where the hands of the watch are snapped back to zero at the end of each element.

The mathematical methods of determining the number of observations required to determine time values at a desired level of confidence are based on the following formula [8] p. 273:

$$\sigma_{\bar{x}} = \frac{\sigma'}{\sqrt{N}}$$

where

$\sigma_{\bar{x}}$=standard deviation of the distribution of averages
$\sigma'$=standard deviation of the universe for a given element
$N$=actual number of observations of the element.

In summary the difficulties associated with using time study or observations to measure nursing services include

- issues of definition — at what point does an activity start or finish?
- representativeness — do the number of activities included in the measured sample represent the norm? Were they randomly selected? Are there differences between patient populations based on for example degree of nurse dependency or applicable clinical specialty?
- the universe of nursing — do the sum of the activities equal the universe of nursing services?
- issues of numerosity — which activities or parts thereof, should be measured? For example bed making consists of many individual activities and combinations thereof, similarly wound care or activities associated with hygiene, mobility etc. To what detail should nursing practice be analyzed and measured?
- estimating the degree of accuracy.
- relating the time values to output.

One way to overcome these limitations is to make use of a self-timing methodology where those providing direct care record all interactive and associated indirect care activities per patient. Activities identified are those known to be significant components of nursing workload, such as mobility, hygiene, nutrition, elimination, observations, medications, technical activities, communication, documentation, ward related tasks. It is more realistic to adopt the patient as the unit of analysis rather than discrete nursing interventions. Patients can also be identified by well defined types so that the analyst is able to differentiate caring norms. This is the methodology adopted by Lowe for the development of TrendCare®. Every hospital using the TrendCare® patient nurse dependency system uses the established international evidence based time values for patient categories for each patient type. All hospitals using the system are invited to participate in international time studies targeted at specific patient types using this well developed and tested standard time study method. The results of these studies then contribute to the data pool to enable benchmarked category time values to be updated to reflect changes over time.

## Self-recording of nursing activity

Self-reporting may be used for various data collection methods. As the observer is the same person as the observed this method is the least expensive method in dollar terms for data collection. This method can be very time consuming and costly relative to data collation and analysis for studies involving a large sample size. The self-recording method requires nurses to record, either continuously or intermittently, how they spend their time. This may include timing or estimating the time taken to perform individual activities or listing activities performed and their frequency of occurrence or by listing the activity being undertaken at various random or predetermined time intervals. When recorded intermittently, the data are the result of a self-recorded sample as distinct from a continuous log.

Much debate continues regarding the accuracy and validity of this type of data. The method is favored because it is thought to be a low cost method at the data collection point and results are more likely to be accepted by the workers themselves due to their participation. This method can be open to manipulation, distortions, omissions and gross inaccuracy if the study is not well organized and managed. Much depends on the commitment by the staff involved in data collection, the co-ordination of the study and the controls that are in place during data collection. The potential for bias is increased if the recorder has a high degree of latitude in choosing not only when to record but also in deciding when an activity commenced and finished. Self-reporting by timing each individual task is essentially a job analysis technique which is useful for some purposes. The method is contra indicated for the measurement of high volume nursing work where precision is required unless the study's design includes appropriate risk minimisation techniques.

TrendCare® timing studies are conducted using a self-recording methodology but individual tasks are not timed. Nurses/midwives and nursing/midwifery support workers record the time utilized for each patient contact on a group of targeted patients in a ward/department for three 8 hr periods of the day (the patient's day, evening and night periods). To minimize the risk of staff providing incomplete and/or inaccurate data a strict process of risk mitigation is implemented. This includes the following:

- The appointment of a hospital project co-ordinator who oversees the study and works in collaboration with an appointed experienced TrendCare® consultant.
- All staff working in the ward/department during the period of the study are educated on the process and have access to the TrendCare® Timing Study booklet outlining the study purpose, objectives, methodology, feedback process.
- A maximum of nine patients are flagged for inclusion in the study in the ward/department for any 8 hr period.
- All staff wear a timing device and are instructed in its use.
- Scannable forms are color coded for each 8 hr period of the day and clearly define when the time period starts and ends.
- A comprehensive list of grouped activities, (with codes allocated to each individual activity), is attached to each timing record.
- Each patient contact is timed and recorded in minutes and seconds and may include a number of activities which are recorded by codes beside each contact time (individual activities are not timed).
- All staff providing care are included in the timings so that a reasonable average skill mix is represented in the study including; qualified and non-qualified staff, highly skilled and less skilled staff, very organized and less organized staff.

- A check list at the bottom of the data collection form contributes to determining if a sample is accepted or rejected for the study.
- All activity codes can be matched against patient care plans. The indicators selected on the patient dependency system provide a method for measuring the level of compliance and accuracy and for determining acceptance or rejection of samples into the study.
- Contact timings for each patient are compared to the category timings in the TrendCare® Patient Episode Acuity report and contribute to the ongoing development and validation of the TrendCare® system.
- Paid staff breaks and personal interruptions are excluded from the timings and are added in as a controlled adjustment for all patient types.

## Work sampling methodology

Work sampling is a work measurement technique usually applied to groups of people or machines. It was first used by Tippett in the British Textile Industry in 1934 and has increasingly been applied to areas not previously measured [8]. Work sampling, also referred to as work measurement sampling, requires an observer to record the actual work being done at the moment of observation.

Work sampling consists of a large number of observations ($N$) taken at fixed or random intervals. Predefined categories of activity pertinent to the work situation and purpose of the study, that is, chosen nursing taxonomy data elements are incorporated on a data collection form unique for each study. This permits observations to be made relative to these predefined activities which collectively make up the universe of work performed in the area under study. Work observations may be recorded relative to any number of variables depending on the purpose of the study, and the questions to be answered. For example, defined nursing activities may be recorded relative to the patient for whom they are performed or relative to the category of staff performing each activity.

The underlying theory of work sampling is, that the percentage of observations ($p$) for any activity provides an estimate of the percentage of time actually spent on that activity, to a known degree of accuracy. Statistically this theory is based on the laws of probability and uses Bernoulli's theorem, random variable and distribution laws and the law of large numbers [8, 16, 17]. The laws of large numbers require mutually independent random variables. In work sampling these variables are the observations made per activity under study. Every activity is mutually exclusive from another. The sum of these mutually independent random variables divided by the total number is "as close to unity as we please" [17] p. 96). Sampling observations may be made by people with no previous work study experience provided they are suitably trained to do so. Observer bias was found to be negligible during Hovenga's studies. Observations made at regular fixed intervals achieve the same results as observations made at random [18] due to the variable nature of nursing.

The mutually independent random variables in work sampling are the observations made per activity under study. It needs to be understood that each observation within a data set of observations has the same value, whereas an individual observation viewed as a proportion of the data set, frequently takes on a value far removed from its mean. The arithmetic mean of a large number of observations viewed as a proportion of the data set expressed as a percentage of occurrence, behaves differently to the mean of a very small number of observations within the same data set. The larger percentage of occurrence is a far more accurate estimate of the actual mean than the small percentage of occurrence.

To make effective use of these random variables there is a need to identify as precisely as possible their laws of distribution. For example if we want to know the range of time taken for wound care we would assume that some average range exists. The difference between the actual and average range may be referred to as the error, whose magnitude will vary from one action to another. This will depend on a number of other variables which act independently of one another. The final error represents the total effect on the time taken to provide wound care. All such errors are approximately distributed according to normal laws. This law was discovered in the middle of the last century by Chebyshev, a Russian mathematician [17].

This knowledge enables the calculation of the number of observations ($N$) required to obtain work sampling percent occurrences ($p$) for a desired relative accuracy to yield a confidence interval of 95%. The relationship between these three variables is such that the smaller the desired standard error and the percentage of occurrence ($p$) the greater the number of observations ($N$) required in the total study.

From the proportions of observations ($p$) made regarding each activity, inferences are drawn concerning the total work under study. When measuring ward work, observations may be made of all nursing and other staff allocated to that ward on every round, or by observing a randomly selected staff member every 2 min (or at some other time interval) or by observing patients for direct patient/nurse interaction for any chosen period of time. Work sampling works because a smaller number of chance occurrences tends to follow the same distribution pattern that a larger number produces [16] p. 3–47).

This sampling technique does not assume that the momentary observation is continued throughout the intervening observation interval. It is based on the fact that the number of times an activity is observed being performed is closely correlated with the total time spent on its performance. For example if an activity is observed 10 times out of a total of 100 observations then it is assumed that the activity consumed 10% of the total time made available during that observation period. The 10% denotes the percentage of occurrence ($p$) of all observations ($N$) made. A sample taken at random, such as nursing work relative to a defined patient population, tends to have the same pattern of distribution as the total patient population. If the sample is large enough, the characteristics of the sample will differ little from the characteristics of the group [8] p. 406).

The formula [8] p. 412) used to determine the sample size (number of work sampling) observations for a confidence interval of 95% or approximately two standard deviations either side of the proportion being estimated and to determine the standard error for each activity once measured is

$$S_p = \sqrt{\frac{p(1-p)}{N}}$$

where

$S_p$ = desired accuracy (standard error)
(an accuracy of ±5% or 0.05 of the proportion estimated is usually considered satisfactory)
95% Confidence Interval = $p \pm 1.96 \, S_p$
$p$ = percentage occurrence of an activity or delay being measured, expressed as a percentage of the total number of observations or as a decimal, that is, 15% = 0.15
$N$ = total number of random observations (sample size)

This means that to obtain a ±5% degree of accuracy for any activity that takes up just 1% of all time allocated there will need to be 158,400 work sampling observations. The number of observations required reduce, as the percentage of time spent on any measured activity increases. Table 1 provides some examples one could use to determine the number of observations for a given degree of accuracy and value of $p$, 95% confidence level or to determine the degree of accuracy of any work sampling result based on the total number of observations recorded.

**Table 1  Table for determining the number of observations for a given degree of accuracy and value of $p$, 95% confidence level**

| Percent of total time occupied by activity $p$ | Degree of accuracy | | | | |
|---|---|---|---|---|---|
| | ±1% | ± 3% | ± 5% | ± 7% | ±10% |
| 1 | 3,960,000 | 440,000 | 158,400 | 80,800 | 39,600 |
| 2 | 1,960,000 | 217,800 | 78,400 | 40,000 | 19,600 |
| 3 | 1,293,300 | 143,700 | 51,700 | 26,400 | 12,900 |
| 4 | 960,000 | 106,700 | 38,400 | 19,600 | 9600 |
| 5 | 760,000 | 84,400 | 30,400 | 15,500 | 7600 |
| 6 | 626,700 | 69,600 | 25,100 | 12,800 | 6270 |
| 8 | 460,000 | 51,100 | 18,400 | 9380 | 4600 |
| 10 | 360,000 | 40,000 | 14,400 | 7340 | 3600 |
| 12 | 293,300 | 32,600 | 11,700 | 5980 | 2930 |
| 15 | 226,700 | 25,200 | 9070 | 4620 | 2270 |
| 20 | 160,000 | 17,800 | 6400 | 3260 | 1600 |
| 25 | 120,000 | 13,300 | 4800 | 2450 | 1200 |
| 30 | 93,300 | 10,400 | 3730 | 1900 | 935 |
| 40 | 60,000 | 6670 | 2400 | 1220 | 600 |
| 50 | 40,000 | 4440 | 2500 | 815 | 400 |

This information should also be used when developing the nursing taxonomy containing all data elements to be measured. This should include only those activities that are engaged in frequently, that is it amalgamates detail into higher level categories of work, such as all technical activities, or direct patient/nurse interaction. Greater detail provides some indication of what is actually included but measuring these less frequently occurring activities require a larger number of total observations to be recorded to obtain valid measures. The number of observations to be made to meet the study's purpose, determines the length of time observers need to be employed and the overall cost of the study.

It's important to provide information fact sheets to be distributed to all stakeholders during the study and to train all observers prior to commencing data collection. During the study one also needs to test observer reliability. All data collected for each data set must relate to the same observation periods to ensure validity when data from the various datasets are linked as part of the data analysis process.

Using work sampling for purposes of work measurement requires that the work sampling study is related to a defined observation period within which the total actual hours and the number of units produced are noted. The personnel being sampled are usually performance rated. This could be used to account for students for example. The formula for arriving at a standard time value is as follows:

$$\text{Standard time} = \frac{\text{Actual hours} * \% \text{of occurrence} \, (p) * \text{average performance rating}}{\text{Number of units produced (patient days/occupied bed hour)}}$$

In nursing studies the work unit may be a defined as a patient day expressed in hours. It is important that unusual circumstances are avoided during the study period to ensure that the results obtained are at the desired degree of accuracy. A formula is used to ascertain the daily limits of error relative to the percentages of occurrence of interest. It needs to be emphasized that the percentage of error reduces as the number of observations increase. An estimate of the number of observations which may be made on any 1 day together with the total number of observations required determines the duration and cost of a work sampling study. The daily number of observations possible is dependent upon the number of staff to be included in the study and the duration and frequency of each observation round. In addition the study period should be at least as long as the longest period of any cyclical behavior or characteristic being studied. In the case of nursing this means for all days of the week and at least all day shifts, desirably every full 24 hr period for 7 days. The sampled population (staff and patients), from which inferences will be drawn, must be similar to and representative of the population to which the results will be applied [8, 16].

Contrary to popular belief, the work sampling methodology does in fact include the measurement of sophisticated cognitive processes used by nurses during the course of their working day. It does this by accounting for all time spent by nurses caring for patients. As such cognitive processes are not usually readily observable as a separate activity, they are not listed in the work measurement taxonomy used for the PAIS studies. These cognitive processes tend to occur concurrently with observable activities although the latter then usually take longer to perform than when these same activities are performed in the absence of such cognitive processes. The only time both observable and cognitive activities are truly concurrent is when the observable activity is a routine one performed by an expert, such as bed making. Hovenga's work sampling studies differed from other reported work sampling studies by concurrently relating the sampling observations of direct care activities to individual patients [4]. Work sampling results of a suitably designed study are able to answer a variety of research questions regarding the distribution of work relative to what, where, why and by whom the work is performed. Intermittent observations made by external observers are non-intrusive.

As a cost-effective and useful methodology, work sampling warrants more in-depth exploration of the various techniques involved to ensure nurse managers, clinicians and researchers appreciate the complexities of the approach and its potential to contribute to an understanding of nursing work [19]. Work sampling reported in the literature lacks a standardized approach [20]. Due to the many ways one can group nursing activities for the purpose of measurement, it is difficult to share or make comparisons between different studies' results.

Unlike the recent Belgian study [12], most studies do not indicate the total number of observations or the percentage of occurrence of the activity being measured. As a consequence the accuracy of the results cannot be ascertained. The work sampling technique, long used in industry, is recognized as "a reliable procedure well suited to the quantitative evaluation of non repetitive irregularly occurring hospital activities which occur on a nursing unit" [18] p. 22). The original PAIS study resulting in standard time values per patient category was based on a total of 41,727 work sampling observations and covered a total of 2371 patient days (excl. Night duty). The degree of accuracy for all four categories measured was ±3% for categories A, B, C and ±5% for category D [1].

Work sampling has the advantage of permitting the calculation of the degree of accuracy of the results. It is cost effective in terms of study period and sample size required, from which inferences may be made with a degree of confidence, relative to the population studied. It is a very useful method to identify possible inefficiencies from which individual aspects of nursing work may be evaluated. Work sampling may be used in conjunction with normal supervisory duties in some instances. Studies may be designed to answer any number of research questions. Work sampling lends itself more readily to be related to a nursing output measure as all nursing work is easily included in the measurement.

## Professional judgments/estimates

In its simplest form, technical experts can estimate the amount of time required for a qualified individual to produce a technical work unit. For example a surgeon can estimate the time required for a given operation, an experienced OR nurse can estimate the time required, not only by operation type but also by surgeon. This information is used daily to schedule the theater lists. Reaching agreement regarding notional times or relative values between defined units of work is referred to as using professional judgment. Professional judgments may also be used to estimate the frequency of occurrence of certain work activities.

Another simple technique for workload estimation is to analyze statistics and/or administrative databases. This historical base represents the time it has taken on average over a specific time period to perform the work. For example the operation register includes start and finish times of all operations performed. This historical data can be analyzed to arrive at an average time. In a sense this same data are used by the technical experts who base their estimates on past experience. Rosters can also be analyzed relative to the number of patients and types of patients serviced during any time period. These are fairly crude measures but good indicators of historical resource usage.

Reaching agreement regarding notional times or relative values, for either nursing activities or a typical patient within a group as described by a prototype classification is another method used to quantify nursing workload. Professional judgments are also used to estimate the frequency of occurrence of certain nursing activities where a standard time value per activity is used. An example of a work study project guideline is presented in Box 1.

---

**BOX 1  One example of work study project guidelines for medical/surgical wards**

**Work study objectives**
- To identify the type of work activities undertaken by the "Registered Nurses, Advanced Care Support Workers and Care Support Workers" in medical/surgical wards.
- To measure the intensity of work for specific groups of tasks according to the scope of practice of (i) RN's, (ii) Adv CSW, (iii) CSW, (iv) Clerks in each clinical area.
- To identify the skill mix requirement for each clinical area by shift.

**Process**
- Develop a data collection tool which delineates the scope of practice for RN's, Adv CSW and CSW.
- Review activities on the data collection tool and update if required for each scope of practice.
- Develop directions for the use of the data collection tool.
- Inform all wards involved in the study on the use of the tool.
- Collect data from each ward.
- Analyze data to determine work activities attended by each type of staff (RN, Adv CSW, CSW, Clerical) in each ward.
- Estimate the average time spent by the "different types of staff" on specific types of activities, e.g.
  - i Ward management
  - ii Advanced clinical activities

---

---

**BOX 1  One example of work study project guidelines for medical/surgical wards—cont'd**

---

    **iii** Technical activities
    **iv** Patient activities of daily living
     **v** Environmental tasks
    **vi** Catering tasks
   **vii** Other
- Identify variances in work activities undertaken by "different types of staff" across all wards included in the study.
- Identify any deviation in relation to scope of practice.

**Method**
Distribute data collection sheets, the objectives of the study and the directions on the use of the data collection tool to all wards involved in the study.

*Timeframe*
The data collection period is to be continued across all days of the week for a 2- to 4-week period depending on the size of the ward.

*Analysis of data*
All completed forms are to be collected by the Project Co-ordinator. These will then be reviewed and the data collated. The Project Co-ordinator will then submit the findings which can be used to re-engineer work processes, FTE establishments and skill mix.

*Directions for the use of the work study data collection form*
This form is to be completed by each "Registered Nurse, Advanced Care Support Worker and Care Support Worker" working over the period of the study.
*It is recommended that "each staff member":*

1. Identifies activities completed during the first 8 hr period of the day, the second 8 hr period of the day and the third 8 hr period of the day by ticking beside the appropriate activity on the form. Activities not accounted for on the form can be listed under the appropriate heading or under "other."

**N.B.** *Activities undertaken should be identified at regular intervals during the shift to enhance the accuracy of time estimates at least 2 hourly (preferably 1 hourly).*

2. Ensures that all work estimates (the approximate time spent for each group of activities) have been documented on the form by the end of each 8 hr period. The total time spent should add up to the total time worked.
3. Ensures that all details at the top and bottom of the form have been completed.

## Conversion of work measurement data to a workload measure

There is a need to convert work measurement data into a staffing formula relative to identifiable patient characteristics and associated nursing service demand. Frequently the use of professional judgments clouds the use of empirical data. When evaluating the various techniques used to quantify nursing practice, a distinction needs to be made between the work measurement techniques themselves and the methods employed in applying these techniques to measuring nursing practice and nursing workloads.

From an industrial engineering perspective actual measured time needs to be converted to a standard time to be used for staff allocation, budgeting or performance evaluation. Time values arrived at from work measurement studies are either applied to individual nursing activities or to a patient type defined by a classification method. Patient classification methods aim to describe various patient groups in a structured manner. Unlike any other industry, every patient cared for is a unique "product." Classifying patients is particularly useful as we have found that time distributions between the many activities undertaken by nurses vary according to patient population types. Patient classification is also desirable from a service management perspective.

As a result of continuing research using a standard work measurement methodology over many years, it has been possible to differentiate between patients for the purpose of defining patient types on the basis of average nursing service use. By 2014 TrendCare® had identified 106 different patient types for whom standard nursing time values had been measured in 829 wards located in 51 hospitals (34 public and 17 private) across 4 countries covering a total of 5,152,708.09 patient days/24 hr [21]. These data were used to establish a new international benchmark range of Nursing Hours Per Patient Day for each patient type (using bed utilization hours as the denominator). When importing data from a wide scope of wards across 51 hospitals it is expected that the sample size of some patient types in individual wards is too small to provide a reasonable average. This relatively small number of outliers could potentially distort the normal distribution. These were excluded from the benchmarking calculations resulting in standard Nursing Hours Per Patient Day for each type of patient.

Research results from work sampling studies undertaken in 1982 in Victoria were compared with 1995 studies from general medical/surgical wards undertaken in Queensland, using the same methodology. This includes the identification of numerous significant indicators of nursing workload presented as percentage of occurrence (*p*) from these work measurement studies as shown in Table 2. This shows the distribution of all nursing work activities during each study period. This also provides some insight into which activities are significant indicators of nursing service demand. These are supportive activities of daily living, all types of technical activity, work associated with medication administration, and all types of direct supportive, instructive and education communication. As such these are good indicators of dependency and can be used accordingly to differentiate between patients based on their most probable nursing service needs. The greater the number of such indicators applicable to individual patients, the greater their nurse dependency. Variations in percentage of occurrence for some activities between studies in Table 2 represents changes over time and/or differences between States or in patient mix.

**Table 2 Percentage of occurrence relative to nursing activities (as reflected by the taxonomy used shown in Fig. 1)**

| Significant indicators — direct and indirect patient related activities | 1982 — 41,727 total observations, *p* = % | 1995 — 52,688 total observations, *p* = % |
|---|---|---|
| Communication between staff and departments, incl. handovers | 17.35 | 19.75 |
| Documentation | 8.99 | 15.87 |
| Technical Activities incl. wound care and IV therapy | 8.76 | 8.78 |
| Mobility assist/bedrest repositioning support | 7.04 | 6.23 |
| Medication administration | 4.91 | 8.13 |
| Direct communication | 4.70 | 3.22 |
| Hygiene assist/full care | 4.65 | 4.66 |
| Elimination, incl. pan room activities | 4.59 | 2.36 |
| Nutrition-assist/feeding | 4.24 | 1.27 |
| Observations | 3.30 | 3.58 |
| Admission/discharge | 1.83 | 1.88 |
| **Total direct and indirect patient related** | **70.36** | **75.73** |
| Meal breaks, personal time, miscellaneous | 29.64 | 24.27 |
| Total study | 100% | 100% |

A comparison between the 1982 and 1995 work sampling studies.

Indirect activity observations can either be distributed equally across all patient categories or distributed in a weighted manner based on dependency by category. These analytical activities require the retention of all data relativities at all times. Percentages of occurrence can then be calculated for any desired configuration. These are then converted to time values by using the total hours recorded during the study period. Activity related time values for all patients are then divided by the number of patient days, calculated according to occupied bed hours (fractional bed days) during the study period, as recorded during the entire study to arrive at an average time value per patient day.

When converting these actual hours into a standard time value one needs to replace actual time spent on meal breaks and personal time with a 15% allowance. Meal times represent unpaid time. These calculations and final standard time values to be used as benchmarks, are shown as an example from a 1996 study [6, 22] in Table 3.

### Table 3  Standard time values calculation

| PAIS Dep. Cat. | Relative value units | Direct minutes per patient day | Indirect minutes per patient day weighted distribution | 15% allowance | Total minutes per patient day | Hours per patient day and degree of accuracy |
|---|---|---|---|---|---|---|
| A | 1.00 | 52.56 | 39.26 | 22.75 | 174.45 | **2.91 ± 4%** |
| B | 1.37 | 72.02 | 53.79 | 27.85 | 213.53 | **3.56 ± 3%** |
| C | 1.91 | 100.43 | 75.02 | 35.30 | 270.61 | **4.51 ± 3%** |
| D | 2.48 | 130.25 | 97.29 | 43.11 | 330.52 | **5.51 ± 4%** |
| E | 3.61 | 189.97 | 141.90 | 58.76 | 450.50 | **7.51 ± 5%** |
| F | 4.49 | 235.93 | 176.20 | 70.81 | 542.84 | **9.05 ± 6%** |

For ease of use it is strongly recommended that the minimum number of key indicators are used for classification purposes, although nurses have a tendency to want to include everything they do. Only those indicators that have been shown to consistently make nursing more time consuming should be included. The base line hours for each patient type include the base routine care expected for that patient type. Work measurement studies show that statistically the identification of every possible technical procedure makes little difference to the overall desired degree of accuracy of the final nursing workload calculation. The key criteria are about the ability to reliably differentiate between patients from the perspective of nursing service needs.

The 1993 NSW nursing costing study reported typical patterns of care by PAIS category [23] p. 75 as shown in Table 4. These continue to be relevant.

### Table 4  Typical patterns of care in PAIS categories A, B, C, D, E, and F (range and number of indicators) pertaining to each category

| Category A (0–3) | Category B (4–6) | Category C (7–9) |
|---|---|---|
| Age<br>Admission/discharge/death<br>Drug administration | Age<br>Observation >4 hourly<br>Patient education<br>Fluid balance chart<br>Drug administration | Age<br>Technical activity<br>Drug administration<br>Patient education<br>Mobility<br>Wound care<br>Observations >4 hourly<br>IV line<br>Fluid balance chart |

**Table 4** Typical patterns of care in PAIS categories A, B, C, D, E, and F (range and number of indicators) pertaining to each category—Cont'd

| Category D (10 − 12) | Category E (13–15) | Category F (16+) |
| --- | --- | --- |
| Age | Technical activity ×3/4 | Age |
| Technical activity | Wound care | Wound care |
| Drug administration | Drug administration | Technical activities × 5 |
| Patient education | Observations >4 hourly | Drug administration |
| Mobility | Hygiene [1] | Observations >4 hourly |
| Wound care | Fluid balance chart | IV line × 2 |
| Observation >4 hourly | IV line | Fluid balance chart |
| IV line | Patient education | Mobility |
| Fluid balance chart | Mental health care | Hygiene [2] |
| Hygiene [1] | Incontinence | Nutritional assistance |
| | Pressure care | Mental health care |
| | | Sensory deficit |
| | | Pressure care |

The study design, work measurement methods adopted, data standards defined and used plus the data collection methods adopted, determine the scope for data analytics and subsequent use of the results. A variety of tools are available to assist with these processes, including stop watches, movement and location sensors, random and statistical tables, various terminologies to describe the domain being studied, video recordings, electronic data retrieval methods, spreadsheets, the use of databases, reporting software and statistical packages. Work measurement study designs need to be able to link various data from five different data sources to be able to create new knowledge, and explain variations. Results obtained from the data analytical processes adopted are ideally applicable for multipurpose use to gain maximum benefits from substantial work measurement data collection investments.

## Making use of study results

Studies undertaken by these authors have been able to demonstrate changes in acuity over time. Most commonly such changes were the result of length of stay reductions which in turn reflect changes to treatment and care patterns. Fig. 4 shows a graph representing 27 surgical patients in 2005 compared with the same group of surgical patients in 2015. The difference is 0.6 NHPPD for 27 patients which equals 16.2 hr/day for the ward as a whole. This equates with 5913 hr/year or 3.6 FTE (productive) nursing positions.

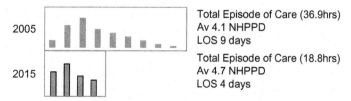

**Fig. 4**

Average length of stay (LOS) — surgical. *From the TrendCare® General Acuity PowerPoint Presentation.*

A similar change was identified for post natal mothers resulting from a change in length of stay from an average of 4 days to 2 days. This equated to an increase of 0.9 HPPD for 28 patients equalling 25.2 hr/day. This equates with an additional 5.6 FTE required to meet this additional workload. It is imperative that changes such as these resulting from changes in treatment and care methods are monitored and quantified. Such changes over time explain why the nursing profession frequently feels that nurses are being exploited, as there tends to be an expectation that they can simply absorb these workload changes by working smarter.

Information generated as a consequence of undertaking comprehensive work studies, not only of nursing and midwifery work but also of other staff categories provides intelligent data and new knowledge which can contribute to the enhancement of numerous operational processes. Nursing and midwifery work measures enable the development and continuing use of standard time values, however data generated can also be used for an extensive list of purposes. These include

- Identifying high volume work items and linking these to quality measures.
- Re-designing work methods for high volume work to improve efficiency.
- Identifying patterns of care by degree of dependency for clinical specialties.
- Calculating the percent of the nursing/midwifery workload generated for each skill level.
- Providing a link to material and surgical supplies for internal and external supply chain management.
- Enhancing additional health service scheduling for each clinical specialty.
- Linking with patient assessment and care planning strategies.
- Providing a patient history and contributing to handovers and discharge planning to ensure continuity of care.
- Supporting nursing/midwifery research.
- Contributing to "big data."
- Linking treatment and care processes to patient outcomes.
- Comparing patient mix based on nursing resource usage between departments and/or organizations.
- Contributing to the development of staff roster patterns that match peaks and troughs in demand.
- Making staffing adjustments in real time to meet unplanned service demands.
- Monitoring trends regarding changes in acuity over time.
- Contributing to workforce planning and continuing professional development strategies.
- Recommending realistic staff establishments and allocating fair workloads to nursing/midwifery personnel.
- Planning nursing/midwifery assignments.
- Demonstrating effective personnel utilization.
- Charging patients for nursing/midwifery services.
- Improving patient safety through more appropriate patient placement.
- Allocating equipment and supplies.

- Monitoring nursing/midwifery dependency levels for all patient episodes.
- Monitoring the progress of all patients during hospitalization.
- Providing nursing/midwifery intensity measures for episode of care costing.
- Contributing to the establishment and maintenance of international benchmarks for nursing and midwifery services.

### Using workload measurement systems with established time standards

To successfully implement any of the above using the data generated from work measurement studies, decisions need to be made regarding the methods to be adopted for the continuing use of a nursing workload measurement system. There are a number of options, all of which enable further analytics to be undertaken to support decisions regarding resource management. The more comprehensive the adopted method is, the more useful the work measurement system's collected data becomes. Benefits extend to the use of the data for performance monitoring and planning, improved operational processes, and enhanced outcomes for patients, staff and the healthcare organization.

**For example:** the following options for data collection use pertaining to nursing and midwifery services:

Option 1 supports operational and strategic resource management decisions.

(a) Predict nursing/midwifery workload requirements for each patient prospectively 24 hr in advance to estimate resource (number and skill mix) requirements for oncoming shifts on a continuous basis.
(b) Update all predicted and unpredicted patients in real time (includes unplanned admissions, transfers etc.).
(c) Update any changes to staff supply and hours available to deliver care in real time (escorts, redeployment, absenteeism, etc.).
(d) Review and update prospective shift predictions of patient acuity and nursing/midwifery hours available for care at the end of each shift/period to reflect the actual demand for nursing/midwifery hours and the supply available.

Outcomes of data use from data collection Option 1:

- Identifies capacity to care in real time.
- Enables staffing adjustments to be made in real time in response to changes in demand.
- Provides extensive use of real time data for additional clinical processes, e.g., dashboard displays, shift handovers, patient/workload allocation, etc.
- Demonstrates peaks and troughs in demand across a week, for months of the year and across a number of years.
- Provides extensive data for reporting on work management, bed management, clinical data, patient episode data, staff work history, productivity and efficiency.

- Facilitates re-engineering of rosters to minimize variances between demand and supply of resources.
- Provides comprehensive clinical and non-clinical data for developing staff establishment profiles.
- Generates extensive data for real time performance monitoring at department, service, hospital and hospital group level.
- Provides nursing and midwifery intensity measures for service and episode of care costing.
- Generates extensive data to assist with effective budget development.
- Provides extensive data for benchmarking and trending changes in patient acuity, productivity and efficiency.

Option 2 is to retrospectively complete acuity measurements at the end of each shift/period on a continuous basis supporting strategic resource management decisions.

Outcomes of data use from Option 2 data collection method:

- Provides retrospective data for measuring demand.
- Demonstrates peaks and troughs in demand across a week, for months of the year and across a number of years.
- Facilitates re-engineering of rosters to minimize variances between demand and supply of resources.
- Provides comprehensive clinical data for developing staff establishment profiles.
- Provides statistical clinical data for budgeting and costing.
- Provides data for benchmarking and trending changes in patient acuity, productivity and efficiency.

Option 3 is to collect a random sample of data for a defined period (e.g., 5 days, 2 weeks) quarterly or annually to enable trend analysis in any setting as is undertaken in Belgium [24, 25] for example.

Outcomes for Option 3:

- Provides a snapshot of demand for the defined period.
- Provides comparison snapshots of demand across a week, for months of the year and across a number of years to identify trends.
- Provides data for estimating staff establishment profiles and associated cost estimates based on trending snapshot periods only.

Successful adoption of any of the above options requires staff education and annual reliability testing to ensure data integrity is maintained. The choice of options is to a large extent dependent upon the functionality of the selected patient acuity and workload measurement system, the way the system is used and its interoperability with other organizational information systems in place. This is explained in more detail in Chapter 9.

## Nursing workload measures' validity

Most nursing/midwifery workload monitoring system use average time values and rely on normal distributions. Prior to their use one needs to seek the degree of accuracy of the time values associated with, for example, each classification category or the method used for nursing resource allocation. To do this one needs to examine the work measurement technique used, whether the universe of nursing/midwifery was measured and the manner by which these data are used by the system. The actual nursing/midwifery resources used or required may be anywhere within two standard deviations of the mean provided the sample used for work measurement consisted of a normal distribution of occurrences for that patient population.

The validity and usefulness of using activity based nursing/midwifery workload measures or nursing/midwifery oriented patient classification models to reflect nursing/midwifery resource usage, is dependent upon the following:

(a)  Accuracy of work measurement, i.e., source data; Work study principles' compliance.
(b)  Ability for the classification/task based model and associated time values to represent the universe of nursing/midwifery work required to be undertaken for any specified group of patients.
(c)  Consistency of data collection in accordance with the nursing/midwifery workload model in use.
(d)  Ability to identify which categories of staff the workload measures relate to.
(e)  Ability of the classification model to discriminate between patient categories on the basis of nursing/midwifery resource usage — reliability.
(f)  Ability of the classification model to represent a normal distribution of occurrence for each class or patient type.
(g)  Technical accuracy of the analysis of the data upon which the staffing formula and hence nursing/midwifery costs are based.
(h)  Data retention as part of the patient's record and stored in a format permitting integration with hospital management information systems.

Standard time values are either applied to individual activities, to categories for each patient type determined by weighted activities to defined units of work or to output measures. Validity, in terms of how well do the resultant time values reflect actual nursing resource requirements, must be demonstrable. The degree of accuracy achieved must be acceptable to the purpose for which the resultant information is used.

Although work measurement studies measure actual labor resource usage, this value is commonly expressed by each workload model as the staff time required. Thus it is assumed that the actual time measured was appropriate and adequate to meet the objectives of the provision of services measured and that these values have a predictive quality. When these time values are later used as a proxy for actual workload and as a basis for costing services, it is further assumed that patients/clients did actually receive the services identified as required to produce the desired outcomes.

The Finnish RAFAELA system described in the previous chapter [26–28] is not based on actual work measurement, thus requiring other methods to determine validity. Linear regression analysis, one-way analysis of variance, *t*-tests and correlation analysis were used, as well as parameters of distribution of the data on data reflecting optimal nursing care intensity from eight Finnish hospitals. This indicated system reliability [28]. A later Norwegian pilot study identified a need to undertake intensive training to ensure that nurses were able to reliably classify patients.

A follow up larger study tested inter-rater reliability and validity by making use of the benchmarked value for each patient category [29] to predict the average assessed value for the same calendar day. Simple linear regression analysis was then undertaken to quantify to what extent the predictor variable explained the variations in values of the dependent variable was accounted for [30]. These results were comparable to those obtained from the Finnish studies but did indicate a number of areas in need of further adjustments. This evidence was considered to indicate satisfactory reliability and validity. A 2016 Finish study evaluated the RAFAELL system's predictive validity by examining whether hospital mortality can be predicted by the optimality of nursing workload using negative binomial regression analyses [31].

## Nursing work measures in use

Various stakeholders are able to make use of these time standards and nursing workload measurement strategies for multiple purposes.

Actual nursing/midwifery work measures commonly form the foundation of various workload measurement systems. These include workload management systems developed from nursing/ midwifery activity based data, patient classification models as well as nurse/patient ratios and midwifery patient ratios derived from an analysis of administrative systems. Many of the systems developed and reported in the literature during the 1970s, 1980s, and 1990s, including GRASP®, QuadraMed/Medicus, Excelcare and TrendCare® have been commercialized in a variety of ways. This represents the evolutionary result of an increasing use of computers during the 1990s. It is difficult to evaluate their efficacy. A Canadian study undertaken early 2004 identified that 70% of healthcare facilities studied made use of the GRASP® or the QuadraMed/Medicus tool as their nursing workload management tool [32].

GRASP uses time standards for direct and indirect nursing activities which are derived from a combination of consensus, published standards, hospital and nursing policy and time studies and rounded off to the nearest one-tenth of an hour. South Australian nurses widely adopted Excelcare, an activity based (units of care) workload monitoring system, which was described as a computer assisted nursing management decision support system for planning, documenting, evaluating, staffing and costing nursing resource usage based on clinically determined patient care standards [1]. ExelCare interventions and the associated timings were found to differ from site to site due to variable on-site updates that were not shared [33].

One critical review of nursing workload measurement methods primarily in use in Australia, highlighted the implications of key stakeholders (health service executives, funders) not fully understanding the impact of excessive nursing workloads on patient and nurse safety outcomes [34]. Methods referred to in this review included the Patient Assessment and Information System (PAIS), Diagnostic Related Groups (DRG) nurse costing models [1], hours of care/ patient day, nurse-patient ratios and mandated ratios, and commercial software packages such as ExcelCare, *E*-care and TrendCare®. Results of the use of commercial software tends not to be made available in publicly available literature, however evidence of their use can be found in health department reports, press releases for health care magazines, internal hospital reports and in conference papers [35].

In 2014 TrendCare® was reported as being used in 70% of Queensland Health hospitals, and in the majority of large Private Hospitals in Queensland. In 2017 TrendCare® was used in all Singapore acute and sub-acute hospitals and in 2018 was selected as the national system for the New Zealand public sector [2]. Ireland made use of the TrendCare® system for their safe staffing/skill mix study to validate national nursing staff establishments [36–38]. It's final report has, among others recommended that [3]:

> *a national workforce planning and workload management IT system be introduced to assist in decisions on nurse staffing and skill mix. This system must be capable of capturing all Framework components. It is also key that such a system integrates with organisational level patient information management systems to enable the development of nursing intensity weight based costing relative to patient Diagnostic Related Groups.*

Actual nursing work measures used as the foundation for any of these nursing workload management systems vary considerably. Measures used are often difficult to identify or evaluate for degree of accuracy or appropriateness. Independent evaluations of each system's validity tend not to be reported in the literature. One comprehensive review of the TrendCare® system, undertaken in 2005 [39], included a comparison of its use relative to the nurse/patient ratio staffing methodology used in Victoria, Australia. This review found that TrendCare® provided a fairer and more equitable workload at a lower cost that the Victorian mandated nurse/patient ratios. The key question when evaluating individual studies or workload monitoring systems is whether the underlying work measurement technique chosen was the most appropriate for the purpose. There is no consensus regarding an acceptable degree of accuracy or indeed on how to assess the validity of nursing workload measurement systems. Yet this is of fundamental importance to those who need to make decisions based on their use.

The effective use of these systems requires those stakeholders who are likely to benefit, to develop the most appropriate data collection and reporting strategy. All strategies combined can then form the basis for information system design and development to ensure all potential beneficiaries are able to receive value and improve their decision making accordingly. Many of these potential uses are explored in greater depth in other chapters. With the digital transformation now in progress, many new opportunities for nursing and/or obstetric care data use become available as shown via the use of not only the TrendCare system but also the RAFAELA system [40]. Reaching full potential does require nursing data to be integrated with other systems as discussed in Chapters 6 and 7.

## *Patient classification principles*

Classification theory identifies what variables are related to one another and the nature of these relationships, which gives it the power to explain. Theories identify the underlying principles used to explain the various relationships within the phenomena being studied. Thus, construction of the theoretical basis of patient classification systems permits testing and review of such systems to advance scientific knowledge regarding this subject [1].

The purpose of nursing patient classification schemes has been identified as being concerned with relating nursing resource usage to patients within defined groups. Medically oriented patient classification schemes are based on diagnosis and disease severity. By using medical diagnosis one may be able to predict on a probability basis, nursing resources required per diagnosis, due to an inherent relationship, although the same diagnosis for different individuals may well lead to significantly different nursing/midwifery service needs. Prediction may be enhanced where the diagnosis is combined with other related variables such as some measure of disease severity or age. The latter may be more discernible from a nursing/midwifery classification scheme.

Although many concepts describe patient phenomena that can be used for classification purposes, it is usually possible to use only some of these for the purpose of allocating patients to a defined classification. The inter-rater reliability is best when clearly defined characteristics or factors are used for classification purposes. The documentation accompanying a patient classification model is the instrument used to communicate these rules to the classifier. The validity of patient categories in terms of their power to discriminate on the basis of resource usage is critical. An important feature of any patient classification model is the crystallization of the rules by which patients are classified. Ideally there should be little overlap between the categories. That is, the actual nursing resource usage for patients classified within a group has a range which differs from that of other categories. A patient's category is usually identified on a daily, per shift or defined period basis but can also be determined regularly for short periods to identify trends.

Work measurement techniques may be employed to determine the relative values between classes or elements of care, as well as actually quantifying resource usage. Health services use polythetic classification by focusing on similarities to achieve homogeneity within groups. Within a polythetic scheme, patients classified into a specific class are not identical, they are more similar to each other in certain characteristics, than they are to non-members of the class. When in use, patterns of similarity emerge for each category. Clearly in any classification system there is a trade-off between the number of classes used, precision, ease of use and cost of data capture for particular purposes. When the system is used as a basis of costing nursing services by patient or diagnosis related group (DRG), then a high degree of precision is desirable.

Another problem for researchers and managers wishing to use nursing classification data are that in addition to the myriad of systems in use, the data used to classify patients were, until recently, not retained as a part of the patient's medical record. A composite category reflecting nursing resource usage over the length of hospitalization is generally not available. An iterative review of the literature that aimed to identify current practices in an effort to determine if there was a "gold standard" patient classification system that could be adopted more widely, revealed [41]:

- Difficulties with measuring workload remain an overarching theme
- Definitions and descriptions of nursing work continue to be deemed inadequate
- There is insufficient evidence of reliability and validity testing
- There is still a need to identify nursing sensitive performance indicators and outcomes

Two separate case studies, not previously published in academic literature, are presented as appendices. These detail the development, use and associated experiences of two different systems, using patient classification principles, the Patient Assessment and Information System (PAIS) developed by Hovenga [1, 4, 42–46] and TrendCare® developed by Lowe [47–50]. These case studies include lessons learned. Each of these systems have enjoyed many years of use. The TrendCare® system has been commercialized and continues to be used extensively across many hospitals located in six countries. PAIS was used by around 100 hospitals, data generated was used for the development of nursing service weights attached to diagnosis related groups(DRGs) in the first Australian national clinical costing study [51]. Both nursing systems were used for later updates of nursing service weights by DRG.

## Developing national nursing service weight measures

Hospital nursing/midwifery service costs constitute a very significant component of its overall operational budget. Where these costs are identified as an average cost per patient the economic value of nursing/midwifery services is hidden and distorted [52]. Undervaluation of nursing/midwifery care within the costing and reimbursement structure is best addressed by the use of patient acuity/classification systems where individual patient acuity data can be linked to billed or funded patient outcomes measures in use. The International Classification of Diseases (ICD) terminology has too many categories to be used exclusively for this purpose. ICD codes plus other criteria are used to group patients according to casemix system in use, such as Diagnosis Related Groups (DRG) or Healthcare Resource Groups (HRG).

Patient acuity/classification systems can be used to form the basis for the development of nursing service weights per DRG/HRG or some equivalent casemix outcome measure. Health economists are interested in identifying which aspects of patient phenomena best explain resource usage and costs of health services provided. Thus a patient acuity/classification system used for this purpose of identifying casemix outcome measures, must be able to demonstrate homogeneity of resource usage within patient categories relative to average length of stay. There needs to be a reasonable trade-off in terms of the number of categories, where each must represent clinically distinct types of patients.

The capture of staffing costs needs to be based on actual staff rostered and linked to the payroll system. Such systems provide data regarding skill mix (based on funded nurse categories), hours and shifts worked, cost of each shift, work location and leave provisions are derived from actual costs. Nursing overhead costs such as nursing administration/management non-clinical hours need to be allocated to specified cost centers within each hospital.

Once a hospital has both systems in place, it is possible to link the patient acuity/classification data per patient to financial and payroll systems as well as to these outcome measures. Service weight measures can also be identified for other hospital services by casemix measure, such as operation room, including anesthetic services, pharmacy, critical care, allied health services, pathology, imaging, prothesis and supply. A variety of measurement techniques are used to derive at these service weight measures relative to individual patients. These measures are collectively used to identify an average cost per patient type treated based on the casemix system in use. This is known as a clinical costing approach. Nursing service weight costs constitute the greatest cost for any in-patient type.

The development of nursing service weights is dependent upon the use of patient acuity/ classification data and an appropriate clinical costing system. The data used must be complete and accurate, and there must be a sufficiently large number of hospitals contributing data in order to obtain a representative sample of cases. Standard service weight measures can be developed for use by other hospitals.

## Evidence of acuity link with patient outcomes

Commonly missed nursing care activities resulting from high acuity and heavy workloads are patient comfort, counseling and education [53]. Salford Royal Foundation Trust reported their outcomes as a result of implementing the TrendCare® system in a paper to NHS Trusts. These included; a reduction in patient harms from the 4th percentile to the lowest percentile, an increase in nurse retention and recurrent cost savings of £2.8 million from their inpatient nursing budgets (40 wards). Other reported benefits were an improved culture around evidence based redeployment processes, a significant vacancy reduction (11% to 3%), excellent staff engagement as a result of improved transparency [54].

One empirical correlational study examined the association between patient satisfaction and the allocation of nursing care hours using a workload management system in one hospital. This found that the patient satisfaction total score statistically correlated with nursing workload management as measured using TrendCare® variables. Their secondary data analysis highlighted that several of these variables influence patient experience, especially respect, communication and involving the patient in the planning of care [55]. According to these researchers these results implied that the wards have positive working environments. Their study confirmed that the allocation of nursing hours according to patient acuity emphasizes the importance of the presence of nurses' caring attributes when providing quality bedside care to patients.

## Future directions

As new technologies continue to be implemented, we also expect continuing changes in nurses and midwifery roles, their boundaries of practice and work activities generally. The impact of such changes on both service demand and staffing needs will need to be monitored to ensure the best possible match between demand and workforce supply is first established and then retained over time. This will require the undertaking of continuing nursing work measurement studies at various time intervals.

The work measurement principles and their past and present applications as explained in this chapter continue to be relevant. The work sampling methodology applied for the PAIS study relied on external observations to be made using a paper and pen approach. New technologies provide new options for work sampling data collection methods. Over the years time study data collection methods as applied for the TrendCare® studies have also improved and become more cost effective over time.

The very extensive unequaled research we've independently undertaken, as detailed in the two case studies (Appendices 1 and 2), plus positions held over the years, have been instrumental in providing us with the wealth of knowledge gained regarding nursing/midwifery resource management. This chapter reflects lessons learned during these processes. Upon reflection it has become evident that the complexity associated with nursing/midwifery practice and the numerous confounding variables influencing their workloads, requires the ability to make use of very large data sets (big data), which is dependent upon the digital transformation, in order to arrive at significant insights able to inform decision makers at every level within the health industry who collectively determine any health service's capacity to care.

Chapter 5 examines how we can or should identify nursing and midwifery human resources' skill mix as the need to match resources with service demand should not only be concerned about numbers and associated costs. Skill mix relates to the capability of providing quality care. Skill development, use and influencing factors are also examined.

## References

[1] Hovenga E. Casemix, Hospital nursing resource usage and costs. [PhD thesis]. University of New South Wales (UNSW), Sydney, Australia; 1995.

[2] CCDM. New Zealand-Care Capacity Demand Management programme. [cited 14 April 2019]. Available from: https://www.ccdm.health.nz/; 2018.

[3] DOH-Ireland. Framework for safe nurse staffing and skill mix in general and specialist medical and surgical care settings in adult hospitals in Ireland: final report 2018. [cited 1 April 2019]. Available from: https://health.gov.ie/blog/publications/framework-for-safe-nurse-staffing-and-skill-mix-in-general-and-specialist-medical-and-surgical-care-settings-in-ireland-2018/.

[4] Hovenga E. A patient classification model (PAIS) and nursing management information system. Melbourne: Hospitals Division, Health Commission of Victoria; 1983.

[5] PAIS. Nursing projects Officer Group Inc. (incl. PAIS User Group). The University of Melbourne Archives; 1983–1993. Available from: http://gallery.its.unimelb.edu.au/imu/imu.php?request=multimedia&irn=5785.

[6] Hovenga EJS, Hindmarsh C. Queensland PAIS validation study. Brisbane: Queensland Health; 1996.

[7] ILO. Introduction to work study. Geneva: International Labour Office; 1979.

[8] Barnes RM. Motion and time study: design and measurement of work. 7th ed. New York: John Wiley & Son; 1980.

[9] Zandin KB, editor. Maynard's industrial engineering handbook. 5th ed. New York: McGraw-Hill; 2001.

[10] Lopetegui M, Yen P-Y, Lai A, Jeffries J, Embi P, Payne P. Time motion studies in healthcare: what are we talking about? J Biomed Inform 2014;49:292–9. https://doi.org/10.1016/j.jbi.2014.02.017.

[11] Zheng K, Guo MH, Hanauer DA. Using the time and motion method to study clinical work processes and workflow: methodological inconsistencies and a call for standardized research. J Am Med Inform Assoc 2011;18(5):704–10.

[12] van den Oetelaar W, van Stel H, van Rhenen W, Stellato R, Grolman W. Mapping nurses' activities in surgical hospital wards: a time study. PLoS ONE 2018;13(4):e0191807.

[13] ASQ. Lean Six Sigma in Healthcare. [cited 27 January 2019]. Available from: http://asq.org/healthcaresixsigma/lean-six-sigma.html.

[14] Kimsey DB. Lean methodology in health care. AORN J 2010;92(1):53–60.

[15] Witt CM, Sandoe K, Dunlap JC. 5S your life: using an experiential approach to teaching lean philosophy. Decis Sci J Innov Educ 2018;16(4):264–80.

[16] Brisley CL. Work sampling. In: Maynard HB, editor. Industrial engineering handbook. New York: McGraw-Hill; 1971.

[17] Gnedenko BV, Khinchin AY. An elementary introduction to the theory of probability. New York: Dover Publication; 1962.

[18] Murphy LN, Dunlap MS, Williams MA, McAthie M. Methods for studying nurse staffing in a patient unit. DHEW Pub.No. HRA 78-3; 1978.

[19] Pelletier D, Duffield C. Work sampling: valuable methodology to define nursing practice patterns. Nurs Health Sci 2003;5(1):31–8.

[20] Blay N, Duffield CM, Gallagher R, Roche M. Methodological integrative review of the work sampling technique used in nursing workload research. J Adv Nurs 2014;70(11):2434–49.

[21] Lowe C. An international patient type hours per patient day (HPPD) benchmarking research study. Brisbane: TrendCare Systems Pty Ltd; 2014.

[22] Hovenga E, Hindmarsh C, editors. Queensland health—PAIS validation study: results and issues for nursing cost capture. Eighth Casemix conference in Australia. Canberra: Commonwealth Department of Human Services and Health; 1996.

[23] Picone D, Ferguson L, Hathaway V. NSW nursing costing study. Sydney: Metropolitan Teaching Hospitals Nursing Consortium; 1993.

[24] Sermeus W, Delesie L, Van den Heede K. Updating the Belgian nursing minimum data set: Framework and methodology. In: al FHRFe, editor. eHealth in Belgium and in the Netherlands. vol. 93. Amsterdam: IOS Press; 2002. p. 89–93.

[25] Sermeus W, Gillain D, Gillet P, Grietens J, Laport N, Michiels D, et al. From a Belgian Nursing minimum dataset to a nursing cost-weight per DRG. BMC Health Serv Res 2007;7(Suppl. 1) https://doi.org/10.1186/1472-6963-7-S1-A6.

[26] Fagerstrom L, Lonning K, Andersen MH. The RAFAELA system: a workforce planning tool for nurse staffing and human resource management. Nurs Manag (Harrow) 2014;21(2):30–6.

[27] Fagerstrom L, Rainio AK, Rauhala A, Nojonen K. Professional assessment of optimal nursing care intensity level. A new method for resource allocation as an alternative to classical time studies. Scand J Caring Sci 2000;14(2):97–104.

[28] Rauhala A, Fagerstrom L. Determining optimal nursing intensity: the RAFAELA method. J Adv Nurs 2004;45 (4):351–9.

[29] Fagerstrom L, Rauhala A. Benchmarking in nursing care by the RAFAELA patient classification system—a possibility for nurse managers. J Nurs Manag 2007;15(7):683–92.

[30] Andersen MH, Lonning K, Fagerstrom L. Testing reliability and validity of the Oulu Patient Classification instrument—the first step in evaluating the RAFAELA System in Norway. Open Journal of Nursing 2014;4:303–11.

[31] Junttila JK, Koivu A, Fagerström L, Haatainen K, Nykänen P. Hospital mortality and optimality of nursing workload: a study on the predictive validity of the RAFAELA Nursing Intensity and Staffing system. Int J Nurs Stud 2016;60:46–53.

[32] Hadley F, Graham K, Flannery M. Assess use, compliance and efficacy of nursing workload measurement tools, workforce management objective A. Ottowa: Canadian Nurses Association (CNA); 2004;

[33] Anonymous. National Service Weights 2002-03 project-final report. Canberra: Australian Government, Department of Health and Ageing, Casemix Costing Section; 2003.

[34] Duffield C, Roche M, Thomas ME. Methods of measuring nursing workload in Australia. Collegian 2006;13 (1):16–22.

[35] Hendry C, Aileone L, Kyle M. An evaluation of the implementation, outcomes and opportunities of the Care Capacity Demand Management (CCDM) Programme—final report. Christchurch, NZ: NZ Institute of Community Health Care; 2015.

[36] Drennan J, Duffield C, Scott AP, Ball J, Brady NM, Murphy A, et al. A protocol to measure the impact of intentional changes to nurse staffing and skill-mix in medical and surgical wards. J Adv Nurs 2018;74 (12):2912–21.

[37] Fitzpatrick T. Sites selected for taskforce ED phase safe staffing/skill mix phase II focusing on emergency settings. World of Irish Nursing and Midwifery 2018. No. 5, June.

[38] Drennan J, Savage E, Hegarty J, Murphy A, Brady N, Howson V, et al. Evaluation of the 'pilot implementation of the framework for safe nurse staffing and skill-mix'—report 3. Ireland: University College Cork; 2018.

[39] Plummer V. An analysis of patient dependency data utilizing the TrendCare system. Melbourne: Monash; 2005.

[40] Liljamo P, Kinnunen U-M, Saranto K. Assessing the relation of the coded nursing care and nursing intensity data: towards the exploitation of clinical data for administrative use and the design of nursing workload. Health Informatics J 2018. 1460458218813613.

[41] Fasoli DR, Haddock KS. Results of an integrative review of patient classification systems. Annu Rev Nurs Res 2010;28:295–316.

[42] Hovenga E. Patient assessment information system (PAIS): a patient classification model based on patient/nurse dependency and a nursing management information system. Melbourne: Health Deartment Victoria; 1985.

[43] Hovenga E. The Patient Assessment and Information System (PAIS) for midwifery patients. Melbourne: EJSH Consulting; 1990 [Unpublished];

[44] Hovenga E. Work sampling at Peter MacCallum Cancer Institute and review of PAIS use. Melbourne: EJSH Consulting; 1988 [Unpublished];

[45] Hovenga E. The Patient Assessment and Information System (PAIS) in extended care. Melbourne: EJSH Consulting; 1990 [Unpublished];

[46] Hovenga E. Casemix and information systems. In: Hovenga E, Kidd MR, Cesnick B, editors. Health informatics, an overview. Melbourne: Churchill Livingstone; 1996.

[47] Lowe C. Trendcare Systems Pty Ltd [cited 22 August 2018]. Available from: http://www.trendcare.com.au.

[48] Lowe C. Validation of an acuity measurement tool for maternity services. Int J Med Health Sci 2015;9 (5):1417–24.

[49] Lowe C. An international patient type hours per patient day (HPPD) benchmarking research study; 2014.

[50] Lowe C, editor. Prospective and retrospective patient nurse dependency system: developed, computerised and trialled in Australia. Nursing Informatics '94, San Antonio. Amsterdam: Elsevier; 1994.

[51] Diers D. Casemix and nursing. Aust Health Rev 1999;22(2):56–68.

[52] Welton JM, Dismuke CE. Testing an inpatient nursing intensity billing model. Policy Polit Nurs Pract 2008;9 (2):103–11.

[53] Lake ET, Staiger DO, Cramer E, Hatfield LA, Smith JG, Kalisch BJ, et al. Association of patient acuity and missed nursing care in U.S. neonatal intensive care units. Med Care Res Rev 2018. 1077558718806743.

[54] Roddy R. E-rostering, acuity and dependency, a presentation to NHS Group. UK: Salford Royal Foundation Trust; 2018;

[55] Goh ML, Ang ENK, Chan Y-H, He H-G, Vehviläinen-Julkunen K. Patient satisfaction is linked to nursing workload in a Singapore hospital. Clin Nurs Res 2017;27(6):692–713.

# Identifying skill mix needs

## Matching available skills with service demands

Any healthcare organization's capacity to care is influenced by not only numbers of staff but also by having the right mix of knowledge and skills available. Identifying those knowledge and skill needs, requires an analysis of each type of service. We have previously identified that the health system employs staff from >70 different clinical specialities [1]. Each specialty has its own set of knowledge and skill requirements. In addition there are specific nursing specialties such as wound management, infection prevention and control, stomal therapy, diabetes educators and many more [2]. Within each specialty there are any number of roles incorporating one to many unique functions. Some of these roles may be defined as sub-specialties. The acquisition of knowledge and skill in any speciality determines career path possibilities and is linked with educational and professional development offerings as well as service delivery innovations and opportunities to participate.

Nurses and midwives bring a diverse set of knowledge and skills to their areas of practice. Even new graduates tend not to have identical skill sets as they vary in life skills when entering their nursing studies and have a variety of mixed experiences throughout their study period. A 3 year degree program does not facilitate the provision of practical work experience in every specialty. Anecdotal evidence suggests that for example, some newly qualified general registered nurses have not experienced pediatric or operating room nursing. Once registered or enrolled many will undertake additional studies, that may lead to formal qualifications or certificates, or they undertake additional learning experiences that simply satisfy continuing education requirements. A nurse may be an expert in one or more areas of practice and a novice in another. Matching skill mix, based on expertise assessment, with service demand is essential for safe quality care. This is especially so for all acute and critical care environments where exposure to risk of harm or adverse events is evident in the day-to-day delivery of complex therapies for patients [3]. In other clinical areas a skill/demand mismatch influences the quality of care provided, the number of adverse events, patient outcomes and patient satisfaction.

It has been recognized from a regulatory perspective [4] that nursing roles and functions are constantly evolving. Every nurse is responsible and accountable for their own breadth of nursing practice. There are numerous anecdotal examples of nurses being legally constrained, far short of their ability and the public need. It is imperative that clinical staff are able to provide their services at a level that best matches their knowledge and skills most of the time. In 2004, the American Association of Colleges of Nursing (AACN) created a new nursing professional, the Clinical Nurse Leader (CNL) prepared at the master's level. The AACN then proceeded to develop a new legal scope of practice and credential for this new nursing professional, and now offers certification [5]. CNL development was initiated to address the critical need to improve the quality of patient care outcomes. This staff category is similar to the Australian Clinical Nurse Specialist/Consultant although the activities undertaken differ. The latter were found to undertake activities that were institutionally, individually and contextually constructed. Some identified themselves as sole practitioners, others as team or clinical coordinators. Unlike the American situation there is no consensus regarding their scope of practice [6].

Economic pressures are known to lead to excessive workloads, inadequate supervision, and a lack of resources. This potentially places the patient at risk and nurses in situations where their capacity to care, deliver services within their legislated and professional scope of practice and code of conduct is compromised. Fig. 1 shows the relationships that collectively influence knowledge and skills availability and use.

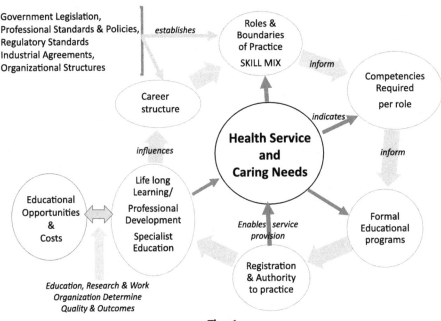

**Fig. 1**
Skill mix development, use and influencing factors.

Evidence of safe and effective nurse staffing practices needs to be measured in a meaningful and preferably, consistent manner to enable continuous improvements to be made. A 2012 focused literature review undertaken for the Canadian Nurses Association [7] noted greater evidence of the complexities associated with the measurement of skill mix. This was the result of emerging new associated topics in the literature such as nursing care delivery models, teamwork and interprofessional collaboration in addition to a stronger focus on skill and competency matching as well as tracking standardized nursing quality indicators to support a business case for investments in nurse staffing. These concepts are covered in depth elsewhere, Chapter 6 explores the models of care and teamwork, Chapter 7 covers rostering, Chapter 10 explores change management and Chapter 11 deals with measuring health service quality.

This chapter provides some historical context to a long standing wicket problem of nurse staffing shortages, how this has been addressed in the past and what has been learned from such events and the provision of non-nursing, unregulated support staff. It then focuses on the metadata required and available to identify knowledge and skill mix needs. These data are also required as contributions for career progression, development and workforce planning as well as for education and training purposes. From an efficiency and effectiveness perspective it is highly desirable for every health worker to be able to contribute their highest level of knowledge and skill within their respective roles. Every health professional needs to be employed relative to their area of expertise and regulated scope of practice.

## Addressing qualified nurse staffing shortages

We have experienced a shortage of nurses since Florence Nightingale days. Every country has dealt with such situations in a variety of ways, most commonly by creating new roles and simply recruiting and training people on the job to assist registered nurses. For example the Australian Red Cross and the Order of St John established Voluntary Aid Detachments (VADs) during World War 1 [8]. Formalized training was provided for this group post war to address the then existing shortage of qualified nurses. A similar situation occurred following the second world war when there was a supply of cheap labor provided by new immigrants, many of whom were war refugees. Nursing Aide Schools were established during the 1950s. Graduates were enrolled by nursing boards as auxiliary nurses. Training courses were upgraded over the years; graduates became recognized as another level of nurse. This continues today. Educational opportunities are critical to the suitable preparation of the caring workforce and need to be in concert with knowledge and skill mix needs at each point in time as relevant to each situation and location.

We need to learn from history. The following case study refers to the 1984 Victorian nursing workforce situation, when the nursing shortage was the result of the introduction of a 38 h week and the transfer of nursing students from the hospital workforce to supernumerary due to the transfer of hospital based nursing schools to Universities [9]. In addition to an overall nurse staffing shortage in terms of numbers, the knowledge and skills of the then available nursing workforce were also insufficient.

---

**BOX 1** Victorian Government case study

---

**Case study**

The 1984 Ministerial Enquiry into Nursing in Victoria undertook a survey of all nurses employed in its 164 publicly funded hospitals to assess the match between post basic qualifications and functional area of practice. This revealed that the proportion of staff, working in a representative group of clinical areas holding appropriate post basic qualifications, did not meet then existing service needs with the exception of obstetric services where >70% of nursing staff were qualified midwives.

**Table B1** Percentage of nurses with appropriate post-basic qualifications in functional areas.

| Functional area of practice | Type of hospital | | | |
|---|---|---|---|---|
| | Metro general % | Metro special % | Country base % | Geriatric % |
| Accident and emergency | 8.0 | 5.4 | 6.0 | |
| Critical care | 30.4 | 38.9 | 24.6 | |
| Geriatric/rehabilitation | 2.4 | 1.9 | 2.1 | 3.3 |
| Obstetric | 79.0 | 74.5 | 70.0 | |
| Operating room | 17.0 | 23.0 | 9.5 | |
| Pediatric | 7.0 | 9.0 | 0.8 | |

---

This case study (Box 1) reveals a most inappropriate match between knowledge and skills, as determined by post graduate qualifications, and clinical specialist health service needs. These results made it possible to influence a policy budget allocation to fund nurses to undertake post graduate studies in any clinical specialty at that time. This funding appears to have been continued as the Victorian Government commissioned a review of post graduate nursing and midwifery education in Victoria in 2015 as a result of a decline in the uptake of training and development grants provided. This review found that the grants provided were highly valued but many barriers to their uptake were reported. Unfortunately the review was unable to identify examples of health services actively monitoring or measuring the perceived benefits of post graduate specialist education to patients or health services, nor did it indicate any percentages of specialist workforce numbers working in their area of practice. These findings were supported by the results of their literature review [10].

These 1984 survey results do suggest that formal specialist post graduate education needed to be examined in the first instance to ensure the available skill mix was the best possible match with service demand. In addition HR systems not only needed to record post registration qualifications but also needed to keep track of content associated with professional development credit points accrued by individual staff members. This is a requirement for all nurses to retain their registration and permission to practice. This study has never been repeated, similar studies were not found in the literature, yet it is important to differentiate between those suitably qualified and those working in an area of specialty.

## Working with a varied skill mix

Shared competencies between any number of health professions also needs to be recognized. Strict adherence to practice restrictions is inappropriate as this stifles flexibility and has the potential to adversely impact patient safety and comfort. Patient centred care requires us to utilize staffing resources in the most productive way including any non-clinical carers. In many situations non-clinical carers are educated to provide very specialized care activities which enables patients to receive the care they require in their home. Such lay carers frequently become the experts in the care of individual patients with very specific caring needs. Collaborative multidisciplinary team practice is highly valued and necessary to deliver health services to those with multiple co-morbidities in a timely manner.

The evidence provided by reported studies pertaining to skill mix and a patient outcomes relationship, were mostly based on administrative data relating to the mix of staff categories, rather than assessed knowledge and skill or formally acquired post registration qualifications in specialty areas of practice. One exception is a very intensive study undertaken by the Centre for Health Economics at the University of York during the early 1990s adopting a multimodal research methodology enabling relationships between health status, nursing care processes, quality of care and patient outcomes to be evaluated [11]. This study differentiated between skill mix and "grade mix," where the latter refers to the number of staff employed at each staff category or pay level. Their literature review revealed several ambiguities and conceptual problems at the measurement level. Its report noted that skill mix is highly political primarily due to financial and organizational staff establishment allocation implications. They concluded that

> investment in employing qualified staff, providing post qualification training and developing effective methods of organising nursing care appeared to pay dividends in the delivery of good quality patient care (p. 144).

A number of studies have reported that a higher proportion of registered nurses, results in a reduction of adverse patient outcomes [12–21]. There are significant limitations when using staff categories/grades as the basis of identifying skill mix. Individuals allocated and paid in accordance with listed staff categories are likely to occupy a variety of roles, where each role requires a specific set of knowledge and skills. Many of these skills are likely to have been acquired via "on the job" training associated with local developments of new roles [22]. New roles and positions requiring less educational preparation continue to be created in response to identified staff shortages. We are now witnessing the introduction of nurse or patient care assistants, potentially as a third level of nurse with a greater division of tasks [23].

## Working to scope

The introduction of new roles changes social and working relationships between clinical staff who collectively are responsible for the care provision of any group of patients. Such changes to the boundaries and scope of practice of existing staff need to be carefully managed although they tend to simply evolve over time. The overall objective is for everyone to be able to work to "full scope," that is to spend less time on undertaking tasks that can be undertaken by lesser qualified staff, often with digital support, and more time to make use of each staff member's highest level of knowledge and skills. This frequently requires significant changes to be made to how the work is organized and how human resources are managed. It may require changes to be made to well established work practices, across multiple departments.

Changes to the available skill mix of staff impacts on patient journeys, care delivery models, teamwork and rostering patterns. Patient acuity or severity of illness patterns can assist with this process of change and its implementation. One significant limitation of focusing only on tasks and skills is the risk of causing fragmented care, rather than holistic patient care. Some flexibility regarding individual boundaries of practice is highly desirable.

## Current skill mix identification methods

Skill mix can be defined in a variety of ways, based on competencies, occupational classification standards, professional practice standards, registration standards or staff categories according to grade/level/band and pay. Any of these essentially define the scope and boundaries of practice associated with each staff category and role.

### Specializations and competencies

Specialization in nursing was first explored by Styles [24] who developed criteria for the designation of a discipline as a nursing specialty. The International Council of Nurses (ICN) provides principles to guide the development of professional regulations across diverse legal, cultural and developmental settings [25]. Both the variety and complexity of clinical specialities relative to roles associated with health services provided, makes it challenging to match service needs with the right skill mix.

The ICN has developed a framework for competencies to apply to a generalist at the point of entry into professional practice-registered nurse [26] and another competency framework for the nurse specialist [27]. This forms the foundation for competencies required to practice in any clinical speciality as well as for any role in other health service delivery, management, research, education or policy development areas. A specialist is a person who has narrowed their focus by acquiring greater knowledge and skills into a niche area of practice to meet the needs of specific types of patients. This includes the ability to operate specialized equipment.

General and specialist nursing practice is delivered in concert with recognized and nationally regulated medical specialties. These plus other regulated health specialties form the basis for the delivery of specific types of health services. Medicine has adopted a strong physiological systems approach as reflected in the naming of clinical specialties [28]. Nursing's focus is on the whole patient within their social and environmental contexts. Holistic patient centred care requires collaborative medical and nursing service delivery strategies.

Countries such as the United States of America, Canada and Australia have two levels of nurses, registered and enrolled or licensed, each with their own boundaries of practice. Registration or enrolment implies that regulatory educational requirements were met. Over the last 15 plus years, Australian Enrolled nurses have been able to obtain advanced skills enabling them to increase their scope of practice. The ability to clearly differentiate between individuals' roles is complex as there is no industrial classification for Enrolled Nurses with Advanced skills and there are variations between Australian States. This has led to role confusion and ambiguity [29]. Delineation needs to be based on each individual's set of formally endorsed and validated set of competencies. Similar issues apply to Registered Nurses who have acquired new specialty and/or post registration competencies.

The nurse specialist role has been in existence in the UK for >20 years, yet it is still surrounded by uncertainty [30]. Mills reported confusion over titles, professional backgrounds, educational requirements and the extent to which such roles afford public protection and pay. This led to the development of the NHS Knowledge and Skills Framework which has six generic competency dimensions to be used as a guide to identify the knowledge and skills applicable to their role [31]. It has frequently been updated and continues to be in use.

A national specialization framework for nursing and midwifery was developed in 2006 as the means to bring "order to the development of specialty areas of practice in Australia" [32]. This policy framework was developed in response to the recognition of an increasing trend of specialization in healthcare. A scan of nursing workforce documentation nationally revealed over 110 areas of both nursing and midwifery "special practice" in use. The framework recognizes 18 national specialties based on specialty recognition criteria, 10 skill domains and 50 practice strands. The latter are defined as areas of practice that did not meet the full criteria for national specialty. Skill domains were defined as areas with common skill groups and common attributes but which may have varied knowledge bases. Nursing Informatics is not included as a recognized specialty in this framework. This work was designed to provide greater national consistency in workforce documentation to enable improvements to be made in workforce planning strategies.

Foundational knowledge and skills for each speciality are most commonly acquired via traditional educational programs. On the job training for some may also be in place. The rate of new knowledge creation and the need to attain new skills to make use of new technologies, requires health professionals to engage in life long learning. Available educational programs are not always up to date and may not be able to cater for each of these clinical specialties. Professional Colleges and national professional organizations develop standards of practice their members need to comply with.

Most countries also have any number of health related competency/training/occupational standards. Australia's National Centre for Vocational Education Research (NCVER) [33] provides a free international research database for tertiary education as it relates to workforce needs, skills development and social inclusion. Organizations such as the Agency for Healthcare Research and Quality (AHRQ) [34], the Cochrane Foundation [35], the Joanna Briggs Institute [36] and many other similar organizations provide evidence based practice guidelines from which required knowledge and skills can be derived.

### Occupational classifications

Occupational classification standards consist of a list of detailed occupations that are combined to form a smaller list of broad occupations further classified into minor and major occupational groups. These codes are used for the purpose of collecting, calculating or disseminating workforce data. The US 2018 Standard Occupational Classification System (SOC) has two major groups applicable to the health industry, (1) Healthcare Practitioners and Technical Occupations and (2) Healthcare Support Occupations. Within these it lists and defines Registered Nurses, Nurse Anesthetists, Nurse Midwives and Nurse Practitioners [37]. Nursing Assistants are classified in the Healthcare support occupations.

There is a four level International Standard Classification of Occupations (ISCO) last updated in 2008 [38]. It's second level lists Health Professionals within which its Nursing and Midwifery Professionals only lists these two categories, this is repeated for the Health Associate Professionals category. National variants are in use in other countries. None appear to include any multidisciplinary occupations such as Health Informatician or Clinical informatician or Nurse Informatician.

The Australian and New Zealand Standard Classification of Occupations, Version1.2 [39] lists four types of Health Professionals (Code 25):

251   Health Diagnostic and Promotion Professionals
252   Health Therapy Professionals
253   Medical Practitioners
254   Midwifery and Nursing Professionals

The Midwifery and Nursing Professionals (254000) are further divided into sub-categories for Midwifery and Nursing Professionals as shown in Box 2. These occupational categories are in use to derive workforce statistics for workforce planning purposes.

It's not clear how Enrolled Nurses, or Advanced Skills Enrolled Nurses or Aboriginal Health Workers need to be classified for statistical purposes. In addition there is a category for Carers and Aides, with further relevant inclusions shown in Box 3.

---

**BOX 2  Midwifery and Nursing Professionals categories as listed in ANZSCO v.1.2**

| | | | |
|---|---|---|---|
| 254111 | Midwife | 254415 | Registered Nurse (Critical Care and Emergency) |
| 254200 | Nurse Educators and Researchers nfd | 254416 | Registered Nurse (Developmental Disability) |
| 254211 | Nurse Educator | 254417 | Registered Nurse (Disability and Rehabilitation) |
| 254212 | Nurse Researcher | 254418 | Registered Nurse (Medical) |
| 254311 | Nurse Manager | 254421 | Registered Nurse (Medical Practice) |
| 254400 | Registered Nurses nfd | 254422 | Registered Nurse (Mental Health) |
| 254411 | Nurse Practitioner | 254423 | Registered Nurse (Perioperative) |
| 254412 | Registered Nurse (Aged Care) | 254424 | Registered Nurse (Surgical) |
| 254413 | Registered Nurse (Child and Family Health) | 254425 | Registered Nurse (Pediatrics) |
| 254414 | Registered Nurse (Community Health) | 254499 | Registered Nurses nec |

---

**BOX 3  Personal Carers and Assistance of relevance to nurses as listed in ANZSCO v.1.2**

423 Personal Carers and Assistants
    4231    Aged and Disabled Carers
        423111        Aged or Disabled Carer
    4233    Nursing Support and Personal Care Workers
        423311        Hospital Orderly
        423312        Nursing Support Worker
        423313        Personal Care Assistant
        423314        Therapy Aide

---

The use of Metadata in both clinical information systems and human resource (HR) management systems makes it easier to undertake a real time analysis of requirements and available knowledge and skills at any point in time. HR systems within any healthcare facility make use of staff categories identified in industry awards and employment agreements. A standard set of Metadata to consistently identify skill mix within any health staffing establishment does not appear to exist. This has an impact on nurse mobility between geographic locations [40].

## Nursing industry awards, agreements and skill mix

A variety of Nursing staff categories are referred to in any industrial award or agreement relative to applicable conditions of employment. The Australian national registered nurse workforce consists of 66 different position titles with a confusing array of titles and practice profiles [41]. These industrial agreements refer to staff categories as defined by their national registration authorities, and where applicable legislative staff categories include; specified periods of time relating to fulltime, part time and casual employees and determine annual "productive hours" for each of these types of employment. Productive hours are calculated by deducting all leave provisions from the effective full time weekly hours per annum. Rates of pay may also vary according to additional graduate qualifications attained and for years of experience. Employment conditions for each staff category determine staff allocations to wards and departments, rostering possibilities and staffing costs for each applicable healthcare facility.

Skill mix staffing establishments for the entire healthcare facility, for each ward and department and for each shift is commonly calculated based on distribution percentages. These need to be carefully considered in terms of types of service demands and a commonly occurring mix of patient types cared for. The results of work measurement studies, as described previously, provide critical information to assist staff allocation and rostering decision making. This is especially important to determine the likelihood of generating sufficient work for non-clinical support staff. Work studies which take into consideration the specific scope of practice for EN's and nursing support staff for an organization, should be conducted for each ward to confirm the appropriate skill mix.

Support services staff are employed under a range of titles and industrial awards. Titles include; Patient Care Assistant (PCA), Patient Services Assistant (PSA), Assistant in Nursing (AIN), Health Care Assistant (HCA), Care Support Workers, (CSW), Rehabilitation Assistant, Physiotherapy Assistant, Diet Aide, Meal Monitor, etc. The hours for the existing roles such as ward clerks/ward secretaries, ward based housekeepers/service assistants, orderlies/porters and couriers have also been extended in order to improve workload management for professional clinicians. The scope of practice of these workers appears to be well defined in most cases, and is generally limited by qualifications, training and the type of work available in the area/service they are working in. A list of activities generally completed by non-nursing support staff, AIN/PCA/PSA/HCA/CSW staff, in acute public and private hospitals include

- Bed making
- Ward tidy
- Cleaning and putting away equipment
- Re-stocking
- Emptying of linen skips
- Porterage
- Tidying of utility rooms
- Patient facial shaves

- Assistance with patient hygiene
- Assistance with mobilizing patients
- Re-positioning patients
- Answering buzzers
- Toileting patients
- Feeding patients (excluding patients with swallowing difficulties or other clinical issues requiring nursing skills for feeding)

Findings from the annual ICN Nurses Wage Survey repeatedly highlight that nurses are generally paid less for their work than professionals in comparable occupations, such as physicians, physiotherapists, teachers and accountants [42]. Overcoming this issue requires nursing work to be evaluated and clearly defined in relation to its value to healthcare delivery facilities and the public at large. Remuneration dimensions tend to focus on five principle factors, (1) knowledge and skills, (2) effort, (3) responsibility, (4) working conditions/environment, (5) freedom to act [42]. Job evaluation underpins the equal pay for work of equal value principle by systematically defining the relative worth of jobs within an organization. The five universal remuneration factors are mental, physical, and skill requirements plus responsibilities and working conditions. Roles are usually evaluated accordingly.

## Job evaluation and skill assessment methods

The staff categories used in any industrial award differentiate according to perceived work value for each classification as described therein. Work value is the value expressed in terms of the salary to be paid. It assumes that all jobs classified into a specific category within any given career structure have equal worth. Work value is considered as a combination of the position occupied, that is the value of the job itself to the organization, and the performance of the position incumbent or the effectiveness with which the individual in the job performs the work as specified. Job evaluation methods [42] useful to consider making use of for nursing are:

1. Non-quantitative
   a. Ranking — based on organizational hierarchies in accordance with level of authority. This reflects organizational structures.
   b. Grading or Classification — this creates job grading based on levels of knowledge and skills or competence in terms of formal qualifications and years of experience.
2. Quantitative
   a. Factor comparison method — pioneered by Edward N Hay during the early 1950s, now widely applied globally, is more expensive to establish but fairer and multidimensional as it considers any number of generic factors and sub-factors that define the level of expertise, importance to a specified role as well as degree of autonomy and authority.
   b. Point rating — most widely used method where points of relative value are allocated to all factors that make up the taxonomy of the work domain. Each role is then evaluated by identifying all relevant factors and sub-factors, all associated points are then summed. The sum of the points for each role provides an index of the relative job value or worth.

The Hay Guide Chart Profile Method of Job Evaluation was well researched by Burton. Its factors and corresponding descriptions are not considered applicable for nursing, but the methodology is [43]. An agreed standardized taxonomy consisting of factors and sub-factors capable of describing nursing work categories or levels based on generic skills relative to the continuum of care is highly desirable. Such a terminology can then be used to describe nursing roles and functions and form the basis for performance evaluation. Given the wide variety of knowledge and skill use relative to individual roles, it is highly desirable for such a taxonomy to concentrate on generic values associated with functions performed within any nursing role. A quantitative factor based, point rating job evaluation method permits comparisons to be made in a systematic and accurate manner; it reduces ambiguity. This method is the most comprehensive, prejudice is minimized, human judgment is more objective, it enables jobs to be placed in distinct categories and is very useful for determining possible career pathways. The chosen job evaluation method needs to be constructively valid and have a high degree of reliability associated with the interpretation and application of the generic definitions attached to each factor and its relevant sub-factors.

### Case Study 1. Nursing job evaluation research study

A nursing career evaluation system (UNCES) was developed for the Australian Private Hospitals sector during the 1990s [44, 45]. This system is applicable to any hospital size, organizational structure, any management style, any existing or new career structure and can define a nursing career structure for each organization. Performance appraisal is an integral part of career path development as it facilitates the identification of staff development needs and assists with career planning. It facilitates matching people with jobs.

This system uses a quantitative position evaluation method classifying hospital nursing work on the basis of work value. The methods used for this development consisted of an extensive literature review, defining boundaries of hospital nursing practice, and identifying generic elements of hospital nursing work based on previous nursing work measurement studies. This information was confirmed/complimented by information obtained from structured interviews and the use of questionnaires. A small group of senior nurses reached consensus regarding the inclusion and ranking of the generic elements of hospital work on the basis of work value. These ranked values were then converted into relative absolute values using the factor comparison method.

Seven major generic job components were identified as factors. These were Expertise, Information processing, Communication, Complexity, Autonomy, Authority and Scope. Each of these were further divided to a total of 45 sub-factors which collectively describe the universe of hospital nursing work. Work value was considered to represent a combination of the position itself and the performance of the position incumbent. Some of the sub-factors had a minimum performance requirement (competency), others had various degrees of progression denoting work value. Where performance was able to be evaluated Benner's model from novice to expert [46] was used.

Positions were evaluated against standard definitions for each sub-factor and translated into a score. The scores for all subfactors denoting all work components of a position were added to arrive at a minimum score reflecting the minimum work value for that position. The maximum score, reflecting the scope for job expansion and maximum possible performance was then calculated for each position (or position type) which had its own score range. This system was evaluated for reliability and validity in four private hospitals ranging in size from 30 to 300 beds, using 81 nursing positions from all levels. There was 80% ($r = 0.806$) correlation between existing grades and UNCES scores demonstrating that 20% of positions in this sample were either over or under valued. There was a 96.7% ($r = 0.967$) correlation between the score for "authority" and the total score. This meant that most positions could be placed in the career structure hierarchy on the basis of that subfactor alone. The greatest level of expertise and the greatest degree of complexity was found to exist in the average charge nurse position.

A further six private hospitals of variable sizes, complexity and service mix trialed UNCES. Implementation consisted of a two-day workshop for members of a local working party established in each hospital to take on the responsibility for implementation. Modifications were made to the implementation manual based on their experiences. Each hospital held educational sessions for their registered nurses who were then able to evaluate their own position. The participation rate ranged from 60% to 80% of all RN staff. Where numerous RNs occupied similar positions, there was overall agreement on which sub-factors applied to that position along with the minimum score for each. Scoring reliability was high, however there was a tendency to identify one's own performance instead of the minimum requirements for any new position incumbent.

The implementation of the system improved communication. It was considered an excellent staff development tool and very valuable by improving clarification of the new nursing career structure. It assisted with the identification of desirable organizational changes. Results were reviewed at another workshop for the local working parties. Generic job descriptions were developed by consensus and the development of performance indicators and methods for staff appraisal commenced.

## Skills Framework for the Information Age (SFIA)

A global well tested job evaluation (factor comparison) method widely used, is the Skills Framework for the Information Age (SFIA) developed for the Information and Communication Technologies (ICT) industry. This is now in its 7th edition [47]. SFIA consist of six generic work categories (factors), each of these have a number of subcategories (subfactors) and each subcategory lists a number of skills which are defined, based on levels of responsibility ranging from 1 to 7. Each level of responsibility is defined based on five attributes, autonomy, influence, complexity, knowledge and business skills. The generically written definitions are applicable to any industry and have been translated into many different languages.

Seven levels of responsibility in nursing could be defined as follows:

1. Works under supervision, basic nursing functions — novice[a]
2. Works under routine direction, understands and makes appropriate use of available technologies and support resources — advanced beginner[a]
3. Identifies and responds to complex issues independently, demonstrates analytical systematic approach to issues and problems — competent[a]

---

[a] Five levels of competency, from novice to expert, in clinical nursing practice referred to above are described in some detail by Benner [46]. This seminal publication continues to be in use today, these levels of competency essentially reflect levels of clinical responsibility.

4. Demonstrates an ability to complete a broad range of complex, technical and professional activities independently — proficient[a]

5. Influences, demonstrates, analyses requirements, aware of practice evidence and industry developments — expert[a]

6. Influences policy and strategy formation, broad business understanding, absorbs complex information — management/administration

7. Authorizes policy, inspires and leads the organization and influences developments within the health industry. Understands, explains and presents the organizations vision and strategic complex ideas — Executive management/administration

Given that nurses are now working in digital environments, it is possible to make use of many of the SFIA categories and subcategories and associated definitions by contextualizing these to suit the health and nursing knowledge and skill domain.

For example; Analytics —

> *"the application of mathematics, statistics, predictive modelling and machine-learning techniques to discover meaningful patterns and knowledge in recorded data. Analysis of data with high volumes, velocities and variety (numbers, symbols, text, sound and image). Development of forward-looking, predictive, real-time, model-based insights to create value and drive effective decision making. The identification, validation and exploitation of internal and external data sets generated from a diverse range of processes."*

Or; Knowledge Management —

> *"The systematic management of vital knowledge to create value for the organisation (patient) by capturing, sharing, developing and exploiting the collective knowledge of the organisation to improve performance, support decision making and mitigate risks. The development of a supportive and collaborative knowledge sharing culture to drive the successful adoption of technology solutions for knowledge management. Providing access to informal, tacit knowledge as well as formal, documented, explicit knowledge by facilitating internal and external collaboration and communications."*

It should be evident from these examples that work value definitions do reflect competency statements. Undertaking a job evaluation activity is similar to undertaking a training needs analysis. The results of either are useful for the development of educational and professional development programs. In addition, such a knowledge and skills framework can form the basis for writing job descriptions, conducting performance appraisals, documenting career pathways and negotiating employment conditions. Using such a framework enables the scope of practice associated with each role to be clearly defined. As there are numerous clinical specialties it is highly desirable that nursing roles in clinical specialty areas are evaluated separately. In most instances any number of knowledge and skill requirements are built on general nursing skills but are unique to the functional requirements of each special area of practice.

Matching skill mix with service demands requires, at a minimum, non-quantitative information obtained from adopting a ranking and/or classification job evaluation method. There is also a need to capture information regarding registration status, qualifications, and evidence of additional knowledge and skills obtained, including some personal attributes such as languages spoken. Not only is there a need to make changes to systems to deal with current knowledge and skill shortages, there is also a need to respond to future skill and competence requirements as new medical technologies are introduced.

HR systems tend to only reflect applicable employment conditions as contained in industrial awards or employment contracts. This may limit opportunities to achieve an optimum match at all times. Given the variable demand for nursing services based on patient types and their various co-morbidities, there are many situations where nurses need advice from another expert who may or may not be working at that time within the same organization. In a digital world such opportunities need to be made available at every point of care via any available communication technology.

There is also a need to consider skill mix issues associated within the nursing and medical professions. High rates of doctors and nurses' skills mismatches have been reported when compared with other professional workers [48]. It is important for all health workers to be able to meet their full potential by applying their knowledge and skills to match the evolving healthcare needs of populations.

A 2018 OECD review of the status of existing surveys [49], was undertaken with an aim to identify evidence on skills requirements, skills use and skills mismatch in healthcare settings. The surveys reviewed were mostly designed to measure qualifications and were not well suited to a system-wide assessment of skills across all health professionals. A general competency framework that includes a set of transversal skills, identified by the OECD as relevant for all health professionals, have been recommended. These common skills are teamwork, communication, socio-cultural sensitivity, awareness of professional and ethical standards, worker's own safety and well being, and adaptive problem solving. This review of health workforce skills assessment found that

> *Increasingly, health care professionals need to apply adaptive problem-solving skills to respond to complex and non-routine patient care issues, while working in complex, multidisciplinary and frequently stressful occupational environments.*

Their findings suggest a need for health workers to be resilient and flexible and to acquire not only technical and clinical skills but also to have cognitive, self-awareness and social skills enabling them to monitor and assess situations, make decisions, take leadership roles, communicate and coordinate their actions with team members to achieve high levels of patient safety and efficiency. There is considerable scope for improving the effectiveness of health professional skills assessment surveys to generate policy-relevant and actionable evidence. There is a need for the health workforce to apply skills such as complex problem solving, critical thinking, judgment and decision making, information processing, information sharing, work autonomously, strong social, interpersonal and analytical skills, engage in active learning and instructing and have the adaptability to work with new technologies [49].

## Education and professional development contributions

Skill characteristics may also be defined as competencies. One could argue that skills reflect personal attributes whereas competencies refer to the performance of a role or set of tasks by using knowledge skills. Health professions make use of competency based or desired graduate outcome statements and professional standards which are used as benchmarks against which individual knowledge and skills are assessed. This is about ensuring that a competent health professional has the attributes necessary for job performance that complies with appropriate standards. Such standards define the level of achievement required for specific areas of practice. Standards are developed and maintained by Professional Colleges/ Associations and serve as the foundation for educational program development and for registration purposes.

Each nation's registration authority publishes their own set of standards. Ideally competency frameworks for each regulated nursing staff category are harmonized and clearly differentiate scopes of practice. Competency standards and educational relationships for any specific regulatory category of staff are shown in Fig. 2. It is possible for any staff category to have not only entry level competencies specified but also additional specialist areas of practice.

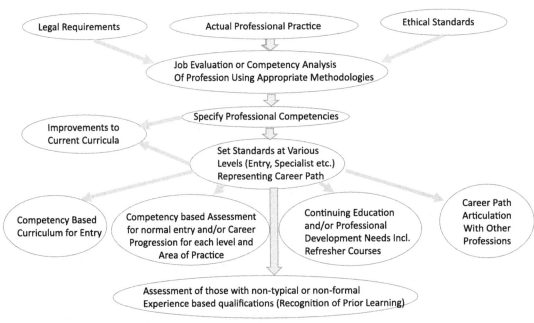

**Fig. 2**

Competency standards development process and relationships. *Source: Adapted from a 1990 Australian Government publication, Gonczi A, Hager P, Oliver L. In: Department of Employment, EaT, editor. Establishing competency-based standards in the professions research paper no. 1. Canberra: Australian Government Publishing Service; 1990.*

A systematic literature review of nursing competency identified 12 consistent broad components representing global community agreement [50]. These competency categories are considered to be useful for the selection or evaluation of qualified nurses. These apply to any registered nurse in various combinations as well as depth and breadth when applied to any clinical specialty at any point of care location. They are defined under the following headings as listed:

1. Personal traits — talent
2. Professional clinical practice
3. Legal and ethical practice
4. Ensuring quality and safety
5. Communication
6. Management of nursing care
7. Leadership
8. Teaching-coaching
9. Cooperation and therapeutic practice
10. Critical thinking and innovation
11. Professional development (life-long learning)
12. Informatics

These broad components defining generic skill requirements apply to any nurse in any specialty. These should be viewed within the context of the 2018 OECD recommendations for transversal skills as described previously [49]. There are a multitude of knowledge and skill competency requirements that need to build on these foundations relative to any clinical specialty as well as within any type of health service for various types of patients.

It's clear that one needs to view health workforce education in the light of workforce knowledge, skill and behavioral requirements based on current and future service demand as well as consider its link with career structures and pay scales. Collectively education providers need to be a step ahead to ensure that the health workforce is suitably prepared. This requires strong relationships between the education providers and health service providers who are able to provide opportunities for experiential learning.

Health systems globally are changing. This requires major reviews and updates to current educational curricula and processes for the health workforce to be better prepared to meet changing future health service demands. The OECD Health Ministerial Meeting held in January 2017 [51] called for

> *a transformative agenda for the health workforce, assessing health professional skills, remuneration and co-ordination, and how these skills and models of care need to adapt in light of digitalisation, wider technological changes, and the evolution of patients' needs.*

These issues will be further explored in the chapter on workforce planning.

## Nursing career pathways

Available nursing pathways differ in every country. In the United States one can commence a nursing career in three different ways. By obtaining a Diploma in Nursing, or an Associate of Science in Nursing or a Bachelor of Science in Nursing (BSN). After obtaining any one of these qualifications they all must pass the National Council Licensure Exam for Registered Nurses to practise. The next step is to become an Advanced Practice Registered Nurse. Many of these programs require a BSN. One can also obtain a Master of Science in Nursing or Doctor of Nursing Practice specializing as a Nurse Practitioner, A Clinical Nurse Specialist, a Certified Registered Nurse Anesthetist or a Certified Nurse Midwife.

Singapore's nursing career pathway starts as a Staff Nurse. Once an Assistant Nurse Clinician, one can continue in Research, or Education, or Management, or Clinical or Informatics from which one can take up a Director of Nursing position before aspiring to become the Chief Nurse. The inclusion of Nursing Informatics as part of the career structure is unique despite the existence of this discipline since the early computing days in the late 1980s.

The first set of nursing informatics competencies were developed in 1987 [52] and have been reviewed and updated in line with health informatics competencies ever since [53]. Informatics roles vary among organizations. The top three informatics competencies for nurses are change management, workflow documentation and information requirements gathering. These skills are associated with the implementation of any health information system in any area of practice. Clinical Informatics has been recognized as a Board-certified Medical Subspecialty by an examination administered by the American Board of Preventive Medicine. This is available to physicians who have primary specialty certification through the American Board of Medical Specialties [54].

Individual educational facilities as well as healthcare organizations have developed their own nursing career pathways covering numerous clinical specialties. Those occupying non-qualified nursing support positions also need to be considered when developing career pathways. Many possibilities exist, however there is no international standard. Career pathway opportunities are closely linked with relevant industrial awards or other employment contractual and prevailing funding arrangements. As a consequence one needs to focus on generic knowledge and skill attributes when developing a system for the purpose of matching service demand with the right skill mix.

Australian nurses undertook major industrial action to achieve a career structure that recognized the value of clinical expertise and experience at the point of care as detailed in the boxed case study. Career structures are closely related to pay scales as people generally like to see their pay increase along with their career advancements. Nurses career structures influence employment opportunities, rostering practices and total nursing costs as well as post registration educational needs.

**Case Study 2. Development of a clinical nursing career structure**

During the mid 1980s the nursing profession in Victoria was concerned about excessive nursing workloads and the lack of expertise and experience at the bedside. Clinical nursing was described as the "Cinderella of the profession. These concerns were the result of a nursing career structure where the only way to be promoted and receive higher salaries was to be appointed to higher hierarchical administrative positions, including education and some rare research positions. At that time just under 80% of all registered nurses were paid at the base rate, the remainder were charge nurses, teachers, supervisors, assistants, deputies and directors of nursing in the public sector. This applied to just under 14,000 effective full time registered nursing positions.

These concerns resulted in several industrial disputes. The first was about obtaining additional non-nursing positions to relieve registered nurses of undertaken routine tasks that could be performed by others. This was resolved on 16th August 1984 and included an agreement that a new career structure recognizing clinical nurses would be developed in consultation with the nurses" union. The first meetings were held on 1st April 1985. The new career structure aimed to reward nurses in accordance with work value. Job evaluation criteria used were knowledge and experience, breadth, interpersonal skills, work complexity, reasoning/creativity and accountability. Position classification standards were clearly defined to be used as benchmarks for implementation purposes. This dispute was resolved on 20th June 1986 when a Clinical Specialist was defined as a:

> Registered Nurse "who is appointed as such to provide a clinical resource, clinical advisory/development role on a full time dedicated basis and undertakes related projects and research and development activities to meet specified clinical nursing needs in the following disciplines of infection control, diabetes education, chemotherapy and stomal therapy" [55].

Further issues were identified during the implementation process. This resulted in a further log of claims on 14th August 1986 followed by landmark strike action resulting in a new decision on 16th September 1986 covering both public and private hospitals and all other healthcare facilities employing Registered Nurses in Victoria.

New South Wales and eventually all Australian nurses have benefited from these landmark events. The 22 July 1986 decision in NSW resulted in the creation of new clinical career positions that provided career advancement opportunities associated with status and monetary recognition. It was the birth of the Clinical Nurse Specialist defined as a registered nurse with specific post-basic qualifications and 12 months' experience, or with 4 years' post basic experience, working in the clinical area of her specified post-basic qualification or working in her specified post-basic experience. This decision listed the 51 applicable clinical specialties [56].

These historical events have resulted in greater recognition and status for all registered nurses in Australia. One current Victorian hospital nursing career structure is available at their website [57].

## Re-engineering clinical services using non-nursing support staff

A global shortage of qualified nurses has had a severe impact on the provision of effective and efficient care in acute health care settings. The extensive use of agency staff, casual workers and overtime, and the increasing incidences of canceled procedures, closed beds and extended waiting lists have all been exacerbated by the nursing shortage. During the past few years there has been a significant increase in the number of support staff assisting with direct patient care in clinical areas. This is most evident in acute medical, surgical and rehabilitation wards where the PCA/PSA/HCA/AIN/CSW roles have been introduced and/or extended to take over work activities/tasks previously undertaken by nursing staff and allied health staff.

The intent outlined by organizations who have re-engineered their clinical services in this way, is not to replace nurses or allied health staff with unqualified staff but to have their workload decreased by others doing the work that is not required to be done by qualified nursing or allied health staff e.g. bed making, manual handling assistance etc. Health care administrators and nurse managers have been forced to examine the work currently done by nurses and where possible support the nursing services with non-nursing staff who can assist by completing non-nursing activities and, under nursing supervision, assist with basic patient care activities e.g. basic hygiene activities, patient reporting, patient mobilization, patient feeding, etc.

These unqualified staff do need to be supervised by Registered Nurses (RNs). This may result in RNs spending more time supervising and less time planning, giving and evaluating care [58]. RNs need to have sound delegation and supervisory skills and apply these with the knowledge of the scope of practice for support staff and recognition that they remain responsible for the outcomes of care. Effective management of skill mix at the point of care can only be achieved safely with the adoption of team work ensuring direct supervision. A systematic literature review regarding the division of labor was inconclusive due to the cultural variety and lack of clarity associated with individual workers' job descriptions [59].

Nurse managers who have embraced this concept and implemented well planned strategies have reported qualitative and quantitative benefits, some of which were planned and others that were unplanned. These include

1. Reduction in nursing time spent doing non-nursing activities.
2. Improved team work within the ward/department.
3. Increased nursing satisfaction related to work content.
4. Increased patient satisfaction.
5. Improved ward efficiency during peak periods.
6. Decreased cost per labor hour.

Hospitals who have reported negative outcomes related to the implementation of these support services generally have

- Not undertaken a review of the work completed by nurses in each ward/department prior to implementation and hence have implemented an inappropriate skill mix for the ward/department.
- Not planned the implementation by ensuring that *all* staff are well informed and supportive of the process.
- Appointed inappropriate employees into these support services positions.
- Not provided appropriate training and/or orientation for these support services positions.
- Not clearly defined the scope of practice and/or work routines for support services staff.
- Not conducted regular feedback sessions with nursing and support services staff in order to resolve initial implementation problems/concerns.
- Not implemented the most effective roster patterns for support services staff i.e. their hours are not concentrated during peak activity periods.

The utilization of support services staff such as; ward clerks, orderlies, patient care assistants, patient services assistant, health care assistants, assistants in nursing, care support workers, allied health care assistants, rehabilitation assistants, meal monitors, diet aides, etc. can improve the effectiveness and efficiency of human resources in clinical areas of acute public and private hospitals. This re-engineering of clinical services requires

1. Effective planning.
2. Comprehensive staff participation from all levels at all stages of the process.
3. Regular staff feedback and debriefing sessions.
4. Clearly defined key responsibilities, scope of practice and work routines for all staff.
5. A committed team approach to change.
6. An effective process to measure associated cost and achieved outcomes.

The success of a skill mix re-engineering project is determined by the level of commitment and the management skills of the unit/ward managers. The percentage (%) of support services staff able to be utilized effectively and efficiently varies according to the case mix and the acuity of patients in a ward. Generally the areas where support services staff have been able to be used in greatest numbers to complete non-nursing tasks and assist with work previously completed by nursing staff, are

1. General medical wards.
2. Surgical wards with an average length of stay (LOS) >4 days e.g. colorectal surgery and orthopedic wards with average LOS >4 days containing joint replacements and spinal surgery.
3. Palliative care units.
4. Medical stepdown units and sub acute units where patients require extensive assistance with activities of daily living.

Areas where support services staff can only be utilized to a limited degree include areas where patients are relatively independent in their activities of daily living e.g. mobility, hygiene, nutrition, etc. These include

1. Day surgery units.
2. Short stay surgical wards including non-complex gynecology and urology.
3. Acute diagnostic cardiology units.

When the correct staffing skill mix is provided for a ward/unit the potential for quality healthcare outcomes and cost efficiency is maximized. Having the required number of staff hours and the correct skill mix enables all health workers to complete their work effectively, remain within their scope of practice, and provide quality patient care. Without the combination of the correct number of staff and the correct staffing mix there is a high risk that patient safety will be compromised.

### Example

The required staffing hours for a 20-bed stroke/palliative care ward for a Monday morning shift are 56 h. This includes

- 1 Nurse Manager
- 1 Ward Clerk
- 3 Registered Nurses
- 2 Healthcare Assistants

This provides a skill mix for patient care of 60% Registered Nurses and 40% Healthcare Assistants. Provided that at least 70% of the care required is basic nursing care that can be completed by a Healthcare Assistant, this skill mix should be adequate, enabling all Healthcare Assistants to work within their scope of practice, and provide quality healthcare to all patients (completing 70% of the basic nursing care required). Registered Nurses should have adequate time to complete Registered Nurse work and some additional time to include some basic nursing care (approximately 30% of their work is generally basic nursing care in a ward environment).

If this skill mix is degraded e.g. no Ward Clerk is supplied, the Registered Nurses and/or the Healthcare Assistants will have to pick up some of the Ward Clerk activities during the shift hence these staff members will have inadequate time to complete their patient care activities. Patient safety and quality care outcomes will be compromised.

If the Registered Nurse/Healthcare Assistant ratio is altered to 25% Registered Nurses and 75% Healthcare Assistants, patient safety and the quality of care will be compromised. In order to complete the Registered Nurse work, Healthcare Assistants will have to complete work beyond their scope of practice, or, Registered Nurses may miss their breaks and work additional hours to complete their work. Regardless of the efforts of the Registered Nurses and the Healthcare Assistants to complete their work, it is highly likely that some care will be delayed, some care may be omitted, and errors could occur.

Each ward/unit has its own work profile and in order to establish a correct skill mix it is necessary to complete a work study. A work study will identify the type of work activities there are in a unit and how long it takes to complete the set of work activities relevant to each shift.

To identify the skill mix required to complete the work activities on each shift it is important to use a tool which can identify the % of work that falls within the scope of practice for each level of skill. In this instance we must identify the scope of practice for each skill level including the Nurse Manager, the Registered Nurse, the Enrolled Nurse, the Healthcare Assistant and the Ward Clerk.

Once the work has been measured for the scope of practice for each skill mix, the ratio or percentage of work that can be completed by the Healthcare Assistant is calculated. The Enrolled Nurse scope of practice includes all health care activities included in the scope of practice of the Healthcare Assistant and additional higher-level care activities such as clinical observations, simple dressings etc. The ratio or percentage of work that can be completed by a Registered Nurse includes the work that can be completed by the Healthcare Worker, and the Enrolled Nurse. The remaining clinical work that is outside of the scope of practice of both the Healthcare Worker and the Enrolled Nurse must be completed by a Registered Nurse. It is important to note that registered nurse work will generally include at least 20% of care that could have been completed by a support worker. This will be demonstrated in the work study data and must be considered when developing the skill mix profile for a ward/unit.

It is important to measure the volume of work that belongs to the role of the Nurse Manager and the Ward Clerk. If adequate resources are not provided to complete the work of the Ward Clerk and the Nurse Manager, some of this work will have to be completed by the Registered Nurses, Enrolled Nurses or the Healthcare Assistants and hence compromise patient safety and quality care.

A common finding during work analysis studies is that some work activities completed by the nursing services team was not funded or accounted for in the operating budget; e.g. care provided to unplanned outpatients. These are often ad hoc visits or phone calls from previously discharged patients and/or their relatives. These additional activities should be measured and monitored on an ongoing basis so that they can be budgeted for/funded. Alternately this work can be moved to another department/cost center. Our objective is to achieve a balance between demand and supply, identify the type of work to be done and to ensure that the appropriate skill mix is in place to complete the work.

If the ratio of Registered Nurses to Enrolled Nurses and Healthcare Assistants is too low, patient safety and quality care is compromised. If the Registered Nurse ratio is too high the cost of care will be increased and there will be less budget to buy additional resources when patient numbers (bed utilization) and patient acuity peaks over budget projections.

Following a review of a range of studies conducted in Australia, New Zealand, Singapore and UK the following recommendations of skill mix could be utilized. Table 1 presents these recommendations to be used as a guide only. These can vary according to patient complexity/acuity and length of stay. If the ward has a mixture of patient types, appropriate adjustments will need to be made to the % of skill mix based on the type of clinical care provided and the scope of practice of the nursing and nursing support staff. Minimum staffing number requirements must also be considered when determining skill mix. It is important to have sufficient numbers of senior nursing staff available after hours to be able to manage workload changes due to new admissions or patient health status deterioration requiring high level knowledge and skills to ensure good outcomes, and to have the capacity to redeploy staff [60].

The highest proportion of non-nursing work and basic ADL care occurs during the hours of 6.00 am to 11.00 am when hygiene and mobility activities peak. The peak activity time for the evening shift for non-nursing activities falls between 4.30 pm and 10.00 pm with basic ADL activities peaking between 5.00 pm and 9.00 pm.

In addition to ward related activities, non-clinical support staff provide first contact services with patients in a variety of clinical service areas. The use of unregulated nursing support workers in acute care nursing units is widespread and growing. They were found to perform many direct patient care activities; more than their licensed/regulated nurse colleagues [61]. Non-clinical support staff are often in situations where they provide a consistency in environments where clinical staff are constantly changing. Everyone who interacts with patients must have excellent "customer service" skills. Exemplary service, including friendliness and empathy, can favorably influence how patients and their families perceive all technical and clinical services provided for them [62]. These authors offer the following five service principles which are of relevance to all front line staff:

1. Create a strong first impression
2. Hire the person, not the resume — good fit with organizational values
3. Prepare performers to perform — training and education
4. Attack needless delay in service delivery
5. Prioritize services that matter most

**Table 1  Recommended skill mix for patient types — Inpatients**

| Ward type | Periods | RN | EN | AIN/HCA |
|---|---|---|---|---|
| General medical/palliative care | 0700–1500 | 60% | 20% | 20% |
| | 1500–2300 | 65% | 20% | 15% |
| | 2300–0700 | 75% | 25% | |
| Cardiology | 0700–1500 | 80% | 10% | 10% |
| | 1500–2300 | 85% | 10% | 5% |
| | 2300–0700 | 85% | 15% | |
| Oncology/hematology | 0700–1500 | 65% | 25% | 10% |
| | 1500–2300 | 75% | 15% | 10% |
| | 2300–0700 | 75% | 25% | |
| Stroke | 0700–1500 | 60% | 20% | 20% |
| | 1500–2300 | 60% | 20% | 20% |
| | 2300–0700 | 70% | 30% | |
| General surgery | 0700–1500 | 70% | 20% | 10% |
| | 1500–2300 | 70% | 20% | 10% |
| | 2300–0700 | 75% | 25% | |
| Major orthopedic and polytrauma | 0700–1500 | 60% | 20% | 20% |
| | 1500–2300 | 65% | 25% | 10% |
| | 2300–0700 | 75% | 25% | |
| Urology/gynecology, laparoscopic surgery | 0700–1500 | 70% | 20% | 10% |
| | 1500–2300 | 75% | 15% | 10% |
| | 2300–0700 | 75% | 25% | |
| Rehabilitation/medical gerontology | 0700–1500 | 50% | 30% | 20% |
| | 1500–2300 | 50% | 30% | 20% |
| | 2300–0700 | 60% | 20% | 20% |
| High dependency medical/surgical, high dependency cardiology, high dependency oncology | 0700–1500 | 80% | 20% | |
| | 1500–2300 | 80% | 20% | |
| | 2300–0700 | 100% | | |
| Day oncology | Day | 80% | 20% | |
| Residential high care | 0700–1500 | 20% | 20% | 60% |
| | 1500–2300 | 20% | 20% | 60% |
| | 2300–0700 | 15% | 30% | 55% |
| Pediatric | 0700–1500 | 70% | 20% | 10% |
| | 1500–2300 | 70% | 20% | 10% |
| | 2300–0700 | 70% | 30% | |
| ICU, CCU, SCU, NIC | 0700–1500 | 100% | | |
| | 1500–2300 | 100% | | |
| | 2300–0700 | 100% | | |

N.B, if EN is not available, 50% of EN work can generally be completed by AIN/HCA.

*Source: From the TrendCare® Skill Mix Recommendations.*

## Future directions for identifying and matching skill mix needs with available staffing resources

It is apparent that there are numerous variations regarding knowledge and skills attained by individuals to enable them to practice in any clinical area. This is evident from the availability of many educational programs, registration standards that are unique for every national jurisdiction, the variety of funding arrangements, industrial awards or employment contracts and subsequent career structures. For now the best available option is to make use of existing staff funding arrangements as these tend to represent work value and scope of practice.

Service demand is changing and becoming more complex due to an increasing number of older patients with multiple co-morbidities requiring long term chronic patient centred care. This requires services to be delivered by teams with a comprehensive combination of knowledge and skills. This is where scope and boundaries of practice for team members need to be able to change to ensure holistic care for individual patients. There is a trend to provide many health services in the home or on an outpatient or day care basis with an aim to reduce lengths of stay for in-patients and in residential care. This also means that patients themselves, their live-in carers and/or social support network need to be educated to contribute to these health service delivery functions.

From a digital transformation perspective there is an absence of a standard set of metadata to describe clinical knowledge and skills. The SFIA example developed by and for ICT Professionals described previously, provides a great foundation for the development of a similar framework to suit the health workforce. It serves as a great example of what can be achieved via global collaboration. ICT Professionals work in all industries, their knowledge and skill requirements are complex, and frequently changing, yet they have managed to evaluate and define these in a generic manner.

There is an increasing need for these digital skills to be integrated with clinical skills to ensure that digital technologies are designed, implemented and used in a manner that ensures data accuracy and patient safety. The desired health, nursing and clinical informatics knowledge and skills can be identified by using the SFIA framework. Its use enables any number of health service roles at various levels of responsibility to be combined as required for individual positions to suit an employee's specific set of knowledge and skills, individual patient centred care requirements, any sized organization and a variety of organizational structures. Clinical knowledge and skills need to be identifiable via its own framework.

The 2018 OECD review [49] identified numerous suitable datasets which collectively could be used as a solid foundation for the development of a suitable framework for a metadata set reflecting health workforce knowledge, skills and behavioral requirements. The availability of such a data set has numerous benefits to the health industry as a whole. The aim is to be able to

1. Identify the collective knowledge and skill needs for any healthcare organization to meet their unique health service demand, quantify workforce capacity need and determine overall staffing establishments and budgets.
2. Reorganize organizational structures to best support individual patient health journeys within physical and locational constraints, identify department workforce capacity needs and relate this to departmental staffing allocations.
3. Identify, define and quantify unique functional roles required to meet service demand in each department or for each service type.
4. Evaluate individual workforce members' set of knowledge, skills, preferences and career pathway desires and negotiate the best fit with available positions to ensure everyone is able to perform at their peak potential and collectively meet all service demand needs effectively.
5. Monitor overall health service demand and workforce availability trends, make staffing adjustments in real-time and more permanently where indicated.
6. Individuals responsible for any point of care function must have access to expert advice regarding any unusual occurrence irrespective of location.

The outcome for nurses who experienced the Nursing Career Evaluation System as described in the case study, was extremely positive as it built self-confidence by formally recognizing the value of functions undertaken by each individual. This is an important feature for nurses who are mostly female, whose work is largely invisible statistically to the public at large and whose services are often taken for granted.

This chapter has included an example of how to re-engineer clinical service staffing arrangements using additional non-regulated nursing support staff. Similar re-engineering strategies need to be adopted to meet the needs of other types of health service delivery. The next chapter explores how to make the best possible use of the available number of staff and their skill mix to deliver safe and high quality care. This requires an examination of nursing and midwifery practice foundations and how their service delivery is organized and managed.

## *References*

[1] AIHW. Metadata Online Registry (METeOR). [cited 5 September 2018]. Available from: http://meteor.aihw.gov.au/content/index.phtml/itemId/181162.
[2] CONNMO. Coalition of National Nursing and Midwifery Organisations. [cited 2018 9 September]. Available from: http://connmo.org.au/.
[3] Rischbieth A. Matching nurse skill with patient acuity in the intensive care units: a risk management mandate. J Nurs Manag 2006;14(5):397–404.
[4] Morrison A. Scope of nursing practice and decision-making framework toolkit. Geneva: International Councilof Nurses (ICN); 2010. [cited 15 October 2018]. Available from: https://www.icn.ch/sites/default/files/inline-files/2010_ICN%20Scope%20of%20Nursing%20and%20Decision%20making%20Toolkit_eng.pdf.

[5] AACN. Clinical nurse leader initiative American Association of Colleges of Nursing; 2004 [cited 24 November 2018]. Available from: https://www.aacnnursing.org/CNL.

[6] Wilkes L, Luck L, O'Baugh J. The role of a clinical nurse consultant in an Australian Health District: a quantitative survey. BMC Nurs 2015;14:25.

[7] Harris A, McGillis-Hall L. Evidence to inform staff mix decision-making: a focused literature review. Canadian Nurses Association; 2012. [cited 7 November 2018]. Available from: https://www.cna-aiic.ca/~/media/cna/page-content/pdf-en/staff_mix_literature_review_e.pdf.

[8] NENA. History of enrolled nursing in Australia. National Enrolled Nurses Association of Australia; 2015. [cited 5 November 2018]. Available from: http://www.nena.org.au/Conference_2015/Presentations/History%20-%20NENA%20Powerpoint%20M%20&%20J.pdf.

[9] McClelland JE. Report of the committee of enquiry into nursing in Victoria. Melbourne: Victorian Government Ministry of Health; 1985.

[10] Darcy Associates CS. Post graduate nursing and midwifery education in Victoria—project final report, Melbourne; Health Department Victoria; 2015. [cited 5 November 2018]. Available from:https://www2.health.vic.gov.au/-/media/health/files/collections/research-and-reports/p/pnmev-final-report-1-oct.pdf.

[11] Carr-Hill R, Dixon P, Ginbbs I, Griffiths M, Higgins M, McCaughan D, et al. Skill mix and the effectiveness of nursing care. York: Centre for Health Economics, University of York; 1992.

[12] Aiken LH, Clarke SP, Sloane DM, Sochalski J, Silber JH. Hospital nurse staffing and patient mortality, nurse burnout, and job dissatisfaction. JAMA 2002;288(16):1987–93.

[13] Aiken LH, Sloane DM, Bruyneel L, Van den Heede K, Griffiths P, Busse R, et al. Nurse staffing and education and hospital mortality in nine European countries: a retrospective observational study. The Lancet 2014;383(9931):1824–30.

[14] Van den Heede K, Clarke SP, Sermeus W, Vleugels A, Aiken LH. International experts' perspectives on the state of the nurse staffing and patient outcomes literature. J Nurs Scholarsh 2007;39(4):290–7.

[15] Butler M, Collins R, Drennan J, Halligan P, O'Mathúna DP, Schultz TJ, et al. Hospital nurse staffing models and patient and staff-related outcomes. Cochrane Database Syst Rev 2011;(7):CD007019.

[16] Duffield C, Roche M, Twigg DE, Williams A, Rowbotham S, Clarke S. Adding unregulated nursing support workers to ward staffing: exploration of a natural experiment. J Clin Nurs 2018;27(19–20):3768–79.

[17] Griffiths P, Recio-Saucedo A, Dall'Ora C, Briggs J, Martuotti A, Meredith P, et al. The association between nurse staffing and omissions in nursing care: a systematic review. J Adv Nurs 2018;74(7):1474–87.

[18] Hendry C, Aileone L, Kyle M. An evaluation of the implementation, outcomes and opportunities of the Care Capacity Demand Management (CCDM) Programme—final report. Christchurch, NZ: NZ Institute of Community Health Care; 2015.

[19] Kane RL, Shamliyan T, Mueller C, Duval S, Wilt T. Nursing staffing and quality of patient care. Evidence report/technology assessment no. 151 (prepared by the Minnesota Evidence-based Practice Center under contract no. 290-02-0009.), Rockville, MD; 2007. Contract no.: AHRQ publication no. 07-E005.

[20] O'Halloran S. Framework for safe nurse staffing and skill mix in general and specialist medical and surgical care settings in adult hospitals in Ireland: final report and recommendations. Ireland: Department of Health; 2018.

[21] Shin S, Park JH, Bae SH. Nurse staffing and nurse outcomes: a systematic review and meta-analysis. Nurs Outlook 2018;66(3):273–82.

[22] Ball J. Guidance on safe nurse staffing levels in the UK. London: Royal College of Nursing; 2010.

[23] MacKinnon K, Butcher DL, Bruce A. Working to full scope: the reorganization of nursing work in two Canadian community hospitals. Glob Qual Nurs Res 2018;5:1–14.

[24] Styles MM. On specialisation in nursing: towards a new empowerment. Silver Spring (MD): American Nurses Association; 1989.

[25] Barry J. Regulatory board governance toolkit. Geneva: International Council of Nurses (ICN); 2014. [cited 15 October 2018]. Available from: https://www.icn.ch/sites/default/files/inline-files/2014_Regulatory_Board_Governance_Toolkit.pdf.

[26] Alexander MF. ICN framework of competencies for the generalist nurse: report of the development process and consultation. Geneva: International Council of Nurses; 2003.

[27] Affara F. ICN framework of competencies for the nurse specialist. International Council of Nurses; 2009. [cited 15 October 2018]. Available from: https://siga-fsia.ch/files/user_upload/08_ICN_Framework_for_the_nurse_specialist.pdf.

[28] List of specialties, fields of specialty practice and related specialist titles. Medical Board of Australia; 2018. [cited 15 October 2018]. Available from: https://www.medicalboard.gov.au/Registration/Types/Specialist-Registration/Medical-Specialties-and-Specialty-Fields.aspx.

[29] Cusack L, Smith M, Cummins B, Kennewell l DL, Pratt D. Advanced skills for enrolled nurses: a developing classification. Aust J Adv Nurs 2015;32(4):40–6.

[30] Mills C, Pritchard T. A competency framework for nurses in specialist roles. Nurs Times 2004;100(43):28–9.

[31] NHS. Simplified Knowledge and Skills Framework (KSF). [cited 27 May 2019]. Available from: https://www.nhsemployers.org/your-workforce/retain-and-improve/appraisals/simplified-ksf; 2019.

[32] NNNET. A national specialisation framework for nursing and midwifery. Melbourne: National Nursing & Nursing Education Taskforce; 2006. [cited 27 May 2019]. Available from: http://www.dhs.vic.gov.au/nnnet/downloads/recsp_spec_framework.pdf.

[33] VOCEDplus. NCVER's international tertiary education research database. [cited 8 October 2018]. Available from: http://www.voced.edu.au.

[34] AHRQ. Guidelines and measures. [cited 13 September 2018]. Available from: https://www.ahrq.gov/gam/index.html.

[35] Cochrane. [cited 8 October 2018]. Available from: https://www.cochrane.org/.

[36] The Joanna Briggs Institute. [cited 8 October 2018]. Available from: http://joannabriggs.org/.

[37] Standard Occupational Classification System. Bureau of Labor Statistics, United States Department of Labor; 2018. [cited 8 October 2018]. Available from: https://www.bls.gov/soc/2018/major_groups.htm.

[38] ISCO. International Standard Classification of Occupations. International Labour Organisation; 2008. [cited 8 October 2018]. Available from: http://www.ilo.org/public/english/bureau/stat/isco/isco08/index.htm.

[39] ANZSCO. Australian and New Zealand Standard Classification of Occupations, 2013, Version 1.2. Australian Bureau of Statistics; 2013. Available from: http://www.abs.gov.au/ausstats/abs@.nsf/mf/1220.0.

[40] Benton DC, González-Jurado MA, Beneit-Montesinos JV. Professional regulation, public protection and nurse migration. Collegian 2014;21(1):53–9.

[41] Gardner G, Duffield C, Doubrovsky A, Adams M. Identifying advanced practice: a national survey of a nursing workforce. Int J Nurs Stud 2016;55:60–70.

[42] ICN. Job evlauation guidelines.[cited 15 October 2018]37 p. Available from: http://www.who.int/workforcealliance/knowledge/toolkit/22.pdf; 2010.

[43] Burton C, Hag R, Thompson G. Women's worth: pay equity and job evaluation in Australia. Canberra: AGPS Press, Australian Government Publishing Service; 1987 .p. 155.

[44] Hovenga E. Origins and development of the Universal Nursing Career Evaluation System (UNCES)—part 1; Melbourne: EJSH Consulting Services; 1992 [unpublished].

[45] Hovenga E. Implementation of Nursing Career Evaluation System (NCES)—part 2; Melbourne: EJSH Consulting Services; 1992 [unpublished].

[46] Benner P. From novice to expert. California: Addison-Wesley; 1984.

[47] SFIA. Skills for the information age framework (SFIA) 7th ed.: SFIA Foundation; 2018 [cited 17 October 2018]. Available from: https://www.sfia-online.org/en.

[48] Schoenstein M, Ono T, LaFortune G. Skills use and skills mismatch in the health sector: what do we know and what can be done? In: LaFortune G, Moriera L, editors. Health workforce policies in OECD countries: right jobs, right skills, right places. OECD Publishing; 2016. p. 163–83.

[49] OECD HDT. Feasibility study on health workforce skills assessment: supporting health workers achieve person-centred care. Paris: Division H; 2018.

[50] Liu Y, Aungsuroch Y. Current literature review of registered nurses' competency in the global community. J Nurs Scholarsh 2018;50(2):191–9.

[51] OECD. The next generation of health reforms. In: Ministerial Statement following OECD Health Ministerial Meeting, 17 January; 2017.

[52] Peterson H, Gerdin-Jelger UE. Preparing nurses for using information systems: recommended informatics competencies. New York: New York National League for Nursing; 1988.

[53] Hovenga E, Grain H. Learning, training and teaching of health informatics and its evidence for informaticians and clinical practice. In: Ammenwerth E, Rigby M, editors. Evidence-based health informatics, studies in health technology and informatics. vol. 222. Amsterdam: IOS Press; 2016.

[54] AMIA. Clinical informatics becomes a board-certified medical subspecialty following ABMS vote. American Medical Informatics Association; 2011. [cited 20 October 2018]. Available from: https://www.amia.org/news-and-publications/press-release/ci-is-subspecialty.

[55] Victoria IRC. Registered nurses award. Melbourne: Victoria State Government; 1986.

[56] Staunton P, Moait S. Briefing notes on the new public hospital nurses (state) award operative from 27th June 1986; 1986.

[57] RCH. Nursing career pathway The Royal Children's Hospital; 2018 [cited 19 October 2018]. Available from: https://www.rch.org.au/nursing/nursing_opportunities/Nursing_Career_Pathway/.

[58] Duffield C, Gardner G, Catling-Paull C. Nursing work and the use of nursing time. J Clin Nurs 2008;17(24): 3269–74.

[59] Lavander P, Meriläinen M, Turkki L. Working time use and division of labour among nurses and health-care workers in hospitals—a systematic review. J Nurs Manag 2016;24(8):1027–40.

[60] Henderson J, Willis E, Toffoli L, Hamilton P, Blackman I. The impact of rationing of health resources on capacity of Australian public sector nurses to deliver nursing care after-hours: a qualitative study. Nurs Inq 2016;23(4):368–76.

[61] Roche MA, Friedman S, Duffield C, Twigg DE, Cook R. A comparison of nursing tasks undertaken by regulated nurses and nursing support workers: a work sampling study. J Adv Nurs 2017;73(6):1421–32.

[62] Berry LL, Deming KA, Danaher TS. Improving nonclinical and clinical-support services: lessons from oncology. Mayo Clin Proc 2018;2(3):207–17.

## Further reading

[63] Gonczi A, Hager P, Oliver L. Establishing competency-based standards in the professions research paper no. 1. In: Department of Employment, EaT, editor. Canberra: Australian Government Publishing Service; 1990.

# CHAPTER 6

# Nursing and organizational models of care

Previous chapters explored nursing practice from the perspectives of service demand, nursing workloads, work measurement and skill mix. Evidence regarding the nurse staffing and improved patient outcomes relationship, as explored in previous chapters, was found to be too weak for specific practice recommendations [1]. There is a need for a better understanding of "how" nursing knowledge and practice, improve patient outcomes. This chapter explores the theoretical professional foundations of the nursing discipline, nurses' professional status and the way these are interpreted and applied at the point of care. All of these factors collectively contribute to options for and effectiveness of the model of care adopted that represents the capacity to care. This publication as a whole is focusing on the management of nurse staffing, rostering and service delivery in acute hospitals or residential aged care facilities. Nursing practice in acute care is most complex. The principles covered in this chapter apply equally to any other area of nursing practice.

Nursing is a practice discipline whose realm of expertise is the diagnosis and treatment of the human response to any event with consequences for an individual's health and well being. Nurses themselves are the key agents for action — that is, the worker is inseparable from the work [2]. A literature review of conceptual models underpinning the nursing discipline found that all emphasize individual (patient)-centered nursing practice processes including assessment, planning, intervention and evaluation. All encompass compassionate care across the life span from an holistic perspective [3].

This differs greatly from the medical discipline or medical practice model which focuses on categorical biological system specific specialties with a perspective for understanding various biological system based conditions. Physicians apply their distinctive knowledge of the disciplines of anatomy, physiology, biochemistry, histology, pathology etc., with an emphasis on addressing abnormalities. There is a danger that nurses and others confuse these related and complementary disciplinary underpinnings as evidenced by discussions and actions that discount the value of nursing [3]. Nurses make a significant contribution to healthcare. Nursing services cannot be considered in isolation of all other health services provided by a great variety of health professionals who all need to work as a team to care for any one individual patient.

Nursing models of care not only relate to the philosophical underpinnings that guide the caring methods adopted, but such models of care also refer to how nursing work is organized,

Measuring Capacity To Care Using Nursing Data. https://doi.org/10.1016/B978-0-12-816977-3.00006-X

distributed between those on duty during any particular shift, and delivered. The latter needs to be referred to as "organizational models of care" to clearly differentiate them between the more patient centered care delivery models. There are numerous "models of care" mostly developed to suit specific clinical scenarios or specialties. This makes it difficult to find relevant references in the literature regarding organizational models of care and creates a great deal of ambiguity regarding their interpretation.

There is no agreed definition, although there is general consensus that models of care provide the framework for the healthcare delivery process in any setting [4]. Many models of care consider individual personalized patient needs, others are focused on how to best deliver nursing services to a group of in-patients. It is apparent that two types of models of care can be implemented, one with an organisational focus, the other based on a nursing theory. Whatever model is chosen it determines communication patterns, who has responsibility for what and who is accountable for outcomes achieved at the end of each shift or episode of care.

Most units or wards have adopted a standard care delivery method although modifications are made on a per shift basis by the nurse in charge to ensure the available collective knowledge and skills are optimally distributed based on known care requirements. Nursing tasks may shift from one group of workers to another during the course of the day to meet patient needs [5]. Unexpected changes due to patient deterioration or new admissions or a request to undertake an unanticipated unit based medical procedure requiring nursing support, plus providing adequate cover during meal and refreshment breaks, may require changes to be made during the shift regarding the original planned distribution of nursing work.

Units (wards) in acute care hospitals organize their care delivery based on their usual mix of patient types, available staff skill mix on any one shift, agreed patterns of skill use, work environment design and the availability of support equipment and services. Due to the number of confounding variables one cannot adopt any specific standard model. One study identified four associated themes associated with the distribution of nursing service delivery activities, the ownership of tasks, managing the workers, ancillary nursing personnel and the struggle to change. It found that nursing boundaries needed to be flexible regardless of model of care adopted [5].

Many nurses work on a part time basis including as members of casual or agency staff. Matching a constantly changing service demand with daily variations in available nurses' skill mix is complex. These factors influence opportunities to provide holistic, well informed patient centered care. One Australian study explored whether nurse staffing, experience and skill mix influenced the model of nursing care in medical-surgical wards. It found that the most frequently reported model of care was patient allocation and team nursing, and that skill mix, nurse experience, nursing workload and factors in the ward environment significantly influenced the model of care in use [6].

Traditional organizational models of care include total patient care, patient allocation, task assignment or functional work distribution, team, and primary nursing. There is little empirical evidence regarding the effectiveness of these traditional models of care in any given situation [7, 8]. A 2017 Canadian study showed that enrolment in the newer team-based primary care practices within in-patient units, was associated with lower rates of post discharge emergency department visits and death [9]. Newly emerging models of care, such as family centered care, nurse practitioner led care and the patient centered medical home or hospital in the home models of care, primarily refer to non-acute community care settings.

There is also a need to explore if there is a relationship between nursing theories, such as the adoption of evidence based practice or multidisciplinary health care, and nursing models of care adopted? How do nursing models of care reflect recognized professional practice or support the application of the nursing care process? How do nursing models of care best relate to patient-centered care? It may be argued that nursing theorists anticipated that models of care would enable those providing nursing care to become more autonomous and accountable in their clinical decisions and organization of care delivery, while boosting the development of nursing as a professional discipline [10].

Fig. 1 details a number of factors used to inform and most likely influence, consciously or unconsciously, decisions made regarding the organizational nursing model of care to be adopted. As situations change so will the distribution of nursing work and the adoption of the nursing model of care. The variable nature of service demand, staff mix, and staff availability at

**Fig. 1**
Factors influencing nursing model of care.

various points in time throughout any patient's journey, makes it difficult to obtain definitive evidence regarding impact of any one or combination of these factors. Collectively these factors provide the context within which nurse managers need to organize the delivery of nursing services. Nursing models of care have a direct relationship with the quality of care provided. Models of care in use influence patient satisfaction, care outcomes as well as staff satisfaction and staff turnover rates. Tools and processes supporting decision making regarding nursing work allocations continue to be developed [11] to assist this process.

Our focus is fairly pragmatic based on organizational models of care, although other models of care based on various health service and theoretical nursing disciplinary concepts, also need to be considered within this context.

## Factors known to influence nursing models of care

Given the numerous confounding variables that need to be considered when making decisions regarding how best to distribute and deliver nursing work, it isn't surprising to learn that the literature has tended to focus on anecdotal reports or single site evaluations. An analysis of traditional nursing care delivery model definitions suggests that their designs were based on professional or organizational value systems or philosophical or theoretical constructs. These determine control mechanisms, expectations, value recognition and professional practice characteristics as well as degree of independence within anyone organizational culture and hierarchical or governance, and operational structures.

Health care organizations have numerous interrelated departments that need to achieve effective coordination and work integration to support individual patient journeys throughout any episode of care. Nurses assume pivotal roles regarding these dynamic interactional matters. There are many interrelationships and interdependencies that collectively make up the nursing work environment. As a consequence there is no single model of care or method to effectively deliver nursing services at the point of care in acute care hospitals [12, 13]. Within such "social systems" it is desirable to also consider that nurses self-esteem is imagined and perhaps actualized in their work environment [14]. In addition nurses' working environments provide the economic basis required to support chosen lifestyles, thus collectively these factors influence staff satisfaction and turnover rates.

One extensive integrative literature review [15] that analyzed the characteristics of nurses' work process in different countries, found that nurses in every country need to deal with social and technical divisions (nursing skill mix and hierarchical subordination). It was noted that nurses have depreciated the value of their managerial roles which are "inseparably composed of healthcare-managerial tasks." Nurses' managerial tasks include coordinate, control, supervise and guide the work process flow in nursing and healthcare. These activities determine the model of care delivery adopted and ensure the smooth operation of health units/wards and the continuity of care while controlling costs for health organizations, yet the value of this type of work undertaken by nurses is reduced. Leal et al. [15] noted that nurses' ideologically view their work as healthcare and that "these professionals mischaracterise the nature of their own work." Nurses bridge the gap between nursing professionals and other health practitioners on behalf of their patients. Evidence was also found that nurses in some countries now undertake specialized procedures previously exclusively performed by physicians. It was noted that

> *As nurses assume tasks previously attributed to doctors, and because the value of their work is lower than the doctors', health service provision becomes cheaper, something that interests not only employers, but also every government, given the increasing maintenance costs of healthcare systems.*

These attitudes contribute to the devaluing of the holistic caring components that belong to the nursing profession. One could argue that by nurses assuming tasks previously undertaken by doctors, do not make the nursing profession socially valued as this scope expansion has not usually granted nurses additional monetary gain or greater social acknowledgement for their work.

These all constitute influencing and contextual factors when determining care delivery models to be adopted by nurses for any one working shift. A 2012 literature review concluded that empirical evidence linking care delivery models and quality of nursing care remains sparse [11]. There is a close relationship between the nursing practice environment and models of care. Such models are primarily differentiated by who is allocated responsibility and control of nursing work and who is held accountable for what and to whom within the many national health system, organizational and available workforce constraints. One systematic literature review regarding nursing work environments concluded that

> *nurses as frontline patient care providers, are the foundation for patient safety and care quality. Promoting nurse empowerment, engagement, and interpersonal relationships at work is rudimental to achieve a healthy work environment and quality patient care [16].*

Every registered nurse will, at various times, be given the responsibility to organize the care for one or more patients requiring them to delegate various nursing tasks to others. Anecdotal evidence suggests that University prepared nurses tend not to be well prepared for this area of practice. This can have negative effects on patient safety and care. One research study explored how newly qualified nurses learn to organize, delegate and supervise care in hospital wards when working with and supervising healthcare assistants [17]. Five delegation styles were identified:

1. The do-it-all nurse who completes most of the work themselves.
2. The justifier who over-explains the reasons for decisions and is sometimes defensive.

3. The buddy who wants to be everybody's friend and avoids assuming authority.
4. The role model who hopes that others will copy their best practice but has no way of ensuring how.
5. The inspector who is acutely aware of their accountability and constantly checks the work of others.

The Magnet Hospital model, based on organizational factors influencing nurses' working conditions, emerged during the 1980s in the United States. This demonstrated the importance of managing these factors well to achieve patient satisfaction, improve outcomes as well as nurse staffing satisfaction, staff retention rates and lower levels of emotional exhaustion. Related empirical multisite research based evidence has demonstrated that the organization of nurses' work and workforce management are major determinants of patient and staff welfare [18–20]. One limitation identified is that possible Magnet effects on other healthcare professions appear not to have been investigated [21]. Internationally the Magnet framework needs to be viewed within the context of cultural differences and each country's health system, in particular funding arrangements. National funding policies are either a major facilitator or inhibitor of what is possible.

The Nursing profession has struggled over the years to attain professional status and value recognition. The transfer of hospital based schools of nursing to Universities was viewed as an important step toward gaining professional recognition. The nurse staffing workforce issues that resulted in the original "Magnet" research that made use of the then available Nursing Work Index (NWI), continue to be present globally. Effective leadership, autonomy and collaborative relationships are important attributes of professional nursing practice [19]. The Magnet hospitals consistently demonstrated three distinct core features of a professional nursing practice model. These were [18]

1. Professional autonomy over practice.
2. Nursing control over the practice environment.
3. Effective communication between nurses, physicians and administrators.

The Magnet nursing oriented organizational model has been adopted nationally by around 8% of US hospitals; it is being promoted via the Magnet Recognition Program® administered by the American Nurses Credentialing Centre (ANCC). The model has been tested for applicability in Europe [21], and in Australia where three hospitals are known to have achieved Magnet recognition [22]. The development of a 46 item measure of professional practice models made use of the original NWI and became known as the NWI-Revised [23] used for further research in New Zealand [24] and Korea [25]. A Queensland study made use of 30 of the original 31 practice environment scale (PES-NWI) and found this to have both construct validity and reliability [26]. Another examination of the validity of nursing practice environment measures using three different factor analytics models found that none of the models were a good fit with the data. It was concluded that not only is the nursing practice environment complex, it has been inadequately examined using factor-analytic approaches [27]. Despite these findings the use of the NWI-R practice environment scale was found to be growing across different clinical settings and countries [28].

Despite context and cultural variations, consistent findings are that health services need to provide organizational flexibility, professional autonomy, continuing education and a progressive career structure for nurses. The NWI-R: with Australian modifications tool [29] was found to be a valid and reliable measure of the Magnet features regarding the nursing practice environment. The Magnet items shown to be particularly related to job satisfaction and intention to leave were those which gauged:

- The existence of a nursing philosophy in the organization.
- Whether care was based on a nursing model.
- The extent to which the organization valued clinically competent nurses.
- The importance of continuity of care.
- The existence of active quality assurance processes.
- The existence of continuing education and preceptor programs within the organization [22].

Another initiative taken several years ago to address the critical need to improve the quality of patient care outcomes was the development of the certified Clinical Nurse Leader staffing category. This nursing category of staff refers to a master's prepared nurse, able to practice across the continuum of care within any healthcare setting to oversee care coordination, provide direct patient care in complex situations and put evidence based practice into action. Their objective is to ensure patients benefit from the latest innovations in care delivery, evaluate patient outcomes, assess cohort risk and have decision making authority to change care plans when necessary. They are leaders and active members of interdisciplinary healthcare teams [30].

This exploration of factors known to influence models of care, has highlighted the need for health service decision makers to appreciate the complexity and critical impact at the point of care of the many interactive organizational variables on nursing care models adopted and associated performance outcomes. Nursing practice features and theoretical foundational factors constitute additional influencing variables. These reflect the well documented nursing process and in particular its nursing care planning component.

## *The nursing process — Conceptual base for nursing practice*

Nurses make use of a problem based scientific method to assess, plan, implement and evaluate a patient's nursing care requirements as shown in Fig. 2. This systematic framework for processing pertinent patient information guides clinical decisions and the organization and delivery of all clinical nursing services. It is formally referred to as the nursing process; it is used globally in nursing education. In some countries two additional steps are added,

(1) nursing diagnosis is inserted in between the assess and plan steps and (2) revise the care based on the results of the evaluation.

**Fig. 2**
The nursing process.

Each step requires the nurse to think critically and to reason accurately [31]. The planning phase requires the nurse to consider the desired outcome(s) both long term and relative to each specific task to be undertaken. Nursing is recognized as a professional practice when a high percentage of care is delivered by registered nurses who are able to exercise considerable discretion and innovation when carrying out their work [32]. This is more likely to result in greater nurses' work satisfaction and improved patient outcomes. It is imperative for everyone providing patient centered care to have sufficient knowledge about their patient's health status before proceeding with any activity and to enable them to recognize significant changes.

Outcome evaluation can be anyone of four possibilities, desired outcome is met, partially met (stable, little change) or not met or the patient's condition has worsened. Such outcomes need to be considered relative to realistic timeframes before any changes to planned services are made. Outcomes are not exclusively due to nursing services provided. Patient deterioration may well be expected and may or may not be corrected, in which case the nurse needs to focus on maintaining patient comfort and obtain assistance as required. Collectively these steps form the foundation for nursing decision making, judgments made, the identification and implementation of all patient related nursing activities.

The USA's nurse education accreditation agencies also require the incorporation of the eight critical thinking steps to be incorporated in the nursing curriculum. These steps are clarity, accuracy, precision, relevance, depth, breadth, and logic. These critical thinking concepts need to be applied to every step of the nursing process. In addition many experienced nurses also make use of their clinical knowledge and intuition or they detect patterns when observing patients at any time, that guide their clinical decision making [33]. All judgments, decisions, observations and activities undertaken for and on behalf of patients need to be documented in each patient's medical/health care record.

The adoption of the nursing process enables nurses to think critically and systematically, taking into account the patient's response at all times. It guides the nurse to gather objective and subjective patient data from which a care plan is formulated and implemented. Nurses need to become informed about each patient's reason for admission and specific treatment plan, then assess their patients following admission to their area of practice and when taking over care from another nurse at the start of every shift within that context.

## Nursing care plans

Care plans need to completely and accurately document all treatment and nursing care requirements using concise statements. They are legal documents and need to be included in medical records. Nursing care plans tend to incorporate everything that is happening for any one patient, they provide direction for individualized care. The care plan serves as a guide to assign staff to care for patients, supports continuity of care, communication and the organization of actions provided by constantly changing nursing staff. Care plans can be multidisciplinary or shared, they can include patients in the development of the plan, which is especially useful to nurture self care, and for long term and/or complex conditions. Care plans can be standardized for commonly occurring conditions. In such instances nurses need only to document variations to the standard plan for individual patients as required, thus reducing the time required to document. Care plans reflect the model of care adopted for each individual patient.

All planned interventions need to be evidence based. This requires the integration of the best available research evidence with clinical knowledge, expertise and patient values. Best available evidence is made available in the literature and via published practice guidelines. Research that focused on multidisciplinary interventions for medical conditions found that despite this availability such new knowledge is not always incorporated in routine clinical practice [34]. Nursing care interventions were not independently identified but were included. This study's conclusion identified *a need for national agreement on clinical standards and better structuring of medical records to facilitate the delivery of more appropriate care* [21]. Many nurses do stick to old habits and well established nursing practices that are not necessarily evidence based. The adoption of systematic processes to advance the adoption of evidence based practice is required to improve outcomes [35].

The formal use of the nursing process and care planning is extremely valuable for teaching purposes, but neither are popular among nurses in the real world of practice. When time is short, nurses rely on memory, personal notes, nurse-patient interactions and observations and they make use of various other documentation to guide the provision of their services.

Planned interventions may be documented in a variety of different locations or communicated verbally. One literature review identified evidence of global and Australian differences in nursing care plans and documentation in use [36]. Only one in five nursing care plans captured the patient's need in their clinical audit.

Australian nurses tend to make use of a narrative style nursing documentation, using standardized care plans with minimal visibility of the nursing process. Nurses report that a lack of time is a major barrier to complete nursing plan documentation. One study of the application of the nursing process by nurses in one hospital reported several factors that impeded its use [37]. These were a lack of sufficient enthusiasm regarding its use, a shortage of nursing staff, no format for writing, a lack of follow up and monitoring, a lack of time, a lack of attention to its importance, deficient in clear instructions, inadequate education and a disbelieve in applying nursing care according to the nursing process. Anecdotal evidence supports these findings.

In addition and adding to the complexities encountered in acute care settings, neither the medical nor nursing profession tend to have adopted an agreed standard medical or nursing language to be used for this documentation. Their natural language, including the use of abbreviations, varies between healthcare facilities and information systems in use. Familiarity with operational and computer systems, their user interfaces, system navigation processes and documentation in use adds to the overall unit's performance efficiency and effectiveness.

It is at the shift handover that the nurse in charge of the shift allocates and distributes nursing work among those coming on duty. At that point all patient conditions and planned activities for the new shift need to be considered within the context of preferred models of care in the light of

- staffing policies such as ratios
- individual nurse's workloads based on patient acuity, including emotional demands vs nurse's capacity
- available staff numbers, mix and known individual nurses' strengths, weaknesses, scope of practice, any nurse's knowledge about or report with individual patients due to having cared for those patients in the past, and a desire for continuity of care
- patient locations and number of patients to be cared for by one nurse, to minimize nurses time and energy required
- service priorities and schedules

In practice these managerial decision making steps tend not to be well documented nor are these contextual factors incorporated in medical records. It is common to only see documentation about actions taken relative to individual patients. This makes it difficult for policy makers and health service executives to fully appreciate the complexity of managing nursing resources relative to continual changes in service demands within the many constraints imposed on those at the point of care.

This is the point in time where "the rubber hits the road" in terms of the collective staffing's ability to provide a high quality nursing service. This is where nurses may make use of different theories of nursing practice and apply this for personalized patient centered care. For example the participator action research (PAR) process based on Orem's self care model that also incorporated the relevant medication protocol, was found to improve the nursing care provided for acutely ill older patients [38]. This process is being promoted as a collaborative means to improve nursing care delivery for patients in varied clinical practice settings [39].

The efficacy of nursing resource management at every level within a healthcare organization, including staff allocations, rostering practices, this daily and per shift focus on nursing work distribution, determines not only the quality of care but also the nurses satisfaction with their chosen profession, value recognition and compensation. Individual nurses have little control over their own work environment as a rule. The factors shown to have the best outcomes were reported [18] as:

- flattening of management layers
- altering the makeup of the care delivery team
- cross training to provide multiskilled personnel
- decentralizing service to the unit or patient room level
- best available physical environment
- augmenting information technology, including the use of bedside data collection devices to enhance patient care and documentation

### Functional or task allocation

This model of care dominated from the post war era to the late 1970s. This was the result of bureaucratic structures, incentives to be efficient, to serve managerial interests and to best accommodate and make use of non-regulated nursing support staff. This situation was viewed as detrimental to nurses' professional image as Registered Nurses were given semi-professional status [40]. Nurses were also subject to physicians' professional dominance, a situation that continuous today in some locations [41].

Functional or task allocation is about delegating the responsibility for undertaking a set of specified tasks to each individual staff member on duty. For example this may include the responsibility for emptying urine bags or for someone else to take responsibility for checking all intravenous transfusions or to take responsibility for all wound care or medication administration. This meant that no one else ever observed or looked for the need to do so for any one patient. As a consequence no one nurse was able to observe the patient holistically to monitor and evaluate progress, detect errors or prevent adverse events; the quality of care was frequently compromised. Task allocation to non-nursing support staff should primarily consist of non-nursing indirect support activities but can include providing assistance with lifting heavy patients. Such staff hours need to be additional to required nurse staffing hours however measured.

The possibility of ineffective task delegation practices is likely to contribute to the duplication, omission or delays of any number or types of nursing activities. This model of task differentiation was the result of adopting what was known as "scientific management principles." This results in workers becoming expert in very specific tasks to the exclusion of all other tasks that collectively make up the total product. This suits the manufacturing industry but not the health industry where people need to be observed and cared for holistically.

In most instances task allocation occurred within teams, which improved the quality of care provided tasks were delegated by one registered nurse responsible for the total care of individual patients. This model of care is again becoming popular due to the skill mix of available staff when all concerned need to work within their own boundaries of practice. The result is fragmented care. The role of qualified nursing staff is critical to patient safety and wellbeing, this is especially so in acute hospital settings.

> *...I'm only dealing with hygiene and elimination and nutrition and I have no idea what's happening to them medically [42].*

An Enrolled Nurse's comment working in an acute setting

A Canadian study found that functional models of care made greater use of licensed practical nurses and assistive staff (orderlies) to deliver nursing services. This resulted in registered nurses perceiving their work environments as less supportive of their professional work [32]. Duffield examined the use of nursing time and argued for work designs that enable registered nurses to do what they do best and to delegate other activities [43]. This was examined in Victoria, Australia during the 1980s when a major industrial dispute resulted in the development of additional non-nursing positions as discussed in previous chapters. Many subsequent studies referred to by Duffield indicate poor nursing support service development and implementation in many hospitals leaving nurses to fill the gaps.

### Patient allocation

This very popular model of care [6] is where a registered nurse is allocated the responsibility to care holistically for one or more patients by meeting all patient care needs on a per shift or daily basis. It is difficult to allocate equitable workloads using this model of care. It minimizes supervision and restricts performance feedback. With this model of care there are less opportunities to support practice development for new graduates and advanced beginner nurses as nurses tend to work in isolation. Senior registered nurses are often overloaded resulting in shorter or missed meal and refreshment breaks. There needs to be recognition that one nurse does need the support of another to provide meal break relief and for nursing activities, such as checking the administration of restricted medications or blood transfusions, repositioning or turning or bathing some patients. As a result patient allocation is frequently made use of within a nursing care team.

## Primary nursing

This model of care has some similarities with the patient allocation model although in this instance a primary care nurse takes responsibility for the care of one patient for the duration of that patient's care episode. Its benefit is care continuity but it is difficult to implement in an acute care hospital setting where patients require 24/7 care, over variable lengths of stay and it requires an all Registered Nurse staff allocation. In essence this model requires nurses to be at the proficient or expert level of knowledge and skills. Depending on patient service demand it can generate an imbalance in work allocation and it restricts the ability to provide clinical supervision. This model of care of best suited to community care.

## Team nursing — A collaborative model of care

Team nursing is becoming more popular in response to a changing mix of available staff. Teams are groups of people who are mutually dependent on one another to achieve a common goal. Team nursing is a method of care delivery whereby a group of nurses and ancillary staff (under the guidance of a registered nurse team leader) are collectively responsible for all of the care requirements for a group of patients. Available skill mix, fiscal restraints and a drive for improving the quality of care are the general motivations to making changes to your model of care. The increasing pressures faced by nurses and nurse managers are challenging nurses to critically examine the way in which nursing care is both planned and delivered to ensure the best possible care is delivered to patients in the most efficient and cost effective way.

How should the effectiveness of team nursing be measured? What are the key characteristics of successful teams? Teams can be very effective but team failure can have major adverse consequences. Team performance is measured by its outcomes, irrespective of how these were achieved. It's important for all staff to learn how to best function as team members as well as team leaders. This requires good managerial, interpersonal and communication skills. Any team member can be expected to lead in situations where their specific knowledge and skills exceed those of other available members of the team. This also requires delegation skills. Introducing a team nursing model of care requires a well planned and structured approach incorporating clearly defined job descriptions and individual roles taking into account levels of expertise and scope of practice among staff members plus the time-consuming nature of various tasks to ensure a fair division of the workload across staff levels [42].

A framework of the key dimensions of what teamwork is was developed by Salas, Sims, and Burke [44]. This can be used as a learning guide for staff on how to manage, develop, and build effective teams. Five core components that promote team effectiveness were identified.

These are team leadership, mutual performance monitoring, backup behavior, adaptability and team orientation. Mutual performance monitoring is influenced by leadership, and team orientation, these factors also influence back-up behavior. Other overarching influencing factors are mutual trust and shared mental models. The latter influence adaptability. Collectively these determine team effectiveness. All of these factors are dependent upon closed loop communication which is a significant facilitator [44]. It's about maintaining mutual trust and respect as team members are interdependent upon each other.

One research study based on this framework of teams incorporating nursing assistive personnel (NAP), found that the Registered Nurse's understanding of the NAP role was not shared by NAP study participants. This plus the results of another study suggests that role clarification is an important component of any team building strategy with the goal of a shared mental model to improve teamwork in acute care settings [45, 46]. A similar research study [47] focused on measuring the shared mental model among nurses and nursing assistants using the TeamSTEPPS Teamwork Perceptions Questionnaire (T-TPQ) composed of five constructs, team structure, leadership, situation monitoring, mutual support and communication. Again significant differences between nursing assistants and staff nurses were found indicating a need to make significant efforts to build cohesive and highly functioning nursing teams to enable them to improve patient safety.

Pearson et al. [48] are of the view that team characteristics should include accountability, commitment, enthusiasm and motivation. Another study explored the influence of unit characteristics, staff characteristics and teamwork on job satisfaction using a previously developed 33 item questionnaire with a Likert-type scaling system [49]. This demonstrated that teamwork plus perceptions of staffing adequacy are important contributors to staff satisfaction. Their Nursing Team Survey (NTS) tool has been used by subsequent researchers [50]. Its use in Iceland demonstrated its solid theoretical base and international suitability [51]. One systematic literature review found that the team nursing model of care also resulted in significantly decreased incidence of medication errors and adverse intravenous outcomes, as well as lower pain scores among patients [52]. Yet another systematic review [53] concluded that it was not possible to determine whether organizing nursing work in a team nursing or total patient care model is more effective in terms of staff well being in acute care settings. These authors found that neither a team nursing or total patient care model had a significant influence on nurses' overall job satisfaction, stress levels or staff turnover. Their recommendation was to provide clear definitions of nursing roles.

A collaborative model of care is where an appropriately skilled nurse takes responsibility for leading a team. In 2010 the New South Wales (NSW) department of health identified the need for a more collaborative nursing model of care. Their ways of working (WOW) project [50] is about providing support to enhance patient care by

- using their nursing resources more efficiently
- improving communication between staff

- improving support for staff new to an area of practice
- improving capacity to effectively use available skills within nursing teams

This project has resulted in the development of clear guidelines, communication and reporting structures enabling teams to function efficiently. This project has resulted in a major set of educational tools to prepare nurses as leaders and members of a care team to deliver patient centered care encompassing the needs of patients, families and staff. This is based on practice development principles of inclusiveness, respect for each other, valuing individual contributions and connecting [50].

Effective teamwork requires access to role clarity tools and activities. These include

- Clear, informative position descriptions for all team members.
- A capability framework to assist team leaders to identify skills and competency among the team members.
- Documentation that explicitly addresses the key role differentiation between differently skilled team members.

Team leaders need to have sound delegation and supervisory skills. All nursing staff need to fully comprehend and embrace professional accountability and responsibility to enhance delegation and supervision outcomes. The Victorian Government's Department of Health has developed a nursing and midwifery delegation and supervision framework [54], consisting of required skills and knowledge pertaining to regulatory, organizational and individual elements with a list of associated recommended tools and activities plus supporting activities, to be used for professional development and guidance purposes. This came about in recognition of the complexity associated with the delegation and supervision of nursing and midwifery care and the need for these activities to be undertaken safely and effectively when working with a variable mix of skills among team members.

## Small team nursing

Coupled with workforce changes, nurses have had to adapt to an increase in patient acuity, increased patient turnover and advanced clinical care requirements, which have significantly increased nursing's scope of practice. Implementing a model of nursing care which provides patient care via a small team, may offer the most practicable approach to provide appropriate supervision and development of a diversely skilled workforce, mentoring new staff, providing comprehensive care for patients and managing care provision in a clinically and fiscally responsible way. Small team nursing is specifically designed to recognize the diversity in clinical competency and to maximize each individual nurse's contribution to patient care. The model facilitates nurses working in a cohesive and supportive environment where patient care can be maximized. The teams can be comprised of a variety of competencies and qualifications.

Small Team nursing offers the opportunity of mentoring and nurturing of nurses at all levels through the process of role modeling. This model of care has the potential to create skilled, passionate and productive nurses by increasing their expertise, confidence, organizational skills and work satisfaction. New graduates, advanced beginner practitioners and nurses new to the ward or speciality will learn and develop more quickly and effectively in a small team environment. Small Team nursing will facilitate a positive culture change in a workplace, improve retention rates of new graduates in the nursing industry, decrease staff turnover in challenging wards and significantly improve the effectiveness and productivity of a nursing service. Patient satisfaction rates will increase due to increased patient carer contact time with improved clinical outcomes being a consequence of this. Financial benefits can also be realized through improved productivity and efficiency resulting from the small team model. These benefits can be major "selling points" when discussing with staff the need to the Small Team model of care.

### The benefits of small team nursing

1. Increases patient observation with more than one person caring for a patient.
2. Facilitates a team approach to clinical decision making (not one nurse in isolation).
3. Facilitates a collaborative approach to patient care.
4. Provides patients with a dedicated team of nursing staff to comprehensively meet all of their care requirements.
5. Supports advanced practice roles and clinical leadership for RNs.
6. Decreases the incidents of missed care and facilitates patient safety due to the increased carer/patient contact with a team of carers caring for each patient.
7. Improves patient outcomes and improved patient satisfaction, through increased supervision and responsiveness to patient needs.
8. Facilitates the allocation of equitable workloads as work is shared across the team.
9. Facilitates nurses to have a more defined role within the nursing team, potentially maximizing the use of the skill sets within the team.
10. Allows diversification of the skill mix, e.g., a group of nursing staff who have varied levels of skills, qualifications and registration levels, including unregulated nursing support workers.
11. Facilitates improved ways of working, for example, medication checking and meal relief within the team to ensure that the patients have a nurse from their team on the ward at all times.
12. Facilitates open communication between team members.
13. Provides a supportive environment with enhanced opportunities for learning by role modeling for new graduate nurses and novice practitioners.
14. Facilitates improved support for staff new to an organization or area of practice.

15. Provides opportunities to improve competence and confidence for all levels of staff through role modeling and mentoring.
16. Enables more effective and efficient use of nursing resources compared to patient allocation.
17. More productive and cost effective than the patient allocation model of care.

**The Unit/Ward Nurse Manager** of the ward provides an ongoing vision for the model of care project. He/she requires a clear understanding of the small team model and clearly articulates the concept to all staff. The person in this role leads the model of care project, manages the change process and is responsible for developing an appropriate roster that will ensure that the team model is successful.

**The Registered Nurse in charge of the shift is** accountable for allocating the patients to each of the small teams. This should be done by using the measured acuity for each patient allocated so that each team has a fair and equitable workload.

**Registered Nurse/Team leader** is responsible for organizing his/her team on a shift by shift basis. To fulfill this role, the senior Registered Nurse must have strong communication skills, knowledge of his/her team's clinical competence levels, the ability to provide direction and delegation, confidence in decision making and a high level of clinical competence. The person in this role needs to function as a member of the team but also provide the overall management of the team on the shift. The Registered Nurse must know the condition and clinical care needs of all patients assigned to the team and ensure that the patients care needs are met through the appropriate delegation of care activities. The team leader is accountable for reporting to the Unit/Ward Nurse Manager.

**Role of the Enrolled/Licensed Nurse or equivalent** will vary between countries, professional bodies, organizations, wards/departments and individual nurses. It is important to have a clearly documented role and scope of practice guideline for Enrolled Nurses within the organization. Clinical skills mapping may be useful if the scope of enrolled nurse practice is not clearly defined. Completing a work analysis will assist in understanding the amount of enrolled nurse activity there is for a shift on an individual ward.

**Role of the unregulated nursing support worker** role and scope can vary significantly depending on the same variables identified above for the Enrolled Nurse. A nursing support worker needs to demonstrate respect for the Team Leader, support the team decision making process, work as a team member, clearly communicate any change in patient condition to the Team Leader and may undertake an activity if its performance:

o Will achieve the desired patient outcome.
o Is supported by the organization as documented in local policies/guidelines/protocols and it is defined as within the scope of practice of the person taking on this role.

o  Has been delegated by the Registered Nurse Team Leader who takes direct responsibility for the care given, is available to provide the required level of supervision, support and education if required and both the worker and the RN team leader are confident that the worker has the required skill level to complete the activity.

o  Is undertaken by a competent worker as evidenced by the successful completion of the necessary education, experience and skill acquisition.

### Leading the change

It is recommended that the Unit/Ward Nurse Manager leads the change process for implementing a model of care. In order for this to be successful, the Ward Nurse Manager must have a clear vision of how the change will be implemented, as well as how the model of care will work in practice within his/her ward/unit. It is essential that the Ward Nurse Manager can clearly articulate the vision to his/her nursing team and develop a plan of action to actively and effectively implement the change. The action plan must address the potential barriers to change and address any concerns the members of the nursing team may have. Issues and concerns which are dealt with up front and in the beginning stages of the project are most likely to be successfully managed and overcome. The Ward Nurse Manager needs to communicate clearly with his/her staff that change takes time and commitment and that the benefits of change are only realized if the change process is properly planned and implemented.

The team model may be undermined if there is:

*   Lack of leadership from the Unit/Ward Nurse Manager.
*   Unaddressed concerns among the nursing staff.
*   Unstable staffing patterns/under resourcing.
*   High levels of casual/bank/agency staff.
*   Lack of leadership from registered nurses within the teams.
*   Lack of clear communication and direction on the team model of nursing.
*   Insufficient time allocation for team planning and communication at the beginning and throughout the shift.
*   Unclear expectations of individual roles within the team.
*   Team sizes larger than 3 staff members, may result in the care becoming fragmented and task orientated.

Key activities required to implement a small team care model are as follows:

*   Develop a vision for the nursing service's desired model of care.
*   Develop objectives, strategies and a timeline for achieving each objective. E.g:
    o  Ensure the small team model of care is implemented in a collaborative and co-ordinated way.
    o  Establish a model of best practice for small team nursing.
    o  Address all staff concerns and respond to valuable feedback.

- Identify key staff for actioning each agreed strategy and monitor achievements.
- Undertake a ward/unit work analysis to aid in the determination of an appropriate skill mix as discussed in the previous chapter.
- Identify the roles, scopes of practice and responsibilities for all levels of staff within the team nursing model. Ensure these are clearly documented and in alignment with organizational policy.
- Prepare Registered Nurses for skills related to delegation, supervision and mentoring. This includes providing a supervision and delegation framework if not already established.
- Ensure there are policies and procedures in place which support the team model of nursing care. The policies should relate to supervision, delegation and roles and responsibilities.
- Ensure that scope of practice definitions are clearly documented and are in alignment with organizational policy.
- Map current scopes of practice for all workforce groups and identify new roles and responsibilities within the team, undertake clinical skills mapping if required, this can be a component of a work analysis study.
- Implement a competency assessment framework for nursing support workers if not previously defined.
- Develop comprehensive shift routines for nursing teams for each shift of the day.
- Ensure rostering practices and patterns support the small team model of care.
- Simultaneously roll out the small team nursing model, new skill mix and roster patterns.

An implementation process overview is provided in Box 1.

---

**BOX 1  Small team implementation process overview**

Vision

- Small team nursing clearly articulates the benefits and outcomes to be achieved.
  E.g., By implementing and maintaining a small team model of care, the nursing service will achieve optimal patient outcomes, provide fair and equitable workloads, increase staff competency levels, accomplish high levels of staff satisfaction and improve the productivity and efficiency of the nursing service.
- The Unit/Ward Nurse Manager understands, articulates and shares this vision with the unit's nursing staff.

Communication

- Involve all levels of staff and union delegates in the planning phase and implementation process.
- Adopt a clear communication strategy understood by all staff. Facilitate cohesion by keeping all staff informed and involved throughout the change process.
- Maintaining ongoing communication on a shift by shift basis is integral to success.
- Hold regular staff 'huddles' at least two hourly during the shift.
- Allocate time at the beginning of each shift for staff handover and team planning.

---

*Continued*

---

**BOX 1** Small team implementation process overview—cont'd

---

Role definition

- Clearly define team member roles and ensure all team members are clear about their role.
- Appoint a Registered Nurse as team leader. Teams may constitute different levels of nurses and/or support workers.
- Coach and mentor staff into their new roles.
- Organize and conduct change workshops to be attended by all staff.

Decision making

- Discuss and establish methods for effective team decision making.
- Team decision making decreases staff isolation, supports junior staff and maximizes patient outcomes.

Participation

- Fully involve everyone in the change process.
- Fully involve all team members in the delivery of patient care.
- Establish and maintain a positive working environment for change.

Manage barriers to change and conflict

- Encourage all staff to express concerns.
- Manage all concerns and negative feelings early during the change process.
- Establish trust and open communication by encouraging all staff to participate in the change process.
- Manage team cultural issues that may be a barrier to change.

*Source: From the TrendCare® Small Team Nursing Model.*

---

## The shift routine example

It is the Unit/Ward Nurse Manager or senior Registered Nurse's role to ensure all changes in patient treatments and plan of care are communicated to the appropriate team if they attend medical staff rounds or are advised of these changes. It is the responsibility of the person in charge to ensure that every team has a fair and equitable workload, are functioning in an effective and harmonious way, and that any concerns with team cohesion are addressed in a timely manner. The Unit/Ward Nurse Manager assumes overall responsibility for the teams on the morning shift and will regularly check in with each team independent of the second hourly "huddles." The senior Registered Nurse assumes overall responsibility on the evening and night shift.

Each Registered Nurse leader coordinates the team and with input from each member identifies the team goals for the shift. This includes planning the key activities such as patients returning from the operating theater, expected admissions and planned discharges. The team leader needs to:

o   Assign/delegate specific responsibilities to team members dependant on the clinical competence and qualification level of each team member.
o   Ensure individualized patient care is prioritized, e.g., feeds, daily hygiene, treatments, etc.

o   Delegate work activities at the beginning of the shift.

o   Ensure that the delegated workload is fair and reasonable for each team member.

o   Regularly communicate with each team member, conduct a "huddle" at least second hourly and re-evaluate the workload of each team member and re-allocate work activities if required as the shift progresses.

o   Provide learning opportunities to team member when delegating work. This can be done by allocating a complex task to two team members for the specific purpose of enabling one to teach/supervise the other.

o   Allocate/negotiate meal and refreshment break times.

o   Report to the Unit/Ward Manager or the Registered Nurse in charge of the shift if the workload becomes too great for the team.

### Evaluate success of team nursing implementation

Team effectiveness in terms of patient safety and quality of care delivered is based to a large extent on team communication enabling staff, irrespective of role, with the most experience to be heard. Team members need to monitor each other's performance and help out in the patient's interest where indicated. It's important for team members to understand each member's role, give and receive feedback to each other. It needs to be noted that teams with a culture of openness and reporting may record a greater number of errors or near misses, than teams where such incidents are hidden; this tends to occur in cases of authoritarian leadership [55]. New graduate nurses were found to have benefited greatly from changing to the small team nursing model of care [56].

Evaluation is an integral step in implementation of a new model of care. Measuring the success of the implementation will help identify areas which may need further development. Having a clear vision statement and objectives in the beginning will facilitate evaluation of the change as success can be measured against the objectives. The before and after measures of success are:

*   Nurse satisfaction levels using staff absenteeism and turnover statistics as well as surveys reporting on completeness of care, workload satisfaction and quality of care delivered, by shift.
*   Patient quality of care indicators and outcomes measures based on clinical pathways and care plan requirements variance to pre-set goals/outcomes for surgical procedures or medical conditions as well as results of patient surveys.
*   Nursing productivity and efficiency measures based on standard reports and graphs.

## Inter and multidisciplinary models of care

In addition to the nursing models of care described so far, options also exist to implement multidisciplinary models of care of which nursing care models are a component. Such teams may refer to their care models as blended, co-ordinated or collaborative and/or refer to patient centered integrative case management. Such models of care are viewed as effective, efficient, and cost-effective care strategies [57]. Making a well coordinated team of health professionals available to provide patient care is deemed to be beneficial as they collectively provide greater insights and a wider range of knowledge and skills when compared with any individual carer. A well co-ordinated multidisciplinary team approach based on accountability, commitment, enthusiasm, motivation and social support by team members to healthcare delivery, was found to improve outcomes [48].

Clinical Nurse Leaders/Clinical Nurse Specialists/Consultants frequently occupy positions where they are expected to provide team leadership, management and collaboration with other health professional team members. They assume accountability, assimilate and apply evidence based information to design, implement, and evaluate patient care processes and models of care delivery [1]. The clinical nurse leader integrated care delivery/practice evidence based model was identified as consisting of four core team practices:

1.  Facilitate effective ongoing communication, including the creation of multimodal communication tools and rounding structures.
2.  Strengthen intra and inter professional relationships by establishing a network of multiprofessional microsystem partners.
3.  Create and sustain teams by bringing people from all disciplines and departments affected by care processes to work together and improve them.
4.  Support staff engagement via an ongoing supportive presence, the provision of resources based on in-the-moment needs, and by empowering staff to perform to their full scope of practice and identify and create solutions to meet patient care needs [1].

This integrated care model is patient centered and has many similarities to the case management model of care. No clear and consistent definition of case management appears to exist. It is defined by the Case Management Society of America as:

> *a collaborative process of assessment, planning, facilitation, care co-ordination, evaluation and advocacy for options and services to meet an individual's and family's comprehensive health needs through communication and available resources to promote quality and cost-effective outcomes [58].*

Multidisciplinary case management teams are especially useful for the delivery of service to those with multiple co-morbidities and chronic care needs. It enables the provision of continuous, well transitional and patient centered care while simultaneously reducing hospital admissions. It assists transitions from hospital to community services. Healthcare providers, especially nurse managers and nursing leaders, need to be made aware of facilitators of and barriers to case management for populations with chronic illnesses [58]. These researchers concluded that to successfully implement case management, information about this model of care needs to be made available to patients. In addition case managers need to be allocated sufficient time and access to relevant resources to enable them to design interventions that best support these patients and their caregivers.

A recent scoping review identified a lack of theoretical frameworks, lack of standard guidelines in case management practice, lack of precise process measures, and limited reports of the explicit role of nurse case managers and role confusion by nurses [57]. Case management can be provided within hospitals although this model of care is more commonly found for home care. Case managers' workload and workload measurement tools are also needed. These need to account for the unplanned and unpredictable nature of case management work and assist with the distribution of more equitable caseloads among case managers and home care teams' [59]. Case managers need to work with a well developed care plan that incorporates activities to be provided by each multidisciplinary team member throughout a patient's health journey.

Integrated patient centered care plans are especially useful when developed as a standard using evidence based practice guidelines suited to specific high volume or costly clinical conditions as identified using any medical diagnosis. These refer to disease specific integrated care models. Such pathways, however titled, detail all essential steps to be undertaken, together with service roles including nurses, a timeline and desirable sequence, toward achieving the desired patient outcomes in one multidisciplinary record [60]. Many of these standard care plans are published and made available by local authorities or professional organizations.

Evidence based standard pathways or care plans need to be contextualized to suit individual patients and the care delivery environment. The plan can include both in-patient and out-patient or home care points of care delivery for any specific episode of care. It's aims are to:

- Facilitate the introduction of clinical guidelines, systematic and continuing audit into clinical practice.
- Improve multidisciplinary communication and care planning with all relevant care providers.
- Manage and improve the quality of care delivery.
- Improve practice consistency, clinician-patient communication and patient satisfaction.
- Support research and development opportunities.

## Organizational models of care influencing patient outcomes

The World Health Organization (WHO) has adopted a global strategy on people-centered and integrated health services [41]. Europe has developed a framework for action for integrated health service delivery [61]. This is viewed by many as a possible solution to the growing demand for improved patient experiences and health outcomes for those with multiple co-morbidities requiring long term care. This requires governing, financing and support service integration at various levels in addition to clinical practice integration. Despite this apparent large scale adoption of integrated care there is no consensus regarding its definition due to the many and varied professional perspectives [62]. The WHO has published a working document that provides an overview of integrated care models based on a scoping literature review [63]. These types of care models, along with interdependencies between primary care, social services and hospitals, need to be considered when developing digital transformation strategies as their success is very much dependent upon effective digital communication and system connectivity.

Many jurisdictions provide various leadership initiatives that aim to improve outcomes of care. In doing so a variety of inter and multidisciplinary models of care are examined and being promoted along with various resources to support care planning activities. One example is the New South Wales Agency for Clinical Innovation which, based on its strategic plan, is about connecting people with ideas to make a difference and to recognize valued partners working on how best to address shared goals. This organization has numerous multidisciplinary networks, a process flow chart for developing a model of care and published models of care pertaining to a large number of clinical specialties [64].

Models of care and patient outcomes are also influenced by organizational policies and procedures such as those pertaining to admission and discharges and the provision of weekend or after hours services. For example it is well known that acute hospitals employ different staff and larger numbers of staff to provide clinical services from Monday to Friday when compared with weekend staff. This may adversely influence discharge preparation for patients and their subsequent quality of life when discharged at weekends [65].

## Success factors

Given the available skill mix it is most appropriate to adopt the small nursing team model of care on a per shift basis in acute care and residential care healthcare facilities. Effective team leadership is critical in managing risk. Interpersonal issues and conflict between team members are known to be detrimental to team effectiveness. Team leaders need to be skilled people managers able to manage and resolve conflict as and when this occurs using interaction strategies and/or skills. Sources of conflict may be due to individual people characteristics, interpersonal factors or organizational factors. Conflict may be, due to poor communication, conflicting roles, horizontal among equals, intergenerational, cultural, disciplinary or ethical perspectives, different levels of experience or status relationships [66]. Collaboration strategies based on mutual respect within a culture welcoming diversity, need to be adopted to enable every team to function effectively. Efficient, accurate and timely communication between team members can be achieved by adopting interdisciplinary patient rounds taking place at the bedside. Not only does this benefit patients but it also improves nurses' job satisfaction [67].

In response to a recognition that many health systems are fragmented, a 2010 WHO report [68] identified the need for interdisciplinary collaborative practice to move health systems to a position of enabling them to meet unmet health needs. This requires interprofessional healthcare teams to be able to recognize and optimize the skills of their team members, share case management and improve the delivery of health care services and health outcomes. This requires the health and education systems to work closely together.

A recent analysis of 20 years of qualitative international research on interprofessional teamwork in hospital settings, identified that interprofessional teamwork was largely absent in acute care. This was found to be influenced by systems perpetuating power imbalances, organizational practices that interfered with interprofessional interactions, representations of teamwork and leadership [69]. In particular healthcare, hospital and insurance systems were found to consolidate medical power and institutionalize their leadership of healthcare teams. This review also points to differentials in status, autonomy and the value associated with certain types of knowledge.

Leaders at all levels were found to play a critical role in facilitating teamwork. These reviewers concluded that existing structures are currently preventing successful in-patient practices in hospitals [69]. This requires politicians, high level decision makers, health service executives, educators and many health professionals to change the existing culture of medicine and healthcare, starting with effective interprofessional education and collaborative practice [68, 70]. Meanwhile from a nursing perspective the best model of patient centered care is to adopt the small team approach led by the most senior available registered nurse. Such models need to be built on whatever treatment care plan is chosen and can be incorporated within inter and multidisciplinary care models.

## *References*

[1] Bender M, Spiva L, Su W, Hites L. Organising nursing practice into care models that catalyse quality: a clinical nurse leader case study. J Nurs Manag 2018;26(6):653–62.
[2] Brennan PF. A discipline by any other name. J Am Med Inform Assoc 2002;9(3):306–7.
[3] Fawcett J. Thoughts about nursing conceptual models and the "medical model" Nurs Sci Q 2016;30(1):77–80.
[4] Hudspeth RS, Vogt M, Wysocki K, Pittman O, Smith S, Cooke C, et al. Evaluating models of healthcare delivery using the model of care evaluation tool (MCET). J Am Assoc Nurse Pract 2016;28(8):453–9.
[5] Rhéaume A, Dionne S, Gaudet D, Allain M, Belliveau E, Boudreau L, et al. The changing boundaries of nursing: a qualitative study of the transition to a new nursing care delivery model. J Clin Nurs 2015;24 (17–18):2529–37.
[6] Duffield C, Roche M, Diers D, Catling-Paull C, Blay N. Staffing, skill mix and the model of care. J Clin Nurs 2010;19(15–16):2242–51.
[7] Tiedeman ME, Lookinland S. Traditional models of care delivery: what have we learned? J Nurs Adm 2004;34 (6):291–7.
[8] Wells J, Manuel M, Cunning G. Changing the model of care delivery: nurses' perceptions of job satisfaction and care effectiveness. J Nurs Manag 2011;19(6):777–85.
[9] Riverin BD, Li P, Naimi AI, Strumpf E. Team-based versus traditional primary care models and short-term outcomes after hospital discharge. Can Med Assoc J 2017;189(16):E585.
[10] McCrae N. Whither nursing models? The value of nursing theory in the context of evidence-based practice and multidisciplinary health care. J Adv Nurs 2012;68(1):222–9.
[11] Harris A, McGillis-Hall L. Evidence to inform staff mix decision-making: a focused literature review. Canadian Nurses Association; 2012. [cited 7 November 2018]. Available from: https://www.cna-aiic.ca/~/media/cna/page-content/pdf-en/staff_mix_literature_review_e.pdf.

[12] Meyer RM, O'Brien-Pallas LL. Nursing services delivery theory: an open system approach. J Adv Nurs 2010;66(12):2828–38.

[13] O'Brien-Pallas L, Meyer RM, Hayes LJ, Wang S. The patient care delivery model—an open system framework: conceptualisation, literature review and analytical strategy. J Clin Nurs 2011;20(11–12):1640–50.

[14] Naisbitt J. Megatrends. New York: Warner Communications; 1982.

[15] Leal JAL, Melo CMM. The nurses' work process in different countries: an integrative review. Rev Bras Enferm 2018;71(2):413–23.

[16] Wei H, Sewell KA, Woody G, Rose MA. The state of the science of nurse work environments in the United States: a systematic review. Int J Nurs Sci 2018;5(3):287–300.

[17] Magnusson C, Allan H, Horton K, Johnson M, Evans K, Ball E. An analysis of delegation styles among newly qualified nurses. Nurs Stand 2017;31(25):46–53.

[18] Havens DS, Aiken LH. Shaping systems to promote desired outcomes. The magnet hospital model. J Nurs Adm 1999;29(2):14–20.

[19] Scott JG, Sochalski J, Aiken L. Review of magnet hospital research: findings and implications for professional nursing practice. J Nurs Adm 1999;29(1):9–19.

[20] Aiken LH, Clarke SP, Sloane DM, Sochalski JA, Busse R, Clarke H, et al. Nurses' reports on hospital care in five countries. Health Aff (Millwood) 2001;20(3):43–53.

[21] Heitmann A, Čišić RS, Meyenburg-Altwarg I. From magnet-hospital to the hospital of the future. Nurs Health 2013;1(4):78–87.

[22] Joyce-McCoach JT. Enabling the transferability of the magnet hospital concept to an Australian context. http://ro.uow.edu.au/theses/3932 Wollongong; 2010.

[23] Aiken LH, Patrician PA. Measuring organizational traits of hospitals: the revised nursing work index. Nurs Res 2000;49(3):146–53.

[24] Finlayson M, Aiken L, Nakarada-Kordic I. New Zealand nurses' reports on hospital care: an international comparison. Nurs Prax NZ 2007;23(1):17–28.

[25] Kim C-W, Lee S-Y, Kang J-H, Park B-H, Park S-C, Park H-K, et al. Application of revised nursing work index to hospital nurses of South Korea. Asian Nurs Res 2013;7(3):128–35.

[26] Parker D, Tuckett A, Eley R, Hegney D. Construct validity and reliability of the practice environment scale of the nursing work index for Queensland nurses. Int J Nurs Pract 2010;16(4):352–8.

[27] Cummings GG, Hayduk L, Estabrooks CA. Is the nursing work index measuring up? Moving beyond estimating reliability to testing validity. Nurs Res 2006;55(2):82–93.

[28] Warshawsky NE, Havens DS. Global use of the practice environment scale of the nursing work index. Nurs Res 2011;60(1):17–31.

[29] Joyce J, Crookes P. Developing a tool to measure 'magnetism' in Australian nursing environments. Collegian 2007;25(1):17–23.

[30] AACN. Clinical nurse leader initiative American Association of Colleges of Nursing; 2004 [cited 24 November 2018]. Available from: https://www.aacnnursing.org/CNL.

[31] Huckabay LM. Clinical reasoned judgment and the nursing process. Nurs Forum 2009;44(2):72–8.

[32] Dubois C-A, D'Amour D, Tchouaket E, Rivard M, Clarke S, Blais R. A taxonomy of nursing care organization models in hospitals. BMC Health Serv Res 2012;12:286.

[33] Melin-Johansson C, Palmqvist R, Rönnberg L. Clinical intuition in the nursing process and decision-making—a mixed-studies review. J Clin Nurs 2017;26(23–24):3936–49.

[34] Runciman WB, Hunt TD, Hannaford NA, Hibbert PD, Westbrook JI, Coiera EW, et al. CareTrack: assessing the appropriateness of health care delivery in Australia. Med J Aust 2012;197(2):100–5.

[35] Hanrahan K, Wagner M, Matthews G, Stewart S, Dawson C, Greiner J, et al. Sacred cow gone to pasture: a systematic evaluation and integration of evidence-based practice. Worldviews Evid Based Nurs 2015;12(1):3–11.

[36] Johnson L, Edward K-L, Giandinoto J-A. A systematic literature review of accuracy in nursing care plans and using standardised nursing language. Collegian 2018;25(3):355–61.

[37] Bakken S, Hyun S, Friedman C, Johnson SB. ISO reference terminology models for nursing: applicability for natural language processing of nursing narratives. Int J Med Inform 2005;74(7):615–22.

[38] Glasson J, Chang E, Chenoweth L, Hancock K, Hall T, Hill-Murray F, et al. Evaluation of a model of nursing care for older patients using participatory action research in an acute medical ward. J Clin Nurs 2006;15 (5):588–98.

[39] Glasson JB, Chang EM, Bidewell JW. The value of participatory action research in clinical nursing practice. Int J Nurs Pract 2008;14(1):34–9.

[40] Brannon RL. Professionalisation and work intensification: nursing in the cost containment era. Work Occup 1994;21(2):157–78.

[41] WHO. WHO global strategy on people-centred and integrated health services, interim report. World Health Organisation; 2015. [cited 24 November 2018]. Available from: http://apps.who.int/iris/bitstream/handle/ 10665/155002/WHO_HIS_SDS_2015.6_eng.pdf?sequence=1.

[42] O'Connell B, Duke M, Bennett P, Crawford S, Korfiatis V. The trials and tribulations of team-nursing. Collegian 2006;13(3):11–7.

[43] Duffield C, Gardner G, Catling-Paull C. Nursing work and the use of nursing time. J Clin Nurs 2008;17 (24):3269–74.

[44] Salas E, Sims DE, Burke CS. Is there a "big five" in teamwork? Small Group Res 2005;36(5):555–99.

[45] Bellury L, Hodges H, Camp A, Aduddell K. Teamwork in acute care: perceptions of essential but unheard assistive personnel and the counterpoint of perceptions of registered nurses. Res Nurs Health 2016;39 (5):337–46.

[46] Dahlke S, Baumbusch J. Nursing teams caring for hospitalised older adults. J Clin Nurs 2015;24 (21–22):3177–85.

[47] Enzinger IH. Teamwork perceptions of nurses and nursing assistants in a community hospital [doctoral]. Walden University; 2017.

[48] Pearson A, Porritt KA, Doran D, Vincent L, Craig D, Tucker D, et al. A comprehensive systematic review of evidence on the structure, process, characteristics and composition of a nursing team that fosters a healthy work environment. Int J Evid Based Healthc 2006;4(2):118–59.

[49] Kalisch BJ, Lee H, Rochman M. Nursing staff teamwork and job satisfaction. J Nurs Manag 2010;18 (8):938–47.

[50] Thoms D. Ways of working nursing Sydney. NSW Health; 2011. [cited 17 November 2018]. Available from: https://www.health.nsw.gov.au/nursing/projects/Publications/ways-of-working.pdf.

[51] Bragadóttir H, Kalisch BJ, Smáradóttir SB, Jónsdóttir HH. The psychometric testing of the nursing teamwork survey in Iceland. Int J Nurs Pract 2016;22(3):267–74.

[52] Fernandez R, Johnson M, Thuy Tran D, Miranda C. Models of care in nursing: a systematic review. Int J Evid Based Healthc 2012;10:324–37.

[53] King A, Long L, Lisy K. Effectiveness of team nursing compared with total patient care on staff wellbeing when organizing nursing work in acute care wards: a systematic review. JBI Database System Rev Implement Rep 2015;13(11):128–68.

[54] McMillan A. Delegation and supervision guidelines for Victorian nurses and midwives. Melbourne: State Government Victoria, Department of Health; 2014. [cited 6 December 2018]. Available from: http://www. health.vic.gov.au/__data/assets/pdf_file/0011/887654/Delegation-Guide-Nurses-Midwives.pdf.

[55] Firth-Cozens J. Cultures for improving patient safety through learning: the role of teamwork. Qual Health Care 2001;10(Suppl. 2):ii26.

[56] Fairbrother G, Jones A, Rivas K. Changing model of nursing care from individual patient allocation to team nursing in the acute inpatient environment. Contemp Nurse 2010;35(2):202–20.

[57] Joo JY, Huber DL. Scoping review of nursing case management in the United States. Clin Nurs Res 2017;27 (8):1002–16.

[58] Joo JY, Liu MF. Experiences of case management with chronic illnesses: a qualitative systematic review. Int Nurs Rev 2018;65(1):102–13.

[59] Fraser KD, Garland Baird L, Labonte S, O'Rourke H, Punjani NS. Case manager work and workload: uncovering a wicked problem—a secondary analysis using interpretive description. Home Health Care Manag Pract 2018. 1084822318803099.

[60] Campbell H, Hotchkiss R, Bradshaw N, Porteous M. Integrated care pathways. BMJ 1998;316(7125):133–7.

[61] Europe W. Strengthening people-centred health systems in the WHO European region: framework for action on integrated health services delivery. World Health Organisation, Regional Office for Europe; 2016. [cited 24 November 2018]. Available from: http://www.euro.who.int/__data/assets/pdf_file/0004/315787/66wd15e_ FFA_IHSD_160535.pdf?ua=1.

[62] Armitage GD, Suter E, Oelke ND, Adair CE. Health systems integration: state of the evidence. Int J Integr Care 2009;9:e82-e.

[63] WHO. Integrated care models: an overview, [cited 24 November 2018]. Available from: http://www.euro.who. int/__data/assets/pdf_file/0005/322475/Integrated-care-models-overview.pdf; 2016.

[64] ACI. Agency for Clinical Innovation. Models of care: NSW Government, [cited 24 November 2018]. Available from: https://www.aci.health.nsw.gov.au/resources/models-of-care.

[65] Kilkenny MF, Lannin NA, Levi C, Faux SG, Dewey HM, Grimley R, et al. Weekend hospital discharge is associated with suboptimal care and outcomes: an observational Australian stroke clinical registry study. Int J Stroke 2018. 1747493018806165.

[66] Brinkert R. A literature review of conflict communication causes, costs, benefits and interventions in nursing. J Nurs Manag 2010;18(2):145–56.

[67] Gausvik C, Lautar A, Miller L, Pallerla H, Schlaudecker J. Structured nursing communication on interdisciplinary acute care teams improves perceptions of safety, efficiency, understanding of care plan and teamwork as well as job satisfaction. J Multidiscip Healthc 2015;8:33–7.

[68] Gilbert JHV, Yan J. Framework for action on interprofessional education & collaborative practice. Whorld Health Organisation; 2010. [cited 23 November 2018]. Available from: http://apps.who.int/iris/bitstream/ handle/10665/70185/?sequence=1.

[69] Petit dit Dariel O, Cristofalo P. A meta-ethnographic review of interprofessional teamwork in hospitals: what it is and why it doesn't happen more often. J Health Serv Res Policy 2018;23(4):272–9.

[70] Gilbert JHV, Yan J, Hoffman SJ. A WHO report: framework .for action on interprofessional education and collaborative practice. J Allied Health 2010;39(3):196–7.

# CHAPTER 7

# Staffing resource allocation, budgets and management

## Using demand side organizational nursing and midwifery workforce planning methods

Previous chapters have explored service demand and various nursing and midwifery staffing models and service delivery aspects that have collectively indicated the complexity associated with workforce management whilst meeting constantly changing service demands. The focus of previous chapters was very much on operational impacts. This chapter's focus is a top down approach making use of demand-side organizational nursing and midwifery workforce planning methods. Budgets determine funds made available for the provision of nursing and midwifery services within any healthcare organization. How are such budget's prepared? What information is required and used to calculate the total nursing and midwifery staffing requirements? What methods are in use to distribute the available nurse/midwife staffing establishment across the many different departments providing nursing and midwifery services? How are departmental rosters designed and developed?

Healthcare facilities have organizational structures based on the mix of health services provided. Each department has its own service delivery hours across days of the week and hours within any 24 hr period. Service demand varies between the types of health services provided, thus departmental staff allocation and rostering patterns need to be developed in a manner that best matches service demands at any point in time. This requires sufficient and appropriate historical information from which patterns of demand can be identified.

## Professional and government nurse staffing initiatives

Here we provide several case studies representing different time periods and countries as examples. No doubt there are many other similar examples. These demonstrate the very significant influence on nurse staffing numbers and budgets over which the nursing profession has limited control or influence as a rule, unless industrial action is taken by them.

Measuring Capacity To Care Using Nursing Data. https://doi.org/10.1016/B978-0-12-816977-3.00007-1

The Victorian Government department initially had legislation requiring a central authority to allocate nurse staffing establishments for every health facility it funded. This task was undertaken by senior nursing advisors including this author. The introduction of a 38 hr week (a 2 hr reduction for every fulltime staff member employed) plus a transfer of nursing education from hospital programs to Universities had resulted in nursing shortages (student nurses were no longer part of the workforce). The need to resolve many subsequent nursing workload issues led to the development of the PAIS patient acuity system as described previously. The nurses' union (ANMF) had resolved that PAIS be introduced and used by all Victorian hospitals by July 1985 [1]. The 2015 Victorian legislation [2] mandates the adoption of nurse:patient ratios as discussed in a previous chapter. Similar actions have been taken elsewhere following this landmark initiative [3] as previously identified, for example in California [4, 5].

In the absence of more appropriate data, bed use and lengths of stay statistics have been used by Government's, Boards and Funders as the basis from which to calculate nurse staffing establishments and funding arrangements for healthcare facilities. This has in the past, and continues to represent a ballpark figure frequently leading to industrial disputes over nursing workload. A Ministerial Enquiry into the staffing, organization and administrative structures of nursing in Victoria had been established late 1983. This committee's final report included instructions of how to calculate safe nurse staffing requirements for every hospital department, guidelines for the use of nursing support staff, rostering, continuing professional development, student supervision, how to reduce staff turnover and absenteeism and prepare to address the impact of technological change [1]. A similar but far wider public inquiry was undertaken by the UK's National Health Service Foundation's Trust. It's report was presented to Parliament in 2014 [6].

The NHS Scotland has, in 2013, mandated the use of 12 nursing and midwifery workload and workforce planning tools covering 95% community, mental health, theaters, emergency departments, neonatal, maternity, specialist nurses and children's services [7, 8]. It's adult inpatient tool determines nursing staffing levels using an acuity-dependency approach based on a staff to bed ratio and average bed occupancy level. This includes a 22.5% predictable absence allowance. It's 2015 toolkit edition includes provision for 'releasing time to care', this is about increasing direct patient-facing time in order to better meet patient need. This toolkit noted that there was no national system for gathering nursing workforce information at ward/unit level limiting the scope for comparative studies to be undertaken. The use of these tools has resulted in an increased use of agency and bank staff to maintain quality and ensure safe service provision [9, 10]. This highlights the need to improve the development of organizational staffing establishments and incorporate the likely need to employ casual and agency staff within allocated budgets to meet periods of high service demand.

The 2018 Health and Care (Staffing) Scotland Bill has placed this existing but enhanced workforce planning method on a statutory footing [11]. The mandated use of staffing tools are intended to establish permanent staffing requirements. This Bill makes it more explicit that the analysis resulting from the application of these tools, is used to ensure better decision making in relation to staffing across health settings. Additional requirements included are that:

- Staff are engaged in the process and informed of outcomes
- Transparent risk-based prioritization and decision making is carried out
- There is the provision of senior clinical professional advice (in addition to use of the Professional Judgment tool).

This Bill did not clarify how compliance would be monitored or what the impact of non-compliance would be, nor were additional costs regarding review and improvements mentioned. As a consequence of this Government mandated requirement, there has been an increased use of agency and casual 'bank' staff to maintain quality and ensure safe services. In addition the Scottish Government is now looking for a digital solution to support observation studies and maintain the validity of existing tools as models of care evolve.

The UK's Royal College of Nursing (RCN) has been actively reporting on and campaigning for safe nursing and midwifery staffing levels following the growing nursing shortages across the UK. The RCN has developed a set of principles for the development of high level objectives. They are arguing the need for legislation, statutory instruments and guidance along with sufficient funding in every country in the UK. It's first objective is a governance framework that details responsibility and accountability for ensuring an adequate supply of registered nurses and nursing support staff is available throughout the health and social care system to meet the needs of the population [12].

The UK's Health Foundation has undertaken research and published a good overview of all the nurse staffing policy related activities undertaken within or for the NHS over recent years in a quest to look for evidence based tools for establishing the staffing needs of each service [13]. This report also includes an overview of approaches adopted in the USA and Australia. It concluded that the current evidence base on nurse staffing and outcomes, although improving, is weak, narrow and incomplete (p. 14).

The NHS England does not favor a mandatory legislative approach or support the use of standardized systems being advocated by and in other UK countries. The NHS England's approach is focused on local flexibility. This means that local management capacity needs to include knowledge and skills to analyze, determine, implement and monitor what is 'safe' and to respond appropriately to staffing issues in a timely manner. Local management also needs to network with local teams to identify effective safe staffing tools and implement systems that enable the effective and consistent use of local data systems [14].

*The available evidence from NHS in England is too fragmented to provide a solid platform on which to base any universal approach to determining nurse staffing, or even to underpin detailed guidelines on best practice in the use of available tools in most care environments [13] p. 9*

The UK's National Institute for Health and Care Excellence (NICE) established a Safe Staffing Advisory Committee (SSAC) in 2014 that has published guidelines for midwifery and adult in-patient wards safe staffing and developed a safe staffing App which allows managers and staff to see an accurate, live staffing position from ward to board [15, 16]. One of the expert papers presented to this SSAC detailed the New Zealand's experience as at 2014 which began in 2005 following a difficult round of industrial negotiations between the nurses union and their 21 District Health Boards [17]. NICE endorsed the TrendCare toolkit on safe staffing for nursing in adult inpatient wards in acute hospitals and maternity in 2019 (https://www.nice.org.uk).

New Zealand established a national Safe Staffing Healthy Workplaces unit with a couple of full time staff under joint governance. By mid 2009 they had adopted a whole of system approach that included an acuity based staffing methodology using acuity data from the TrendCare patient nurse dependency system, social structures to support monitoring and change, the development of monitoring metrics and a new and more sophisticated way of managing variance between demand and capacity. In 2015 the NZ Ministry of Health funded an independent review of the Care Capacity Demand Management (CCDM) program [18], and as a consequence, the approach has now been rolled out throughout the whole of NZ. It does not specify minimums or maximums (except where these are service specifications such as having two nurses minimum on a night shift to keep a service operational regardless of patient utilization). This work has been extended to also cover allied health staffing. A national approach driven by the Ministry of Health, adoption by all relevant stakeholders, system consistency between organizations and the ability to generate high quality timely data, were identified as key success factors.

## Rostering fundamentals

The WHO health workforce models document includes a glossary of planning terms, occupational categories, work locations and more [19]. WHO developed and field tested a rational method that can be used for setting activity (time) standards for health personnel and translating these into workloads for setting staffing levels in health facilities. This high level Workforce Indicators of Staffing Needs (WISN) planning tool has now been in use since the 1990s, it was computerized in 2010 and is being made available through the WISN community [20]. It's software user manual includes a list of terms used with definitions.
Instructions regarding its use, findings and lessons learned were published in 2016. This indicated that [21]:

- a range of methods can be employed to arrive at valid and reliable workload activity standards for various departments or types of health services;
- involvement from all relevant stakeholders yields better and more reliable results;
- the activity should not be a one-off activity, but be integrated into the existing health system enabling the monitoring of trends over time;

- it is imperative that roles and responsibilities be clearly defined for each staff category based on competencies and skills.
- Staff need to work in accordance with their scope of practice all or most of the time.

Preparing effective staff rosters that enable the provision of 24/7 care provision all year round is a challenging task. There are many aspects, including numerous constraints, that need to be considered as part of this process. Rosters impact nursing staff's family life, general life style, social and personal activities, sleep patterns and work environments. Staff evaluate rosters based on ability to meet patient needs, their own family needs and personal objectives. Rosters are instrumental in contributing to staff satisfaction, high quality care outcomes and the healthcare organization's financial performance.

In recognition of the complexity of calculating the right staffing levels required to provide safe care, Governments/funders have made various attempts to provide tools and guidance for healthcare facilities to establish the best possible staffing mix to provide nursing services [14]. This chapter explores past and current developments of nurse and midwifery staffing establishments and associated funding initiatives, how departmental staff establishment requirements are calculated, followed by methods used to develop rosters, determine staffing numbers and skill mix for any working day and to meet the service demand all year round, along with performance objectives to be achieved.

Rosters need to balance the conflicting needs of patient care, staff requests and budget restraints. Other budget considerations to be explored are, shift work impacts, various possible rostering patterns, patient turnover rates, variable or an unstable patient type mix, staffing instability due to high staff turnover rates and/or the use of casual or agency staff, and their impacts on patient outcomes [22]. All of these factors influencing nursing workload in some manner need to be budgeted for. We'll finish this chapter by exploring the use of zero based and activity-based costing/funding, the use of case-mix classifications as hospital 'products' and outcome measures against which cost can be identified.

## Data variables required to calculate nurse staffing needs

As indicated in previous chapters, there are few if any standard data definitions for the many variables one needs to make use of to calculate service demand and staffing requirements. The number and types of patients cared for represent service demand for each service type. For in-patients the number of patients are usually identified relative to patient days or occupied bed days. There is also a need to consider overhead nurse staffing requirements to cover all organizationally determined nursing managerial and supervisory positions. When developing budgets for nursing services it is necessary to consider the nursing costs associated with patient transfers, which may be required to provide patients with access to services not available within the healthcare facility where they have their first point of contact. There are likely to be instances when nurses are required to accompany some patients during the transfer.

**Table 1 Service demand and staffing data variables and definitions**

| Variable | Definition |
|---|---|
| Number of patients | • Number of persons treated/cared for relative to any time period or location or service type or casemix |
| Patient days | • Occupied bed days — as defined by the midnight census for in-patients<br>• Bed utilization — the total number of patients cared for based on beds used across the 24 hr day<br>• Factorial bed days — based on episode hours, or occupied bed hours where 24 hr = 1 patient day |
| Length of stay | • Inpatients — the number of overnight stays — each count as one day<br>• Day only patients — measured in hours, or, one episode is counted as one day |
| Health facility type | • Tertiary/teaching or regional or rural or remote acute care facility<br>• Sub-acute care facility<br>• Residential long term care<br>• Community/home care |
| Service type | As per organizational and departmental arrangements, for example;<br><br>• Medical services<br>• Nursing services<br>• Midwifery services<br>• Allied Health Services<br>• Community services<br>• Diagnostic services — organ imaging, pathology, etc.<br>• Non-clinical support services<br>• Environmental services |
| Patient type | • As per diagnosis/health status and treatment<br>• As per any one of many patient nurse dependency and other clinical classification systems<br>• As per health service type, e.g., pediatric or orthopedic or maternity patient<br>• As per cost center allocation or casemix grouping |
| Nursing intensity | • Each patient type has a known pattern of nursing service demands per day of stay, specific to that patient type<br>• Each patient type has an associated benchmark range indicating the expected nursing intensity in hours per patient per day (HPPD) |
| Discharge status | • Patients are discharged to home or to another service<br>• Patients are cured or have an improved health status<br>• Patients have a deteriorated status or are deceased |
| Full/whole time equivalent hours (FTE/WTE) | • The productive hours available from one full time staff member across a one-year period, to complete the work within a unit<br>• These hours do not include; training, orientation, sick leave, annual leave, maternity/parental leave or any other type of leave or absenteeism |

Table 1    Service demand and staffing data variables and definitions—Cont'd

| Variable | Definition |
|---|---|
| Hours of Service | • Hours by days of the week the service is being provided<br>• A 365 day 24/7 service equals 8,760 hr. This requires 5-6 full time equivalent staff (FTE/WTE) for one person to be present at all times |
| Productive hours | • Staff hours worked and paid to provide a service including clinical, non-clinical, department-based educator and management/supervisor hours<br>• Productive hours do NOT include, orientation, training, any type of leave or absenteeism<br>• These hours can be divided by a range of denominators relevant to the type of service to measure productivity and efficiency for example;<br>  o  Ward — productive hours divided by patient days (HPPD)<br>  o  Emergency — productive hours divided by attendances (HPPA)<br>  o  Outpatients — productive hours divided by occasions of service (HPOS)<br>  o  Community — productive hours divided by patient visits (HPPV)<br>  o  Operating theater — productive hours divided by operating minutes (HPOM), procedures (HPP), cases (HPC) |
| Non-productive hours | • Training, orientation, sick leave, annual leave, maternity/parental leave or any other type of leave or absenteeism |
| Skill mix | • Staff categories (determined by qualification, skills, experience) employed to complete work as described in their employment contract or industrial award associated with their scope of practice<br>• Each of these staff categories has an associated cost used for budgeting purposes. Refer Chapter 5 for further details |
| Overhead staffing | • Staff hours utilized by another service such as infection control, operational management, human resource administration, quality control, clinical or information governance, Information technology, project management, staff development which are used to support a department and are costed to that department |

Table 1 provides some definitions in use. This demonstrates the variability regarding all statistics produced using these parameters.

Accounting for all the different types of leave, staff training and development and various types of absenteeism, appears to be one significant aspect impacting on the calculations in use to determine actual productive hours available for each nurse employed. Variations in actual productive hours have major repercussions on the number of FTE staff that need to be allocated as the staffing establishment for any unit or department employing nurses. FTE positions can be shared by dividing the FTE into two or more part time/casual positions based on hours worked, where the total combined hours of the part time/casual staff equal the hours of one FTE. This is an important consideration for rostering purposes.

## Projecting nursing service demand and workforce requirements

Past experience has informed the nursing profession that nursing workloads have tended to be severely underestimated by health service executives, politicians and funders alike. Cost containment measures undertaken by administrative staff, for example, reducing the availability of support services, equipment or supplies such as drugs, linen or meals, can have significant workload implications on nursing staff, especially when the non availability occurs after hours [23].

It is critical to have appropriate nursing information systems that will enable realistic service demands to be identified and measured in real time, as this provides useful retrospective trend data from which future projections of workforce requirements can be made. A lot of nursing work is officially and statistically invisible, yet it needs to be accounted for in budget allocations. For example, patient care that is not budgeted for is frequently provided to outpatients who go to a ward post discharge for follow-up care, such as; wound/stoma review, removal of packs/sutures/staples, or postnatal advice. These patient visits are often not scheduled, and the patients may or may not be advised to go to the unit/ward by medical staff. There are many other activities generally not measured or accounted for such as; patient escorts, waiting in the radiology department with a patient, talking to concerned post discharge patients or relatives on the phone, providing specialist advice to other departments, providing specialist care to patients in other departments, calibrating equipment, ordering clinical supplies, restocking, and providing supervision to new and novice practitioners, to name a few.

In many acute hospitals hospital managers, finance managers, and nurse managers have traditionally used, and are still using, the midnight census count as a demand indicator and as the denominator for calculating nursing/midwifery hours per patient day (NHPPD/HPPD). Midnight census counts, and in fact any census count at any time of the day will not capture the true count of all patients cared for in busy acute wards with a high patient turnover. If nursing and midwifery workloads are to be accurately measured all patients who have received care must be included. This includes all patients who were discharged or transferred out of the ward. Measuring bed utilization for three 8 hr periods of the day and taking an average for the day will give a more accurate measure of patient numbers.

The implementation of electronic health/medical record systems won't achieve their potential to improve patient safety and organizational efficiency unless they include accurate and timely nursing data enabling minor adjustments to be made operationally to reflect the real time experience. In addition, given the projected shortage of registered nurses, there needs to be alternative strategies in place to reduce nursing service demand. Strategies to achieve this may include;

- reducing the number of late discharges by ensuring that medical staff process discharges in a more timely way,
- reducing the amount of non-nursing work that nurses are currently having to complete by conducting a work analysis and providing the most suitable skill mix,

- reduce the number of one on one/specialled patients in the ward/unit by reviewing one on one/specialling data, and related protocols and where appropriate introducing the concept of co-horted watch,
- investing in health information systems that will make nursing documentation, care planning, patient assessments, medication administration, diet ordering, observations, and rostering less time consuming.

To initiate the implementation of some of these strategies may indicate a need for health system and service delivery reform.

Previous chapters have highlighted the many data and system constraints that need to be overcome to more accurately project nursing service workforce requirements. It's been demonstrated that the use of an evidence-based patient acuity system associated with well defined patient types overcomes many of these constraints. Patient acuity systems can provide service demand data for every unit/ward/department, day of the week and time period within any 24 hr day relative to patients treated. There is also a need to have data about nurses engaging in non-clinical activities such as professional development, ward related project work or meeting attendance or sick leave so that accurate productive service delivery can be calculated. Such data are required not only to project staffing establishments by skill mix but also for rostering purposes to ensure there is the best possible match between service demand and staff availability. This data should also be used for budgeting and activity based costing purposes. Developing roster patterns that accurately match peaks and troughs in workloads for days of the week and periods in each day prevents over or under staffing, hence maximizing efficiency and productivity and minimizing waste. Patient acuity systems also enable more accurate information to be collected to demonstrate compliance with legislative ratio or hours per patient day (HPPD) requirements.

Projecting organizational nursing service demand and workforce requirements needs to be undertaken using a bottom up approach. Every healthcare organization delivers a suite of health services as required by its organizational structure. The service demand for each service type or unit/department forms the foundation for projecting staffing requirements by skill mix. Ideally each unit/departmental manager takes responsibility for preparing their own service demand projections for the following year(s) as they are best placed to know and understand trends. However this is only possible if they have the relevant data.

## Calculating departmental/unit nurse staffing requirements

Chapter three explored the many different methods in use to measure nursing care demand. Of these the most prevalent are the use of nurse to patient ratios, full-time equivalent (FTE) nursing staff per patient day, Nursing Hours Per Patient Day (NHPPD) and the use of patient acuity systems [24]. It was demonstrated that the use of midnight census as the basis for identifying the number of patient days did not adequately indicate nursing service demand. Variations regarding calculations used to identify productive hours worked by nursing staff was identified as another issue of concern. It was noted that for 'productive nursing hours', 'patient days' and 'occupied bed days' to be meaningful and consistently interpreted, there would need to be standard definitions for these concepts. In the absence of adopting such standards, the degree of accuracy of service demand calculations is unknown.

Many hospitals have stopped using the midnight census as the indicator for calculating the count of patients per day for inpatient areas and are using one of the following:

1. Fractional bed days where each patient's length of stay is measured in hours and one patient day equals 24 hr.
2. Multiple bed census counts done at different times of the day from which an average number of patients per day is calculated.
3. Patient days by bed utilization for three equal 8 hr periods of the day capturing all patients cared for during the 8 hr day, evening, and night periods.

The result of these new counts generally is a higher patients per day rate than the midnight census count with the exception of option [1] which, depending on admission and discharge times may in some instances calculate a lower patient count than the midnight census method. The third option, Patient days measured by averaging bed utilization for three periods of the day is the most accurate measure as it identifies the total number of patients actually cared for during the 24 hr period. This calculation will match the number of patient categorized in a real time patient acuity system. This calculation must be used as the basis for calculating nurse staffing requirements if the total nursing demand is to be captured.

NHS Scotland [8] uses a triangulated approach based on three main workload indicator sources for calculating unit/ward based staffing:

1. Outcome of the specialty specific workload measurement tool
2. Outcome of the Professional Judgment tool
3. Clinical quality indicator evidence or evidence from local quality dashboards or scoreboard.

> *The success of workload measurement tools is related more to the willingness of decision-makers to use this information for staffing decision-making than the comparative merits of the tools themselves [8], p. 25*

This unit/ward based assessment also needs to make use of the funded staffing establishment that formed the basis of providing the unit's original staffing budget. The budgeted staffing establishment then needs to be compared with the actual staff use including supplementary staffing such as the employment of casual or agency staff. This learning toolkit [8] made no mention of how the original unit/ward's staffing establishment and budgets are arrived at. The 2018 Bill [11] indicates that this is reviewed annually by the NHS Board based on the information provided by each unit. The inclusion of six guiding principles for staffing was to mitigate the risk that resources may be diverted away from staff groups and settings not covered by these existing workload and staffing tools. These guiding principles are (p. 24):

1. Taking account of the particular needs, abilities, characteristics and circumstances of different service users,
2. Respecting the dignity and rights of service users,
3. Taking account of the views of staff and service users,
4. Ensuring the well being of staff,
5. Being open with staff and service users about decisions on staffing, and
6. Allocating staff efficiently and effectively.

Similarly the United Kingdom's Skills for Health has published a nursing workforce planning tool to assist with the identification of service demand for various departments with skill mix suggestions using a spreadsheet [25]. This makes use of five methods, professional judgment, nurses per occupied bed, activity quality, time task/activity and regression based systems. Tool user guides and training regarding their use is provided.

### Use of nurse:patient ratios to capture FTE/WTE measures for clinical care

Chapter 3 also included some examples of legislative nurse to patient ratio requirements, in particular the Victorian example mandating ratios for various types of hospitals and in-patient care units for each of three shifts. In this instance all healthcare organizations that need to be compliant with this legislation need to calculate their nursing staffing establishment accordingly. Such calculation is usually based on the average number of patients per day per unit in order to identify the total number of patients cared for in any one year/financial period. This does not reflect the nursing workload associated with each patient. Where patients per day is calculated based on midnight census there will be significant inaccuracies. Other variables to be considered are the variable number of hours included in the length of each shift and any number of employment contractual requirements that determine the total productive hours for any FTE/WTE funded position and the number of shifts any one FTE/WTE provides per week/fortnight.

For example; an acute care general medical/surgical unit with an average midnight census of 25 occupied beds/patients, needing to care for a higher number of patients during the morning and afternoon shifts. In addition, the number of staff actually required to meet the ratio is rounded up or down to the closest whole number, thus if there are 25–27 occupied beds/patients on the day and evening shifts, there is a need for 7 nurses to comply with the; 1:4 ratio on the day shift, 7 nurses to comply with 1:4 ratio on the evening shift and 3 nurses to comply with the 1:8 ratio on the night shift.

- If one FTE/WTE productive hours equal 1371.6 (based on 6 weeks annual leave, 2 weeks paid sick leave, 4 days mandatory training and work 38 hr/week) or 180.5 days (7.6 hr/day), and
- the requirement is to cover 16 hr of the day (morning and evening shifts), with $7 \times 2$ FTE/WTE plus 3 FTE/WTE nurses on the 10 hr night shift.
- then 17 FTE/WTE nurses need to be available on any one 24 hr day.

Where one FTE/WTE annual productive working days equals 180.5 days, then the total nurse staffing establishment to be included in the budget will equal $17 \times 365$ divided by $180.5 = 34.38$ FTE/WTE. Assuming that each day of the week is staffed the same, this may be required for general medical wards where patient numbers remain constant across the total week. This equals 5.16 productive NHPPD utilizing midnight census calculation of patients per day. NB training, leave and absenteeism hours are not included in productive hours.

The Victorian Safe Patient Care legislation [2] includes specific ratios for a number of patient types and special services. It also mandates that there may be no >20% enrolled nurses included in mandated ratios for acute ward or a general medical or surgical ward. Ratios also differ by level of hospital. There are variable mandated staffing requirements for only 12 patient types:

- Aged high care residential wards
- Emergency departments
- Coronary care units
- High dependency units
- Palliative care inpatient units
- Rehabilitation and geriatric evaluation management
- Operating theaters
- Post anesthetic recovery rooms
- Special care nurseries
- Neonatal intensive care units
- Antenatal and post natal wards
- Delivery suites

This legislation includes many additional clauses and exemptions making it difficult to interpret for the purpose of setting an organizational staffing establishment whilst ensuring legislative compliance. It's up to individual nurses or midwives, with or without union support, to notify hospital management of an alleged breach of the ratio or a ratio variation, and follow a local dispute resolution process in the first instance.

### Use of Nursing (Care) Hours Per Patient Day (NHPPD)

Nursing Care Hours Per Patient Day is a measure developed by using the historical nursing hours worked to achieve safe staffing and dividing it by the average patients per day in a ward/unit [26]. When using Nursing Hours Per Patient Day (24 hr) as the service demand indicator from which to calculate staffing establishments, there is a need to first identify the various types of patients and/or wards/departments for whom such a standard measure has been developed. Patient days is again one of the measures used however defined.

To calculate the nursing staffing establishment, one needs to:

- multiply the NHPPD by the number of patients per day = total nursing hours required per day,
- then multiply by 365 = total nursing hours per year,
- then divide by the number of productive hours per FTE/WTE = the number of FTE required for one year.

Thus a general medical/surgical unit with an annual average daily occupancy of 25 patients and a required 5.0 NHPPD requires 125 hr per day, 45,625 hr per year and a staffing allocation of 33.26 FTE/WTE (where 1 FTE/WTE contribute 1371.6 productive hours per year).

### Use of patient acuity data

Patient acuity data provides an average daily time value for every patient type included in the acuity/patient nurse dependency system in use. This method accommodates the different mixes of patient types in each ward, and the different levels of acuity/patient dependence for each patient type. It provides historic data which reflects the actual care requirements of patients over a selected period of time. When patient acuity systems are used on a continuous day to day basis, they provide invaluable evidence of seasonal changes in demand when patient numbers and their associated nursing service demands fluctuate. Some acuity systems have been developed specifically to be used for only a limited time each year with the objective to obtain a snapshot of nursing demand for establishing nursing staffing establishment. Below is an example of the type of data required from an acuity system to develop an accurate staff establishment related to inpatient care for nursing and midwifery services.

To calculate nurse staffing needs for a unit requires an estimate of that unit's projected patient type mix for the following year. The acuity system should provide evidence of the mix for previous years, showing the occurrence rate of each patient type as a percent of the total number of patients. This data should be used for the calculation of FTE/WTE unless there is a known change in ward structure or purpose. In this case the percentage for relevant patient types can be adjusted for the purpose of the next years staff establishment calculations. It is highly recommended to make use of benchmark HPPD values for each patient type to cross reference nursing intensity for each patient type.

The report example in Fig. 1 for a surgical ward, generated from a computerized patient acuity/ patient nurse dependency system, shows the required hours for care/demand for nursing hours/ HPPD required per Patient Type. This data explains peaks in demand (increase in HPPD required) due to a change in patient types in the ward. For example, an increase in the percent of palliative care patients in this ward would increase the average demand for nursing hours/HPPD required. Similarly, an increase in patients requiring one on one care/specialling because of post-surgical delirium, would have a significant impact on the demand for nursing hours/HPPD required. Surgical wards with a shorter length of stay display a higher measure of acuity and required HPPD.

### Ward Acuity HPPD per Patient Type

Printed: 19/05/2019
8:42:45 AM

**Month:** March, 2019
**Ward:** Ward 4 Bee

**By Patient Days**

| Patient Type | | Average Patient Days with % of days | | 'Actual' Hours Required by Acuity | 'Actual' HPPD Required by Acuity | 1:1 Hours for Part of Shift Included in 'Actual' |
|---|---|---|---|---|---|---|
| SUR | Surgical | 706.00 | (83.5%) | 3037.55 | 4.30 | 8:00 |
| | -1:1 Care | 1.00 | (.1%) | 28.00 | 28.00 | |
| ORT | Surgical - Orthopedic | 39.00 | (4.6%) | 175.25 | 4.50 | |
| PCU | Medical - Palliative Care | 33.67 | (4.0%) | 217.39 | 6.46 | 2:00 |
| | -1:1 Care | 1.33 | (.2%) | 36.00 | 27.00 | |
| U/G | Surgical - Urology/Gynecology | 28.33 | (3.4%) | 112.05 | 3.96 | |
| MED | Medical | 16.67 | (2.0%) | 68.15 | 4.10 | |
| HDS | Surgical - High Dependency | 14.00 | (1.7%) | 109.08 | 7.79 | 28:00 |
| | -1:1 Care | 5.33 | (.6%) | 146.00 | 27.38 | |
| SDO | Day Only - Surgical | 0.33 | (.0%) | 1.10 | 3.50 | |
| | **TOTALS:** | 845.67 | | 3931:37 | 4.65 | 38:00 |

*\*\*Number of Patients Not Categorized for the Period:* 4.00

**Fig. 1**

A 28 bed ward example of monthly acuity data. *Produced from the TrendCare® Software Version 3.6.*

This example of a 28 bed ward with a 95% bed utilization (26.6 patients/day) using an acuity system has an extensive range of patient types for medical and surgical patients. The calculations for two wards with bed utilization rates of 34.645 patients per day and 34.4 patients per day respectively are as shown in Tables 2 and 3.

**Table 2** Using patient acuity data and actual bed days — Surgical ward — (annual average bed utilization of 34.645)

| Patient types | Patient days/year | Annual average HPPD/patient type | Hours per year |
|---|---|---|---|
| Major orthopedic surgery (18%) | 2,276.51 | 4.7 HPPD | 10,699.60 |
| Minor orthopedic surgery (66%) | 8,347.18 | 4.2 HPPD | 35,058.16 |
| Medical (11%) | 1,391.26 | 4.4 HPPD | 6,121.54 |
| Infectious surgical and orthopedic (5%) | 632.36 | 4.8 HPPD | 3,035.33 |
| | | | |
| Total patient days/year | 12,647.31 | 4.34 HPPD | 54,914.36 |
| Other clinical hours for escort, outpatient care, X-ray (100%) | | 0.08 HPPD | 1,011.78 |
| Ward total clinical HPPD requirement | | **4.42 HPPD** | 55,926.14 |

Required staff = 55,926.14 hr per year/1,371.6 hr = **40.77 FTE/WTE** — (assuming 1 FTE productive hours value/ 1 year = 1,371.6 hr).

**Table 3** Using patient acuity data and actual bed days (by bed utilization) — Medical ward — (annual average bed utilization of 34.4)

| Patient types | Patient days/year | Annual average HPPD/patient type | Hours per year |
|---|---|---|---|
| Cardiology (31%) | 3,892.36 | 4.42 HPPD | 17,204.23 |
| Medical respiratory (24%) | 3,013.44 | 4.6 HPPD | 13,861.82 |
| General medical (18%) | 2,260.08 | 4.59 HPPD | 10,373.77 |
| Medical gastric (16%) | 2,008.96 | 4.3 HPPD | 8,638.53 |
| Stroke (6%) | 753.36 | 5.2 HPPD | 3,917.47 |
| High dependency medical (5%) | 627.8 | 7.5 HPPD | 4,708.5 |
| | | | |
| Total patient days/year | 12,556 | 4.67 HPPD | 58,704.32 |
| Other clinical hours for escort, telemetry, outpatient care, X-ray (100%) | | 0.12 HPPD | 1,506.72 |
| Ward total clinical HPPD requirements | | **4.8 HPPD** | 60,211.04 |

Required staff establishment = 60,211.04 hr per year/1,371.6 hr = **43.9 FTE/WTE** (assuming1 FTE productive hours value for 1 year = 1,371.6 hr).

### Using patient demand measures to calculate staff establishments

Tables 2 and 3 demonstrate how to use HPPD measures to calculate FTE/WTE requirements for 1 year for a surgical ward with an average bed utilization of 34.645 patients per day and a medical ward with an average of 34.4 patients per day. The mix of patient types is shown as a percentage of patient days for one year. The calculations include:

1. The total required hours for each patient type per year.
2. The total hours for all patients.
3. The number of FTE required for the ward using the organizations' value for productive hours per FTE/WTE as the denominator to calculate. This is calculated based on the total

hours per year minus all leave provisions, including non-productive orientation and professional development/study days for new staff (subject to turnover rates). This varies by hospital.

The two examples are based on the same ward size in terms of the number of beds available and the same value for productive hours per FTE/WTE was used. Each ward has its own unique bed utilization rate and patient type/acuity mix. These latter differences impact on the total staff establishment required for each ward, for the surgical ward it is 40.77 FTE/WTE and for the medical ward it is 43.9 FTW/WTE.

If nursing service staff establishments (FTE/WTE requirements) are to be accurately predicted, patient demand measures must be used for the calculations. The retrospective data used must include acuity measures for all patients who received care, accommodating the fact that numerous beds can be occupied/utilized two or three times by different patients due to discharges, transfers in and out of the ward, admissions and deaths. Patient days measured by bed utilization captures this churn factor which is very common in acute medical and surgical wards today. The shorter the patient length of stay in a ward/unit/department the greater the churn factor becomes. If this churn factor/bed utilization is not accounted for the demand for nursing services will be significantly underestimated. Bed days measured by this method can be up to 30% higher than the midnight census measure in large (30 bed) wards that are perpetually full and have an average length of stay of 3 days per patient.

Patient days measured by bed utilization reflects reality in nursing services. Although this larger denominator decreases the HPPD measure, it can significantly increase the hours required for care due to the greater volume of patients captured using the annual bed utilization statistic. The measure of patient acuity in a ward includes all of the care requirements for inpatient care, but does not include other clinical activities completed by nurses on the ward such as; monitoring telemetry on patients external to the ward, taking phone calls to give advice to patients previously discharged from the ward, completing patient escorts out of the ward, waiting for patients in X-ray, providing care to outpatients in the ward area, etc. These activities will vary from ward to ward and hospital to hospital. These hours are often "invisible," yet are necessary for service delivery and must be measured and added to the clinical work demand for that ward/unit/department.

## Staffing needs for other service types

### Day only departments

Over the past decade many surgical procedures and medical procedures that previously required one to three over-night stays in hospital, are now being admitted as day only patients (admitted and discharged on the same day). To calculate the staffing establishment for day only areas, the same methodologies can be used as described above for inpatient areas, with the exception that each attendance in the day only area is counted as one patient day. In day only areas there is generally only one acuity measure for the day compared to three measures a day for inpatients. Acuity measures have been established for day only areas to accommodate short consultations, varied lengths of stay up to 24 hr, and an over-night stay.

### Obstetric services

Midwifery services differ from general acute hospital nursing services. As such, different methods need to be employed to calculate service demand and staff FTE/WTE requirements. Major influencing factors for staffing requirements are average length of stay, the average birth rate, number of inductions, the percentage of caesarean sections and complicated deliveries. Some of these factors relate to prevalent clinical protocols adopted as well as the distribution of deliveries managed by midwives versus obstetricians. Other factors to be considered are the type of ante-natal and post-natal services provided, the availability of neonatal intensive care and/or specialist pediatric facilities, as well as whether the facility is used for midwifery and/or obstetric training.

The availability of speciality services determine the need for patient transfer and transport requirements for high risk pregnancies which may have staffing and budget implications. Special care nurseries, neonatal intensive care units and delivery suites each requires their own specific daily staff establishments based on projected service demands. The use of a comprehensive valid and reliable patient acuity system in these speciality areas will provide valuable data to assist with developing realistic staffing establishments and associated budgets. Acuity patient types for maternity services include: maternity acute assessment, antenatal short stay, antenatal, labor, labor preterm, labor assist to independent midwife, elective caesarean, maternity high dependency, postnatal caesarean, postnatal vaginal birth, postnatal day only, and postnatal readmit.

Acuity patient types for neonates include: neonatal ICU postnatal, neonatal ICU surgical, high dependency special care, special care, special care long stay, special care rooming in.

### Geriatric, disability and rehabilitation residential services

The residential patient types and their associated nursing work demands are the primary determining factors for staff requirements. The model of care adopted is another factor that can have major workforce implications. For example, it is highly desirable to adopt a model of care that supports self-management with supervision providing moderate or minimal assistance. This takes far more time than if the carer were to undertake the same activity on the patient's/resident's behalf, such as wheeling them to the dining room in a wheelchair, or serving meals at the bedside versus assisting them to walk to the dining room. There needs to be active promotion of mobility, activity and exercise for these types of patients/residents for them to retain as much independence as possible. It is well known that the self-management model of care has far better outcomes but is frequently unattainable due to staff shortages. There may also be a need to incorporate a palliative care service within some of these types of facilities.

Setting a staffing establishment for any type of residential service needs to follow the same principles as described previously. There are significant differences in terms of possibilities as staffing establishments for residential aged and disability services are commonly influenced by legislative and policy initiatives as described in Chapter 12. These include the mandatory use of specified resident assessments which identify care needs and are usually linked to funding based on regular activity reporting. Most funding models are based on care requirement assessments, many international models are linked to care plans [27].

Various acuity systems are in use for the purpose of day to day management of staff, developing and maintaining effective roster patterns, identifying service demands and calculating FTE/WTE requirements. The TrendCare acuity system provides 10 patient types for rehabilitation services and 8 patient types for residential care. Setting staffing establishments within overall budget allocations is the responsibility of organizational management who will need access to comprehensive data on patient/resident acuity and ward/department bed utilization.

The preference is to provide this group with home and community care, therefore long-term residential care is only suitable for those who can no longer be supported by such alternative care approaches as they require high-level nursing care. This is a common occurrence for those people who are living alone. Service demand for these types of residents need to be met by multidisciplinary teams. Some residential facilities also provide respite care of 2–4 week duration enabling home carers to be relieved of caring responsibilities for these time periods.

## Operating theaters

The Victorian Safe Patient Care legislation [2] requires each operating theater to be allocated three nurses, to cover three roles, instruments, circulating and anesthetics. Access to additional roaming staff to provide assistance with positioning patients, moving equipment and other support activities is highly desirable. Recovery rooms need to have a registered nurse for every unconscious patient and a suitable ratio of patients per nurse depending on the type of procedures being performed. Similar regulations/standards apply to other Australian States and Territories. Total staffing for any operating room suite on any one day is very much dependent on;

* the number of available rooms,
* operating lists/schedules,
* type of surgery performed,
* education services provided,
* emergency after hours projected need and available support services,

- the need for registered nurses to undertake a surgeon's assistant role for some surgical lists or procedures,
- the need for nursing staff to conduct pre-admission clinics and/or attend to patient preparation prior to surgery.

Daily staff requirements over 365 days should be calculated considering all of the relevant points above and utilizing Nursing Hours per procedure, and nursing hours per operation minute for each speciality and each specific type of procedure to determine the overall staffing establishments that needs to be allocated to an operating theater department.

## Accident and emergency departments

Service demand varies significantly with geographical location, surrounding industries and population demographics, construction work, likelihood of road trauma as well as the availability of primary care facilities, especially after hours. It's important to evaluate historical data in terms of patterns of activity by time of the day and day of the week as well as a percentage of attendees who need to be admitted or transferred to another facility. The latter is directly related to medical staff and specialist service availability locally or accessible via an established telehealth infrastructure. There may also be seasonal variations due to holiday makers, severe heat or cold snaps, or the probability of communicable disease outbreaks, natural disasters such as fires, floods, hurricanes, cyclones or earthquakes. Peak staffing levels need to match peak patient attendance periods.

Every emergency department needs to have a triage facility to determine the urgency for clinical intervention. Workload is measured by the number of attendees and the use of an acuity system that determines various patient types however defined. The number of emergency attendances may be influenced by funding arrangements, such as free public emergency access versus the need to pay for a primary care consultation, and accessibility. The Agency for Healthcare Research and Quality has published an Emergency Severity Index (ESI) [28], a five level triage algorithm that provides clinically relevant stratification of patients into five groups from least to most urgent based on acuity and resource needs. Other similar triage systems are also in use, such as the five level Australasian Triage Scale (ATS). This has been modified to become the Canadian emergency triage and acuity scale (CTAS) to better suit that population [29, 30]. The TrendCare system provides 12 different emergency patient types and incorporates the Australasian Triage scale, the Manchester Triage scale and the Singapore Triage scale together with their adaptions for pediatrics and mental health.

From a staffing skill mix and rostering perspective there is a need to collectively consider Patient Triage levels, complexity/acuity, time in the emergency treatment area, waiting times, admissions from emergency, transfers to other facilities and the workload generated using a valid and reliable patient acuity and workload management system. One comparative study found that the ESI has better resource discrimination ability than the PACS making it a more useful tool for resource management [31]. Another study in 2018 in Ireland identified that the TrendCare system was more comprehensive and gave a more realistic nursing intensity measure than the BEST system.

### Specialist outpatient departments

There are many and varied specialist outpatient clinics each requiring their own mix of staff. Each of these clinic types need to undertake their own work measurement study and collect patient throughput statistics for a variety of patient types. Future staffing requirements can then be accurately projected.

### Supervisory and administrative clinical staff

Every clinical unit needs to have a manager responsible for the operational management during normal working hours. An allocation of 0.2 HPPD should be budgeted to account for ward management hours five days a week. This will ensure that wards of 26 or more patients per day will have approximately 7–8 hr of management per day Monday to Friday. Smaller wards with low acuity and low patient numbers of 20 or less patients per day, and low to moderate acuity may be allocated a small clinical component each day which will enable them to assist with meal relief and some clinical care.

Each ward/unit should have a registered nurse named in-charge when there is not a ward/unit manager working, including evening, night, weekends, and public holidays and when the manager is on leave. Time should be allocated for ward co-ordination activities relevant to the type, size and throughput of the ward. Ideally the nurse in-charge on busy after-hours shifts should not have a patient allocation and should complete their clinical component by assisting other nurses with their workloads and/or assisting with unplanned admissions, transfers, and clinical procedures. An allocation of.1 HPPD should be budgeted to account for after-hours shift co-ordination on the evening and weekend day shifts. Large wards with huge turnover and admission rates may need more than this allocation to co-ordinate the ward activity.

Every hospital facility needs to employ senior staff to assist with administrative work concerned with, for example, maintaining accreditation, quality management, supporting information technology projects/system use, providing professional development or supporting the adoption of evidence based practice. Setting the nursing staff establishment for this area is very much dependent upon the hospital facility's size, type, organizational structure and strategic directions.

Every in-patient facility needs to have sufficient senior leadership/management positions covering evening, night and weekend services with responsibility for and ability to manage any emerging local or hospital wide issue or crisis, patient flows or staff reassignment as required to meet unexpected high service demand [23]. There tends to be minimal staffing provision and service availability after hours, which has the potential to compromise patient safety. Senior nurses occupying these positions need to represent and assume any of the roles occupied by unit managers, clinical specialists, directors and hospital administrators during week days, on an as required after hours basis primarily for problem solving purposes. They need to monitor hospital wide activity, have access to areas such as the kitchen or pharmacy, equipment/supplies store, know who to call in as and when required, and support any individual staff member or individual patients and their families on an as required basis.

One literature review [32] found a greater incidence of various adverse events over weekends and during the night suggesting that not only is it imperative to have sufficient staff after hours but also that the presence of suitably qualified senior nursing staff is likely to have a direct impact on patient safety. These positions need to be included in the overall nurse staffing establishment.

## Significant variations resulting from method used

If the midnight census measure of patient days was used for the above calculations the required hours for clinical care of inpatients would be significantly reduced, and hence the nursing workloads for these wards significantly underestimated. This may explain why so many ward budgets are in deficit, because, as stated earlier most financial managers develop the nursing budgets and use midnight census measures to predict patient days and to develop staffing budgets/and efficiency targets.

The NHPPD method is based on nursing hours worked, not patient demand/acuity measures, and generally uses midnight census data from the patient administration system as the denominator for calculating NHPPD. The risk of perpetuating under and over staffing is a concern when this method is used, as worked hours are used instead of demand. The Nurse to Patient ratio method also uses nursing hours worked but generally uses census data from 3 points of time in the 24 hr period as the denominator for calculating HPPD. Computerized, interoperable, real-time patient acuity systems use actualized required hours for care and use patient days by bed utilization differentiating HPPD measures between patient types.

The variations between the three methods described above have budget, projected nursing workload and skill mix implications. The methodology of these three different approaches to estimating demand for nursing services and calculating FTE/WTE staffing establishments can be compared, but the accuracy of the calculations of these methods depends on a set of variables unique to each methodology. These include;

- NHPPD method — the staffing establishment at the time the data is collected and its variance to the work demand, patient length of stay and ward churn,

- Ratio method — the variance in patient acuity between patients across each shift, length of stay and ward churn,
- Patient acuity/patient nurse dependency method — the level of completeness of the acuity data and inter-rater reliability measures.

All these approaches can be applied to any type of in-patient ward/unit, rehabilitation unit, sub-acute ward or long stay residential service. The only differences are the types of patients/ residents cared for and the different types of activities included in other clinical hours when calculating the clinical FTE/WTE calculations. Each type of patient requires its own mix of nursing work and associated skill mix. The more accurate the patient type differentiation and known associated nursing services, the better the ability to project staffing requirements. Given the ad hoc nature of service demand, there is a need to identify patterns of low and high demand and to allow for real time adjustments to be made to cover unexpected high demand within the staffing budget allocation. The latter is based on degree of probability which can also be calculated and included in the budget provision.

Another major factor is the ward/unit size in terms of number of beds and bed utilization relative to the types of patients accommodated. For example; it's efficient to have high dependency complex patient types in smaller units/wards, and it is more efficient to accommodate low acuity patients such as diagnostic cardiology and laparoscopic surgery in large wards of at least 30 beds.

The total number of patients in one unit/ward impacts on the amount of work generated for non-nursing support staff. In smaller units with acute care patient types, it's more difficult to keep such staff fully occupied/productive all the time. Another consideration for all acute care wards/ units is the need to have a pre-defined minimum staffing level for each ward which should be maintained irrespective of the amount of work generated during low activity, low bed utilization and low acuity.

Nurse:patient ratios and HPPD per patient type will differ for different types of health services based on the average complexity of the types of patients cared for and the average length of stay for each patient type. This is where the adoption of a patient acuity approach enables a more accurate reflection of the real world across any nation's health system.

## *An international patient type HPPD benchmarking research study*

In 2014, a major benchmarking study of HPPD per patient type was undertaken by Lowe utilizing data from hospitals using the TrendCare acuity/patient nurse dependency system [33]. Given the large amount of data collected it was possible to establish international benchmarks for 106 patient types. Data were collected in four countries across 51 hospitals (34 public and 17 private) relating to 106 well defined patient types accommodated across 829 units/wards. The sample covered a total of 5,152,708.09 patient days (24 hr bed occupancy) of which 35.39%

were excluded as these had incomplete data or could not verify that the ward submitting the data had completed inter-rater reliability during the 12 months prior to the data being collected, leaving a balance of 3,329,023.19 patient days. Work measurement was undertaken across 15,135,720.27 Patient acuity shift categories representing 24,915,186 nursing hours required [34].

Fig. 2 provides one example of the results which demonstrate the HPPD range for required hours (demand measured by acuity) associated with the ICU patient type. The graph shows the (1on 1)/specialling hours for each ICU unit in the study (Purple (black striped in the print version)) and the statistical HPPD range within 2 standard deviations from the mean (green shading (rectangular box indicated by solid line in the print version)). Each column is color coded to identify the hospital and numerically coded to identify each individual ICU unit/ward. It is evident from this graph that the large tertiary hospitals included in the study have numerous ICU units with high numbers of patients requiring (1on 1)/specialling, and higher HPPD requirements as shown by the use of one color for each hospital. Smaller hospitals with only one ICU unit and fewer patients requiring (1on 1/) specialling have a lower HPPD range.

Following this benchmarking study the range of ICU patient types were extended to include specialities such as medical, surgical, cardiothoracic surgery, neurosurgery, orthopedics, head and neck surgery, trauma, vascular surgery and medical cardiac. The international benchmarking study using 2019 data will include this wider range of patient types.

Table 4 displays a sample of all collected benchmarking data, for a small selection of patient types included in the study, before filtering out wards that could not confirm Inter-rater reliability results of 90% or higher. These data show that there are very few outliers outside of the one to two deviations from the mean for almost all patient types. In this example the patient type of cardiology has 3005 extreme outliers, however this is only 1.78% of a large sample size (168,572.25). 98% of samples fell within 1 standard deviation of the mean. Following this study a high dependency Cardiology patient type was developed, as it was identified that these outliers were mainly related to cardiology speciality hospitals. The ICU values shown in Tables 4 and 5 do not include patients requiring 1:1 specialling, therefore the HPPD values shown apply only to other types of ICU patients.

Note the high dependency Maternity patient type is a rare occurrence as there were only 22.99 patient days for that patient type, however the nursing hours required for this patient type are high, ranging from 9.5 to 10.79 at one standard deviation, averaging 10.195. This patient type is most commonly found in tertiary maternity units, and hence increases their overall acuity HPPD.

Table 5 includes only data that is confirmed as having 90% or greater inter-rater reliability. The table displays the scope of acuity measures across 1, 2 and 3 standard deviations from the mean for each patient type. These data are used to develop international benchmark ranges for the acuity measures for each patient type which can be used to develop staff establishments when

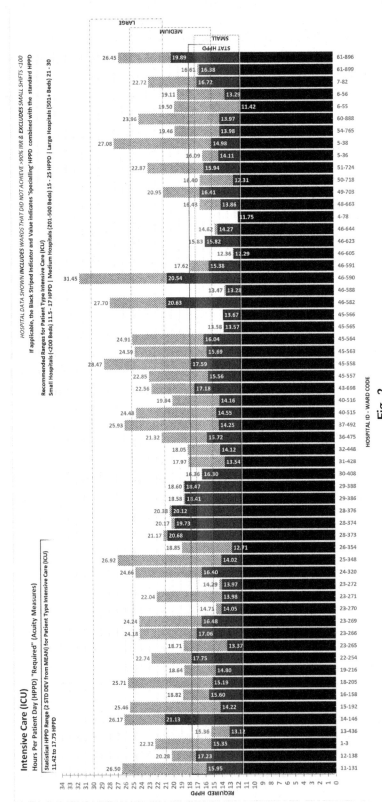

**Fig. 2**

HPPD range for intensive care by hospital. *From the TrendCare® Benchmarking Report.*

## Table 4 Patient acuity HPPD within 1, 2 and 3 standard deviations showing sample size and outliers excluded from HPPD range calculations

| Patient type | 1 standard deviations | | | | | 2 standard deviations | | | | | 3 standard deviations | | | | | Out of range |
|---|---|---|---|---|---|---|---|---|---|---|---|---|---|---|---|---|
| | HPPD range | | | Samples | | HPPD range | | | Samples | | HPPD range | | | Samples | | |
| | 1σ Min | 1σ Mean | 1σ Max | 1σ Included | 1σ Excluded | 2σ Min | 2σ Mean | 2σ Max | 2σ Included | 2σ Excluded | 3σ Min | 3σ Mean | 3σ Max | 3σ Included | 3σ Excluded | Excluded |
| Antenatal short stay | 2.79 | 3.419 | 4.2 | 1350.96 | 210.99 | 2.79 | 3.662 | 6.09 | 1429.62 | 132.33 | 2.79 | 4.006 | 7.79 | 1561.95 | | |
| Babies — Medical up to 24 months | 4.47 | 5.554 | 6.94 | 37265.93 | 5198.31 | 3.67 | 5.792 | 9.1 | 41739.91 | 724.33 | 3.67 | 6.027 | 9.63 | 42464.24 | | |
| Babies — Surgical up to 24 months | 4.75 | 5.74 | 7.01 | 2755.29 | 143.32 | 4.26 | 5.636 | 8.14 | 2889.28 | 9.33 | 4.26 | 5.826 | 8.87 | 2898.61 | | |
| Baby — Postnatal | 2.94 | 3.264 | 3.7 | 13613.29 | 15165.97 | 2.82 | 3.316 | 4.18 | 28772.6 | 6.66 | 2.82 | 3.391 | 4.45 | 28779.26 | | |
| Cardiology | 3.5 | 4.163 | 4.88 | 79460.72 | 6328.57 | 2.86 | 4.147 | 5.65 | 82762.65 | 3026.64 | 2.55 | 4.146 | 5.67 | 82783.64 | | 3005.65 |
| Cardiothoracic surgery | 4.03 | 4.6 | 5.19 | 12735.93 | 5053.98 | 3.01 | 4.529 | 5.97 | 17745.91 | 44 | 3.01 | 4.65 | 6.58 | 17789.91 | | |
| Colorectal surgery | 4.15 | 4.89 | 5.62 | 11837.23 | 16807.28 | 3.35 | 4.889 | 6.2 | 28633.18 | 11.33 | 2.86 | 4.841 | 6.2 | 28644.51 | | |
| Coronary care | 9.03 | 10.261 | 11.12 | 7900.3 | 1882.31 | 7.85 | 10.426 | 12.65 | 9782.61 | 244.65 | 7.85 | 10.426 | 12.65 | 9782.61 | | |
| CVA | 4.26 | 5.041 | 5.71 | 22462.87 | 1736.94 | 3.48 | 5.077 | 6.62 | 23955.16 | 9.33 | 2.97 | 5.049 | 7.45 | 24199.81 | | |
| Day of surgery admission | 1.21 | 1.66 | 2.59 | 19351.33 | 9.33 | 1.21 | 1.66 | 2.59 | 19351.33 | 9.33 | 1.21 | 2.917 | 9.2 | 19360.66 | | |
| Day oncology/hematology | 2.07 | 3.742 | 5.9 | 8613.65 | 63.33 | 2.07 | 4.725 | 9.64 | 8676.98 | | 2.07 | 4.725 | 9.64 | 8676.98 | | |
| Day Only — Cardiology | 2.82 | 3.127 | 3.41 | 132.32 | 435 | 2.82 | 3.415 | 4.28 | 567.32 | | 2.82 | 3.415 | 4.28 | 567.32 | | |
| Day Only — Medical | 2.93 | 3.989 | 4.59 | 413.64 | 1000.65 | 1.84 | 4.141 | 7.57 | 1414.29 | | 1.84 | 4.141 | 7.57 | 1414.29 | | |
| Day Only — Surgical | 1.54 | 3.063 | 3.84 | 11993.62 | 2255.66 | 0.92 | 2.762 | 3.84 | 13602.62 | 646.66 | 0.92 | 3.52 | 8.85 | 14249.28 | | |
| Discharge/transit lounge | | | | | | | | | | | | | | | | |
| Emergency — Adult | 3.31 | 3.62 | 4.12 | 27969.99 | 1265 | 1.8 | 3.256 | 4.12 | 29234.99 | | 1.8 | 3.256 | 4.12 | 29234.99 | | |
| Emergency — Adult Trauma | 2.29 | 3.755 | 5.22 | 4844.66 | | 2.29 | 3.755 | 5.22 | 4484.66 | | 2.29 | 3.755 | 5.22 | 4484.66 | | |
| Endoscopy | 1.53 | 1.67 | 1.81 | 1489 | 8.33 | 1.53 | 2.15 | 3.11 | 1497.33 | 9.33 | 1.53 | 2.15 | 3.11 | 1497.33 | | |
| Hemodialysis day only | 2.09 | 2.385 | 2.68 | 11182 | | 2.09 | 2.385 | 2.68 | 11182 | | 2.09 | 2.385 | 2.68 | 11182 | | |
| Hemodialysis inpatient | 7.23 | 8.63 | 10.03 | 6745 | 161 | 2.84 | 6.7 | 10.03 | 6906 | | 2.84 | 6.7 | 10.03 | 6906 | | |
| Hepatic surgery | 4.02 | 4.674 | 5.13 | 3325.29 | 335.33 | 3.54 | 4.649 | 5.65 | 3660.62 | | 3.54 | 4.649 | 5.65 | 3660.62 | | |
| High dependency — Maternity | 9.6 | 10.195 | 10.79 | 22.99 | 9 | 9.6 | 11.127 | 12.99 | 31.99 | | 9.6 | 11.127 | 12.99 | 31.99 | | |
| High dependency — Medical | 5.6 | 7.237 | 9.33 | 18676.7 | 7150.89 | 4.28 | 7.267 | 11.25 | 23342.95 | 2484.64 | 4.28 | 7.48 | 12.18 | 25816.26 | | 11.33 |
| High dependency — Mental health | 2.69 | 5.45 | 6.89 | 2554.65 | 1367.66 | 2.69 | 7.58 | 18.23 | 3922.31 | | 2.69 | 7.58 | 18.23 | 3922.31 | | |
| High dependency — Oncology hematology | 6.73 | 7.366 | 8.1 | 191.99 | 1170.99 | 5.78 | 7.102 | 8.1 | 1333.65 | 29.33 | 5.78 | 7.791 | 11.93 | 1362.98 | | |
| High dependency — Surgical | 4.94 | 6.936 | 9.45 | 33473.07 | 7657.95 | 3.53 | 7.052 | 12.02 | 40772.02 | 359 | 3.53 | 7.123 | 13.38 | 40859.02 | | 272 |
| High dependency — Mental health adolescent | 6.52 | 11.255 | 15.99 | 53.66 | | 6.52 | 11.255 | 15.99 | 53.66 | | 6.52 | 11.255 | 15.99 | 53.66 | | |
| Intensive care | 13.12 | 15.107 | 17.59 | 58430.57 | 9741.31 | 11.42 | 15.463 | 20.68 | 68018.88 | 153 | 11.42 | 15.609 | 21.13 | 68171.88 | | |
| Intensive care burns | | | | | | | | | | | | | | | | |
| Interim care | 3.17 | 3.451 | 3.89 | 11046.93 | 5689.65 | 2.87 | 3.514 | 4.24 | 16628.25 | 108.33 | 2.87 | 3.571 | 4.59 | 16736.58 | | |

*From the TrendCare® Benchmarking Report.*

## Table 5  Benchmarked patient type acuity HPPD — Standard deviation ranges per patient type — summary table

| Pt type | Benchmarked patient type | 1 std deviation | | | 2 std deviations | | | 3 std deviations | | |
|---------|--------------------------|-----|------|-----|-----|------|-----|-----|------|-----|
| | Calculated HPPD ranges | Min | Mean | Max | Min | Mean | Max | Min | Mean | Max |
| ANS | Antenatal short stay | 3.41 | 3.56 | 3.83 | 2.99 | 3.50 | 4.20 | 2.99 | 3.50 | 4.20 |
| BBM | Babies — Medical up to 24 months | 4.71 | 5.32 | 5.81 | 4.34 | 5.26 | 6.10 | 4.34 | 5.33 | 6.65 |
| BBS | Babies — Surgical up to 24 months | 4.86 | 5.67 | 6.30 | 4.43 | 5.51 | 7.01 | 4.43 | 5.51 | 7.01 |
| BAB | Baby — Postnatal | 2.94 | 3.28 | 3.70 | 2.82 | 3.35 | 4.18 | 2.82 | 3.35 | 4.18 |
| CAR | Cardiology | 3.50 | 4.19 | 5.03 | 2.94 | 4.16 | 5.65 | 2.94 | 4.25 | 6.51 |
| CRS | Cardiothoracic surgery | 4.03 | 4.60 | 5.19 | 4.03 | 4.78 | 5.97 | 3.19 | 4.66 | 5.97 |
| COL | Colorectal surgery | 4.25 | 4.89 | 5.57 | 3.59 | 4.97 | 6.20 | 3.42 | 4.91 | 6.20 |
| COR | Coronary care | 9.74 | 11.47 | 12.77 | 7.85 | 10.87 | 12.77 | 7.85 | 11.12 | 14.88 |
| CVA | CVA | 4.44 | 5.07 | 5.52 | 4.06 | 5.05 | 6.01 | 3.40 | 4.95 | 6.01 |
| DOS | Day of surgery admission | 1.21 | 1.42 | 1.79 | 1.21 | 1.72 | 2.59 | 1.21 | 1.72 | 2.59 |
| DYO | Day pncology/hematology | 2.07 | 2.12 | 2.17 | 2.07 | 2.12 | 2.17 | 2.07 | 2.12 | 2.17 |
| MOD | Day Only — Medical | 1.91 | 1.99 | 2.06 | 1.91 | 2.74 | 4.25 | 1.91 | 2.74 | 4.25 |
| SDO | Day Only — Surgical | 1.30 | 2.28 | 3.55 | 0.92 | 2.49 | 3.84 | 0.92 | 2.49 | 3.84 |
| EMD | Emergency — Adult | 1.80 | 2.75 | 3.69 | 1.80 | 2.75 | 3.69 | 1.80 | 2.75 | 3.69 |
| END | Endoscopy | 1.53 | 1.67 | 1.81 | 1.53 | 1.67 | 1.81 | 1.53 | 1.67 | 1.81 |
| HDO | Hemodialysis day only | 2.09 | 2.39 | 2.68 | 2.09 | 2.39 | 2.68 | 2.09 | 2.39 | 2.68 |
| HDI | Hemodialysis inpatient | 2.84 | 6.44 | 10.03 | 2.84 | 6.44 | 10.03 | 2.84 | 6.44 | 10.03 |
| HEP | Hepatic surgery | 4.23 | 4.92 | 5.13 | 3.69 | 4.82 | 5.65 | 3.69 | 4.82 | 5.65 |
| HDM | High dependency — Medical | 5.85 | 7.18 | 9.00 | 4.87 | 7.19 | 10.24 | 4.87 | 7.42 | 12.18 |
| HMH | High dependency — Mental Health | 4.23 | 5.89 | 6.87 | 4.23 | 8.98 | 18.23 | 4.23 | 8.98 | 18.23 |
| HOC | High dependency — Oncology Hematology | 7.03 | 7.14 | 7.24 | 5.78 | 6.68 | 7.24 | 5.78 | 6.68 | 7.24 |
| HDS | High dependency — Surgical | 4.98 | 6.97 | 9.81 | 3.53 | 7.13 | 11.23 | 3.53 | 7.23 | 13.38 |
| ICU | Intensive care | 13.12 | 14.93 | 17.18 | 11.42 | 14.77 | 17.75 | 11.42 | 15.28 | 21.13 |
| INT | Interim care | 3.17 | 3.46 | 3.89 | 2.87 | 3.53 | 4.24 | 2.87 | 3.59 | 4.59 |
| LAB | Labour | 12.74 | 13.72 | 15.05 | 12.74 | 13.72 | 15.05 | 8.01 | 12.91 | 15.05 |

NB — Includes only data from wards with IRR >90% and sample size >100.
*From the TrendCare® Benchmarking Report.*

there is no other accurate acuity data available. Variations between hospitals in relation to HPPD acuity measures are mainly due to hospital type and their related variances in patient complexity and length of stay. Variable models of care, patient demographics, cultural differences and care protocols can also cause variations in HPPD measures.

## Rostering methods

A nursing roster shows all on duty and off duty periods for each member of staff within a defined area of practice. Rosters need to reflect anticipated workload variations, such as a high nursing workload on surgical operating days, or potential trauma admissions on certain days of the week. Rosters need to match expected workload peaks and troughs to ensure sufficient staff with the right skill mix are available when required. Only then is it most likely that 'safe' care can be provided. Automated rostering support systems are now being made available. These technologies may be accessible via workers' mobile devices. Its important that the software underpinning these Apps is able to accommodate the many variables to be considered as part of the rostering process and for these to be linked with the organization's human resources and payroll systems.

The staff establishment provides the number of FTE/WTE staff required to be employed and allocated to the unit. The roster profile identifies who is working full or part time and identifies any specific contractual arrangements that not only need to be complied with but which also may have rostering cost implications. The allocated staffing establishment reflects the staffing budget and if it is not realized with full or part time staff the financial consequences of backfilling with casual/bank or agency staff may be a budget deficit.

There needs to be some provision in the budget for employing casual or agency staff as required at short notice for sick leave, other unplanned special leave and for unpredicted peaks in acuity and/or activity that significantly increase the nursing workload. Known seasonal service demand variations should alert the unit manager to the number of staff who need to be on leave at specific times of the year to best match projected workload demand. This should be planned to ensure staff know about leave options so they can plan their own holiday schedule.

### Foundations for roster development

Roster development should start by identifying peak service demand times and days of the week to determine required skill mix and number of staff for different time periods and days of the week based on historic patient acuity data, daily unit/ward activity routines, cyclic service demand patterns and future projections. Consideration of the time patient activities start and finish and the impact they have on the use of resources is critical. Regular patient care activity trends generate increased workload at specific times within a shift. Examples include:

- Admission and discharge trends.
- Scheduled elective surgery.
- Patient transfer patterns.
- Routine patient care activities.
- Doctor's rounds.
- Routine patient care programs, clinics or services conducted.

An analysis of ward work studies reveals that in most medical surgical wards the 4 hr of the morning shift is where workload peaks and hence additional nursing hours should be provided during this period of the shift. The workload in the afternoon peaks mid shift, between 1700 and 2100 hr. During this peak period nurses also require a meal break. This increases the workload for those nurses who remain on the ward if additional hours are not rostered for this period. The peak activity period for the night shift falls during the last 2 hr of the shift when patients are generally all awake as shown in Fig. 3. This determines the number of staff and skill mix required for each day and various time periods. Such information forms the basis for developing a rostering template.

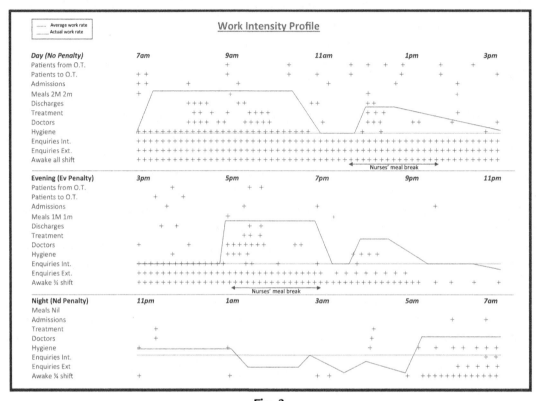

**Fig. 3**

Nursing work intensity profile across a typical 24 hr in-patient period. *From the TrendCare® Clinical Training Booklet Version 3.6.*

Shifts can be of any number of hours duration between a minimum of 2 hr and a maximum of 12 hr shifts. The average Australian nurse works 33.4 hr per week as around 50% work part time. Short shifts suit some people. Long (10 or 12 hr) shifts are popular with some and not others, a lot depends on an individual's external factors. The number of shifts individuals like or need to work per week will vary based on the number of weekly normal hours.

For example those working 3 × 12 hr shifts work a 36 hr week, those working 4 × 10 hr shifts and those working 5 × 8 hr days work a 40 hr week. In Australia rosters need to consider an extra day off every 4 weeks for full time staff to account for a 38 hr week. France has a 35 hr week, and England has a 37.5 hr week. These variables indicate numerous rostering options in terms of number of shifts each person works per week as well as the number of hours per shift.

Staff working a 9.5 hr night shift cover 4 shifts/week. Those rostered to work 8 hr day and evening shifts but required to work 38 hr/week, can only cover 19 shifts over a 4 week period. There needs to be an overlap of hours between shifts in every 24 hr period to enable nurses to handover patient care from one shift to the following shift. Accuracy is subject to the rounding factor required to account for required ratios.

When considering the length of shifts one needs to be careful not to build in periods of over staffing relative the service demand generated. For example; the use of five nurses all working 12 hr shifts is in many cases inefficient, as there are periods during any 24 hr timespan where a greater number of staff would be beneficial or where three nurses are not required. The known daily peaks of service demand should be used to make decisions about how many staff at what skill level are required to provide optimum care. This forms the basis for deciding the best possible arrangement of lengths of shifts. This also needs to be negotiated with staff members allocated to the unit/ward as some prefer to work more frequent short shifts, where others prefer less frequent long shifts. Variable shifts do influence clinical handover processes that also need to be considered as do ward/unit routines. Rosters can be re-engineered as part of an overall change management process.

The most effective method of managing the workload during peak activity times is to utilize short shifts. Part time staff can be routinely rostered on short shifts e.g. short 4 hr day shift (0700 to 1100) and short 4 hr evening shift (1700 to 2100). The night duty workload during peak activity time is most effectively managed by rostering one or two day shifts to start 1 or 2 hr earlier than the normal day shift. This provides additional work hours for the last 1 or 2 hr of the shift.

If these peak activity periods are not accommodated nurses will become fatigued and patient care activities will not be completed thus compromising patient care. The impact of this is magnified when unpredicted activity occurs in the second half of the shift with nurses being unable to complete their shift on time. Rostering staggered full shifts will also relieve the workload during these peak periods but will be less financially effective.

In critical care areas, peaks and troughs are generally not as consistent on a shift by shift basis. A creative combination of long and short shifts will generally produce the required roster objectives. Twelve hour shifts can be effective in critical care units in achieving roster objectives. Staff working in critical care areas have reported high levels of satisfaction with these shifts. To achieve patient continuity, staff satisfaction and cost efficiency, the following principles should apply;

- The number of 12 hr shifts utilized should be limited to 60%–70% of rostered hours.
- 12 hr shifts may make up the core staffing but need to be complemented by a mix of 8 hr, 6 hr and 4 hr shifts.

- Any one nurse should only work a maximum of three 12 hr shifts in a week.
- Nurses working three 12 hr shifts in a week should have their shifts rostered together and days off not split as work fatigue is minimized when *days off* are grouped together.

In order to provide the right skill mix for each shift, managers must be aware of the type of work that has to be done. Case mix within the ward, average patient length of stay, skill level and scope of practice of staff, and the model of care delivery must be considered. Work studies demonstrate that higher numbers of enrolled nurses and non-licensed carers can be included in the roster in non critical care areas. There is less opportunity to use such staff in wards with high acuity, short length of stay and rapid turnover. The skill mix should be carefully balanced to ensure;

- Patient safety is not compromised.
- Staff are able to complete all work which lies within their scope of practice.
- Staff receive fair and equitable workloads.
- Staff new to the environment, novices and students are orientated and supervised.
- There is a balance between hands-on clinical hours and non clinical hours.
- The proportion of registered nurses to enrolled nurses suits the case mix and the type of work in the ward.
- The level of support staff is proportional to the amount of work available relative to their scope of practice.

If a roster is to accurately provide human resources over a period of time it is critical that the creator of that roster is fully aware of all activities occurring during that period. The type and volume of work in a department can be identified by conducting a work study, which involves staff completing a form for each shift they work over a one month period. An appropriate roster skill mix can then be developed.

In order to match clinical work requirements to nursing staff skills, it is essential that the skill level of each nurse is evaluated and nurses are identified at particular skill levels. Nursing skill levels may be defined as;

- Team leader (able to be in charge on a shift).
- Competent (requires no direct supervision).
- Advanced beginner (requires guidance in complex situations).
- Novice (new practitioner requiring some supervision and guidance).

Some automated roster systems enable each nurse to be rated at a level of competence. This enables identification of the best skill mix from the nurses available to be made. If set parameters for staffing are not met the system will alert the user.

If the model of care used in the ward to deliver patient care is *primary nursing* or *patient allocation to individual nurses*, the roster will require a skill mix of nurses who are all at a competent level and above. If the model of care is *patient allocation with team nursing* a range of nurse levels and nursing support staff can be incorporated into the roster.

Once these factors are known and a suitable template developed, then the next step is to allocate the required off duty days for each full time member of staff in accordance with their employment contract and by considering any special requests. Care must be taken to ensure that no staff member is allocated to work an extended number of consecutive shifts. For example; no one should work consecutively for 10 days or more. Most wards/units limit the number of consecutive shifts to 6 or 7. It is useful to have rotating rosters covering a 6 or 8 week period, as it provides staff with some certainly to enable them to plan their own social and family life. There are other rostering methods that are used for rostering nursing services.

## Cyclic rostering

The cyclic rostering method has pre-set roster patterns developed around industrial rules, occupational health and safety requirements and organizational policies. Each line represents a nurse roster and is given a unique number and skill level. The allocation of nurses to the set roster patterns can be done on a fortnightly, monthly, six weekly or two monthly basis by the nurse manager or by the nurses themselves using self rostering concepts. This method of rostering can be very popular when implemented with skill and equity.

Benefits are:

- Industrial rules, occupational safety requirements and ward rostering policies are easily incorporated and maintained.
- Staff are able to predict their work pattern in advance.
- Can be accommodated on simple spreadsheets or sophisticated roster systems.
- Roster development time is short.
- Ability to optimize shift skill mix from available resources.
- Fair and equitable shift distribution.
- Can accommodate self rostering concepts.
- Shifts can be swapped within the finalized roster.

Disadvantages are:

- Multiple requests within a week may not be accommodated easily.
- Many nursing staff reject this rigid process.
- Some staff become inflexible and resist moving from one cycle to another to accommodate others.

## Self rostering

During the past 15 years the concept of self rostering has become popular. One of the main objectives of this method is to increase staff satisfaction and to attract and retain skilled nursing staff. Self rostering works best in units where there is a variety of personal circumstances amongst staff and they have different preferences about work patterns. It is critical that all nurses concerned fully understand the *rules of rostering*, the outcome objectives for the ward and the minimum requirements for staffing each shift for each day of the week.

Benefits are:

- Improved staff understanding of rostering.
- Staff have more control over hours worked.
- Flexibility is improved if staff are prepared to work additional shifts, change starting times, etc.
- Improved staff satisfaction and reduced absenteeism.
- Staff roster needs and ward staffing needs can be matched.
- Staff can better plan out of work activities.
- Encourages teamwork.

Disadvantages are:

- Requires set protocols.
- Requires significant staff education.
- Difficulty in matching ward requirements and rules to staff needs.
- Industrial rules and occupational health and safety requirements may be compromised.
- Requires regular staff satisfaction review.
- Some staff end up with only *left over* shifts.
- Additional work for the unit manager when staff ignore rules.
- Problems with staff meeting the deadline for posting the roster.
- Requires up to 4 hr of management time to adjust the roster so that it meets all requirements.
- Nurses can become fatigued when they select shifts for social reasons and neglect to consider work fatigue.

### Request focus rostering

Still the most common method of rostering is to make use of weekly or fortnightly requests. This method attempts to accommodate as many staff requests as possible and involves developing a roster from the *ground up* each time. This method is time consuming and is best generated by computerized roster systems. The most common staff complaint related to this model is that some staff are considered to be *favored* as other staff generally receive *left over* shifts after the *favored staff* have been given their requests. This method requires nurse managers to invest significant time to meet staff requests, provide an adequate skill mix for each shift and obtain a cost effective staffing outcome.

Benefits are:

- The ability to meet specific needs of staff members.
- The ability to incorporate a range of shift lengths, shift start and finish times.

- Able to be generated by a computerized rostering system.
- Staff who continually have requests accommodated are satisfied.

Disadvantages are:

- Time consuming process (6 to 8 hr).
- Industrial rules, occupational health and safety requirements and ward roster policies are often compromised.
- Some staff feel disadvantaged.
- Difficult to obtain optimum skill mix from the staff available and also accommodate all staff requests.
- Cannot easily accommodate self rostering.

If nursing staff are to be involved in roster development, they must be educated in:

- Industrial rules.
- Occupational health and safety requirements.
- Policies relating to roster requests, annual leave, rostered days off, etc.
- The competence level of each nurse.
- The skill mix profile for each shift.
- The nursing intensity required for patient care (measured by acuity for each shift).
- Identifying peaks and troughs in workload.
- Financial outcome targets relating to ward salaries.
- Staff training requirements.
- Staff attendance requirements at meetings, etc.

### Rostering process

All of the above possible variables to be considered indicate that developing rosters based on the number of FTE positions allocated to each unit is not a simple matter. In essence the rostering process consists of several steps.

*Step 1*: Consider all 'must comply with' constraints. These should be readily evident in the rostering template.

*Step 2*: Consider all 'if possible' constraints such as staff requests, these are penciled in for starters but may need to be modified, this requires negotiation with staff to determine significance or importance of meeting such requests, as well as ensuring equity.

*Step 3*: Allocate all remaining staff based on number and skill mix requirements by day of the week and time period. This is about filling in the remaining blanks.

*Step 4*: Reach agreement with staff regarding a suitable roster rotating period and a rotational pattern to be adopted. This may require the development of anywhere between four to eight consecutive rosters that can then be continued throughout the year. Frequent reviews and updates are necessary to accommodate staff and service demand changes.

*Step 5*: Establish a process for staff to negotiate shift exchanges between other staff who represent the same skill mix.

### Rostering principles

Rosters need to balance patient care, employee and organizational needs. Principles to be considered when developing a roster are as follows:

- Ensure sufficient staff with an appropriate skill mix are available at all times to provide patient care and meet nursing service demands.
- Enable a fair and equitable distribution of workload.
- Facilitate suitable coverage to be provided for meal, refreshment and personal breaks.
- Conform to all relevant regulatory, legal and policy requirements.
- Meet individual staff requirements and requests by incorporating sufficient flexibility.
- Facilitate staff supervision, training and clinical handovers.
- Facilitate any planned external commitments such as staff meetings, workshop attendance, annual leave.
- Have a strategy in place to meet ad hoc additional staffing requirements to cover any unexpected absence, sick leave or to meet unusual high service demands.

Efforts to contain healthcare costs may result insufficient numbers of senior staff rostered after normal business hours when the overall in hospital staffing is significantly reduced, this can have severe adverse consequences [23]. Given the many variables and constraints that need to be considered when developing rosters, it is clear that the rostering process is a complex task. A good roster is crucial to a unit's ability to provide safe and high quality care as well as maintain staff well being and satisfaction. Suboptimal rostering leads to poor use of available human resources and adds unnecessary costs resulting from an increased use of casual and agency staff. Some of these additional costs can minimized by employing permanent part-time staff.

There needs to be recognition that the nursing workforce is predominantly female who predominantly continue to have primary family responsibilities. All members of the health workforce will have various personal lifestyle considerations, most commonly child care arrangements, and many others such as weekly choir practice or any other community activity. This is unique for every individual. Roster changes may have a flow on effect amongst other family members or impact family budgets. Staff may have partners working shift work, or who are away from home on business from time to time, aging parents etc. Such personal constraints change throughout any individual's life cycle.

There are periods in life when individuals prefer to work every weekend, others may not be able to do so. The same applies to evening shifts or night duty which is particularly difficult for single parents to undertake. Thus adopting a policy where everyone must work their share of out of hours (unpopular to some) shifts is frequently not conducive to achieving a collaborative and supportive staffing culture. Rostering protocols frequently determine a willingness for staff to be allocated to be part of that unit's staffing establishment. It's also a major reason for nurses to choose to be part of the casual staff pool or to work for an agency, or it influences the type of health service they wish to be employed in, as that's the only way they can control when and where they work. As a result of increasing industrial democracy, staff have obtained more input into their roster design and are generally better protected and provided for by industrial rules.

The type and number of shifts allocated to staff can have a substantial effect on staff satisfaction as these impact on an individual's;

- Economic circumstances.
- Time and energy spent with families and in activities other than work.
- Health.
- Degree of satisfaction achieved from work.
- Opportunities to learn and develop.

The rostering process is ideally aided by the use of suitable information systems able to make use of the best available data representing the many variables and constraints that need to be considered and which have incorporated best possible algorithms. Many past attempts have floundered due to the poor or non availability of suitable data. Experienced unit/ward managers tend to have a good 'feel' about when they are likely to have busy and quiet times, they prepare their rosters accordingly.

### Evaluating the suitability of rosters

A roster should be regularly evaluated to assess its suitability to the staff and effectiveness in achieving roster objectives. Rosters plan the working life of people and their feedback is critical. Staff satisfaction surveys, performance reviews, staff exit interviews or questionnaires can be used for this. The nurse manager should ascertain if the roster:

- Meets staff needs.
- Provides adequate staffing numbers.
- Provides an adequate skill mix.
- Matches workload peaks and troughs.
- Provides an equitable mix of shifts.
- Provides an adequate skill mix so that novice practitioners, new staff and students are appropriately supported.
- Provides opportunity for staff professional development.

Other indicators of roster suitability include:

- Patient outcomes, e.g. adverse events.
- Sick leave rates.
- Staff deployment rates.

- Staff turnover rates.
- Efficiency outcomes related to budget.

Ongoing evaluation should be conducted to measure the degree to which the roster:

- Meets workload demand.
- Provides equitable treatment for all staff.
- Meets staff personal preferences.
- Satisfies occupational health and safety requirements.
- Complies with industrial and legal requirements.
- Minimizes staff fatigue.
- Maximizes efficient resource use.

The Australian Nursing and Midwifery Federation has published a comprehensive rostering policy [35] that emphasizes the need for balance between service demand, patient safety and employee well being. It also indicates the need for adequate clinical handover time, relevant staff rest breaks during and between shifts, advance posting to support nurses and midwives' work-life balance, equitable distribution and facilitate staff access to professional development opportunities.

The New South Wales (NSW) Ministry of Health has published rostering resource manual [36] to inform, guide and educate staff with rostering responsibilities to plan, develop, maintain and operate rosters that meet patient, staff and organizational needs. This includes an outline of all mandatory obligations that need to be considered. Similar documents are available in other Australian States. It is highly recommended that every healthcare facility prepares a similar document to ensure regulatory compliance and rostering consistency.

A 2017 survey revealed widespread nurse dissatisfaction with NSW health rosters which were described as short term rosters where staff did not know their roster >2 weeks in advance. More than 78% of nursing staff indicated that they wanted their rosters to be published 4 weeks or more in advance. Four week rosters applied only to 18.92% of those survey participants [37].

### Roster reengineering

When the type of patients, or the type of treatment changes, the resources required also need to change. Patient acuity data clearly demonstrates that certain wards generally require higher nursing hours per patient per day than other wards. This explains why some surgical wards of the same size and occupancy require more nursing hours and have higher costs. As the case mix, length of stay or treatment methods or work routines change within a unit so must the rosters to reflect the altered ward activity, acuity and skill mix requirements. This process of *roster re-engineering* should be undertaken on an annual or as required basis. The worksheet demonstrated in Fig. 4 is generated by a computerized acuity/patient dependency system for the general purpose of roster re-engineering.

# Ward Roster Re-Engineering - Clinical Hours Required (Including 1:1 Care)

Printed: 07/05/2019 10:08:23 AM

**Ward: Medical Ward**  Date Range: 01/05/2018 to 30/04/2019  Roster HPPD: **4.31**

| Shift | Monday | Tuesday | Wednesday | Thursday | Friday | Saturday | Sunday |
|---|---|---|---|---|---|---|---|
| **DAY** | RNs, RMs<br>ENs, PNs, MCs<br>OTH<br>Avg Pts: 38.00<br>Avg Hrs Req: 71:09<br>FTE Shift: 8.89<br>Bed Util HPP: 1.87 | RNs, RMs<br>ENs, PNs, MCs<br>OTH<br>Avg Pts: 37.15<br>Avg Hrs Req: 69:46<br>FTE Shift: 8.72<br>Bed Util HPP: 1.86 | RNs, RMs<br>ENs, PNs, MCs<br>OTH<br>Avg Pts: 37.15<br>Avg Hrs Req: 69:35<br>FTE Shift: 8.70<br>Bed Util HPP: 1.87 | RNs, RMs<br>ENs, PNs, MCs<br>OTH<br>Avg Pts: 36.65<br>Avg Hrs Req: 69:19<br>FTE Shift: 8.66<br>Bed Util HPP: 1.89 | RNs, RMs<br>ENs, PNs, MCs<br>OTH<br>Avg Pts: 36.48<br>Avg Hrs Req: 68:22<br>FTE Shift: 8.55<br>Bed Util HPP: 1.87 | RNs, RMs<br>ENs, PNs, MCs<br>OTH<br>Avg Pts: 33.21<br>Avg Hrs Req: 63:13<br>FTE Shift: 7.90<br>Bed Util HPP: 1.90 | RNs, RMs<br>ENs, PNs, MCs<br>OTH<br>Avg Pts: 34.77<br>Avg Hrs Req: 67:00<br>FTE Shift: 8.38<br>Bed Util HPP: 1.93 |
| **EVE.** | RNs, RMs<br>ENs, PNs, MCs<br>OTH<br>Avg Pts: 39.46<br>Avg Hrs Req: 58:12<br>FTE Shift: 7.27<br>Bed Util HPP: 1.47 | RNs, RMs<br>ENs, PNs, MCs<br>OTH<br>Avg Pts: 39.11<br>Avg Hrs Req: 57:40<br>FTE Shift: 7.21<br>Bed Util HPP: 1.47 | RNs, RMs<br>ENs, PNs, MCs<br>OTH<br>Avg Pts: 38.37<br>Avg Hrs Req: 56:46<br>FTE Shift: 7.10<br>Bed Util HPP: 1.48 | RNs, RMs<br>ENs, PNs, MCs<br>OTH<br>Avg Pts: 37.83<br>Avg Hrs Req: 55:49<br>FTE Shift: 6.98<br>Bed Util HPP: 1.48 | RNs, RMs<br>ENs, PNs, MCs<br>OTH<br>Avg Pts: 36.21<br>Avg Hrs Req: 53:49<br>FTE Shift: 6.73<br>Bed Util HPP: 1.49 | RNs, RMs<br>ENs, PNs, MCs<br>OTH<br>Avg Pts: 34.35<br>Avg Hrs Req: 52:45<br>FTE Shift: 6.59<br>Bed Util HPP: 1.54 | RNs, RMs<br>ENs, PNs, MCs<br>OTH<br>Avg Pts: 36.37<br>Avg Hrs Req: 55:56<br>FTE Shift: 6.99<br>Bed Util HPP: 1.54 |
| **NGT.** | RNs, RMs<br>ENs, PNs, MCs<br>OTH<br>Avg Pts: 35.75<br>Avg Hrs Req: 31:10<br>FTE Shift: 3.90<br>Bed Util HPP: 0.87 | RNs, RMs<br>ENs, PNs, MCs<br>OTH<br>Avg Pts: 35.15<br>Avg Hrs Req: 31:59<br>FTE Shift: 4.00<br>Bed Util HPP: 0.87 | RNs, RMs<br>ENs, PNs, MCs<br>OTH<br>Avg Pts: 35.15<br>Avg Hrs Req: 31:51<br>FTE Shift: 3.98<br>Bed Util HPP: 0.91 | RNs, RMs<br>ENs, PNs, MCs<br>OTH<br>Avg Pts: 34.85<br>Avg Hrs Req: 31:51<br>FTE Shift: 3.98<br>Bed Util HPP: 0.91 | RNs, RMs<br>ENs, PNs, MCs<br>OTH<br>Avg Pts: 34.90<br>Avg Hrs Req: 30:18<br>FTE Shift: 3.79<br>Bed Util HPP: 0.87 | RNs, RMs<br>ENs, PNs, MCs<br>OTH<br>Avg Pts: 32.31<br>Avg Hrs Req: 29:18<br>FTE Shift: 3.66<br>Bed Util HPP: 0.91 | RNs, RMs<br>ENs, PNs, MCs<br>OTH<br>Avg Pts: 36.54<br>Avg Hrs Req: 33:05<br>FTE Shift: 4.13<br>Bed Util HPP: 0.91 |
| **24 Hr** | Avg Pts/Day: 37.74<br>Avg Hrs Req: 160:31<br>FTE Shift: 20.06<br>Bed Util HPPD: 4.25<br>Bed Util %: 99.31% | Avg Pts/Day: 37.23<br>Avg Hrs Req: 159:25<br>FTE Shift: 19.93<br>Bed Util HPPD: 4.28<br>Bed Util %: 97.98% | Avg Pts/Day: 36.79<br>Avg Hrs Req: 157:12<br>FTE Shift: 19.78<br>Bed Util HPPD: 4.30<br>Bed Util %: 96.81% | Avg Pts/Day: 36.79<br>Avg Hrs Req: 155:26<br>FTE Shift: 19.43<br>Bed Util HPPD: 4.26<br>Bed Util %: 95.95% | Avg Pts/Day: 36.46<br>Avg Hrs Req: 151:30<br>FTE Shift: 18.94<br>Bed Util HPPD: 4.33<br>Bed Util %: 92.11% | Avg Pts/Day: 35.00<br>Avg Hrs Req: 148:26<br>FTE Shift: 18.55<br>Bed Util HPPD: 4.38<br>Bed Util %: 89.19% | Avg Pts/Day: 35.89<br>Avg Hrs Req: 156:01<br>FTE Shift: 19.50<br>Bed Util HPPD: 4.35<br>Bed Util %: 94.45% |

NOTES:

\* This report / work sheet identifies the average required hours for each shift on each day of the week for the selected period.

\*\* The report facilitates the development of a roster profile based on acuity trends and the set skill mix profile in staff maintenance.

\*\*\* The Built-in unpredicted value of 12.5% has not been included in the Day and Evening shift totals.

\*\*\*\* Ward FTE Shift Values:  Day: 8:00  Evening: 8:00  Night: 8:00

v3.6  Copyright © 1993,2019 Trend Care Systems Pty Ltd

Page 1 of 1

**Fig. 4**

A computerized roster example. *Produced from the TrendCare® Software Version 3.6.*

Indications for initiating roster re-engineering include;

- Introduction of new services.
- Change in ward case mix.
- Change in surgical or treatment activity.
- Significant decrease in average length of stay.
- Change in patient acuity.
- Introduction of a new staff skill mix.
- Need to improve efficiency.

Benefits are:

- More timely response to patients' needs.
- Improved morale.
- Decreased fatigue.
- Improved satisfaction.
- More efficient utilization of human resources.

Roster patterns in patient care areas cannot remain static. They must be regularly evaluated and appropriately re-engineered to match the fluctuating needs of patients, staff and the organization over time. The roster should be recognized by nurse managers as the most important human resource management tool at their disposal to match demand and supply. Effective rostering is critical to achieve optimal operational outcomes for any health care service.

In order to evaluate the effectiveness of the methodology used to develop staff roster profiles, managers should have a process in place to monitor the daily and shift by shift variance between demand and supply. Variances due to predictable peaks and troughs in demand can be identified if bed utilization is also monitored. The graph displayed in Fig. 5 is used to monitor how well the roster pattern is matching the demand for clinical care. Roster re-engineering is indicated when consistent patterns of miss-matches occur.

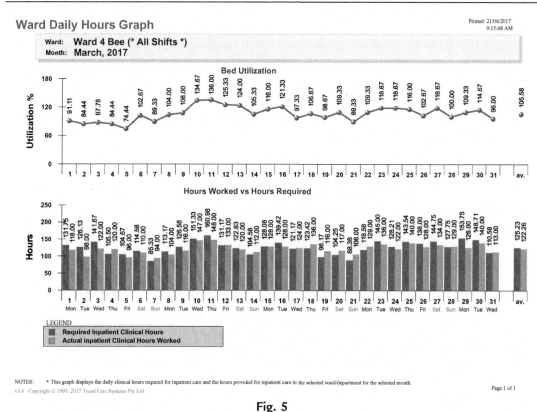

**Fig. 5**

An example of daily and shift by shift variance between demand and supply. *Produced from the TrendCare® Software Version 3.6.*

## *Workforce availability*

Employment contracts determine how many hours one full time employee is able to contribute in any 1 week, beyond which additional hours are classified as overtime. Such contracts also determine leave provisions and any number of other employment requirements for full time employees, employers need to adhere to the conditions of the contract as well as being aware of any cost implications. Other employment contracts cover casual or permanent part time or agency staff. There are also legislative requirements to be considered such as anti-discrimination, health and safety and organizational policies. It is important to remember that rosters also have a medico-legal role and can be challenged in court.

Changes to any of these employment parameters have very significant staff establishment and cost implications. When a 40 hr week changed to a 38 hr week, it was agreed that the duration of rostered shifts remained the same but that a reduction of 2 hr/week equated to a full 8 hr shift once every 4 weeks. This became known as an Accrued Day Off (ADO) in Australia. In the United Kingdom the working week changed to 37.5 hr, this enabled nurses to work three 12 hr shifts in any 1 week, and a 6 hr shift once a month. These instances demonstrate the impact that employment contracts have on rostering practices and possibilities.

## Financial management

Identifying required staffing establishments is one critical part of calculating the capacity to care. Also to be considered are the financial implications. Financial management is an important management skill [38]. Responsible resource management requires access to relevant, accurate and timely cost information associated with all available resources, especially the allocation of human resources, and their relationship with service demand. Those managing the workflow at the point of care need to be included in the development, management and review of budgets as they have intimate knowledge about likely impact of any funding or service demand changes. Strong collaboration between finance and nursing is highly recommended [39–41].

We have all too often witnessed situations where this has not been the case. There is a tendency for some financial directors to take control of nursing budgets. Given that these tend to be the biggest cost component in hospitals, such budgets are often made use of to support other departments with the expectation that nurses will somehow manage. We have observed numerous instances where Directors of Nursing have not been provided with all the necessary information. The same is true for unit/ward managers. In addition many individuals allocated to take on these positions do not necessarily have the relevant financial management skills, preventing them from assertively requesting the information they require [42, 43].

In such situations it may be necessary to provide nurse managers with access to a non-nurse data analyst comfortable in reading, interpreting data and communicating with nurse managers to enable them to make cost effective management decisions leading to significant savings without compromising patient safety or outcomes [44]. Alternatively, healthcare organizations could consider employing a senior nurse with the appropriate financial management skills, in the position of Chief Nursing Finance Officer who liaises between finance and unit/ward/ departmental managers to manage the nurse staffing budget. Five core competencies shared by all healthcare leaders, as identified by the United States Healthcare Leadership Alliance (HLA), continue to be relevant; these are communication, relationship management, professionalism, knowledge of the healthcare environment, leadership and business acumen [41].

### Roster budgeting processes based on service demand

A roster is a plan to utilize human resources. In health care facilities make up a large proportion of the budget. It is most important therefore that the developed rosters remain within the pre-set budget targets such as $ per month for labor costs, and efficiency measured in HPPD. Before posting rosters, managers should check that these targets have not been over run. To calculate the average daily clinical hours per patient day from a roster profile, the following steps should be taken;

NB. In the following example one FTE/WTE shift = 8 hr. This should be adjusted to fit with the relevant service standards.

*Step 1*: Total the average number of full time equivalent (FTE) shifts rostered for each shift of the week (A) shifts.

*Step 2*: Calculate the average hours per week (A x 8) hours.

*Step 3*: Calculate the average hours per day (A x 8)/7 hr per day (B).

*Step 4*: Calculate the average number of patients per day (C) patients.

*Step 5*: Calculate the average number of hours per patient per day B/C hours per patient per day (HPPD).

*Step 6*: Add casual/agency hours (e.g. 5%) to the total HPPD for backfilling sick leave and to accommodate unpredictable peaks in workloads (Table 6).

**Table 6 An example of using HPPD to calculate a ward staffing costs**

| **A. Calculating clinical HPPD in a roster** — Example: | |
|---|---:|
| 1. Total FTE shifts/week includes | 102.5 FTE |
| 2. Total hours/week (average) (@ 8 hr/shift) | 820 hr |
| 3. Total hours/day (average) (7 days/wk) | 117.14 hr |
| 4..Total patients/day (average) | 26 patients |
| 5. Total hours/patient/day (HPPD) (average) | 4.50 HPPD |
| 6. Additional 5% casual/agency | 0.23 HPPD = 5.98hr/day |
| **Total hours/patient/day** | **4.73 HPPD** |

| **B. Calculating the cost of clinical hours in a roster:** utilizing Australian dollars as an example and the average cost of one productive hour (65%RN, 15%EN, 20% AIN/HCA) with on costs = $45 | |
|---|---|
| **Total hours/patient/day (HPPD)** | **(Permanent/substantive staff)** |
| Total $ cost for permanent staff/day @ $45/hr × 117.14 hr | = $ 5,271.3/day |
| Total for one month roster (30 days × $ 5,271.3) | = $158,139/month |
| Total for additional 5% casual hours (5.98 hr/day × $60/hr) | = $ 358.8/day |
| 1 calendar month (30 days) clinical roster cost including casual | = $168,903/month |
| *top up* staff cost | = $(5,271.3 + 358.8)/day × 30 days |

If this same pattern was maintained for a year (365 days), *annual cost of roster for clinical hours only = $2,054,986.5* to provide nursing services for a daily average of 26 patients.

There may need to be a contingency to cover CPI (consumer price index) salary increases.

N.B. *By allowing 5% casual staffing additional to the roster flexibility can be maintained* within a roster and average sick leave covered.

*Continued*

Table 6  An example of using HPPD to calculate a ward staffing costs—cont'd

| C. Calculating non-clinical staff HPPD included in roster: | |
|---|---|
| Ward clerk (8 hr/day) 5 days/week | 0.23 HPPD |
| Orderly (4 hr/day) 7 days/week | 0.15 HPPD |
| PCA non-clinical (4 hr) 7 days/week | 0.15 HPPD |
| Ward manager hours (8 hr) 5 days/week | 0.23 HPPD |
| Ward supervision after hours | 0.17 HPPD |
| Staff training (2%) | 0.09 HPPD |
| Staff orientation | 0.03 HPPD |
| Quality improvement (meetings/audits) | 0.01 HPPD |
| **Total non-clinical hours/month** | **1.06 HPPD** |

| D. Total roster HPPD and allowance for casual/agency top up | |
|---|---|
| Permanent/substantive clinical hours | 4.48 HPPD |
| Additional 5% casual/agency clinical hours | 0.22 HPPD |
| Ward manager hours | 0.23 HPPD |
| Ward supervision/shift co-ordinator after hours | 0.17 HPPD |
| Staff training | 0.09 HPPD |
| Staff orientation | 0.03 HPPD |
| Quality improvements — meetings/audits | 0.01 HPPD |
| Ward clerk | 0.23 HPPD |
| Orderly | 0.15 HPPD |
| PCA non-clinical | 0.15 HPPD |
| **Total roster and additional casual top up** | **5.76 HPPD** |

The above roster components need to be included to ensure that all requirements can be met within the overall unit/ward budget allocation. These time values may also be used to calculate service demand generated for each ward for non-clinical support services by staff category/role.

### Staffing establishment budgeting processes

Budgetary restrictions place enormous pressures on nurse managers by requiring them to consistently provide quality care to a growing number of more complex patients. Nurse managers are often presented with a budget target without their involvement in its development. This target is usually presented as a financial figure (e.g. $ 2.2 million per annum) or in labor hours (e.g. 5.1 hr per patient per day — HPPD). In order to achieve a pre-set target nurse managers must:

- Clarify exactly what is included in the budgeted target.
- Clarify the average cost per productive hour for each staff type and all *on costs* that are included in the target (e.g. penalty rates, superannuation, etc).
- Clarify the projected patient throughput for the next 12 month period.
- Identify peaks and troughs in workload according to patient dependency/acuity data.
- Decide on a suitable staff mix.
- Collect retrospective data on absenteeism rates for the past 12 months.
- Identify the number of hours required per month for staff to attend meetings.

- Calculate the number of nurse management hours required.
- Calculate the number of supernumerary hours required for ward supervision on the evening, night and weekend shifts.
- Clarify the allowance for training.
- Identify hours required for orientation.
- Calculate casual and agency usage as a percentage of total clinical hours worked.
- Identify the number of hours required per month for clerical, orderly and other support staff.

### Zero based budgeting

An alternative and highly recommended method is the adoption of the Zero Based Budgeting process where income minus expenses equals zero. This process enables nurse managers to make use of known service demand, modified to include future service demand projections and to consider associated proposed staff rostering patterns developed to optimize the match between service demand staff numbers and skill mix. This should form the basis for any unit's staff establishment and budget allocation. This budgeting process allocates funding based on program efficiency and necessity rather than budget history [45]. It requires every program and expenditure to be considered at the beginning of each budget cycle and each line item needs to be justified in order to receive funding.

McKinsey [46] indicates that zero-based budgeting is a repeatable process that can be used to rigorously review every dollar in the annual budget, manage financial performance on a monthly basis, and build a culture of cost management amongst all employees. Adopting the process described in the previous section as the means of determining staffing requirements to best meet service demand, provides transparency and visibility into all cost drivers. It builds an organizational culture of cost management by using a structured approach to facilitate cost visibility, cost governance, cost accountability and associated incentives. This process provides opportunities for staff to participate and meaningfully debate options aimed at meeting projected service demands in the most cost effective and value driven manner.

Adopting this process requires a central co-ordination team to develop deep visibility into existing costs and set desirable targets for the next budget cycle. The organization needs to have systems and processes in place to support the process and facilitate its introduction. When applied to variable costs, as is the case with demand driven health service provision, budgets may need to be adjusted accordingly and reported monthly. This enables new operational actions to be considered to either curtail or increase demand, better address other cost drivers or better manage unproductive costs by improving overall productivity and/or outcome performance.

### Activity based costing (ABC)/funding (ABF)

Traditional cost methods simplistically assume homogeneity across patients and providers. Evidence indicates this does not reflect reality. To deal equitably with issues of effectiveness, quality and efficiency in patient care it is necessary to merge the financial with all medical (clinical) information. Only then would it become possible to trace the relationship between clinical decisions made and their impact upon resource usage and costs [47]. This requires all input and output factors to be quantified.

The budgeting processes described represents the Activity Based Costing approach. This is used to provide accurate patient-level cost information independent of service delivery processes in use. This enables costs comparisons to be made regarding service delivery processes in use relative to patient outcomes. That is, cost variations can be identified by variations in the use of service delivery protocols. Such variations can then be examined to explore possible changes to be made to improve efficiency without adversely impacting on outcomes.

Patient acuity systems identify actual nursing and midwifery staffing costs associated with patient types that can also be linked to patient outcomes however identified. Other services also need to be able to be identified on a per patient type basis to enable the identification of total costs per discharged patient treated. This method captures how and why clinical processes, activities and protocols vary from one patient type to another, including amongst patients who present with the same clinical condition but who may vary in other person centered aspects. In addition, activity based costing provides information about the actual mix of resources used to treat individual patients. An accurate determination of actual costs relative to services provided for any type of patient and associated outcomes, is viewed as an ideal mechanism for assessing value [48].

This activity based costing method has been in use in healthcare since the late 1980s [49] and is viewed as representing a more useful method for healthcare facilities than the use of traditional cost-accounting methods, but is considered difficult to implement and resource intensive. The adoption of this costing method requires healthcare organizations to structure their financial systems to enable clinical costing processes to be linked to the financial system. This can happen at various levels of detail to predominantly support a bottom up approach. Where this is not possible a top down approach is used.

Clinical costing in Australia commenced in 1986 at the then Queen Victoria Medical Centre (now Monash Medical Centre (MMC) in an effort to computerize information flows through the hospital's various functional departments and to improve management information. The first step in clinical costing is to capture the use of services of the primary and secondary components (wards and departments) by each patient. That is by means of a workload indicator such as drugs dispensed, theater minutes used and nursing intensity (patient/nurse dependence). The second step is to determine the total cost for each ward or department. This requires the identification of individual cost centers to which all labor (from the payroll) and non labor (from the general ledger) costs are allocated. Tertiary costs (overhead charges) also need to be allocated to these cost centers, using a predetermined allocation method, to arrive at a total departmental cost. The third step is to distribute these total costs amongst all patients serviced by the department in question.

This requires a costing model. Such models may range from the most general to the most specific. The most general model simply calculates an average cost per patient and the most specific model records actual labor time and consumables used per patient. The latter was perceived to represent an expensive costing model to operate, although the aim was to move as far as possible towards the adoption of the most specific costing model. The use of relative value units is advocated where the relative value is assessed based on the relationship of resource utilization of any one workload indicator to the mean resource utilization of all workload indicators. There needs to be a costing model for each unique clinical department which spells out the relative value for each workload indicator [47]. Costs are then distributed on this basis. For example, the departmental salary costs are distributed on a relative value basis whereas overhead costs may be distributed equally amongst all patients.

The alternative top down costing method employed in Australia is usually referred to as 'cost modeling'. This method does not require the detailed data collection described above, instead the relationships between costs, production processes and hospital products are modeled. This method requires the following input data [47]:

- Expenditure by cost center, classified by type of expenditure, for example, salaries and 'other', and by type of cost center, for example, 'overhead', 'patient care'.
- Proportion of total resources used in each patient care cost center which are associated with inpatient activities — the inpatient fraction of each of these cost centers.
- A statistic for each overhead cost center such as total costs, staff hours, floor space, bed numbers, bed days and admissions, which can be measured for all relevant cost centers and used to allocate expenditure by overhead cost centers to all patient care cost centers.
- Number of patients discharged and the number of patient-days, by patient care cost center, for each patient class/type.
- Measures of relative resource use (service weights) by patient class/type for specified patient care and ancillary services.

The first Australian service weights, including nursing service weights, were empirically developed by using details stored in a database obtained from Maryland, USA. These nursing service weights have been reviewed and updated a number of times since then using available patient acuity data obtained from Australian hospitals. Other updates were based on the adoption of new versions of the ICD-10-AM coding system and changes to the patient classification rules over time. Given the complexity and numerous different ways one is able to allocate costs, it is of no surprise that there is little consistency between the many activity based costing applications in use.

A major 2011 health reform policy initiative undertaken in Australia aimed to improve health outcomes as measured relative to casemix. This resulted in the establishment of the

Independent Hospital Pricing Authority charged with determining the National Efficient Price (NEC) for public hospital services, allowing for the national introduction of activity based funding. This authority has published a set of standards and business rules applicable to the patient costing process that each Australian jurisdiction now needs to comply with. They then report their cost data annually to this authority [50]. These data underpin Australian Hospital Patient Costing Standards which by early 2018 was in its 4th version.

One systematic review [51] of its use in healthcare found that applications were often constrained by organizational boundaries rather than spanning the full cycle of care for a medical condition. It also became apparent from the literature that process maps, resource inclusion and time estimates were developed using variable methods, each demanding different amounts of resources and providing variable levels of accuracy. These authors noted that if the purpose of the cost analysis is operational improvement, then only resources around which the initiative is focused need to be included. Indirect costs, described as such in traditional accounting systems, need to be assigned to specific output measures using methods based on cause and effect to more accurately reflect real costs.

## Casemix definitions (hospital 'products')

Product costing methods were first explored during the late 1970s when it was determined that it was nearly impossible and too costly to administer such methods in hospitals because of the many different 'products' (patient types and associated outcomes). As a result decisions were made to group patients in meaningful ways by representing homogeneity in terms of resource usage and thus reduce the number of 'products' (discharged patients) to be accounted for. This was the birth of what is now known as casemix costing and funding using benchmarked standard costs per patient group thus defined. A number of different hospital patient classification (casemix) systems have been developed since then and are in use for a variety of purposes.

The National Health Service (NHS-England) makes use of their Health Resource Groups (HRG4+) as their casemix measures [52] to describe NHS healthcare activity in England. These classifications underpin their reimbursement system from costing through to payment and support, local commissioning and performance management. Casemix based funding in New Zealand underpins its National Minimum Data Set, and applies mainly to medical, surgical, obstetric, and neonatal inpatient services. Other services, such as; outpatient services, emergency department, mental health, rehabilitation, disability support and health of older people, are funded by different methods [53].

California's office of Statewide Health Planning and Development [54] leads data collection and dissemination of information about California's healthcare infrastructure and makes use of a Case Mix Index (CMI). This represents the average DRG weight, calculated by summing the Medicare Severity-Diagnosis Related Group (MS-DRG) weight for each discharged hospital in-patient and dividing the total by the number of discharges to reflect the patient diversity, clinical complexity and resource needs.

Casemix was first introduced in Victoria, Australia during the mid 1980s. This generated a lot of research and development activities associated with the adoption of activity based costing. It formed part of the 1988 Medicare Agreements following which an Australian Government funded casemix development program was established [55]. There is a high correlation between length of stay and total costs or measures of casemix complexity. Standard costs are determined by those hospitals who are able to employ clinical costing methods. A variety of casemix systems continue to be regularly reviewed, updated and in use to primarily support costing, payment or funding arrangements.

Australia and many other countries continue to make use of casemix systems to classify patients treated. The Australian Refined Diagnosis Related Group in-patient classification, groups patients by using patient demographic data. These include; age, sex and length of stay, principal diagnosis plus procedure(s) undertaken plus any additional diagnosis, using the International Classification of Diseases (ICD) classification system and its nationally recognized procedure terminology, the Australian Classification of Health Interventions (ACHI), to arrive at an AR-DRG code for each discharged patient. These data are used not only for activity based costing and a casemix funding model but also for a number of national data collections. ICD coding accuracy influences hospital funds allocated.

Australia's casemix model [56, 57] uses AR-DRG cost weights which are modified for four types of hospital stay, extended, typical, short and same-day and overnight stay. Cost weights increase with level of complexity and cost. Australia has adopted an Episode Clinical Complexity Score (ECCS) based on the Diagnosis Complexity Level (DCL) which is a complexity weight estimating the relative cost of each diagnosis within a particular Adjacent DRG (ADRG). The ECCS is the measure of the cumulative effect of DCLs for a specific episode [58]. Australia has developed and makes use of six standard classifications, as described by the IHPA [59] in an effort to provide a nationally consistent method of classifying all types of patients, their treatment and associated costs. South Australia appears to have adopted different activity based funding models applicable to a number of different health service types [60].

### Use of casemix classifications and nursing service costs

The use of nursing focused patient acuity systems as previously described, constitute a very significant cost component of any hospital casemix system in use. Australian nurses embraced casemix and took advantage of various early research findings that showed the necessity for a per case nursing weight and the promise of using existing nursing patient classification systems for allocating nursing resource use by DRG. Diers [61] has described this research history, including the use of DRGs by nurses, and noted that the then widely adopted PAIS classification system made it possible to develop the original Australian nursing cost weights during the early 1990's [62]. Later updates were based on data obtained from hospitals using the PAIS, TrendCare and South Australian Excelcare nursing workload measurement systems.

Casemix has now moved into the financial landscape as most citations are about costing, cost modeling, service weights or grouper adjustments. A lack of clinical input often results in senior decision makers not fully understanding service delivery work associated with individual AR-DRGs. Clinicians have the knowledge and skills to investigate and explain outlier cost variations. The use of the casemix system increases the transparency of clinical practice. This visibility enables management to influence decisions regarding admissions, treatment options, discharges and length of stay. It enables a variety of performance variations between any number of entity types to be highlighted. It is imperative however for Clinicians to be included in such decision making processes. They need to see their own work reflected in casemix information for continuing engagement.

> *...until clinicians – nurses and physicians – can find, within casemix information, their own work, they will remain on the periphery of the policy arguments that center on costs and funding [61].*

Few nursing managers regularly receive a breakdown of patients as identified by the casemix system in use and associated data such as average lengths of stay or data regarding outliers, service costs or comparisons to State or National benchmarks to enable them to explore service areas that may benefit from some in-depth evaluation. One study undertaken around 2001 [42] found that few nurse managers demonstrated comprehensive knowledge of, or were regularly provided with or were in positions to make use of casemix data. Anecdotal evidence suggests that this has not changed despite a greater use of activity based funding processes and new information systems.

Many Australian health services no longer make use of a nursing workload measurement system since the advent of mandating or using nurse:patient ratios and their continuing use of the midnight census to calculate the average number of patient days. This change is limiting the opportunity to accurately include current nursing service costs as these relate to individual AR-DRGs within financial or clinical costing information systems. This in turn negatively impacts on the ability to accurately reflect nursing service costs in AR-DRG funding models in use [63]. Nursing service costs are not mentioned as a unique entity in the set of standards and business rules applicable to the patient costing process that each Australian jurisdiction now needs to comply with [59].

It is beneficial to develop and implement similar service workload measurement systems to account for costs incurred by other services provided, such as pathology, organ imaging, allied health etc. Total costs also need to include various organizational overhead expenses. Service weights, a measure of the mean cost of the specified service for any patient type relative to the mean cost for all patient types, were calculated for every health service including nursing and may be developed as recognized standards by pooling data from several hospitals. These are then used within various funding algorithms in use by Governments.

## Connectivity requirements for nursing resource management

Table 6 in Chapter 3 listed all input, process and output variables associated with the capacity to care. These formed the basis for the identification of metadata standards and included some analysis of available metadata that relate directly to standard terminologies. The focus of health service delivery is patient-centered care, it is therefore useful to adopt a patient perspective when analyzing the current state of data collection and systems in use relative to future needs. Anyone with a health concern relies on services provided by one or more healthcare providers who need to have access to information about their past history, and current situation in a chronological relevant manner. Any string of health encounters may occupy a short period, several hours, days, weeks, months or years depending on the individual's overall health status and functional ability. This indicates the need for teamwork, multidisciplinary collaboration, multi-provider and multidisciplinary communication needs. Patients should not have to rely on their memory, decide what is relevant or essential information to share and/or repeat themselves over and over again.

Such patient journeys are also good indicators of the variety of internal and external healthcare providers and support organizations that nurses, within any one healthcare facility, need to be able to potentially liaise with or consider. This includes insurance companies, registries, researchers, and various Government departments. This information may be used as a guide for establishing your desirable information system requirements, current capabilities and gap analysis. Human resource management relative to service demand by type, are fundamental to assessing your capacity to care. There is a need to be able to link and make use of data from multiple systems. The greater the connectivity and interoperability options within the healthcare facility, the easier it is to produce the information required for efficient and effective resource management and many other supportive functions such as:

- Care capacity management
- Clinical pathway tracking and multidisciplinary care planning
- Care plans and outcomes reporting by patient type
- Multidisciplinary clinical handovers
- Diet ordering and reporting
- Rostering and leave planning
- Patient risk assessment and reporting
- Patient nurse dependency and workload management
- Workforce planning
- Qualifications, registrations, training and competency registers
- Staff health registers

- Employee and payroll management
- Data flow, multidisciplinary communication
- Communication with external healthcare service providers
- Productivity efficiency and quality reporting, including meeting accreditation requirements

This requires links between patient master index/administration, human resource management, diet & nutrition, patients' clinical record, diagnostic services, therapeutic services, incident/ adverse events and financial information. This provides a great variety of automated reporting and trend monitoring opportunities to inform decision makers at all levels in a timely manner. Successful system connectivity design and interoperability schema development requires the full cooperation of all stakeholders, including vendors of legacy systems and organizational ICT departments, who are major influencers regarding system use, maintenance, security and privacy compliance.

Clinical nurse specialists with ICT/Informatics skills are well positioned to assist with, and/or manage system implementation, user training and use. Additional system implementation contributor roles need to have an understanding of data/information needs for executive strategic developments and vision realization plus the ability to customize systems to suit National, State, Regional and unique facility requirements. This includes data tagging and the ability to match fields and identifiers in accordance with information governance, standards compliance, organization and location needs.

### *Linking electronic health records with nursing resource management*

The implementation and use of EHR systems are growing exponentially. Such systems store a lot of episodic point of care data that could be collected and linked with other data for secondary use including nursing resource management. Every type of clinical record system has been designed in accordance with its own proprietory data structure or reference information model. This means that every system's design and methods used to link with nursing resource management data is fairly unique although there are many similarities. The International Organization for Standardization (ISO) has the following definitions associated with different types of health records as follows [64].

1. An integrated care electronic health record (EHR):
   *'a repository of information regarding the health status of a subject of care, in computer processable form, stored and transmitted securely and accessible by multiple authorized users, having a standardized or commonly agreed logical information model that is independent of EHR systems and whose primary purpose is the support of continuing, efficient and quality integrated health care'*
2. Health record:
   *'a collection of data and information gathered or generated to record the clinical care and health status of an individual or group'.*

3. Healthcare record (synonym medical record, clinical record):
   *'a health record produced for and used within a healthcare organisation or by a healthcare provider'*.

4. EHR:
   *'a health record with data structured and represented in a manner suited to computer calculation and presentation'*.

5. Personal health record:
   *'a health record controlled by the person, or a representative of the person to whom it pertains'*.

Episodic healthcare records used within healthcare organizations are most commonly in use. Nursing resource data needs to be linked to these types of records as described in Chapter 6. One systematic review was able to confirm the potential of this technology to aid patient care and improve clinical documentation. Common negative impacts include changes to workflow and work disruption. Concerns regarding the accuracy and completeness of records, interoperability with related systems, privacy and security remain. It was concluded that current systems have been designed to automate clinical documentation rather than transform and innovate new models of care to add value for patients and clinicians.

These findings also suggest a lack of socio-technical connectives between clinicians, patients and the technology and a lack of consideration for future developments for patient-accessible EHRs, when developing and implementing EHRs [65]. Another national survey found that *'with the current EHR systems physicians are not able to conduct their work in an efficient way'* and that:

> *'EHR and related IT systems should serve a single physician but also their work with numerous other parties since clinical processes are characterised with a high degree of communication and cooperation [66]'*

The same comments apply to nurses who spend even more time than physicians interacting with EHR/EMR systems. Such systems need to be designed in a manner that supports not only physicians but also nursing and allied health workflow and information needs.

> *'Adverse events associated with health IT vulnerabilities can cause extensive harm and are encountered across the continuum of health care settings and sociotechnical factors. The recurring patterns provide valuable lessons that both practicing clinicians and health IT developers could use to reduce the risk of harm in the future'* [67].

One systematic review found that end-user support or lack thereof was the most important factor in both successful and failed implementations, respectively [68]. These authors suggest that following the Expanded Systems Life Cycle management model during implementation, instead of a traditional project management approach, may contribute to greater success over time. Another review of user satisfaction evaluation methods resulted in the development of a guide to assist future evaluators [69].

At a minimum any hospital or residential aged care facility needs to focus on those systems that facilitate effective nursing resource management. Undertaking this type of digital transformation by any healthcare provider is a major project requiring a considerable amount of business/clinical pathway analysis, requirements specification preparation, system design and development and the adoption of a major change management program. When doing so it is beneficial to also consider additional use of data collected and used at the points of care or service delivery. This may guide you to identify the required data standards to facilitate data entry once only, and making use of such data many times, to suit multiple purposes at higher levels within the national health system. Adopting a data-centric approach maximizes health data use. Further details regarding digital transformation requirements enabling data and technical connectivity able to support nursing/midwifery resource management are provided in Chapter 9.

## Capturing and using the data operationally

Acuity systems need to be able to predict/estimate future nursing service demands to enable nurse resource allocation planning to occur. The system then needs to be able to capture real time modifications to reflect actual workloads generated and serviced. This ensures:

- The system's ability to reflect real time service demands on any dashboard.
- Support is provided for operational decision making at any level within the organization as required due to unplanned events.
- The provision of accurate and timely information that represents clinical operational realities, including dietary, treatment, admissions, transfers and discharge changes.

A link with the payroll system ensures every nurse or midwife can be paid according to actual hours worked. Chapter 10 includes a detailed description of patient acuity data collection methods and nursing data uses. The following chapter explores the many issues associated with workforce planning to ensure that each healthcare facility is able to employ the number of staff with the required knowledge and skill mix to meet its projected service demands.

## References

[1] McClelland JE. Report of the committee of enquiry into nursing in Victoria. Victorian Government Ministry of Health: Melbourne; 1985.
[2] Safe Patient Care (Nurse to Patient and Midwife to Patient Ratios) Act. No. 51 Melbourne 2015, [cited 30 November 2018]. Available from: https://www2.health.vic.gov.au/health-workforce/nursing-and-midwifery/safe-patient-care-act; 2015.
[3] NNU. National Campaign for Safe RN-to-Patient Staffing Ratios. [cited 24 August 2018]. Available from: https://www.nationalnursesunited.org/ratios; 2018.
[4] Aiken LH, Sloane DM, Cimiotti JP, Clarke SP, Flynn L, Seago JA, et al. Implications of the California nurse staffing mandate for other states. Health Serv Res 2010;45(4):904–21.

[5] Bolton LB, Aydin CE, Donaldson N, Brown DS, Snadhu M, Fridman M, et al. Mandated nurse staffing ratios in California: a comparison of staffing and nursing-sensitive outcomes pre- and postregulation. Policy Polit Nurs Pract 2007;8(4):238–50.

[6] NHS. Hard truths: the journey to putting patients first UK: the stationary office limited on behalf of the controller of her majesty's stationary office, [cited 30 November 2018]. Available from: https://assets.publishing.service.gov.uk/government/uploads/system/uploads/attachment_data/file/270103/35810_Cm_8777_Vol_2_accessible_v0.2.pdf; 2014.

[7] NHS-Scotland. Nursing and midwifery workload and workforce planning tools, NHS Scotland Workforce Planning-Resources: NHS Scotland, [cited 30 November 2018]. Available from: http://www.knowledge.scot.nhs.uk/workforceplanning/resources/nursing-and-midwifery-workload-and-workforce-planning-tools.aspx; 2011.

[8] Ferguson C, Moore R. Nursing and midwifery workload and workforce planning learning toolkit: NHS Education for Scotland, [cited 30 November 2018]. Available from: https://www.nes.scot.nhs.uk/media/248268/nursing_midwifery_workforce_toolkit.pdf; 2013.

[9] News B. NHS Scotland reports staffing at record high: BBC News, [cited 30 November 2018]. Available from: https://www.bbc.com/news/uk-scotland-scotland-politics-36471626; 2016.

[10] NHS. Use of locum, agency and bank staff Scotland, NHS Ayrshire & Arran: NHS Ayrshire & Arran, [cited 30 November 2018]. Available from: https://www.nhsaaa.net/news/latest-media-responses/use-of-locum-agency-and-bank-staff/; 2018.

[11] Jepson A. The health and care (staffing) (Scotland) bill: The Scottish Parliament-SPICe, [cited 30 November 2018]. Available from: https://digitalpublications.parliament.scot/ResearchBriefings/Report/2018/9/4/The-Health-and-Care–Staffing–Scotland–Bill; 2018.

[12] Borneo A. Staffing for safe and effective care. London: Royal College of Nursing (RCN); 2018.

[13] In short supply: pay policy and nurse numbers-Nurse staffing: can the new guidelines make a difference: The Health Foundation; 2017. [cited 30 November 2018]. Available from: https://www.health.org.uk/sites/default/files/Workforce%20pressure%20points%202017_nurse%20supplement%20FINAL_0.pdf.

[14] Fenton K, Casey A. A tool to calculate safe nurse staffing levels. Nurs Times 2015;111(3):12–4.

[15] NICE. Safe staffing advisory committee, terms of reference and standing orders: National Institute for Health and Care Excellence, [cited 1 December 2018]. Available from: https://www.nice.org.uk/media/default/Get-involved/Meetings-In-Public/Safe-Staffing-Advisory-Committee/SSAC-terms-of-reference.pdf; 2015.

[16] NICE. Safe staffing guidelines 2016, [cited 1 December 2018]. Available from: https://www.nice.org.uk/guidance/service-delivery–organisation-and-staffing/staffing#panel-pathways.

[17] Lawless J. Safe staffing: the New Zealand public health sector experience: NHS England, NICE Safe Staffing Advisory Committee, [cited 1 December 2018]. Available from: https://www.nice.org.uk/guidance/sg1/documents/safe-staffing-guideline-consultation11; 2014.

[18] Hendry C, Aileone L, Kyle M. An evaluation of the implementation, outcomes and opportunities of the care capacity demand management (CCDM) programme-final report. NZ Institute of Community Health Care: Christchurch, NZ; 2015.

[19] WHO. Human Resources for Health: models for projecting workforce supply and requirements, Geneva: World Health Organisation; 2001.[cited 17 December 2018]. Available from: https://www.who.int/hrh/tools/models.pdf.

[20] WHO. Workload indicators of staffing need tool and software., Geneva: World Health Organisation; 2010. [cited 18 December 2018]. Available from: https://www.who.int/workforcealliance/knowledge/toolkit/17/en/.

[21] WHO. Workload indicators of staffing need (WISN): selected country implementation experiences, Geneva: World Health Organisation; 2016.[cited 17 December 2018]. Available from: http://apps.who.int/iris/bitstream/handle/10665/205943/9789241510059_eng.pdf?sequence=1.

[22] Duffield CM, Roche MA, Dimitrelis S, Homer C, Buchan J. Instability in patient and nurse characteristics, unit complexity and patient and system outcomes. J Adv Nurs 2015;71(6):1288–98.

[23] Henderson J, Willis E, Toffoli L, Hamilton P, Blackman I. The impact of rationing of health resources on capacity of Australian public sector nurses to deliver nursing care after-hours: a qualitative study. Nurs Inq 2016;23(4):368–76.

[24] Min A, Scott LD. Evaluating nursing hours per patient day as a nurse staffing measure. J Nurs Manag 2015; 24(4):439–48.

[25] Skills-for-Health. Nurisng workfore planning tool: skills for health, [cited 13 December 2018]. Available from: https://tools.skillsforhealth.org.uk/nursing_planning/.

[26] NQF. Nursing hours per patient day 2012, [cited 23 August 2018]. Available from: http://www.qualityforum.org/WorkArea/linkit.aspx?LinkIdentifier=id&ItemID=70962.

[27] McNamee J, Poulos C, Seraji H, Kobel C, Duncan C, Westera A, et al. Alternative aged Care assessment, classification system and funding models final report.: Centre for Health Service Development, Australian Health Services Research Institute. In: University of Wollongong. 2017.

[28] Gilboy N, Tanabe T, Travers D, Rosenau A. Emergency severity index (ESI): a triage tool for emergency department care, version 4. Implementation Handbook 2012 Edition: AHRQ Publication No. 12-0014. Rockville, MD, [cited 5 December 2018]. Available from: https://www.ahrq.gov/sites/default/files/wysiwyg/professionals/systems/hospital/esi/esihandbk.pdf; 2012.

[29] Bullard MJ, Musgrave E, Warren D, Unger B, Skeldon T, Grierson R, et al. Revisions to the Canadian emergency department triage and acuity scale (CTAS) guidelines 2016: Canadian Association of Emergency Physicians, [cited 5 December 2018]. Available from: https://www.cambridge.org/core/journals/canadian-journal-of-emergency-medicine/article/revisions-to-the-canadian-emergency-department-triage-and-acuity-scale-ctas-guidelines-2016/E2CB3E2063C54E11259313FA4FEAE495; 2016.

[30] Bullard MJ, Melady D, Emond M, Musgrave E, Unger B, van der Linde E, et al. Guidance when applying the Canadian triage and acuity scale (CTAS) to the geriatric patient. CJEM 2017;19(5):415.

[31] Fong RY, Glen WSS, Mohamed Jamil AK, Tam WWS, Kowitlawakul Y. Comparison of the emergency severity index versus the patient acuity category scale in an emergency setting. Int Emerg Nurs 2018;41:13–8.

[32] Weaver SH. Exploring the administrative supervisor role and its perceived impact on nurse and patient safety: Rutgers, the Sate University of New Jersey; 2016.

[33] Lowe C. TrendCare. TrendCare benchmarking report 2012/2013 data; 2014.

[34] Lowe C. An international patient type hours per patient day (HPPD) benchmarking research study unpublished. TrendCare Systems Pty Ltd: Brisbane; 2014.

[35] ANMF. Rostering policy Melbourne: Australian Nursing and Midwifery Federation, [cited 6 December 2018]. Available from: http://anmf.org.au/documents/policies/P_Rostering.pdf; 2015.

[36] Health NMo. Rostering best practice, system performance support branch, HSSG Sydney: NSW Ministry of Health, [cited 27 November 2018]. Available from: https://www.health.nsw.gov.au/Performance/rostering/Publications/rostering-resource-manual.pdf; 2016.

[37] NSW-NMA. Short-term rosters a bugbear for many Sydney: NSW Nurses & Midwives Association, [cited 27 November 2018]. Available from: http://www.nswnma.asn.au/short-notice-rosters-a-bugbear-for-many/; 2017.

[38] Noh W, Lim JY. Nurses' educational needs assessment for financial management education using the nominal group technique. Asian Nurs Res 2015;9(2):152–7.

[39] Esposito-Herr MB, Persinger KD, Regier A, Hunt SS. Partnering for better performance: the nursing-finance alliance. Am Nurse Today 2009;4(4):29–31.

[40] Madigan CK, Harden JM. Crossing the nursing-finance divide: strategies for successful partnerships leading to improved financial outcomes. Nurse Lead 2012;10(4):24–5.

[41] Waxman KT, Massarweh LJ. Talking the talk: financial skills for nurse leaders. Nurse Lead 2018;16(2):101–6.

[42] Blay N, Donoghue J. The provision and utilisation of casemix and demographic data by nursing managers in seven hospitals. Aust Health Rev 2003;26(1):209–18.

[43] Bai Y, Gu C, Chen Q, Xiao J, Liu D, Tang S. The challenges that head nurses confront on financial management today: a qualitative study. Int J Nurs Sci 2017;4(2):122–7.

[44] Douglas K. Taking action to close the nursing-finance gap: learning from success. Nurs Econ 2010;28 (4):270–2.

[45] Deloitte. Zero-based budgeting: zero or hero?, [cited 7 December 2018]. Available from: https://www2.deloitte.com/content/dam/Deloitte/global/Documents/Process-and-Operations/gx-us-operations-cons-zero-based-budgeting.pdf; 2015.

[46] McKinsey. Five myths (and realities) about zero-based budgeting, [cited 7 December 2018]. Available from: https://www.mckinsey.com/business-functions/strategy-and-corporate-finance/our-insights/five-myths-and-realities-about-zero-based-budgeting; 2014.

[47] Casemix HE. Hospital nursing resource usage and costs. [PhD thesis]Sydney, Australia: University of New South Wales (UNSW); 1995.

[48] Goldberg MJ, Kosinski L. Activity-based costing and management in a hospital-based GI unit. Clin Gastroenterol Hepatol 2011;9(11):947–9 e1.

[49] Chandler IR. The Yale cost model, In: The second international conference on the management and financing of hospital services, Sydney, Australia; 1988.

[50] IHPA. Independent Hospital Pricing Authority Canberra 2011, [cited 26 March 2019]. Available from: https://www.ihpa.gov.au/; 2011.

[51] Keel G, Savage C, Rafiq M, Mazzocato P. Time-driven activity-based costing in health care: a systematic review of the literature. Health Policy 2017;121(7):755–63.

[52] NHS. National Casemix Office London: NHS Digital, [cited 10 December 2018]. Available from: https://digital.nhs.uk/services/national-casemix-office.

[53] The New Zealand Casemix System. An overview Wellington: Ministry of Health New Zealand, [cited 10 December 2018]. Available from: https://www.health.govt.nz/publication/new-zealand-casemix-system-overview-0; 2015.

[54] OSHPD. CHHS open data, Casemix index: Office of Statewide Health Planning and Development, [cited 11 December 2018]. Available from: https://data.chhs.ca.gov/dataset/case-mix-index.

[55] H.E. Casemix and Informaion Systems. Hovenga E, Kidd MR, Cesnick B, editors. Health informatics, an overview. Melbourne: Churchill Livingstone; 1996.

[56] VSG. Casemix funding, [cited 10 December 2018]. Available from: https://www2.health.vic.gov.au/hospitals-and-health-services/funding-performance-accountability/activity-based-funding/casemix-funding.

[57] ACCD. AR-DRG Version 9.0-Final Report, [cited 11 December 2018]. Available from: https://www.ihpa.gov.au/sites/default/files/publications/ar-drg_v9.0_final_report.pdf; 2016.

[58] ACCD. Classification Information Portal (CLIP): Australian Consortium for Classification Development, [cited 2018 10 December]. Available from: https://www.accd.net.au/.

[59] IHPA. National activity based funding for Australian public hospitals: Independent Hospital Pricing Authority, [cited 11 December 2018]. Available from: https://www.ihpa.gov.au/.

[60] Gov't-SA. Casemix funding for South Australian Public Hospitals, Adelaide: Government of South Australia; 2018.[cited 10 December 2018]. Available from: https://www.sahealth.sa.gov.au/wps/wcm/connect/cd968c40-4851-4899-82ab-8193fb305b0b/CFMmethodology1819_FCS.pdf?MOD=AJPERES&CACHEID=ROOTWORKSPACE-cd968c40-4851-4899-82ab-8193fb305b0b-mts.qLK.

[61] Diers D. Casemix and nursing. Aust Health Rev 1999;22(2):56–68.

[62] Picone D, Ferguson L, Hathaway V. NSW nursing costing study. Sydney Metropolitan Teaching Hospitals Nursing Consortium; 1993.

[63] Heslop L. Status of costing hospital nursing work within Australian casemix activity-based funding policy. Int J Nurs Pract 2012;18(1):2–6.

[64] ISO. ISO/TR 20514:2005 health informatics-EHR definition, scope and context; 2005.

[65] Nguyen L, Bellucci E, Nguyen LT. Electronic health records implementation: an evaluation of information system impact and contingency factors. Int J Med Inform 2014;83(11):779–96.

[66] Kaipio J, Lääveri T, Hyppönen H, Vainiomäki S, Reponen J, Kushniruk A, et al. Usability problems do not heal by themselves: national survey on physicians' 2019; experiences with EHRs in Finland. Int J Med Inform 2017;97:266–81.

[67] Graber ML, Siegal D, Riah H, Johnston D, Kenyon K. Electronic health record-related events in medical malpractice claims. J Patient Saf 2019;15(2):77–85.

[68] Gruber D, Cummings GG, Leblanc L, Smith Dl. Factors influencing outcomes of clinical information Systems implementation: a systematic review. Comput Inform Nurs 2009;27(3):151–63.

[69] Nahm E-S, Vaydia V, Ho D, Scharf B, Seagull J. Outcomes assessment of clinical information system implementation: a practical guide. Nurs Outl 2007;55(6):282–8 e2.

# CHAPTER 8

# Workforce planning

## Nursing and Midwifery Workforce Statistics

Health workforce planning is critical to a sustainable health system. A projected shortfall requires health services to consider and make changes to staff availability but also changes to the way health service delivery systems are organized. Workforce planning strategies are based on projected service demand associated with projected changes in population demographics as well as projected workforce demographics. Every nation needs to monitor trends regarding these population and workforce changes and make policy adjustments accordingly. There is a global trend toward increases in health service demand accompanied by a decreasing health workforce. The demand for health services is influenced by an aging population living longer with increasingly complex health problems due to increases in chronic diseases and multi-morbidities [1].

The World Health Organization (WHO) has developed global strategic directions for strengthening the nursing and midwifery workforce [2]. Nursing roles are considered critical in achieving global mandates as nurses respond to the health needs of people in all settings and throughout their lifespan. This WHO publication provides a flexible framework for policy makers, practitioners and other stakeholders at every level of the health care system, for broad-based, collaborative action to enhance nursing and midwifery development capacity. Nurses and midwives are well placed to keep the population healthy; this contributes to national productivity and to society generally. The employment of highly skilled nurses and midwives reduces the cost burden of healthcare service delivery. Governments need to acknowledge these positive returns on investments made in nursing and midwifery education, ensure the contribution of nurses and midwives to society are valued and that they are appropriately rewarded so that they remain in the workforce.

There are nearly three times as many nurses and midwives compared to doctors that constitute part of the global health workforce. Nurses and midwives deliver the majority of health services globally yet the nursing profession is absent or poorly represented in all significant policy making committees and governance structures. In 2014 the WHO resolved that the commitment toward universal health care be renewed. This was followed by the development of a global human resources for health strategy that is now aligned with its framework on integrated people-centered health services [3]. The strategy is primarily aimed at planners and policymakers but its contents are useful for all other health industry stakeholders. Policies are required to address health workforce education, the inflows and outflows of staff, maldistribution and inefficiencies and to regulate the private sectors.

Many low and middle income countries are confronted with acute health workforce shortages and inequitable distributions of skilled health workers. From a workforce planning perspective this is due to the supply of qualified workers, recruitment and retention practices, and inadequate knowledge concerning health workforce performance links to outcomes [4]. Global health workforce projections relative to projected demand by 2030, based on the unrealistic assumption that healthcare delivery will continue in accordance with current practices, shows wide variations between regions and a significant global shortage. Our major challenge is to determine future investments influencing both workforce supply and demand factors in order to achieve a healthy population, an equitable distribution of health workers and a sustainable health system.

Workforce planning tools in use will be examined in this chapter, along with service demand and workforce supply statistics, actual workforce participation and turnover rates. Workforce planning strategies need to be employed by individual healthcare facilities, as detailed in Chapter 7, as well as by every government jurisdiction. Traditional nursing and midwifery workforce planning activities have tended to count all nurses and midwives relative to both demand and supply irrespective of type of nurse/midwife or the availability of specialty skill sets. Data availability is a common planning constraint. Many national workforce and service demand statistics represent a questionable degree of accuracy or completeness.

Key policy issues to address nursing and midwifery shortages are workforce participation and retention rates, including the percentage of nurses and midwives working part time, the effective use of the available nursing and midwifery workforce, innovation and reform regarding health service delivery methods, training capacity and supportive employment opportunities for new graduates. There is a need to identify specific issues that dominate in each geographical region, for specific healthcare facilities as well as for individual clinical specialties. This requires us to identify data collection needs to monitor constantly changing trends and to assess the likely impact of re-organization in health service delivery following proposed new policy initiatives that aim to better respond to ever changing health service demands.

Previous chapters have covered a lot of nursing and midwifery service demand detail relative to a great variety of patient types and associated skill mix requirements. Ideally these demand factors are referred to by clinical specialty, in order to identify and quantify service demand by skill mix. Workforce planning is greatly assisted by monitoring statistical trends and in its absence one needs to consider a variety of known factors regarding projected changes in service demand to enable workforce requirements to be projected. Such changes are due to the adoption of new technologies, expanding applications of digital and telehealth and the adoption of new service delivery methods. A future health workforce needs to be organized and capable of responding to the diverse needs of individual persons at different points in time, in a variety of geographical areas, across variable states of health, and throughout progressive life changes [5].

This chapter focuses on the supply component of workforce planning by exploring the types of statistics generated and the efficacy of nursing and midwifery workforce planning models in use. Workforce statistics assist policy makers and healthcare organizations with the ability to contribute to achieving the WHO strategic directions by 2030 [3] which are to:

- Optimize performance, quality and impact of the health workforce through evidence-informed policies on human resources for health, contributing to healthy lives and well-being, effective universal health coverage, resilience and strengthened health systems at all levels.
- Align investment in human resources for health with the current and future needs of the population and health systems, taking account of labor market dynamics and education policies, to address shortages and improve distribution of health workers, so as to enable maximum improvements in health outcomes, social welfare, employment creation and economic growth.
- Build the capacity of institutions at subnational, national, regional and global levels for effective public policy stewardship, leadership and governance of actions on human resources for health.
- Strengthen data on human resources for health for monitoring and accountability of national and regional strategies, and the global strategy.

## Nursing and midwifery's future perspectives

In 2010, the US Institute of Medicine released a publication entitled The Future of Nursing: Leading Change, Advancing Health [6] within which it documented a new healthcare landscape with a patient-centered focus able to provide accessible and affordable care. This was the result of major national changes in healthcare policy. Many other countries have since followed a similar trend. This report addressed the following question: What roles can nurses assume to address in relation to the increasing demand for safe, high-quality, and effective health care services? A very relevant consideration when undertaking a nursing workforce planning activity as this is likely to have an impact on nurses' scope of practice.

Primary and preventative care, where nurses make significant contributions, are seen by many as the central drivers of healthcare now and in the future. One of the key messages in this report was the need for better data collection and an improved information infrastructure to support effective workforce planning. This needs to enable nursing workforce requirement projections by role, skill mix, region and demographics to inform necessary changes in nursing practice and education. It was noted that the power to change conditions to deliver better care, does not rest primarily with nurses. It also lies with governments, businesses, healthcare institutions, professional organizations, other health professionals and the insurance industry. Following this landmark report a campaign for action was launched targeting six major areas, or "pillars" [7]:

- Advancing education transformation
- Leveraging nursing leadership
- Removing barriers to practice and care
- Fostering interprofessional collaboration
- Promoting diversity
- Bolstering workforce data

Recommendations made in the original report are now being implemented and progress is being made. Continued progress was found to require a greater focus and effort in some specific areas. One of the identified barriers to the collection of nursing workforce data includes the lack of national indicators required to provide consistent information. Little progress has been made on building a national infrastructure that could integrate the diverse sources of the necessary data, identify gaps, and improve and expand usable data, not just on the nursing workforce but on the entire health care workforce [7]. This is a consistent story for nations generally. There is a global concern regarding an escalating nursing shortage.

The US Retention Institute at NSI Nursing Solutions Inc., have managed to provide industry insight in their annual reports regarding Registered Nurse (RN) staffing. It noted that according to a March 10, 2017 news release from the Bureau of Labour Statistics, healthcare has added over 30,000 jobs, on average, per month for the past year [8]. This report indicates that the RN labor market continues to tighten with a RN Recruitment Difficulty Index increasing to 86 days at considerable expense to the hospitals concerned. Reported turnover rates were calculated using raw data on all employee terminations excluding temporary, agency and travel staff. It did not account for internal transfers. A slight majority of hospitals include all employment classifications, the others include full/part time staff only. Hospital turnover rates were reported to range from 5.7% to 27.2%. Since 2012 the average hospital turned over 81.7% of its RN workforce over a five-year period.

Australia's 2014 nursing workforce and future requirements report indicates that demand will significantly outstrip supply due to population health trends, combined with an aging nursing workforce and poor retention rates. This significant gap may be reduced to some extend if more nursing students complete their education, there is improvement in the workforce participation rate among new graduates, along with an increase in early career retention. This work applies to the following nursing sectors, acute care, aged care, critical care and emergency, mental health and other [9], indicating a big picture approach devoid from the ability to examine these issues in greater detail. Similar scenarios are reported in other countries.

Major workforce planning issues include poor data, a simplistic view of the nursing profession and nurses' contributions to the health industry as a whole, continuing medical dominance [10] and the lack of a coordinated approach between education providers, governments, employers and the profession. These issues combined with the lag-times in implementing changes, plus broader national economic impacts that have influenced these stakeholders, has resulted in a "boom and bust" cycle in nursing education. Australia has experienced periods of limited employment opportunities for new nursing graduates, while experienced nurses continue to be recruited from overseas [9]. Overseas recruitment is a case of robbing Peter to pay Paul and should be avoided as it often reduces the capacity of providing good health care in under developed countries that have paid for those nurses' education and who are also short of experienced qualified nurses.

New nursing graduates must be provided with opportunities to consolidate their learning to enable them to stay as a member of the nursing workforce. Australia has a three-year bachelor degree program. Other countries have a four- or five-year degree program to prepare Registered Nurses. These programs include one or two years in an internship program where the participants are paid for hours worked. A longer degree program enables a greater amount of supervised clinical practice to be included in the curriculum; its graduates are better prepared for the real world and more likely to make nursing their career choice. The United States is adopting a residential program, similar to the new medical graduates, to address this issue. Ireland has implemented a one year internship for registered nurses.

## Nursing workforce structures and statistics

Workforce planning is usually based on the identification of a nursing workforce structure consisting of the levels of regulated nurses, such as in Australia, Registered nurses, Midwives and Enrolled nurses [9]. Numbers tend to be identified in terms of the number of nurses in or out of the nursing workforce. If participating in the workforce, are they employed in nursing, on extended leave or looking for work? If employed in nursing or midwifery, then they are grouped according to five dominant roles; clinician, administrator, teacher/educator, researcher or other. If not in the nursing workforce then they may be employed elsewhere in a different role and not looking for nursing work, employed as a nurse in another country, unemployed and not looking for work or retired. Those employed as clinical nurses are further identified by their employer's health service type. The greatest number of nurses work in the aged care sector, and the smallest number of nurses work in research, policy and health promotion.

According to the OECD, the number of nurses per 1000 population (headcount) ranges from three to around 17. In 2011 Australia had 10 nurses per 1000 people, the United States had around 12 and the United Kingdom around eight. Mexico had the smallest number and Switzerland the largest number of nurses relative to the population size [9]. The average age of nurses in the workforce is rising as is the number of nurses working part time. Around 90% of the nursing workforce is female. Male nurses tend to mostly work in mental health, followed by Emergency, Management and Critical Care. They are least likely to work in Maternity or Child and Family health.

Another consideration is the geographic nursing workforce distributions. Nurses' demographics also vary by their geographic location. For example, a greater percentage of younger nurses work in cities, whereas a greater percentage of older nurses work in regional, rural and remote areas. There is a significant difference in the number of nurses employed relative to population numbers and retirement rates in various geographic locations. It would also be useful to be able to track individuals longitudinally across their career paths from graduation/first registration onwards, to note migration, job vacancy rates relative to clinical specialty, area of practice and geographical location along with relocation patterns.

## Nursing and midwifery workforce education and professional development

Meeting future health workforce demands requires the provision of education and training opportunities designed to support universal healthcare and the delivery of integrated people-centered health services [5]. The 2018 OECD study on health workforce skills assessment found that health care professionals need to increasingly apply adaptive problem-solving skills to respond to complex and non-routine patient care issues while working in complex, multidisciplinary and frequently stressful occupational environments. This requires resilience and the ability to be flexible with technical, clinical, cognitive, self-awareness and social skills to enable them to monitor situations and respond appropriately [5].

For nurses and midwives to acquire such skills, educational institutions may be required to introduce new institutional educational support systems and instruction methods aligned with country accreditation and regulatory systems and standards. The latter needs to be the result of collaborative efforts aimed at improving workforce competency, quality and efficiency. A 2016 OECD Programme for the International Assessment of Adult Competencies [11] revealed a significant skill mismatch in health care with many doctors, nurses and midwives reported being either over-skilled or under-skilled for the positions they occupied.

Existing programs may need to be reviewed and updated to ensure new graduates as well as the existing workforce are able to acquire all necessary skills. These include interpersonal skills, effective communication, teamwork, self-awareness, and analytical skills including problem-solving to devise customized patient-centered care and effective computer/digital skills.

All health professionals share many core skills [5]. The OECD Health Ministerial meeting held in January 2017 called for a transformative agenda for the health workforce, by adopting drivers of change to address the need for new skills and competencies. This was achieved by assessing health professional skills, remuneration and workforce co-ordination. Workforce skills and models of care need to adapt to, and be well prepared for, digitalization, wider technological changes, and the evolution of patients' needs. As the health workforce is increasingly mobile it is also highly desirable to have internationally comparable skill sets based on a National Qualifications Framework.

A common generic and non-job specific competency framework is proposed [5]. This needs to encompass all the major categories of health professionals and include "areas of focus" or major functions based on a common set of general skills, both cognitive and non-cognitive. These skills are required by every health care team to suit the care needs of specific chronically ill patient profiles. This allows new health professional categories to be included as team members. The effectiveness of work organization and management practices across the healthcare system also needs to be assessed. Policy levers enabling such changes to be made were identified as; payment/financing, organization, regulation, licensing, education and training.

From a workforce planning perspective, it is important to know the number and types of educational programs available and the number of admissions to and completions of these programs that occur per annum. This information enables these programs to be matched with the types of health services that need to be delivered, and their specific workforce requirements so that the programs remain sustainable over time. This information is also required, not only to identify new nurse graduate numbers, but also post graduate courses whose graduates are prepared to take on more senior and specialist roles. All first year courses have high attrition rates, so it's important to know the likely successful completion rates and subsequent registrations enabling graduates to enter the nursing workforce.

All post registration studies lead to additional knowledge and skill acquisition enabling its graduates to work in more senior roles or to work in specialist areas of practice. This area is not well served with workforce planning activities. Few reports have included workforce supply and demand data regarding the many specialty areas, yet attrition rates and/or part time participation rates are known to vary between these areas due to the variable higher emotional and/or physical demands. This is also a result of these areas of practice not being formally recognized via registration processes or the use of occupational category standards. For example; the new discipline of Health or Clinical Informatics is included in many traditional organizational role titles. Dedicated positions for these roles are difficult to identify as these roles have not enjoyed widespread recognition as representing a unique specialty.

It is important to balance the nursing and midwifery supply with projected demand. In response to known shortages universities tend to increase student intakes. However, the supply of new graduates has a reasonably long lead time which needs to be considered. In the interim other measures are taken to minimize the impact such as refresher programs, overseas recruitment and the substitution of qualified positions with non-qualified health care workers. This imbalance has negative consequences. For example; Australia has experienced situations where supply exceeded demand or there was an inability of healthcare providers to increase the number of nurses employed due to budget constraints. This resulted in thousands of new nursing graduates unable to find work in the short term or to consolidate their education [12]. This was exacerbated by the shortage of experienced nurses making hospitals reluctant to employ new graduates while they continued to pursue the recruitment of experienced nurses internationally.

## *Workforce planning models and tools*

The employment of human resources within the health industry to deliver healthcare services constitutes the biggest cost of any healthcare facility. Managing the supply of the health workforce influences the capacity to meet service demand, representing a significant investment. Workforce planning models and tools are required to provide the evidence needed on which to base any number of health service policy decisions. Health workforce planners need to be able to make use of a range of the best available data irrespective of degree of accuracy or completeness. A WHO workforce planning tool includes a list of available methods and tools, describes the processes and resources needed to undertake health workforce planning and describes pathways by which projection results can be optimized to inform policy and program decision making [13]. It lists approaches commonly used as:

- Workforce to population ratio method.
- Health needs method which is a more in-depth approach based on projected changes in population needs for health services.
- Service demand method based on observed health service usage rates for different population groups.
- Service targets methods as an alternative approach specifying service delivery targets by service type.

Each of these approaches has its own data collection and processing needs. Data quality and availability determines the degree of precision and strength of the evidence that can be provided. The choice of variables to be considered should be determined on the basis of their compatibility to policy intervention. This then identifies the scenarios that can be evaluated using the chosen planning model. WHO provides a widely used 234 page instruction document describing models for projecting workforce supply and requirements [14]. Workforce planning open source software is freely available [15].

It is common for workforce planners to start with a model that reflects existing workforce supply trends and service workforce use, based on the assumption that this will continue into the future. This model is then modified to accommodate variable scenarios of change and the likely impacts following policy changes or alternative futures such as changes in models of care, service delivery arrangements, scopes of practice, skill mix, medical or information technologies. Ideally such scenarios are formulated on well informed predictions based on actions taken or changes about to take place. Some of these changes, such as the number of nurses retiring in five or 10 years time, are fairly predictable, action needs to be taken to minimize any adverse impacts resulting from such facts.

The principal method of modeling in use is mathematical simulation modeling making use of a "stock" (current workforce) and "flow" (people entering and exiting the workforce) model, as well as the use of trends or influences on either of these demand and supply factors. The workforce can be broken down into any number of different cohorts, this then allows for different flow rates to be calculated by cohort and year, for each of the input and output factors.

The current workforce changes each year in terms of age and any number of outflows such as retirements, illness/death, career changes, emigration and staff working less hours. Any future workforce is influenced by changes in age of the current workforce plus new graduates entering, re-entries, immigration, late retirements and an increase in average hours worked (workforce participation rates). Various scenarios can be modeled and tested by posing any number of "what if" questions based on related assumptions. These findings are then used to develop new policies or initiatives with the aim of achieving the most beneficial and possible outcomes within any stated time period. Such scenarios need to be considered by every unit/ department or healthcare facility as each of these scenarios has its own characteristics and possibilities. Given the many confounding variables, models can become and should be reasonably sophisticated to provide the best possible information on which to base decisions that will have significant future impacts.

One OECD working paper [16] documents the results of a study of 26 workforce projection models from 18 countries. Most models apply primarily to medical practitioners and are based on the assumption that there is no gap between supply and demand. There are a few more sophisticated models that have made use of information indicating current imbalances. Some relate to the national economic status as this may adversely impact a future workforce supply, current unemployment or under employment for some categories of staff. Only one health workforce planning model had included wages or payment rates as a variable impacting supply and demand for health workers. Few attempts to link health workforce projections with health expenditure projections were found. It was found that different health expenditure growth scenarios can lead to very different conclusions.

One striking example is the expenditure growth in digital health, generating a demand for a new category of health workers, such as health, nursing or clinical informaticians. In the absence of these informatics specialists many clinicians have obtained information and communication technologies (ICT) knowledge and skills. ICT professionals have also obtained some knowledge and skills pertaining to special health service needs as a result of applying these technologies to the health industry. A major constraint to any of these studies is an inability to identify roles that need these new skill sets as the discipline of clinical or health informatics is not officially recognized in any occupational classification standard nor is it identified as a staff category in most surveys. This is a result of the digital health transformation now underway, requiring major changes to be made to all current curricula used for preparing the new graduates and registrants to enter the health workforce. Similar scenarios relate to new clinical specialties or sub-specialties requiring new specialized knowledge and skills.

Canada undertook a study to assess the size of this new skill demand and found that as at 2014 the employment of health informatics and health information management professionals in the public and private sector was estimated to be approximately 39,900 persons with a substantial growth rate, ranging from 6200 to 12,200 over the following five years [17]. This has an impact on health services' ability to optimally manage their nursing resources and on the nursing profession itself. The distribution identified in the Canadian study was 51.4% information technology (IT), 14.3% Health Information Management and 34.3% other professional roles which included changing roles for nurses and midwives.

Demand for these staff categories is directly related to the pace of health ICT expenditure. This also influences changes to health service delivery methods and the boundaries of practice (skill mix required) thus impacting on the required health workforce projections as a whole. For example; there is a new demand for human resources to provide support, make use of and continue the development of digital technologies to optimize their functionality. These new roles tend to be occupied by those already in the health workforce creating an under recruitment at entry level for these roles, leading to a systematic shortage of experienced professionals over the longer term [17].

As was the case in previous chapters, this again leads us into the need for data standards and efficient data collection processes. It's important to identify and record the many existing different databases containing many of the confounding variables that may be used for workforce planning purposes. Many of these are based on data collected via population surveys. Any nation's economy has a strong link with the health of its population. The health industry is a major employer, with its services carrying a significant cost percentage of any nation's GDP for the benefit of the population at large. As such it is highly beneficial to monitor the health industry in a similar manner as economic activity is monitored. Metadata requirements to be considered for modeling purposes are listed in Table 1.

Table 1 Metadata requirements for workforce planning

| Supply | Demand |
|---|---|
| • Available education programs and training pathways<br>• Learner capacity and numbers entering each program annually<br>• Number of graduates per annum, number of registrations for workforce entrants and number of graduates (completions) by clinical specialty and/or area of practice<br>• Immigration, and short term working visa rates<br>• Workforce characteristics by health service type, roles/scope of practice, location and clinical specialty<br>  – Retention and participation rates<br>  – Percentage working part time<br>  – Average weekly hours worked<br>  – Turnover rates<br>  – Vacancy rates by roles, hard-to-fill vacancies<br>  – Unemployment or underemployment rates<br>  – Wages and payment methods | • Population demographics by location, migration and growth patterns<br>• Public health & disease burden by location<br>• Health service types available by location<br>• Health service usage by population demographics, health service type and clinical specialty<br>  – Average length of stay per episode of acute care<br>  – Population percentage with multiple co-morbidities requiring lifelong chronic disease management by age group<br>• Health expenditure projections and associated new workforce roles<br>• Workload indicators of staffing needs relative to every available type of health service |

The United States nursing workforce minimum data set [18] developed by the National Forum of State Nursing Workforce Centre and the National Council of State Boards of Nursing (NCSBN) is used every two years and consists of:

* demographics,
* type of nursing degree, where it was obtained, incl. country,
* highest level of education,
* type of license held and status,
* employment status (full-time, part-time, per diem (casual) not employed),
* reasons for unemployment,
* number of currently employed positions held,
* typical hours/week worked,
* employer service type and location,
* employment specialty of primary and secondary nursing position held,
* position title,
* states in which license is held.

One critical evaluation of nursing workforce forecasting models and content of workforce planning policies for nursing professionals [19], found that current methods are inconsistent, do not sufficiently account for local nursing migration patterns, socioeconomic and political factors that can influence workforce projections and oversimplify the complexity of the nursing workforce. Differences exist between governments and policy makers regarding their use of national economic evidence. Their decisions have a direct impact on the nursing workforce sustainability [20].

Economic factors have a direct impact on workforce supply, demand and the match between these parameters. National health service delivery policy initiatives along with the organization of healthcare delivery structures and the trend toward greater specialization, have an impact on individual patient journeys and the ability of health care professionals to deliver patient centered care. These factors also need to be considered as they influence workforce and skill mix demand by geographical location.

Workforce planners need to obtain far better understanding of the unique features associated with the nursing and midwifery workforce. Workforce planners need to consider not only the numbers of nurses and midwives but also the variety of nursing work together with the level of expertise required [21]. They must also recognize that each health service type has its own workforce demographic, skill mix requirements and use as well as variable participation and turnover rates between service types.

## *Recruitment to the profession*

Unlike previous generations when women were encouraged to take up either a nursing or teaching career, today's secondary school graduates have many more career choices. The university entry score is significantly lower for nursing than other science based degree programs. Consequently students often enter the nursing program and then move to a more favored program in their second or third year of study using the nursing program as a stepping stone to a more favored career. This limits the number of Registered nurses emerging from these programs.

Another limiting factor to meeting the nursing supply needs is the number of available places in nurse education programs. The latter is dependent on the number of suitably qualified nurse educators, clinical sites willing and able to provide clinical student placements, clinical preceptors and budget constraints. Nursing programs need to recruit large numbers from the available pool of secondary school leavers to meet supply needs. This can result in attracting a number of individuals for whom nursing is a second or third choice. These graduates often experience difficulty when transitioning from University to workplace nursing practice. Nursing graduates acquire valuable knowledge and skills that can be applied to many other jobs, making it easy for dissatisfied nurses to obtain other positions in related industries.

Nursing as a profession needs to be an attractive career choice, this is influenced by its public image. The media and social media play a very significant role in projecting this image. Such public images influence public perceptions about the type of work nurses undertake, working conditions, working hours, financial rewards, pay equity, professional recognition/status, continuing education and career path opportunities, gender equity, family friendliness and work-life balance. There is a common view that nurses undertake low level menial work that can be undertaken by anyone. This is wrong and grossly outdated. Unfortunately this view continues to be expressed publicly by politicians and/or their advisors. These factors are viewed comparatively to the roles of other professions when individuals are making career choices. Nursing represents a knowledge based discipline.

The image of nursing that needs to be projected is that nursing is a highly versatile profession of equal status to other healthcare professionals, with enormous diversity and career opportunities. Greater efforts are required to assist nurses with career pathway planning. This may require the provision of various short-term clinical placements within an organization to be part of a professional development strategy. Opportunities also need to be provided to attract those who have left nursing back into the profession by providing comprehensive re-entry programs, including supervised practice. Local economic factors can play a positive role in nurse recruitment as nursing is viewed as a profession most likely to have continuing job opportunities.

The World Health Organization (WHO) workforce reports indicate that globally there are three times as many nurses and midwives than physicians. It has a global code of practice on the international recruitment of health personnel [22] which:

- Discourages active recruitment from countries with official health workforce shortages.
- Encourages countries to develop sustainable health systems that; would allow as far as possible for domestic health services demand to be met by domestic human resources.
- Focuses on policies and incentives which supports the retention of health workers in underserved areas.
- Emphasizes the importance of a multi-sectoral approach in addressing the issue.

## Workforce participation

Job dissatisfaction influences workforce participation rates. Workforce participation is the percentage of nurses working full or part time. Those who are registered to practice but are not in the workforce nor looking to be employed are non-participating. Nurses who are unhappy with rigid or controlling organizational structures or rostering arrangements will opt to work part time to give them greater flexibility and personal control or may leave the workforce entirely or seek employment outside the health industry. During the 1980's Victoria's Nursing Registration Board's annual survey provided evidence that around 30% of nurses and midwives were working part-time. This has changed significantly to more than 50% of nurses working part time today. This is significant from a workforce planning perspective because if a large percentage of part time workers work one extra shift the nursing shortage could be alleviated.

In 2018 Australia had 390,585 nurses and midwives registered. In 2015 nurses worked an average of 33.5 hr/week. Australia's statistics do not indicate the percentage of nurses working part time, they do indicate FTE per 100,000 population by principal area of main job occupied. One Australian study found that nurses' motivators to work part time were complex and that nurses' workforce participation decisions are made in contexts that may be unique to the profession [23]. Reasons for working part time may be voluntary to have time to pursue other interests including further education, meet family caring responsibilities, accommodate poor health, support a better balance between work and lifestyle, escape negative work environments or transitioning to retirement [24].

As at 2017, the most commonly reported initial nursing education of RNs in the United States is the Associate Degree in Nursing (ADN), Bachelor's or graduate degrees were received by 41.8% of RNs, and 17.1% received a Master's degree in nursing. The average age of RNs is 51. This survey reported that 65.4% were actively employed in nursing full-time, 12.1% part-time and 7.1% per diem (casual). The remaining number of licensed nurses were retired, unemployed or working elsewhere. Overall about 16.7% of RNs were employed in two or more nursing positions. More than half of RNs work at least 40 hr/week in their principal nursing position, and another 22.6% work 41 or more hours per week [25].

## Employment characteristics

Industrial awards and employment contracts globally refer to a large variety of position titles associated with variable practice profiles [21] describing employment position characteristics. This lack of a common framework or a standard nursing career structure makes it difficult to do any meaningful workforce comparisons other than using numbers. It also limits the adoption of more detailed workforce planning methods other than to do so locally and/or at organizational levels. Many employment characteristics of relevance to nurses seeking employment are not explicitly included in the industrial awards or employment contracts.

Of importance to most nurses when considering employment is the type of rosters in use in the department, unit or ward in which they would work. Many nurses have caring or case management responsibilities for elderly parents, or family members requiring them to have some flexibility in working hours. Other nurses may prioritize maternity or parental leave conditions, suitable childcare arrangements, reasonable car parking costs, or possible public transport arrangement to meet shift start and end times when considering job opportunities. Some health service locations are in areas where housing is expensive, or the availability of any suitable rental accommodation nearby is limited or too expensive. Other factors to be considered are; the work environment itself in terms of its organizational culture, employment opportunities among local health services and career progression opportunities. There are many influencing factors that determine a nurse or midwife's willingness to accept a full time position, especially when such positions include shift work. The priority of these factors change throughout a nurse or midwife's career due to changing personal circumstances.

## Retention and turnover rates

Staff turnover refers to staff departing from one role or position where there is a need to fill the vacancy created. Nurses/midwives departing may continue to be employed by the same organization in which case they simply change positions from one area of practice, or unit/ward or department to another. Internal transfers may be the result of promotion or a career move. Alternatively nurses/midwives may leave one position to take on another position elsewhere, leave the profession, take a temporary break to study or have a child or leave permanently to retire from the workforce. Many nurses/midwives reduce their hours and work part time. Position changes are often the result of individuals needing to have emotional or physical respite or due to changes in their personal lives requiring them to work different hours which may not be available within the healthcare organization that employs them.

These changes may be either voluntary or involuntary. They may be initiated by the employing organization due to structural or operational changes, or they may be planned rotations to fit with career development or educational needs. Whatever the reason such changes incur, they are a cost to the employing organization and are referred to as staff turnover. The occurrence rate of staff turnover is measured as the percentage of total full time equivalent (FTE/WTE) terminations relative to the budgeted FTE/WTE. These turnover rates will vary by department, unit or ward as well as by service type. It is therefore highly desirable to monitor both planned involuntary turnover rates and unplanned voluntary turnover rates as part of any healthcare organization's human resource management strategy. These rates incur costs that need to be considered as part of the budget as well as for workforce succession planning to ensure long term sustainable service provision. There is no standard definition or method in use to calculate turnover rates relevant to a standard time period, making it difficult to make meaningful comparisons between voluntary turnover rates [26].

A widely used Nursing Turnover Cost Calculation Methodology (NTCCM) was developed during the 1980s. A comparative review of published studies that employed this methodology was undertaken to compare the relative weightings of individual cost items to determine whether some components of this costing methodology has changed over time and across countries [26]. This methodology requires the identification of various cost data items that refer to direct and indirect cost. Direct costs refer to recruitment, human resource management, temporary replacement and where relevant service reduction costs. Indirect costs include orientation, supervision and training of new staff including a decrease in productivity during this induction period, plus unused leave provisions that need to be paid out.

Data items that need to be available to apply this NTCCM methodology are:

- Full time equivalent (FTE) budgeted positions.
- Number of new staff appointed per annum.
- Number of positions vacated, either voluntary or involuntary per annum.
- Number of vacant funded positions at any point in time and/or annual average (vacancy rate).
- Recruitment activities undertaken, including applications received and processed, and all associated costs.
- Pre-employment and hiring costs per position filled.
- Temporary replacement costs.

- Termination costs per employee vacating their position.
- Human resource management and payroll processing costs.
- New employee orientation, supervision, training/education costs.
- Orientation period lower productivity costs.

Clinical staff turnover rates not only have an impact on costs, they may also impact on patient outcomes, remaining staff job satisfaction, continuity of patient care, and overall care delivery efficiency and/or productivity. Management need to make a concerted effort to foster staff retention. This requires a good understanding of the causes of voluntary turnover among nurses. One review of nursing turnover research indicated that the national average is estimated to be 21% and national data indicates that 40% of hospital nurses are dissatisfied with their jobs [27].

As most nurses are female, one significant indicator of intent to stay in the nursing workforce is the degree of work, career or professional orientation and the importance of work for individuals. Anecdotal evidence suggests that many nurses working part-time have a low nursing and/or career orientation. Generational values regarding work and careers are other variances to be considered when promoting nurses' workforce retention [28]. Work environments conducive to staff retention were found to have nurse managers/leaders who were supportive, collegial nurse-physician relationships and collaborative teams [29]. The US 2017 RN staffing report found that while an overwhelming majority of organizations (85.7%) view retention as a key strategic imperative, this was not evident in operational practice/planning. The RN vacancy rate in the United States stands at 8.1%. Vacant nursing/midwifery positions usually require the use of overtime, per diem/casual/bank and agency staff to fill the gaps [8]. Vacancy rates are an indication of recruitment difficulty.

### Causes of dissatisfaction and turnover

There are many reasons for nurses/midwives to become dissatisfied with their work or to initiate a change in employment. These reasons are often very personal, such as desire to study or follow a career path, or to meet new personal requirements. These very personal reasons have nothing to do with the working conditions experienced. One study adopted the social cognitive career theory to identify the antecedents of nurses' professional commitment as this is an indication of the likelihood of leaving the nursing profession. It identifies two career barriers, perceived discrimination and a lack of advancement, and two career supports, human and social capital, that is experience received, that collectively explain career interest. These are that experiences such as perceived discrimination, a lack of advancement, career supports including human and social capital. It is recommended that nurse managers monitor and attempt to lower career barriers while providing strong career supports to enhance nurses' professional commitment as a retention strategy [30].

A poor match between roster or shift work requirements and personal lifestyle desires, difficult personal relationships with managers or co-workers are other frequently occurring reasons for a desire to leave a position. Other reasons are poor graduate preparation or the lack of support for new graduates, a desire to earn more, tiredness due to constantly heavy workloads, working with patients who have behavioral issues due to drugs or mental health issues or negative workplace behaviors (bullying, verbal abuse). Negative workplace behaviors promote a sense of powerlessness [31].

Lateral violence and bullying among nurses is fairly common and has a strong relationship with staff intention to leave. Risk factors for negative interactions were found to be university educated nurses, nurses who only work day shifts, witnessing negative interactions toward other nurses and perpetrating negative interactions. Negative interactions consist of gossip, false information shared with others, not being acknowledged or credited for work done and sarcastic comments. Once such lateral violence becomes bullying behavior the impact on nurses becomes more severe over time and is associated with a greater likelihood of leaving. Stress is a major pre-cursor of this type of behavior. It is highly recommended for nurse managers to acquire communication and conflict management skills enabling them to intervene early, adopt zero-tolerance strategies and codes of conduct, to improve the unit/ward's micro-climate [32]. Hospital units characterized by trust and respect among nurses are less likely to have a culture of bullying [33].

Many nurse graduates are known to leave the profession within their first 2 years of practice. Working environments encountered during those early years must be designed to attract and retain new recruits. This requires a healthy balance between workload demand, complexity and time constraints, respect, the provision of social and professional support for the maintenance of self-esteem and a work/life balance, career opportunities and adequate financial reward. Retention predictors relate to personal dispositional factors, such as self-esteem, general self-efficacy, locus of control and emotional stability, situational workplace factors including opportunities to learn and grow, and new graduate support structures. It's important for nursing management to create satisfying and engaging work environments for new graduates in support of their transition into the nursing profession [34].

An imbalance between effort and reward plus a lack of social support were found to be major influencing factors toward the intent to quit. New nurses' main reasons for changing jobs were found to include a lack of challenges, a desire for change, other career opportunities and dissatisfaction with working conditions [35]. This may differ from one generation to another. Anecdotal evidence suggests that older experienced nurses tend to have higher expectations of new graduates than the graduates are willing or able to deliver.

Some health services represent very stressful work environments often resulting in staff burnout or emotional exhaustion. Symptoms are often not recognized until it's too late. Some nurses are more resilient or better at managing such stresses than others. There are also cultural differences [36] and a number of moral distress factors resulting from having to deal with conflicting demands, unrealistic time frames, ineffective collaboration or having to condone or witness:

- unethical organizational processes or medical procedures,
- breaches of patient confidentiality/privacy,

- an inability to influence medical decisions related to futile care or inadequate management of pain,
- ignoring patient rights and desires,
- a presentation of inadequate information,
- inadequate cooperation with other colleagues,
- chronic understaffing and a resulting inability to provide quality care [31,37].

Some of these instances are likely to be a consequence of nurses not being treated with respect by management or medical staff or a failure to have their professional contributions formally recognized. Power inequalities keep registered nurses in oppressed positions [38]. This may be due to the continuation of working conditions created by the hierarchical nature of the division of labor, as determined by the registered scopes of practice, reporting hierarchies, nurse managers not being given formal decision making authority (delegation of powers) and/or the prevalence of medical dominance in many countries or areas of practice. Medical dominance is often exacerbated by the maintenance of traditional operational processes that minimize opportunities for multidisciplinary teamwork. It is time to dismantle and unveil historical organizational hierarchies within health care organizations that inadvertently propagate oppressive nursing work conditions, as these may act as organizational antecedents to horizontal violence [38].

According to Svensson [39] an increased prevalence of chronic illness has resulted in a change of emphasis from the medical body systems perspective of death prevention to a holistic nursing perspective of handling life. This is further supported by a change from nursing work task allocation to small team nursing which facilitates closer nurse-patient relationships as each nurse has fewer patients to care for at any one time. This in turn enables individual nurses to directly interact with medical staff rather than exchanging their knowledge about each patient via the unit/ward nurse manager during formal ward rounds. Nurses need to be able to interact with medical staff in the prevention of adverse events. For example we have frequently observed experienced nurses informally teaching newly appointed medical staff. This was confirmed by a field study in a large English hospital and is a major incentive for nurse retention as this facilitates care continuity and patient safety [40].

Another form of stress for nurses are organizational protocols and inflexible boundaries of practice between health professionals. This often leads to nurses having to observe patient distress where activities cannot be provided in a timely manner by the person with the primary responsibility for undertaking an activity that could have been undertaken by the nurse within a shorter time frame. High stress eventually adversely impacts on a nurse's health and emotional well-being.

A leading Job Demands-Resources (JD-R) occupational stress model may be used to guide the understanding of specific workplace characteristics that can lead to negative organizational outcomes including high turnover rates. This model acknowledges that stress stems from the individual's response to a perceived imbalance between demands made on them and the available resources the individual has available to deal with those demands [41]. The energy required to manage heavy nursing workloads for long periods is a major cause of role stress and job dissatisfaction leading to an increase in turnover rates. Nursing job demands may be a combination of physical, psychological, organizational, social, or emotional demands. Available resources used to deal with such demands include resilience, autonomy, inner strength, maturity, self-efficacy, optimism, motivation, professional commitment, social and professional support plus interpersonal, role, organizational and professional knowledge and skills. Service demands and the energy required to meet these vary according to health service type. As a result, turnover rates also vary by health service and role type. Emergency and behavioral health services tend to have far higher turnover rates than other services for example [8]. A high demand and high resource mismatch will often lead to higher turnover rates.

A continued high demand and high resource mismatch leads to a depletion of energy and health impairments, and exhausts an individual's physical, emotional and mental resources. This may be exacerbated by an individual's personal lifestyle circumstances such as domestic violence, relationship or family issues. Matching such demands with available resources requires effective recruitment strategies, advanced people management skills, appropriate rostering and the ability of management to generate and maintain a work environment that, together with suitable job characteristics, facilitates work engagement. Criteria for work engagement will vary among individuals and impacts their turnover intention. For example; experienced nurses benefit from greater job autonomy that fosters a sense of responsibility and meaningfulness that subsequently keeps them in clinical practice [42]. Well matched JD-R leads to better patient outcomes, high productivity, low turnover and high retention rates.

There is not usually any one single factor that causes a nurse to leave. Most nurses love the work they do, but need to employ various defense mechanisms to deal with any one of the turnover causal factors identified. Turnover is a complex, dynamic process that unfolds over time diminishing the capacity of nurses to function effectively in the practice setting [31]. The many and varied factors known to influence turnover rates need to be considered operationally by service managers who themselves also need appropriate policy and human resource management support. There is a strong need to foster healthy work environments for student nurses, new graduates, clinical educators and experienced nurses to optimize retention rates.

## Replacement and succession planning

Proactive succession planning is a key strategy that needs to be widely adopted to address nursing shortages. Few if any effective health workforce succession planning programs exist [43]. Those reported in the literature have focused on leadership and internal succession planning, excluding its relationship to external recruiting [44]. Most Healthcare leadership succession planning focuses on senior executive positions and lags behind other industries [45].

Previous chapters have described nursing work from numerous different perspectives. Educational and experiential preparation and development for the many different roles relating to the many and varied health service types and clinical specialties are unique for each role. Succession planning is about ensuring a continuing supply of human resources able to fulfill and meet the service demands of all nursing roles and is an intricate component of workforce planning.

Succession planning benefits from the development of career pathways and programs, including incentives, that actively support individuals to develop their own preferred career pathways. Some individuals actively identify their passions, strengths, values, preferred job characteristics and life directions. This enables them to identify and take advantage of educational and on the job opportunities for professional career development. Others need to be guided and presented with such opportunities as incentives toward greater professional recognition and engagement which in turn increases their willingness to stay in the workforce and thus reduces turnover rates. This is an important required step in any female dominated occupation where its workforce tends to undervalue their own contributions.

Nurse managers play an integral role as creators and maintainers of the overall work environment, reducing turnover rates and providing opportunities for professional and career development of staff. These are key positions for which retention and succession planning is particularly important. Studies show that resource allocation for proactive and deliberate development of current and future nurse leaders is currently lacking. An organization's commitment to identifying and developing those with high-potential promotes the retention of high performing individuals [46]. Such commitment is necessary not only for nurse leaders but also for all nursing roles within an organization. This requires continual knowledge transfer and sharing with other team members. Such a strategy supports new graduates preventing them from leaving within their first years of practice, promotes professional engagement and encourages them to develop a preferred career path. We cannot afford to loose nursing knowledge and experience from the health industry or from academia responsible for maintaining the supply of a well educated nursing workforce.

National workforce planning needs to be supported locally by organizational workforce planning activities to ensure that the projected continuing service demands can be met. The value hospitals place on their people will have a direct correlation to staff commitment, confidence and engagement [8]. Healthcare organizations need to strategically consider how they can best ensure a continuing supply of the talent they need to meet current and future service demands. In doing so they contribute to the national and global health workforce pool. National policies that address education and industrial matters plus specific program initiatives play their part, but it is essential that individual healthcare facilities influence, contribute to and make use of national opportunities for workforce development. A replacement and succession plan is especially necessary for key leadership positions in order to maintain stable and effective leadership. It is also an important element of developing staff to enable them to reach their full potential.

Professional organizations such as the American Nurses Association (ANA) and the Australian College of Nursing (ACN) also need to play a role by providing webinars, scholarships, conferences and/or educational programs. The ACN has developed a comprehensive emerging nurse leaders development program for 50 ambitious go-getters per annum to advance their nursing careers. This program is a self-paced, self-driven blended program consisting of formal education, self-reflection, mentoring and action based learning [47].

Turnover and vacancy rates vary by health service type, position type and department. It is extremely valuable to measure and monitor these. Vacancies that are difficult to fill and high turnover rates need extra attention by identifying the usual human resource supply chain's location, entry level requirements, modes of education/training delivery, practical experience requirements and graduation numbers. This provides a baseline from which to proactively develop a strategy to address these issues. Turnover rates can be reduced by addressing the causes of turnover. This may require focused strategic initiatives using human resource management expertise and support. By identifying role competencies based on job demands the resource demand factors for each role can be identified. This provides a baseline to be used for recruitment and professional development purposes ensuring the best possible JD-R match.

It is important that human resource managers work with people currently occupying key leadership positions to address any staff talent gaps and to enhance retention. Nurse managers are usually highly qualified and experienced clinicians, but in most situations they need to develop people management and administrator skills. Leaders need to identify and develop those with high potential for specific roles or positions. Healthcare facilities need to have documented vision, mission, values, stakeholders, specific health service type objectives and performance criteria. These can utilized as a positive and motivating force in every department by embedding them within leadership roles, position behavioral statements and evaluations to address very specific strategic directions to support retention, replacement and succession planning. In some situations there may need to be organizational or service delivery changes to make better use of the available workforce.

## Meeting future demands

Alleviating a continuing nursing and/or midwifery shortage by making greater demands on those remaining in the workforce is unsustainable as it creates a negative spiral, with an increasing number of dissatisfied nurses/midwives leaving the profession, compromised patient safety and an increase in service costs. Nurses and midwives have a strong focus on the need for safe staffing levels however identified and play a vital role in the prevention of adverse events. Unsafe staffing levels generate enormous stress and compromise patient safety.

The OECD noted that many countries face significant challenges due to increasing health service demands in fiscally constrained environments [5]. In this situation it is important for health service executives to explore new service delivery options and provide services relative to nursing and midwifery resource availability. If surgeons are unavailable, surgery doesn't happen, the same principle needs to apply to nursing services. This means that when nursing workload monitoring systems indicate the need for more nurses, then either more nurses are recruited or services are reduced to accommodate such situations. Addressing the issues identified as causes of nursing/midwifery dissatisfaction and turnover will improve nursing and midwifery working conditions which in turn will improve recruitment, retention and turnover rates. Satisfied nurses/midwives will also improve the public image of nurses and midwives, which in turn further enhances recruitment.

There needs to be a greater collaborative approach to workforce planning activities that make use of health service demand by geographic locations based on public health patterns, to identify knowledge and skill demands, rather than focusing on specific health professional occupations and numbers. Turnover statistics need to be monitored by health service type so that supply needs can be more accurately calculated and made use of. The population numbers requiring any of the many types of health services is also changing. For example a greater number of people living with life-long chronic health conditions will require a greater number of nurses to be able to meet complex clinical needs to work in community and home care [48]. The supply and skill mix of nurses must change to meet this changing demand. Nurses provide support to individuals living with chronic conditions, to improve their self-care and independence. Nurses also support the informal carers.

With the increasing number of specializations it is imperative that workforce planning places a greater emphasis on the need to meet changing skill demands relative to changing roles by health service type. Only then is it able to meet the health system's workforce needs with a sustainable and fit-for-purpose workforce. The future holds many uncertainties. A two-case multiphase mixed method study revealed that a few dominant actors of considerable influence (governments and health service providers) are in conflict over a few critical workforce issues. This study offers a novel approach and guidance for workforce planners and policy makers on the use of complimentary data and methods to overcome the limitations of conventional workforce forecasting. A framework for exploring the complexities and ambiguities of a health workforce's evolution is offered [49].

With current technologies it is possible to ensure the availability of expert advice at any time of day or night from any location. Establishing the means to do so ensures health professionals can fully deliver in accordance with their scope of practice. Health policy initiatives need to accommodate these changing requirements by facilitating network models of care with shared governance and leadership reflecting integrated working practices. This also requires interprofessional education to promote improved understanding and integrated team work.

These potential changes need to be part of the activities undertaken by health service planners who in turn need to work closely with those who undertake workforce planning.

This chapter plus all previous chapters have highlighted the need for data consistency, accuracy, completeness, agreed data structures and frameworks. This level of data is required to undertake data analytics to; determine service demand by health service and patient types, measure workloads, identify workforce knowledge and skill mix, document models of care, manage human resources, plan and prepare the health workforce.

The next chapter focuses on the collection, transfer, linking, sharing and use of data and informatics generally from information system connectivity and interoperability perspectives.

## *References*

[1] Health Workforce Australia. Australia's future health workforce—nurses. Canberra: Commonwealth of Australia; 2014. p. 84.

[2] WHO. Global strategic directions for strengthening nursing and midwifery 2016–2020. Geneva: WHO; 2016.

[3] WHO. Global strategy on human resources for health: workforce 2030. Geneva: Wolrd Health Organisation; 2016. [cited 18 December 2018]. Available from: https://www.who.int/hrh/resources/pub_globstrathrh-2030/en/.

[4] Liu JX, Goryakin Y, Maeda A, Bruckner T, Scheffler R. Global health workforce labor market projections for 2030. Hum Resour Health 2017;15(1):11.

[5] OECD HDT. Feasibility study on health workforce skills assessment: supporting health workers achieve person-centred care. Division H; 2018.

[6] IOM, IOM, editors. The future of nursing: leading change, advancing health. Washington, DC: The Natioanl Academies Press; 2011.

[7] IOM, Altman S-C. In: National Academies of Sciences, Engineering, and Medicine, editor. Assessing progress on the Institute of Medicine report, the future of nursing. Washington, DC: The National Academies Press; 2016.

[8] Colosi B. National health care retention & RN staffing report. NSI Nursing Solutions, Inc; 2017. [cited 30 December 2018]. Available from: https://www.emergingrnleader.com/wp-content/uploads/2017/09/NationalHealthcareRNRetentionReport2017.pdf.

[9] HWA. Australia'S future health workforce—nurses. Health Workforce Australia; 2014. [cited 13 December 2018]. Available from: http://www.health.gov.au/internet/main/publishing.nsf/content/australias-future-health-workforce-nurses.

[10] Willis E. Medical dominance: the division of labour in Australian health care. Sydney: Allen & Unwin; 1990.

[11] Schoenstein M, Ono T, LaFortune G. Skills use and skills mismatch in the health sector: what do we know and what can be done? In: LaFortune G, Moriera L, editors. Health workforce policies in OECD countries: right jobs, right skills, right places. OECD Publishing; 2016. p. 163–83.

[12] ABC-News. Thousands of nursing graduates unable to find work in Australian hospitals: union. Australian Broadcasting Corporation (ABC); 2014. [cited 31 December 2018]. Available from: https://www.abc.net.au/news/2014-05-24/thousands-of-nursing-graduates-unable-to-find-work/5475320.

[13] WHO. Models and tools for health workforce planning and projections. Geneva: World Health Organisation; 2010. [cited 17 December 2018]. Available from: http://apps.who.int/iris/bitstream/handle/10665/44263/9789241599016_eng.pdf?sequence=1.

[14] WHO. Human resources for health: models for projecting workforce supply and requirements. Geneva: World Health Organisation; 2001. [cited 17 December 2018]. Available from: https://www.who.int/hrh/tools/models.pdf.

[15] iHRIS. Open source human resources information solutions: IntraHealth International; [cited 17 December 2018]. Available from: https://www.ihris.org/ihris-suite/download/.

[16] Ono T, LaFortune G, Schoenstein M. Health workforce planning in OECD countries; a review of 26 projection models from 18 countries. OECD Health Working Papers, OECD Publishing; 2013.

[17] CHIMA, CIHI, COACH, ICTC, ITAC. Health informatics and health information management: human resources outlook 2014–2019. Available from: https://www.echima.ca/uploaded/pdf/reports/HI-HIM-HR-Outlook-Report-Final-w-design.pdf; 2014.

[18] HRSA. Health workforce: Health Resources & Services Administration; [cited 19 December 2018]. Available from: https://bhw.hrsa.gov/health-workforce-analysis/.

[19] Squires A, Jylhä V, Jun J, Ensio A, Kinnunen J. A scoping review of nursing workforce planning and forecasting research. J Nurs Manage 2017;25(8):587–96.

[20] Buchan J, Twigg D, Dussault G, Duffield C, Stone PW. Policies to sustain the nursing workforce: an international perspective. Int Nurs Rev 2015;62(2):162–70.

[21] Gardner G, Duffield C, Doubrovsky A, Bui UT, Adams M. The structure of nursing: a national examination of titles and practice profiles. Int Nurs Rev 2017;64(2):233–41.

[22] WHO.Global Health Observatory (GHO) data—health workforce: World Health Organisation; [cited 31 December 2018]. Available from: https://www.who.int/gho/health_workforce/en/.

[23] Jamieson LN, Williams LM, Lauder W, Dwyer T. Nurses' motivators to work part-time. Collegian 2007;14(2):13–9.

[24] Burke RJ, Dolan SL, Fiksenbaum L. Part-time versus full-time work: an empirical evidence-based case of nurses in Spain. Evidence-based HRM: a Global Forum for Empirical Scholarship 2014;2(2):176–91.

[25] Smiley RA, Lauer P, Bienemy C, Berg JG, Shireman E, Reneeau K, et al. The 2017 National Nursing Workforce Survey. J Nurs Regul 2018;9(3).

[26] Duffield CM, Roche MA, Homer C, Buchan J, Dimitrelis S. A comparative review of nurse turnover rates and costs across countries. J Adv Nurs 2014;70(12):2703–12.

[27] Gilmartin MJ. Thirty years of nursing turnover research: looking back to move forward. Med Care Res Rev 2012;70(1):3–28.

[28] Robson A, Robson F. Do nurses wish to continue working for the UK National Health Service? A comparative study of three generations of nurses. J Adv Nurs 2015;71(1):65–77.

[29] Nantsupawat A, Kunaviktikul W, Nantsupawat R, Wichaikhum O-A, Thienthong H, Poghosyan L. Effects of nurse work environment on job dissatisfaction, burnout, intention to leave. Int Nurs Rev 2017;64(1):91–8.

[30] Chang H-Y, Chu T-L, Liao Y-N, Chang Y-T, Teng C-I. How do career barriers and supports impact nurse professional commitment and professional turnover intention? J Nurs Manage 2018;1–10. https://doi.org/10.1111/jonm.12674.

[31] Hayward D, Bungay V, Wolff AC, MacDonald V. A qualitative study of experienced nurses' voluntary turnover: learning from their perspectives. J Clin Nurs 2016;25(9–10):1336–45.

[32] Bambi S, Guazzini A, Piredda M, Lucchini A, De Marinis MG, Rasero L. Negative interactions among nurses. An explorative study on lateral violence and bullying in nursing work settings. J Nurs Manag 2019; https://doi.org/10.1111/jonm.12738.

[33] Arnetz JE, Sudan S, Fitzpatrick L, Cotten SR, Jodoin C, Chang C-H, et al. Organizational determinants of bullying and work disengagement among hospital nurses. J Adv Nurs 2018;75(6):1229–38. https://doi.org/10.1111/jan.13915.

[34] Laschinger HKS. Job and career satisfaction and turnover intentions of newly graduated nurses. J Nurs Manage 2012;20(4):472–84.

[35] Lavoie-Tremblay M, O'Brien-Pallas L, Gélinas C, Desforges N, Marchionni C. Addressing the turnover issue among new nurses from a generational viewpoint. J Nurs Manage 2008;16(6):724–33.

[36] Guo Y-F, Plummer V, Lam L, Wang Y, Cross W, Zhang J-P. The effects of resilience and turnover intention on nurses' burnout: findings from a comparative cross-sectional study. J Clin Nurs 2019;28(3–4):499–508. https://doi.org/10.1111/jocn.14637.

[37] Zolala S, Almasi-Hashiani A, Akrami F. Severity and frequency of moral distress among midwives working in birth centers. Nurs Ethics 2018. 0969733018796680.

[38] Blackstock S, Salami B, Cummings GG. Organisational antecedents, policy and horizontal violence among nurses: an integrative review. J Nurs Manage 2018;26(8):972–91.

[39] Svensson R. The interplay between doctors and nurses—a negotiated order perspective. Sociol Health Illn 1996;18(3):379–98.

[40] Allen D. The nursing-medical boundary: a negotiated order. Sociol Health Illn 1997;19(4):498–520.

[41] Gabel Shemueli R, Dolan SL, Suárez Ceretti A, Nuñez del Prado P. Burnout and engagement as mediators in the relationship between work characteristics and turnover intentions across two Ibero-American nations. Stress Health 2016;32(5):597–606.

[42] Wan Q, Li Z, Zhou W, Shang S. Effects of work environment and job characteristics on the turnover intention of experienced nurses: the mediating role of work engagement. J Adv Nurs 2018;74(6):1332–41.

[43] Lorrie EL. Making the case for succession planning: who's on deck in your organization? Nurs Leadersh 2011;24(2):68–79.

[44] Griffith MB. Effective succession planning in nursing: a review of the literature. J Nurs Manag 2012;20 (7):900–11.

[45] Titzer JL, Shirey MR. Nurse manager succession planning: a concept analysis. Nurs Forum 2013;48 (3):155–64.

[46] Titzer J, Phillips T, Tooley S, Hall N, Shirey M. Nurse manager succession planning: synthesis of the evidence. J Nurs Manage 2013;21(7):971–9.

[47] ACN. Emerging Nurse Leader program: Australian College of Nursing; [cited 31 December 2018]. Available from: https://www.acn.edu.au/leadership/emerging-nurse-leader-program.

[48] Drennan VM. More care out of hospital? A qualitative exploration of the factors influencing the development of the district nursing workforce in England. J Health Serv Res Policy 2019;24(1).

[49] Rees GH, Crampton P, Gauld R, MacDonell S. The promise of complementarity: Using the methods of foresight for health workforce planning. Health Serv Manage Res 2018;31(2):97–105.

# CHAPTER 9

# Digital health ecosystems: Use of informatics, connectivity and system interoperability

## A need to resolve data issues

Chapter 3 focused on the actual data used to measure nursing care demand that defines nursing workloads and the capacity to care. This chapter included a table that identified the many variables known to influence resource usage and in-patient costs and noted that it would be useful to be able to make use of system dashboards, where data from multiple sources are presented on one screen to assist decision making at any level within the healthcare facility. The introduction of precision medicine, new global directions to provide patient centered care, and the increasing use of health information technologies and continuing health IT developments, makes it imperative to consider what any nation's digital health ecosystem should look like. Chapter 3 also explored digital transformation requirements, primarily in terms of data standards and the need to reach agreement regarding metadata standards that relate to those variables known to influence productivity and capacity to care.

Chapter 4 described nursing work and how this could be measured noting that work content relates to outcomes, the result of work performed. Work measurement is about documenting processes that determine how available input resources are used to produce outputs. We need to be able to describe, measure and relate all these factors to determine our capacity to care. Chapter 4 also noted that work consists of productive work as well as many inefficiencies, which either the individual worker or management can minimize. Nursing workload measurement systems are about determining nursing productivity relative to patient outcomes. This needs to be considered within the context of any type of healthcare facility as well as the nation's health system as a whole. Chapter 5 explored one very significant input factor, workforce skill mix.

Chapter 6 explored significant process factors, nursing care processes, nursing and organizational models of care. This chapter noted numerous influencing confounding variables. Chapter 7 explored processes associated with the management of staffing resources including the data variables required to measure actual service demand need. Chapter 8 explored workforce planning that determines the size and availability of the health workforce in any location.

All of these previous chapters identified numerous data issues including the lack of suitable metadata standards, poor data availability, inaccuracies and incomplete data that impede our ability to make well informed decisions at every level. In summary we need to relate these issues to our desire to optimize productivity by improving efficiency and effectiveness. Productivity is the ratio of output to input. It applies to any business or industry or an economy. It is therefore essential to consider all resources used (input), processes or activities undertaken to create a change, and what was achieved (output) in a format that enables optimum use of our new digital world.

Leaders and managers are instrumental in increasing productivity and quality as they are in positions to create a favorable organizational climate where collaboration and co-operation can flourish. Managers are responsible for obtaining the facts, planning, directing, co-ordinating, controlling and motivating staff in order to deliver any health service. They need to ensure that all resources needed (input) to undertake all care delivery processes are made available at the right time and place. The health workforce needs to apply the right knowledge and skills to ensure that every process is undertaken correctly and that available resources are used appropriately. It is now widely recognized that there is a need to transform our current systems to a more integrated approach, to enable more efficient use of available resources, avoid fragmentation, and facilitate information sharing for better and faster decision making.

Digital health ecosystems apply to individual healthcare organizations, networks or regions as well as the nation as a whole. The delivery of the potential of "digital health" requires national or regional digital health initiatives guided by a robust strategy that integrates financial, organizational, human and technological resources [1]. Digital health refers to "the field of knowledge and practice associated with any aspect of adopting digital technologies to improve health, from inception to operation" [2]. Successful implementation requires collaboration. There is a need to think globally and act locally by filling capacity gaps and building on the related current digital health journey.

## What is a digital health ecosystem?

An ecosystem is a complex adaptive system consisting of diverse components that interact locally and are subject to the process of natural evolution and selection within any community of interest such as a national health system. Patterns at higher levels emerge within any ecosystem from localized interactions and selection processes at lower levels [3]. An ecosystem generally refers to groups of interacting entities (representing a "community of interest") that depend on each other [4] as shown in Fig. 1. Within any ecosystem it is necessary to consider all possible people, equipment/supplies, communication and information flows between the

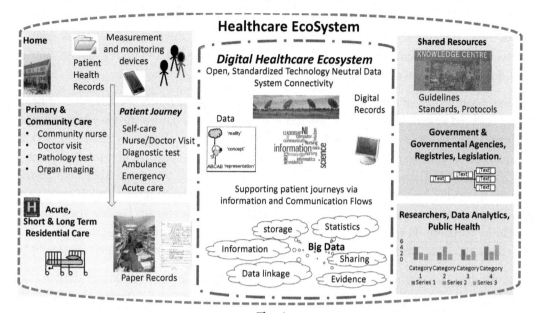

**Fig. 1**
A digital health ecosystem within a healthcare ecosystem [5].

many entities that collectively make up the ecosystem. Every patient journey has its own unique associated communication and information flow needs that must be supported by digital technologies in a manner that facilitates care continuity.

Sustainable health systems need to be able to optimize the use of health data for all decision makers. This is enabled by system wide active involvement of co-operative professional communities, businesses, governments and people generally to co-create value, develop and embrace innovation using digital technologies in an optimal and responsible manner. The WHO/ITU strategy for digital health [1] asserts that the development and implementation of a successful national strategy needs to consider seven components, governance and leadership, strategy and investment, legislation, policy and regulation, services and applications, standards and interoperability, infrastructure and human resources. Similar components are required to develop and implement a successful local digital health strategy that is in line with national initiatives. Wilkinson et al. [6] described the guiding principles for improved data management for data rich research environments, as FAIR (findable, accessible, interoperable, reusable). These guiding principles are highly relevant to the health industry.

The health industry represents an extensive and complex community of stakeholders who need to co-operate to create value and sustainable digital health ecosystems to benefit all. Only then can healthcare facilities ensure efficient delivery of all healthcare services resulting in desired and expected outcomes. Our next question is what factors need to be considered to produce an effective digital health ecosystem? This question will be answered differently by each type of stakeholder as they each value a digital health ecosystem from a variety of perspectives. Stakeholders may be differentiated by grouping them as a producer or user. Producers consist of technology developers, knowledge producers and policy makers/funders. What does such an ecosystem look like?

A digital health ecosystem requires a digital health strategy including the adoption of health service oriented system architectures providing a platform that enables stakeholders to connect with each other to communicate, share, link and process data. Collectively such adoption and use make up an ecosystem supporting knowledge development and leading to significant benefits for all [7]. Romanelli refers to business ecosystem and notes that SMEs (small to medium enterprises) tend to develop information technology infrastructures able to manage big data. They tend to have more flexible IT infrastructures and fewer legacy issues. They can adapt quickly by focusing on innovation to gain higher efficiencies. Other broad groups of ecosystems referred to in the literature are "innovation" and "platform" focused ecosystems [4].

The Australian Government has published a digital service standard to "ensure digital teams build government services that are simple, clear and fast." Its criterion 7 requires such teams to "use open standards and common platforms," criterion 8 states "make source code open (by default)" and criterion 9 states "make it accessible" [8]. These criteria are fundamental to our ability to fully connect every system within a domain's ecosystem. Somehow that has not been translated to be applicable to the health industry and its ecosystem in most countries. Digital health ecosystems need to consider all of the features described in this chapter.

## Essential ecosystem features

The health industry consists of many interrelated organizations that have significant autonomy yet need to be coordinated. This is enabled by a modular oriented architecture where each entity has a large degree of autonomy in how they design, price, and operate their respective modules, as long as they interconnect with others in agreed and predefined ways [4]. This type of ecosystem provides processes, rules, standards and codified interfaces to encourage alignment. In addition to codified interfaces one could also interconnect via open interoperability schema to create the conditions for an ecosystem to emerge. The adoption of connectivity and interoperability schema enables end users to choose access from a set of entities (producers or complementors) who are bound together through some interdependencies, by all adhering to certain standards. The communication medium within a digital ecosystem is a P2P (point-to-point) communications network or a real time data distribution system, TCP/IP (Transition Control Protocol/Internet Protocol) or UDP (User Datagram Protocol)-based networks [9].

For example access to some of these complementarities, such as a terminology server, may be an essential requirement to produce a complex value proposition. There are many such or similar interdependency instances within the health industry challenging the establishment of an effective ecosystem, connecting numerous entities to suit multiple purposes. The strength of ecosystems, and their distinctive feature, is that they provide a structure within which all its entities (complementarities of all types in production and/or consumption) can interoperate to create or extract more value, without the need for vertical integration such as, for example, within the Google/Android ecosystem [4]. This type of ecosystem cannot maximize value for a health industry focused on driving universal access, patient centered service delivery and financial sustainability while producing good population health outcomes.

A digital health ecosystem's design needs to be driven by what drives "value" for each type of user and what is valued most by the "community of interest" and/or groups of stakeholders to be served. What is of benefit and of high value to some users may be detrimental for, or not in the best interest of, other users. Users most likely to benefit greatly are patients and individual healthcare providers. This is likely to be at the expense of healthcare facilities and funders who need to be able to justify such investments. This is where regulation needs to play a role to reduce overall costs while maximizing benefits, by standardizing interdependencies to suit each role or group of stakeholders by type. Ideally every health system and healthcare facility will adopt an evidence based approach for the development and use of such standards and guidelines within their digital health ecosystem.

There is a need to focus on designing technology neutral, ecosystem oriented reference architectures able to address a new type of evolutionary, dynamic, knowledge-sharing, self-organizing, self-controlling, self-reliant architectural model similar to natural ecosystems. Such an open health computing platform differs from a service-oriented architecture (SOA) as there cannot be a hierarchical topology based on interest, or a single point of administration, as the system must be self-configuring and adaptive [9]. Beale [10] describes an open health computing platform as consisting of:

- **openly specified service models**, including for the EHR, demographics, terminology, resource location, medications and other reference data, security services etc. → allows us to share data and knowledge;
- **openly specified information**, consisting of common models of data types, basic data structures, "clinical statements," various kinds of documents, particularly for cross enterprise use: discharge summaries, referrals → allowing us to aggregate data;
- **openly specified semantic models of content and process**, allowing systems to reliably compute on data representing domain level notions like blood glucose measurement, cholesterol result, antenatal examination, prescription;
- **openly specified model of querying**, allowing the locked-in health information to be retrieved for secondary uses as well as point-of care.

A key finding from one systematic review is that semantic data exchange can only be successful with the use of a common platform or system architecture of which the RM (reference model) is fundamental [11]. One component example in use in the health industry is the openEHR Foundation's reference architecture [12–17] that has adopted a two level modeling approach making use of Archetypes freely accessible via its clinical knowledge manager [18]. The openEHR's information model is native (integrated) into the system and used as a reference for the clinical models.

The openEHR standard is the only currently available open standard for the representation of fine-grained structured clinical content that is sufficiently mature and proven at scale. This standard's design represents health information content, context, audit, and versioning, allowing applications to commit data to a coherent longitudinal EHR, as well as read and query it. Its implementation and use support the creation of a digital health ecosystem as it enables the realization of application centric (use of exchange formats and/or APIs (Application Programming Interface) for use with legacy systems) and data centric interoperability. The latter stores data in a vendor neutral format by normalizing the data and using a defined data layer, while APIs provide programming routines or protocols developed to enable software applications to share data.

> Gartner believes that truly effective and sustainable open architectures will need a capability for vendor-neutral data persistence, such as utilizing a common schema or set of archetypes and rules for managing structured and unstructured data. Providing open messaging standards (for example, FHIR®, HL7) for data exchange in specific use cases will only go so far in meeting the architectural challenges of digital citizen-centric care delivery [19].

The management of structured and unstructured data refers to examples such as those currently used for digital images in VNAs (Vendor-Neutral-Archives), or the openEHRs CKM (Clinical Knowledge Manager), or the IHE's (Integrating the Healthcare Enterprise®) XDS (Cross-Enterprise Document Sharing) interoperability profiles, in combination with services for trust/consent, ecosystem governance and oversight, and reuse of data and processes for secondary purposes, such as research and population health.

The transformation from an application view to a highly desirable data-centric view, requires:

- data integration in a future proof manner,
- open data access to normalized data stored in vendor neutral formats where coding to standard terminologies is cross-mapped, and
- innovation.

The IHE-XDS and openEHR work well together, and this combination has been used successfully at scale. XDS handles unstructured and semi-structured data while openEHR handles fine-grained structured data with links between the openEHR's clinical data repository.

Linking legacy systems with new open applications using this transformative approach enables you to build an ecosystem of Apps and applications which share data without additional integration. When Apps commit to accessing and storing all data in, for example, a patient record, you can grow the relative ecosystem, taking advantage of innovation from different vendors. This architecture is fueling the next generation of solutions known as the Postmodern EHR (Electronic Health Records) [19] or ERP (Enterprise Resource Planning) systems.

The continuing maintenance costs and relevant risks associated with the use of health information exchanges reduces the ability for such exchanges to be sustainable. In summary, interoperability standards are used to integrate existing systems and data. Storing the data in an open, vendor neutral format enables an ecosystem of vendors to innovate. This approach is being widely adopted by several openEHR industry partners for a variety of applications [20].

## Healthcare ecosystem connectivity frameworks

Healthcare ecosystem connectivity is about achieving interoperability between distributed systems in accordance with the international standard ISO/IEC 10746 RM/ODP [21], the interoperability reference architecture model, as well as the applicable rules such as the interoperability reference architecture framework (also known as Generic Component Model GCM or GCM Framework). It provides structuring principles that guide system software builds. Connectivity requires consistent adoption and use of standard guiding principles to ensure the same interpretation of technical specifications.

This high-level reference framework (IT architecture), consists of multiple views, including a reference information model view associated with an agreed set of data types.

> *Datatypes are the fundamental building blocks around which the semantics of a given piece of data are built. Formally, a datatype is fully specified when both its semantics (in other words, its formal meaning) and the set of legal computational operations that can be performed on an instance of the datatype are rigorously specified [22].*

Effective interoperability enabling optimum data use, requires the adoption of a standard reference architecture/information model, a standard set of data types, standard terminologies and unique identifiers for patient data matching purposes. There are two different technical pathways to achieving interoperability between distributed systems. One is to provide the link at the application level, as is used for transactions and messaging referred to as "interfacing" [23,24].

The other links at a much lower computing level using for example an object modeling approach [18,25,26], which speeds up processing as concepts are "integrated" and able to work synchronously. This is made possible when systems are designed in accordance with a standard reference. Programming protocol adoption is based on functional needs. Networked systems reliant on interface protocols are more costly and time consuming to maintain with no real-time data synchronization possibilities. Interfaced systems do not share the same databases, so mapping of codes between systems is required.

Blobel [27] has modeled a reference architecture that is applicable to the health industry. His model is similar to all other architectural frameworks, in that it includes multiple reference models that represent views or perspectives at variable levels of granularity. Each of these perspectives represents a domain or view point (VP) within a digital health ecosystem that needs its own connectivity framework relevant to the jurisdiction or healthcare provider network.

A number of additional high level (jurisdictional) reference architectures exist such as the Federal Enterprise Architecture Framework (FEAF) [28] or the whole of government enterprise Australian Government Architecture (AGA) Framework program [29]. There are enterprise specific reference architectures or frameworks such as the Open Group Architecture Framework (TOGAF) v.9.2 [30] or the Zachman Framework [31], representing an enterprise ontology, a set of concepts and categories showing their properties and the relations between them.

These are generic organization centric reference frameworks intended to be applicable to any industry, although no evidence was found regarding their suitability to be applied to a national digital health ecosystem. Such ecosystems need to adopt a framework that applies to patient-centric healthcare delivery models facilitating the delivery of collaborative, multidisciplinary and cross-organizational healthcare services. Ideally such a framework adopts a standard national platform enabling virtual healthcare record sharing and communication among authorized healthcare professionals.

A national strategic interoperability framework needs to be able to create the desired health information integration and interoperability backbone. A health industry's interoperability framework must reflect all possible personal and national data, information and knowledge flows, data element lifecycles and individuals' health care journeys encountered throughout their lifespan. There is a need to adopt a pragmatic blend of existing standards enabling every stakeholder to move forward, by providing an evolving incremental pathway without requiring replacement of current systems.

To this end HIMMS (Healthcare Information and management Systems Society) put a call to action to the USA Department of Health and Human Services and the broader health information and technology community, demanding integration between interoperability approaches adopted [32]. HIMMS is seeking trusted exchange frameworks for the public good. They have recognized the need to:

• Educate the community to appropriately implement existing and emerging standards, data formats and "use cases" (specific situations in which a product or service could potentially be used), to ensure a comprehensive, integrated approach to care,
• Ensure stakeholder participation from across the care continuum, including patients and caregivers,

- Identify the "minimum necessary" business rules for trusted exchange to enhance care coordination,
- Standardize and adopt identity management approaches, and,
- Improve usability for data use to support direct care and research.

Sound knowledge is required about the relationships between desired health data use by a large variety of different users and required system functions to meet purpose of use, as these factors dictate the degree and type of interoperability required between any number of stakeholders. Once these are known, technology decisions regarding the best possible solutions can be made. Ideally all digital data, information and knowledge can be managed independently of the technology in use and:

- can be persistent throughout their lifecycles for all user environments.
- persist for as long as the need to exchange information between the systems continues.
- can be safely and efficiently exchanged, shared and linked.

Strategic and future focused use of information technology has the greatest potential to deliver the desired outcomes. Critical features are compliance with appropriate standards, user incentives, accurate and complete data and documented user guides providing information, knowledge and accountabilities. Healthcare is a global business, therefore it is highly desirable to think global (or national) while acting locally. The focus needs to be on data specification. Achieving the desired level and type of interoperability within any digital health ecosystem requires an agreed national infrastructure to enable optimum use of these assets, consisting of:

- An agreed vision.
- A high-level digital reference roadmap/architecture that represents the vision.
- An agreed open platform.
- Consistent adoption of an agreed set of unique identifiers.
- Consistent adoption and use of:
    - Clinically developed conceptual models,
    - Agreed standard terminologies,
    - An agreed set of data types.
- National health data governance/custodianship arrangements.
- Professional clinical colleges providing leadership regarding clinical data sharing, linking and using data relative to the management of specific health issues and for meeting clinical trial information needs.
- A transition pathway covering a specified time period.
- Legislative changes that will enable the vision to be realized.

There also needs to be changes made to traditional vendor business models. Most vendors like to be recognized as being different, preferring not to change their system's information model. This is reflected in a continuing strong focus on developing system interfaces based on data mapping rather than on integration to become part of a seamless digital ecosystem, as is evident by today's state of the art.

## Today's state of the art

Computing technologies in use today are the result of various progressive developments over a period of around 50 years. Health information systems in use today were originally designed at various points along that continuum based on the latest technologies at the time. Pre-1997 the focus was on "best of breed," from 1998 to 2014 systems focused on integrated Electronic Patient Record (ERP) suites or Electronic Medical Records (EMRs) which are provider-centric, or Electronic Health Records (EHRs) which are patient-centric. Healthcare organizations have any number of stand-alone systems in use that are connected via various network configurations.

From 2015, post-modern ERPs with core and multi-vendor process oriented systems, have considered using Integrating the Healthcare Enterprise (IHE)/HIE (Health Information Exchange) interoperable schema to connect with each other [24]. Many are now gradually converting to using multiple Apps (Applications) via an ecosystem App store and linked via Application Program Interchanges (APIs). API's consist of three key characteristics, a computational view that indicates what functions and procedures can be called in the interface, such as "create_patient," the informational view that specifies functions, returns data associated with "take data" as arguments, and provides a protocol specification indicating in what order API calls need to be made, how exceptions are reported and more [33].

HIE refers to the process of electronically transferring, or aggregating and enabling access to patient information and data across provider organizations or between different types of entities. It's use has increased over time [34]. IHE International works with vendors and healthcare professionals (users) across 17 countries developing and testing, HL7 (Health Level Seven) messaging standards and implementation protocols [23] to meet data interchange requirements for many clinical domains addressing specific clinical needs. Its mission is to improve healthcare by providing specifications, tools and services for interoperability. IHE protocols are used to resolve issues associated with inconsistent interpretation of Health Level Seven (HL7) messaging standards [23].

Many legacy information systems in current use have any number of limitations regarding their ability to interoperate between vendors, technologies, business processes and applications. Most enterprise-based systems are complex, difficult and costly to implement and maintain. Many have out-dated system architectures and interfaces, were not designed to support clinical practice, and/or do not have clinically desirable functionality and/or are time consuming for clinicians to use. Their data quality is questionable, they contain many data errors and data are difficult to find, access, link or process unless specifically programmed to produce pre-determined reports. Current procurement processes in Australia continue to favor software vendors who operate closed development models, selling their software with proprietary licenses resulting in what is referred to as "vendor lock-in" to ensure continuing data access over time.

Proprietary formats requires data migration to new formats every 10–15 years or so when applications are replaced. Application centric data exchange solutions will only go so far in meeting architectural challenges of digital citizen-centric care delivery [19]. Meanwhile we need to manage a transitional period to enable the continuing use of expensive legacy systems. This requires a learning culture, collaboration and trust and the adoption of multiple standards in a manner that best suits the relevant ecosystem. Change requires adaptive leaders with entrepreneurial and co-production mindsets, no single technology or technical group.

Australia, along with most other nations does not have a National Health System Reference Architecture. The lack of an architecture or open platform that supports standardized APIs, as well as current EHR vendor technology and their business practices, are viewed by the USA Government as structural impediments to achieving interoperability [35]. The JASON taskforce report recommends an urgent focus on creating a unifying software architecture to migrate data from legacy systems to a new centrally orchestrated architecture to better serve clinical care, research, and patient uses. There is now a desire to reorient healthcare interoperability away from "siloed legacy systems," toward a centrally orchestrated interoperability architecture based on open APIs and advanced intermediary applications and services. The health sector's major software vendors are increasingly making new APIs available to facilitate and improve data sharing opportunities.

The use of Open APIs is enabling vendors and start-ups to create web and mobile applications for individual patient use by aggregating patient specific data retrieved from various EHR portals. The ONC (Office of the National Coordinator for Health Information Technology) issued a regulation that included API "certification criteria" in 2015 [36–38] via the US HITECH Act and has published a Certified Health IT Product List [39]. In addition to a national health ecosystem reference architecture, there is a need for a data infrastructure across healthcare organizations that supports interoperability. National programs such as shared EHR programs don't deliver "plug and play" clinical decision support rules.

National Digital Health Frameworks are now being developed. The European Commission has adopted a Digital Single market strategy. The Digital Health Society was established as part of the European Connected Health Alliance, to initiate and coordinate European taskforces addressing key challenges including the development of a convergence roadmap based on interoperability standards [40]. As an alternative or for inclusion in a national framework, one could consider adopting the openEHR architectural framework [41]. The United Kingdom's INTEROpen group was established to implement their vision for a health and care community through which digital information seamlessly flows [42]. New Zealand has included this framework in its national interoperability architecture. The Australian Government funded the establishment of a Digital Health Cooperative Research Centre early in 2018 [43].

The openEHR domain is patient centered with a consumer record focus rather than an enterprise or jurisdictional domain focus. Its structural focus is on concepts enabling data integration, machine learning and automation. Its framework is technology independent. Tooling and data exchange schema are influenced by the architectural framework or "platform" adopted. Reynolds and Wyatt [44] argue for the superiority of open source licensing to promote safer, more effective health care information systems. It's viewed as essential to a cost effective rational procurement strategy, especially for underdeveloped countries with poor resources.

The European Union (EU) has developed an innovative technological platform to enable the re-use of EHR data for clinical research via a 4 year project EHR4CR (Electronic Health Records for Clinical Research) [45–47]. This project highlights the need to adopt one common platform and to match legislative and standards requirements to be successful. There are many EU examples of extensive collaborative unification between European experts in different fields (clinicians, biologists, IT-specialists, ethical and legal), a large number of organizations and countries, with impressive results.

### Shadow systems and health data

Large healthcare facilities tend to have multiple legacy systems that may or may not be well connected or integrated within their own facility and are likely to be less connected with external systems. Many "shadow" systems have been developed and are in daily use by individual Clinicians and/or Managers to compensate for large system limitations. There are similarities with what may also be referred to as "Ghost charts" [48] or "feral information systems" or "workarounds" [49]. They may be simple working documents, databases, spreadsheets, or devices containing routinely collected data generated as part of every day healthcare service delivery processes not under the jurisdiction of a centralized information systems department. These systems collect and process data representing any one of the many data variables pertaining to the variety of input, process and output factors as identified in previous chapters.

Based on anecdotal evidence, it is not uncommon for large healthcare facilities to be able to identify hundreds of shadow systems. One research study of three companies external to the health industry concluded that the use of shadow systems is hindering high enterprise wide system integration [50]. Other concerns are that shadow systems may add to a healthcare facility's risk to cybersecurity [51].

Data stored within shadow systems are of great interest for the purpose of supporting local activities, monitoring and managing every day operations and staff resources but difficult to link to other data for the purpose of creating new information useful for every day operational decision making and/or to inform areas in need of process, quality and safety improvement or to undertake predictive risk modeling via a learning healthcare system. Overcoming this challenge requires the adoption of strategies aimed to fill data gaps required to inform pragmatic operational research. This requires extensive data linkage, the development of analytical methods and the collection of more data on a routine basis along every point of a patient's journey through the health system [52]. Analysis of routinely collected data is commonly referred to as "secondary use." Failure to adopt this approach not only limits opportunities for improving productivity, it also limits opportunities for meaningful "Big Data" analytics.

Experience tells us that there are numerous data matching challenges as not every system in use has adopted the use of the same metadata or data standards. Semantic interoperability requires the precise and unambiguous use of terms. Healthcare organizations and software vendors are increasingly using maps to convert data from one code system to another code system where terms from one system don't match terms in use by another system. In response to the increasing use of this mapping methodology, an ISO technical report (ISO TR12300) [53] detailing mapping principles was published and a follow up resource that provides map quality measures (MapQual) [54] has been developed. Failing to achieve a quality match has the potential of compromising patient safety. Every data map is developed for a specific purpose, using it for another purpose has a high probability of failing to meet the requirements for the intended use and/or compromising data retrieval.

The benefits to be realized from an ability to share data and enable the automation for both primary and secondary data use is widely accepted. This is entirely dependent upon the collection of quality "little" or source data, successful interoperable system implementations within healthcare organizations and beyond. One of the major issues slowing down digital transformation is the lack of data and data exchange standardization, including the absence of interoperability between EHRs. System connectivity is dependent upon achieving intra-domain and inter-domain interoperability, requiring connectivity between components to occur at identical levels of granularity [27].

Another related challenge is patient matching in the absence of a standard universal unique patient identifier or when there is non-compliance with the standard, to ensure that data pertaining to a specific patient in one system is accurately transferred to that same person's clinical record. These challenges along with the adoption of variable reference architectural frameworks and/or variable proprietary enterprise systems that don't reference any standard architecture, are significant roadblocks to achieving the many potential benefits of a well connected national digital health ecosystem. Overcoming these variations requires the development of various interoperability schema or approaches. Fundamental to these are the adoption of health data exchange standards, in particular the widely used and well known HL7

(Health Level Seven) messaging standards [23] and the Integrating the Healthcare Enterprise (IHE) protocols.

### Connectivity and interoperability

Healthcare is continuously changing and today's monolithic applications can't cope. The future is multi-vendor, adoption of standard open "platforms" and the adoption of vendor neutral health data repositories. openEHR provides a proven open platform for new niche applications using clinical data. Meanwhile numerous alternative connectivity and interoperability strategies are in place and expanding to continue the use of costly legacy systems. Walker et al. [55] provided the first cost estimates applicable to the United States in 2005 for their level 4 (optimum) interoperability. The roll out costs were estimated to be $276 billion and annual costs were estimated to be $16–20 billion for an annual net value estimate of $33.7 billion. This value did not consider clinical or organizational effects or probable societal or workflow impacts.

We know that there are three essential types of interoperability: syntactic, semantic and pragmatic, and that there are numerous levels within each of these types that determine functionality. Whatever the approach to interoperability adopted, it needs to support the business and/or care requirements relative to the health status of any individual, health provider operational needs, and the jurisdictional desired digital health ecosystem. Connectivity solutions continue to evolve. The focus should be on the difference between system interfaces and system integration and the many shades of gray in between that have resulted from an increasing use of APIs and HL7-FHIR (Fast Healthcare Interoperability Resources) models.

The continuing use of proprietary platforms and system architectures severely limits the possibilities of real time synchronization as their connectivity protocols rely on data mapping and programmed interfaces. Another significant limitation is having no direct access to the source data structure to support querying. No access to the real meaning of the variety of concepts represented by the data extracted that is needed for data analytics based on extracted data. When deciding on a connectivity schema linking any number of systems, one needs to consider the desired functional data use, not only for a specific application but also for potential secondary data uses.

Any interface has two components, content and process. Content is more integrated when the connecting systems share an information model and/or a data dictionary. Both are needed to avoid data mapping. An interface process is the creation of a message based on the data contained in the source system and forwarded to an external system where it is imported and actioned. Most interfaces are about process points of information exchange. The openEHR modeling approach means less data conversion and more effective data reuse based upon the reference information model rather than at the process point of its collection.

The achievement of any level of interoperability within each type determines the potential for actual data use and benefits to be achieved. Every solution is likely to have limitations. An agreed interoperability measurement scale could become a useful tool for decision makers to determine the associated functionality of data exchanged using any proposed technical interoperability schema. Interoperability standards need to indicate how information is exchanged electronically, its usability in terms of relevance and comprehensibility, its application potential in terms of human use and data computability and its potential impact on patient safety, cost savings, care coordination, process and outcomes improvements, user engagement and experience.

Not only do we need to consider interoperability in terms of clinical data exchange and use, it is also critical to be able to link clinical data with the person to whom it belongs. This link must be retained whenever such data is linked or transferred or shared. In the absence of nationally and/or state/jurisdictional governed unique identifiers, health care provider organizations issue their own unique identifiers. These then need to be mapped every time clinical data pertaining to an individual is transferred, linked or shared to ensure context and its relationship to an individual is retained.

Walker et al. [55] noted that:

> *these imprecise processes may generate errors and redundant information, limit the efficacy of clinical decision support, and create information and cognitive overload for clinicians.*

They went on to say that "on-demand, seamless integration of local and remote records, is far more likely to offer clinicians the integrated information they need for providing optimal care." This functionality, their level 4, was defined as:

> *Machine-interpretable data — transmission of structured messages containing standardized and coded data; idealized state in which all systems ex-change information using the same formats and vocabularies (examples: automated ex-change of coded results from an external lab into a provider's EMR, automated exchange of a patient's "problem list").*

This definition matches Elkin's highest level of semantic interoperability measurement scale [56].

## Measuring interoperability

At the 2007 Medinfo held in Brisbane, Elkin et al. [56] presented an interoperability scale with good inter-rater agreement to assist implementers of healthcare standards to better understand the level of interoperability provided by standard specifications being considered for implementation and to mitigate risk of choosing a healthcare standard when developing health IT solutions. This scale essentially details variations in data processing possibilities facilitated by various standards. It enables those faced with the choice of which standards to implement to compare the relative levels of syntactic, semantic and pragmatic interoperability and associated data processing functionality provided by each standard specification.

Another interoperability standards measurement framework published in April 2017 [57] does not measure individual standards' operational (semantic) functionality. The aim of this measurement scale is to understand:

- if specific standards are built into health IT products,
- the use of these standards,
- the method of deployment into production systems,
- the level of conformance or customization during implementation and
- their availability to end users.

Another measurement in frequent use is the HIMSS Analytics Electronic Medical Record Adoption Model (EMRAM). This eight-stage model is used to score hospitals relative to their electronic medical records (EMR) capabilities or functions. This score evaluates levels of achievement regarding the functions made possible as a result of an EMR system's connectivity and ability to share and link data within any individual organization. Hospitals achieving a high level of EMR functionality and interoperability are not necessarily able to electronically interoperate with external organizations. This limits care coordination for individuals receiving care from multiple providers collaboratively addressing one or more concurrent health issues. In other words, such organizations are not necessarily a good fit with any national digital health ecosystem.

These limitations are being overcome by the establishment of health information exchanges established to support individual communities/networks by facilitating access and retrieval of clinical data via the use of an interoperability schema developed specifically to suit that community and the systems in use by its members. These schema are designed to make use of any number or set of compatible health informatics standards in the absence of seamless system integration using a nationally agreed standard platform.

A continuing limitation for many such communities is the lack of the consistent adoption of unique identifiers for individuals for whose benefit such data are transferred, linked or shared. Another limitation to consistency and data accuracy is the need for schema maintenance to ensure a continuing ability to transfer, link or share data as all connecting systems are upgraded at different times over any time period. Such schema are designed to reconcile diverse codes, data structures, and terminologies by developing and using maps. These critical technical interface development and mapping activity factors add to the cost and risks associated with the use of health information exchanges.

## Interoperability standards and schema

Standards development activities undertaken since the early 1980s aimed to provide solutions to these interoperability challenges. The Digital Imaging and Communications in Medicine (DICOM®), standards committee of the National Electrical Manufacturers Association (NEMA), was one of the first to develop a standard for communicating medical imaging information [58], now published by the International Organisation for Standardization (ISO12052:2017). All Health Level Seven (HL7) standards are published by an official national standards organization, the American National Standards Institute (ANSI). It's important to note that HL7 (Health Level Seven) standards [23] focus on technical interoperability (health data exchange) resulting in limited operational (semantic) interoperability [59].

The now very popular Fast Healthcare Interoperability Resources (FHIR®) standard has greatly improved operational information flows [60]. The HL7 FHIR® reference model is defined by a collection of information models (resources). These can be profiled (or not) to generate a clinical information model. FHIR® resources are produced from a superset of data found in legacy systems to be directly used by developers. Strictly speaking these are not "models" since none of the usual inheritance, encapsulation of common elements or typing practices are used. The successful use of the FHIR® standard is dependent upon the degree of agreement reached between the parties. Reaching agreement in this manner is easier to achieve within organizations by a small number of stakeholders, including Clinicians, than agreements to suit more extensive networks. This limits software re-use opportunities [61].

A review of six US Medicare Accountable Care Organisations [62] has again highlighted the need to adopt a standard approach. It was found that those who used a single electronic health record system across their provider networks, were able to share data in real time, enhancing providers' ability to coordinate care. Others had access to robust health information exchanges enabling access to patient data from external providers. Despite this it was noted that the full potential of health IT has not been realized. Reported limitations for those using multiple EHR systems were the result of poor interoperability compromising the capacity to care resulting in:

- Reliance on other means to share data including phone calls and faxes,
- Burdensome and frustrating use of EHRs,
- Physician burnout due to workload associated with EHR management,
- Access to health information exchanges with little or incomplete data creating difficulty regarding care coordination,
- Inability to offer health IT to patients other than online portals to their EHRs,
- Inability to use analytics to customize care to suit an individual patient's needs.

A private sector initiative is currently undertaking a project aimed to advance industry adoption of modern, open interoperability standards known as the Argonaut project [63]. This project's purpose is to develop a first generation FHIR®-based API and Core Data Services specification to enable expanded information sharing for electronic health records and other health information technology based on Internet standards and architectural patterns and styles. This work is being sponsored by numerous vendors including Cerner, Epic and Accenture. The scope of this work is about accelerating FHIR® model development by focusing on more specific FHIR® profiles and documentation including a focus on accompanying security specifications and documentation to enable interoperability to be achieved between legacy proprietary systems, where their vendors contractually control data access. The use of these APIs creates inherent risks for errors and patient safety due to data mapping.

Health care organizations are now in the process of managing the rapid uptake of FHIR® and the greater use of APIs. From a national infrastructure perspective there needs to be the means to facilitate the re-use of some frequently used key FHIR® models. These may be chosen to become approved HL7 standards. Without the adoption of agreed standard models there will be a proliferation of FHIR® models developed to meet local needs, many repeats and inconsistencies limiting overall national progress toward achieving a well-connected digital health ecosystem. FHIR® models do not comply with well-known modeling conventions as demonstrated by Beale [61] who concluded that:

> the FHIR resources appear to be the result of separate committees working with almost no cross-referencing, methodology, or common design basis. The result is that each Resource is something like a "bag-of attributes," presumably due to the application of the so-called 80/20 rule in the committee context.

FHIR® models need to consider a very large and unmanageable number of data points. This limitation is overcome by the adoption of generic models which are perfectly suitable for messaging but not for semantically coherent clinical data storage. Context is integral to how clinicians process information. Context needs to be considered when developing decision support or artificial intelligence systems, when adopting various data analytic strategies, and for any number of relevant standards including those used for data mapping.

The FHIR® resources are said to contain a continuum of ontological levels, they have two characteristics undesirable in a stable information model [64]:

1. *volatility* — clinically specific resources will clearly need to change over time. How this affects dependent profiles is an interesting question;
2. *open-endedness* — one must presume that the set of resources will simply keep growing to accommodate new major clinical information categories.

The consequences of these factors are that the model is "never finished" and that any databases or software based on it will likewise need continued maintenance. There is another issue associated with downstream information-processing for secondary data use. That is an inability to know whether data structures that appear to be nearly the same can be treated the same way; the default with FHIR is that every Resource is its own thing [65]. One international standards expert [66] noted that:

> FHIR® was designed and intended as an API/interchange standard, most major ITC organizations (Google, Microsoft, IBM, etc) and at least one major EHR vendor (Cerner) have embraced FHIR® as a persistence object schema and datastore schema. FHIR® remains underspecified at an international level, as the binding of value sets is relegated to implementation guides (e.g. the US Core FHIR® implementation guide). Given current national predispositions toward national Health Information Technology (HIT) codes (virtually every country has their own darn publication of Procedure codes, and oddly the entire world is not yet using ICD11) this separation is highly pragmatic. The barriers to international HIT interoperability have less to do with technical syntax specifications, and vastly more to do with the nearly intractable tendency of countries to specify their own, sometimes proprietary, coding systems.

Adoption of the openEHR architectural framework requires the use of a more comprehensive modeling approach [67]. These models are able to link with FHIR® models where required, for example the openEHR Adverse reaction risk archetype [18] and FHIR® AllergyIntolerance resource [68] were jointly published and aligned at the end of 2015. Archetypes incorporate the collaborative work undertaken by a large virtual international community of multidisciplinary professionals turning health data into an evidence based electronic (computable) form to ensure universal interoperability within any digital health ecosystem. This approach consists of multi-level, single source modeling within a service-oriented software architecture delineated by a set of specifications published by the openEHR foundation and freely available to anyone.

The openEHR is an open standard specification governed by a not-for-profit foundation, it is freely available to be implemented by any developer. OpenEHR specifications, the result of 25 years of research and development, represent the only serious instantiation of the ISO 13606 Reference Model for electronic health record communication that details a hierarchical structure for clinical information [69]. This standard is being increasingly used in massive projects in United Kingdom, Norway, Finland, Slovenia and Germany. Also China, Chile, Brazil, Italy and the Caribbean have adopted this approach. In fact ISO 13606 was derived from the experience of openEHR, and more recently has been influenced through experience with other standards including FHIR®. The clinical modeling work is more extensive than that undertaken anywhere else previously with a volunteer community of over 2000 people from 93 countries.

The ISO13606 domain clinical models known as "archetypes" [70] are external to the software. Each archetype defines a maximum number of possible datapoints and data groups that apply to the modeled concept. This provides for a maximum number of different contexts to suit all clinical specialities and potential data uses. Templates, consisting of a "recombination" layer of models to define data sets, make use of only those datapoints from a set (any number or selection) of archetypes that are required for any specific "use case" or application, such as any type of clinical assessment. Templates may be used as message definitions for legacy systems, as well as data sets for new applications, including forms. This means that any set of archetypes can be re-used for multiple use cases. It enables significant parts of the software to be machine-derived from the archetypes. Archetypes represent atomic data in an open standard; a critical success factor able to withstand changes in technology, particularly exchange and persistence.

These "Archetypes" (models) are in use by various applications, including in the United Kingdom's National Health Service where they are used for the conversion of shadow systems enabling the use of vendor neutral data repositories. Other users are: Queensland Health and some private hospitals for their infection control systems, a number of Western Sydney Primary Health facilities and the Northern Territory Health's Shared EHR, and openEHR industry partners whose systems are widely implemented in Scandinavian countries. Conforming components and systems are "open" in terms of data models and APIs. Strategically, the openEHR approach has a health record focus well suited to patient centered care. This approach enables a platform-based or "open back-end" software market in which health industry vendors and solution developers interface with each others systems via standardized information models, content models, terminologies and service interfaces.

An OPEN collaboration of individuals, industry, standards organizations and healthcare providers have agreed to work together to accelerate the development of open standards for interoperability in the health and social sector. They have provided a collaborative forum for sharing experiences and providing solutions aimed at overcoming this confusing landscape. In July 2018 they published an overview of what has been described here [42]. It was noted that these many standards are in the process of converging. The FHIR® and openEHR standards do not address the same problem. They both create clinical information models but the openEHR models (archetypes) are data-centric vendor independent across openEHR platforms. FHIR® is application centric as it creates a common model for use across any application to then form the basis of defining interoperability between applications whatever open or proprietary models they were built on. Those making use of application centric interoperability solutions do so to enable the continuing use of expensive legacy systems.

openEHR and FHIR® provide complementary but different approaches to connecting a complex patchwork of applications into a single coherent system [71]. The data centric openEHR approach focuses on normalizing the health data first, and building new systems on top of legacy systems to avoid the interoperability issues all together. This is about defining a data layer within an open system architecture. This is the most important layer as it facilitates optimum data use to improve outcomes, better manage chronic diseases and to enable better population health management. This requires data to be stored in vendor neutral formats as opposed to proprietary formats of most legacy systems. The data centric approach enables digital health data storage and use throughout the lifetime of any patient.

Almost every openEHR-based system is currently developing FHIR® interfaces to ensure they can support the patient's journey between openEHR and non-openEHR systems [71]. openEHR connectivity schema can be built once and shared between all openEHR vendors since they are based on common openEHR archetypes. Applications clustered on an openEHR- based platform don't need exchange solutions such as FHIR® because they are able to communicate directly with each other through shared clinical data repositories. This allows new market entrants to concentrate on developing truly innovative applications and functionally to meet niche clinical requirements. FHIR® capacity is only needed to enable them to communicate with wider ecosystem vendor proprietary applications [71].

## Computing platforms

The New Zealand's Ministry of health has adopted a strategy to pull together its four regional systems and underpin them with a National Health Information Platform (nHIP). Norway has adopted a similar strategy as has Brazil. The adoption of a national standard open platform provides numerous technical and economic benefits for healthcare providers, system or application users, start-ups and software vendors. Using open platforms prevents organizations from being locked into any proprietary information system where the users need to pay the vendor for any desired software changes [72]. Open platforms need to be viewed from both developer and deployment perspectives. Any technical platform provides flexibility and an ability to re-use fundamentals for the development of any number of new systems without having to re-engineer this from scratch. One defining characteristic of an open platform is that its specified interfaces now referred to as APIs, are publically exposed to enable any software developer to make use of it to develop their own system while making use of those standard fundamental characteristics.

Services platforms have what is referred to as "front-end" and "back-end" components making use of implemented platform components. Platform APIs can be defined at many levels. A "back-end" example is a data repository or information-related service of some kind. A "front-end" example is a business oriented application of some kind. Platform examples based around mobile operating systems are Amazon, Apple and Google. Third party developers can build applications using their published platform definitions and then assume that those applications will work on the chosen platform, but the platform owner implements the platform themselves. One could argue that this represents standards based procurement where the platform owners control and publish the standard, but leave it to external parties to build the applications. In these cases, the applications can only be executed by mobile devices.

Product platforms on the other hand refer to private sets of specifications relating to a piece of private IP (intellectual property) that may be published in some way. Such platforms may be best of breed, reflecting the best available defacto starting point for an industry platform or a "trojan horse" platform. It is therefore important to identify API and platform relationships to determine the roles being played. Woland's Cat blog [33] describes the following key roles:

- Platform specifier — defines and publishes the platform specifications.
- Platform mandater — mandates the use of the platform specifications in the target ecosystem.
- Platform implementer — implements the platform — i.e. back-end(s).
- Platform procurer — pays for platform production deployments.
- Platform deployer — actually deploys the platform implementation.
- App developer — develops Apps that talk to the platform.
- App procurer — buys the Apps.
- App deployer — runs the Apps.

Any "platform" may technically consist of many interfaces at many levels. What is referred here as an App for convenience may in fact be someone else's platform service. Every "platform" reflects a particular ecosystem.

Any national health system needs to establish its own ecosystem based on a standard national platform in order to optimize data sharing, linking and processing to benefit all its stakeholders and transform to a sustainable digital health system. Once a jurisdiction mandates or makes available an open platform to suit its ecosystem, its healthcare providers have greater flexibility and the ability to avoid procuring very expensive, inflexible proprietary ERP/EMR/EHR systems that reflect lock-in software solutions. No system is able to meet every possible information need; that's the reason for the development and use of shadow systems. Many of the best innovations come from small companies focussed on solving very specific problems. In the absence of a standard and openly available platform such companies typically implement non scalable or clinically unsafe solutions, issues that are easily resolved with the adoption of an open platform.

The world has recognized that health data is the primary health industry's asset valued as the main business drivers for both healthcare providers and recipients. Adopting open platforms contradicts the traditional software developers' business models. Some are making a change to this business model by partnering and establishing health services platform consortia [73,74], including the Argonaut project [63] mentioned previously, to meet the needs of specific healthcare ecosystem. However these consortia continue to oblige procuring healthcare providers to procure costly, monolithic single-vendor solutions that in essence equates with such providers becoming investors in these vendor driven technologies while depriving them of any influence or control in the long term. This is in opposition to situations where healthcare provider procurers make use of the availability of an open platform designed to suit their ecosystem. This is where they have the power to purchase any set of software applications from multiple vendors that are known to work together. The use of vendor consortia should be viewed as a transitional solution to benefit those healthcare providers who have invested heavily in legacy systems. This interim solution is economically inefficient.

Any open platform for use in the health industry represents an alternative collaborative structure to the product platform consortia enabling cost-sharing, less gross investment needs for some of the main ecosystem components, especially in the back-end space. It provides agility as open source creates an IP ownership structure with few legal or commercial boundaries, enabling developers to simply adopt it. Their primary benefit is compliant implementations of specific interface layers enabling automated sharing results in high level interoperability between such systems.

Open platforms need to be able to solve a number of health information challenges, including data sharing, semantic computability (data merging and computer processing), flexibility and adaptability over time (irrespective of new medical and/or technical advances), ability to meet new requirements, cost effective and efficient implementation pathways and continuing clinical relevance. This also requires a national standards development program that includes the major stakeholders as contributors, and specifies and/or mandates, base and new standards' compliance requirements complete with implementation guidelines. These standards, include API definitions to suit the chosen platform, specify information interoperability solutions, platform mechanics, as well as languages/terminologies and ontology. The latter is a fundamental requirement to trust computers to reason with data for AI (artificial intelligence) purposes.

Ontologies provide the definitional underpinnings of health data to prevent errors regarding the meaning of concept, for example actual blood pressure versus a target blood pressure, an allergic reaction versus an allergy. Base standards are the most critical as the wrong choices can severely compromise the sustainability of any healthcare enterprise. Software no longer represents large parts of health IT, ontologies, terminologies, archetypes, computable guidelines and other "model artifacts," most of which are to a large extent open source, making up a large part of the health IT layer.

In summary a platform with no ontological underpinnings is unlikely to support reliable semantic or meaningful computing [33]. At the end of the day it is the developers who tend to adopt the best available platform unless a platform is mandated for a specific ecosystem by the relevant jurisdiction or other appropriate legal entity. Platforms need to be well documented and easy to understand, easy to start developing with. This can be made possible via available downloads, demonstrator server sites and software developer kits (SDKs). Platforms need the ability to do more than initially desired and provide for programmer usability to generate industry credibility. Open source platforms for the health industry also need to have a suitable governance structure.

A digital health ecosystem's architecture may need to include various digital technologies such as cloud computing, mobile computing, or the Internet of Things (IoT) where each is based on its own platform. This requires interoperation between platforms and the application components that use them. The scope of IoT enabled products allows for the integration of network technologies, devices, sensors, software and distinct infrastructures well suited to support new healthcare services. This presents new cybersecurity challenges and possible threats requiring an ontology-based cybersecurity framework designed to minimize such risks [75].

The Open Group, a global consortium, is leading the development of open, vendor-neutral technology standards and certifications to enable the realization of boundaryless information flows through global interoperability in a secure, reliable and timely manner [30]. This group licenses the use of a number of standard reference architectures although none appear to have been developed specifically to suit the health industry. Its 2014 open platform 3.0™ white paper [76] provides a good insight regarding new and emerging technology trends converging with each other and leading to new business models and system designs. Its public draft standard specifies interoperable application platforms enabling enterprises to gain business benefit from these new technologies [77].

One research study investigated strategies to implement and scale digital platforms in highly regulated settings such as healthcare [78]. Openness on code and content layers were observed as fueling platform growth as suppliers, healthcare providers, insurers, and patients are more likely to use and contribute to the platform. This is where end users can access and potentially modify source code. In opposition, risks such as a lack of or uncertainty about regulations were observed as prompting the provider to close the platform to uphold control which in turn reduces the benefits for potential suppliers. With proprietary platforms the source code is secret, end users can only access and execute the machine code.

The type of license associated with any software purchase does not affect its quality, it determines source code development with long term cost and functionality implications for the purchaser [44]. License arrangements usually include details about source code IP, governance and management arrangements. Adopting open platforms enables connectivity between competing software implementations compliant with the same set of standards fostering innovation to benefit users and consumers. It empowers users to change applications and/or software development teams without losing data or facing conversion costs. This provides protection against monopolies and reduces costs. Large global distributed development communities scrutinise source code, make use of open source platforms, share functionality experiences and problem solving to inform their communities, leading to higher quality software [44].

## Interoperability, clinical needs and secondary data use

Healthcare service demand is driven by the health status, care, diagnostic, treatment and caring requirements of individuals. Healthcare providers are individual clinicians from different disciplines and specialties who provide their services using their knowledge, skills and supporting resources via many and varied organizational structures. The need for interoperability is primarily to support these health service delivery processes in a manner that fits with the desired or possible individual journeys or pathways through any healthcare system to suit any individual's specific health status. It's important that communication and supporting dataflows also includes nursing data.

Clinicians need to play a far more active role in defining their optimum data needs as this determines functional and technical requirements [79]. They can contribute by working as a team to:

1. develop clinical models such as archetypes where evidence based clinical knowledge, including clinical data and process needs, are converted into computable formats,
2. develop FHIR® models,
3. model dataflow requirements based on known diagnostic and treatment protocols for their specialty to support individual patient journeys,
4. set data standards that must be adhered to by all and governed in a manner that enables their desired research to be undertaken and/or data use for performance monitoring.

A national digital health ecosystem with a fully integrated infrastructure should be able to automate all data collections from EHRs and service provider/enterprise systems in real time for use in dashboards where desired [80]. Let's use one example to demonstrate this need using maternity care, a common health service in demand globally. Variables influencing desired dataflow and connectivity requirements are determined by the woman's geographical location, service availability, past obstetric history, any co-morbidities, referrals, individual financial and/or insurance status, family support, individual choices such as public versus private care, shared care, desired home delivery versus, birth center or hospital delivery. Similar generic variables influence any other individual's journey from the first point of access which may be an Ante-natal clinic or independent midwife, an emergency room or community based midwife/Lead Maternity Carer or an Obstetrician/specialist.

Antenatal care is about monitoring progress and early detection of any pregnancy complication(s) so that these can be managed in a timely manner. This requires data contributions to an individual's record from multiple service providers. The variables listed above again apply and need to be considered for connectivity purposes. Time periods for individual patient journeys, vary based on their specific health status and service needs relative to the condition(s) being treated or managed.

The desired outcome for maternity care is a healthy mother and a full-term healthy baby. This is not always achievable. Birthing processes and post natal care add another dimension to dataflow needs to suit all individuals. In the case of newborns, it's an opportunity to establish a new lifelong health record, monitoring developmental goals, vaccinations and any other ongoing health service needs. It's important to consider the need for dataflow that populates one's health record to enable continuity of care throughout one's lifetime based on key data retrieved from episodic health care services. This determines connectivity needs for any digital health ecosystem.

Each of these many variable options and the consistency of adoption (or the lack thereof) is dependent upon agreements reached between all relevant parties during the software and/or modeling development phases. The adoption and use of any interoperability option plus the use and maintenance of associated data and unique identifier maps, determine the extent to which exchanged data and information can be used for both primary and secondary purposes. This in turn determines data processing limitations and impacts upon the potential benefits of digital system implementation as this applies to any ecosystem. Another consideration is the information needs of each specific healthcare provider to support their service delivery processes, including supply chain and resource management.

Secondary data use requires extensive data linking to generate new information and knowledge suited for multiple purpose uses. There is a continuing lack of interoperability among the data resources for EHRs within healthcare organizations and between healthcare providers. This is a major impediment to the effective exchange of health information. These interoperability issues need to be solved going forward, or else the entire health data infrastructure will be crippled [81].

A national (or global) digital health ecosystem needs to be designed according to required and desired national data and information flows to support the health industry as a whole. This requires process and clinical (pragmatic) interoperability as well as semantic and technical interoperability schema that are primarily designed to support the management and processing of health and associated data. Collectively interoperability is about ensuring systems are able to work together as required for an optimally functioning digital health ecosystem.

The American Medical Informatics Association published a white paper in 2007 indicating that the United States requires a framework for the secondary use of health data with a robust infrastructure of policies, standards, and best practices. It is argued that such a framework can guide and facilitate widespread collection, storage, aggregation, linkage, and transmission of health data. Such a framework needs to provide appropriate protections for legitimate secondary use [82]. The same principles are applicable to any nation. Health and nursing data should be viewed as scientific data as it is a primary asset supporting learning environments, enhancing our ability to make any number of improvements regarding, for example, resource use and from which numerous discoveries are made.

An assessment of the quality of data contained within EHRs in terms of completeness, correctness, concordance, plausibility and currency, found that these data quality features were difficult to measure. It was concluded that the clinical research community needs to develop validated, systematic methods of EHR data quality assessment [83]. EHR data quality is essential for any secondary clinical use. Scientific data needs to comply with FAIR data management principles as described in Chapter 3.

## Using source data and information for multiple purposes

The use of electronic handover tools generated by integrated EHR systems provide more and better information to care teams during shift handovers than paper-based handovers [84]. Lessons learned from evaluation studies regarding the use of nursing models and nursing documentation systems in Finland were that different health professionals have different information needs for content and representation. The use of "tailored templates" were recommended. These need to cover essential features to suit specific situations for both data entry and information display, thus enabling context specific applications of the chosen nursing model [85]. It is possible to set up randomized trials or other quasi-experimental designs needed to produce evidence of practice. Qualitative nursing research that explores the relationship between practice and information use could be used as a precursor to the design and testing of, for example, nursing information systems. It is important to involve nursing staff in the design and development of nursing record systems [86].

Electronic nursing documentation, links to EHRs, nursing data requirements and use were detailed in Chapter 6. Here we discuss how such data needs to be linkable to other patient data and potentially used for a variety of purposes. The use of health information exchange in any form does make data available for use to answer clinical or epidemiological, financial, or utilization-based research questions. Such information exchange is dependent upon a consistent use of agreed health informatics standards. The majority of studies using such data are done primarily to evaluate the use and impact of health information exchange on health care delivery and outcomes, although significant barriers to effective health information exchange exist. These barriers include technical infrastructure limitations, business processes limiting secondary use of data, and a lack of participating provider support [87].

The re-use of clinical data was found to be a fast-growing field as potential benefits have been widely recognized [88]. Data re-use continues to primarily rely on data obtained from data warehouses filled from source systems with copied data that has been cleaned, filtered with a modified structure based on the data storage scheme. Data quality from such sources need to be questioned as it is insufficient to simply transfer data into the research database without contextual knowhow of their meaning at that time. The Clinical Data Interchange Standards Consortium (CDISC) works with a global community of experts to develop and advance data standards of the highest quality for research use [89]. Researchers also make use of the Research Electronic Data Capture (REDCap) secure web application to build and manage online surveys and databases [90]. Natural Language Processing is used to extract data from unstructured clinical data in EHRs. Clinical trial research is a multi-million dollar business that stands to gain significantly from the global adoption of agreed standards that optimizes connectivity and interoperability to benefit all stakeholders within the digital ecosystem.

Secondary data use is expected to result in significant time and cost savings, speed up data processing and the acquisition of useful new information and knowledge as well as improve health outcomes. A variety of agreed data sets are in use for this purpose, but there are many data overlaps between such data sets that have not been harmonized. Internationally agreed aims are to:

- Improve the patient experience (increased access or choice, safety, quality of care, patient empowerment, trust or confidence, increasing satisfaction).
- Improve the clinician experience (Increased access to information, increasing trust providing the best care, improved usability, increasing satisfaction, time savings).
- Realize cost savings per capita.

Structuring EHR data, or methods used to do so, have rarely been viewed as an intervention for the purpose of studying its impact of using structured EHR data for secondary purposes [91] within a healthcare facility. Yet this is beneficial for the study of impacts on care processes, productivity and costs, patient safety, care quality or other health impacts. These endpoints tend to be discussed as goals of secondary use and less as evidence-supported impacts, resulting from the use of structured EHR data for secondary purposes [92]. Benefits are expected to be greatest where such data includes nursing data.

The efficiency and effectiveness of secondary data use for any purpose is entirely dependent upon the ability to accurately automate data linkage, transfers, reorganization, extraction and processing [15]. This ability can only be achieved once all source data are collected in a computable form. A variety of roundabout strategies are in use to overcome deficiencies in semantic interoperability, such as data re-entry or manual coding of record content to enable categorization or classification, such as ICD (International Classification of Diseases) coding [93], and further data processing. In such instances context and meaning of data processed is either lost or severely compromised.

To overcome some of the issues associated with data inconsistencies, a large international group of collaborators, have established the Observational Health Data Sciences and Informatics (OHDSI) community [94,95]. Their aim is to improve health outcomes for patients around the world. This community has developed a Common Data Model (CDM). Observational data holders working as OHDSI partners, are required to translate their data to the OMOP common data model (CDM) by mapping each element contained within their database to the approved CDM vocabulary and placing their data in that data schema. This information model explicitly and formally specifies encoded concepts and relationships enabling multicenter, global analyses to be executed rapidly and efficiently using applications or programs developed at a single site [96]. OHDSI provide resources, including common representation (terminologies, vocabularies, coding schemes) enabling the conversion of a wide variety of datasets along with tooling, to make use of the data once converted to the standard format [97]. The data model and its associated vocabulary services is maintained by the OHDSI group.

A primary issue with this approach is the degree of accuracy and retention of meaning of all data mapped to the CDM. Jiang et al. [98] noted that a variety of data models have been developed to provide a standardized data interface that supports organizing clinical research data into a standard structure for building the integrated data repositories. The HL7 Fast Healthcare Interoperability Resources (FHIR®) as discussed are increasingly in use for data exchange between systems. One study designed and assessed a consensus-based approach for harmonizing the OHDSI CDM with the HL7 FHIR® W5 classification system. Despite finding only fair to moderate agreement for model and property level harmonization these researchers concluded that FHIR® W5 is a useful tool in designing the harmonization approaches between data models and FHIR®, and facilitating the consensus achievement [98]. This is an application centric solution to health data interoperability.

Similarly the HL7 international community has a "vocabulary" workgroup responsible for providing "an organization and repository for maintaining a coded vocabulary that, when used in conjunction with HL7 and related standards, will enable the exchange of clinical data and information so that sending and receiving systems have a shared, well defined, and unambiguous knowledge of the meaning of the data transferred." This group works cooperatively with others such as standards development organizations, creators and maintainers of vocabularies, government agencies and regulatory bodies, clinical professional specialty groups and other HL7 workgroups [99]. One of their current projects is a Unified HL7 Terminology Governance Process.

Many countries have recognized the need for a standard terminology and have adopted the SNOMED-CT® terminology as their national ontology-based terminology standard that covers the medicine domain. Ontologies represent knowledge about invariable truths about a particular domain of discussion. Another definition for ontology is a "formal, explicit specification of a shared conceptualisation" [100]. It is critical for the correct code to be identified for each concept to enable data sharing and a consistent interpretation of meaning. Adopting a standard terminology is not the only requirement to enable meaningful secondary data use. For example if organization A assigns the code 75367002/Blood pressure (observable entity) to identify a blood pressure recording, while organization B chooses the concept 163020007/On examination — blood pressure reading (finding), then semantic interoperability will be hampered as both concepts belong to different hierarchies in SNOMED CT. This is critical when using clinical data in decision support systems.

The EU funded a study that undertook a cost-benefit analysis of this terminology [101] and another study that examined its use for large scale e-health deployments in the EU [102]. A central challenge was found to be the difficulty in delineating the benefits of semantic interoperability from the implementation of clinical terminology/SNOMED CT® and distinguishing between speculative or expected or perceived versus the observed or measured benefits. Such benefits can usually only be quantified in a meaningful way in a concrete use case or case studies. Most evaluations undertaken regarding the use of SNOMED CT® were found to have focused on administrative and financial transactions rather than on delivering clinical care. Recommendations made to the EU are as follows [102]:

1. Any decision about the adoption and role of terminological resources, including SNOMED CT®, must be part of a wider, coherent and priority-driven strategy for optimizing the benefits of semantic interoperability in health data, and of the overarching eHealth Strategy of the European Union and its Member States.
2. SNOMED CT® is the best available core reference terminology for cross-border, national and regional eHealth deployments in Europe.
3. SNOMED CT® should be part of an ecosystem of terminologies, including international aggregation terminologies (e.g. the WHO Family of Classifications), and including local/national user interface terminologies, which address multilingualism in Europe and clinical communication with multidisciplinary professional language and lay language.
4. The adoption of SNOMED CT® should be realized incrementally rather than all at once, by developing terminology subsets that address the interoperability requirements for prioritized use-cases, and expanding this set over some years.
5. Mechanisms should be established to facilitate and co-ordinate European Member State co-operation on terminology and semantic interoperability, including common areas of governance across national terminology centers, eHealth competence centers (or equivalent national bodies).

*An elaborate communication strategy with the scientific associations of health care providers (medical sub-disciplines, primary care physicians, allied health personnel [nurses and midwives]) is needed to inform, educate and convince with regard to the necessity of semantic interoperability, well structured electronic health records, performant end user terminologies, suitable international reference and aggregation terminologies, and clinical documentation skills [102].*

The Norwegian health authorities have adopted a trail blazing strategy to transform their system to a patient-centered healthcare service. This research activity aims to "gain knowledge and a common understanding of the consequences of implementation in the transition to ontology-based terminologies in relation to strategic plans for the healthcare sector" [100]. These researchers note that for large-scale communication of data to be useful, its interpretation must enable different actors in different organizations to draw the same conclusions and avoid wrong inferences to be made. This means that data from many sources must conform to a common representation and meaning by faithfully specifying certain contextual aspects. This requires a significant level of semantic interoperability based on common reference models, clinical information models and the adoption of standard clinical terminologies including SNOMED-CT, LOINC (Logical Observation Identifiers Names and Codes) [103], ICD (International Classification for Diseases) [93]. Norway is one of many countries to deploy OpenEHR data centric solutions for this purpose [104].

### *Decision support systems — Using secondary data*

The use of clinical decision support (CDS) systems need to fit with a clinician's workflow. The clinicians (and patient) are always in the position of plotting a course "from here" solving a clinical process any time throughout a patient's journey. Such journeys are full of decisions to determine the next steps as a patient's condition is forever changing. This requires revising the plan of care in real-time. Workflow and decision support are intimately related.
The openEHR Specification program now has Task Planning Model Specification to address this issue [105]. Other requirements may be communication with other care team members who each have their own plan of care. All players operate independently rather that mindlessly following someone else's directions. This has been likened to a GPS system required to constantly recalculate directions.

Unfortunately many care provider users of CDS systems are frequently bombarded with inappropriate and inapplicable CDS that often are not informational, not integrated into the workflow, not patient specific, and with CDSs that may present out of date and irrelevant recommendations [106]. The effect of CDS systems on clinical outcomes, health care processes, workload and efficiency, patient satisfaction, cost, provider use and implementation has found that "both commercially and locally developed CDSSs are effective at improving health care process measures across diverse settings, but evidence for clinical, economic, workload, and efficiency outcomes remains sparse" [107]. Collecting useful real-world timely evidence is only possible via the use of effectively integrated information systems using, preferably internationally agreed, standard data representations.

Decision support systems need to make use of computable code to be able to convey new knowledge to clinicians, or any other potential user, in a way that can easily be used in practice. One tried and tested technology able to convert evidence based clinical or other knowledge, or any source data, into computable code is via the development of archetypes (constraint models). The internationally supported openEHR's Clinical Knowledge Manager (CKM) repository contains evidence-based models that can be used by any software developer for any type of application. The subsequent vendor-neutral data repositories facilitate the use of the

Archetype Query Language (AQL) where unlike other query languages, the syntax is independent of applications, programming languages, system environment and storage models. It's minimum requirement for data to be querying with AQL is for the data to be marked at a fine level of granularity with the appropriate archetype codes and terminology codes [108]. This may be native openEHR-structured data, or legacy system data to which the relevant data markers (mainly archetype paths and terminology codes) have been added [14,15,17,109].

Decision support systems can also use computable code from systems based on relational databases making use of another query language, such as SQL to extract the necessary source data for secondary use. The usability of data contained in any database is limited by the known data field relationships built into the database structures. A DSS can be programmed to evaluate care relative to patient outcomes although this is dependent upon the available access to all relevant source data that collectively meet the necessary criteria [110]. Irrespective of how data are retrieved data quality must be retained. This is particularly important for operational analytics [111].

### National and international health data uses

Given the multitude of actual and possible purposes for the use of secondary data it is helpful to consider the entire national and global health data ecosystems. These are currently badly fragmented. Ideally health data flow from point of care or source, needs to be considered regarding all possible uses across the global, regional and national health data ecosystems. To this end the World Health Organization has published a circular graph that represents this ecosystem and demonstrates the vastness of health data secondary uses [112]. At the center of the WHO circular graph is individual health data. This can take many forms at various levels of granularity. The graph shows four separate sectors:

(a)  Standard data from health services, public health and research,
(b)  Expanded data about the environment, lifestyle, socioeconomic, behavioral and social,
(c)  Capabilities — technological, analytical and policy,
(d)  Stakeholders — individuals, groups, health services, research and academia, healthcare industry, data & ICT industry and Governments [112].

This diagram (Fig. 2) serves as a comprehensive overview of secondary data use at every level within the health system and their relationships. It's a useful framework from which to describe the health data, information and knowledge ecosystem that begins with all personal health source data that may be captured via observations, diagnostic tests and examinations using any number of devices. Such data may need to come from a multitude of locations but needs to be combined into a health record pertaining to any individual. These data may also need to be linked with other data from different systems to generate new information and/or discover new knowledge.

# Evolving health data ecosystem

E. Vayena, J. Dzenowagis, M. Langfeld, 2016

**Fig. 2**

WHO evolving health data ecosystem [112].

Secondary data use needs to be considered as part of any system implementation process. Project managers have a dual responsibility when developing or implementing eHealth systems or infrastructure. First there is a need to meet the objectives of the specific system or initiative, but secondly the system or initiative needs to fit the broader requirements of the Digital Health system into the future. Short term approaches or those which focus on a singular result, are likely to deliver systems, but those systems are less likely to deliver the long term benefits desired. Decision makers need to understand successful implementation is not judged by a system being operational and being used, but by whether that system achieves what was desired of it and for clinical systems, this must include appropriateness to clinical workplace and safety. Only a data-centric approach to interoperability facilitates full reporting automation to suit all of the above data uses within a digital health ecosystem.

*National and international reporting — An example*

The Australian Institute of Health and Welfare (AHIW) has the responsibility to manage Australia's international reporting to WHO. This national agency provides information and statistics on Australia's health and welfare by collaborating with federal and state-level agencies as well as non-government organizations across multiple sectors. Its main activity is to develop national data standards, collect data accordingly, link multiple data sets, undertake data analysis and produce reports. They undertake specific project work with defined deliverables using their data collections. All data are protected to the point where outside researchers can only view data and engage the AIHW staff to undertake data analysis on their behalf. Data use is highly inflexible as none of the raw data is permitted to be copied for further analytical processes required to answer specific research questions. Data use needs to overcome many hurdles via this Agency's secure environment. This severely limits the use of this valuable data asset for the public good. The AIHW reached health data agreements with other jurisdictions back in 1991 to ensure data consistency and the best possible degree of accuracy of data collected.

Australia has many other Government data collections used primarily for funding purposes, such as Medicare, the Pharmaceutical Benefits Scheme, Prescription drug monitoring, Aged care, the Commonwealth Home Support Programme. Every country has similar data collection requirements, some far more comprehensive than others. A number of underdeveloped countries have yet to make use of a national birth and death register.

To achieve the efficient exchange and interoperability of health data requires codes, standards and the standardization of health data terminology. Protecting privacy should be a priority. A certification or accreditation service for organizations processing personal health data may need to be considered as part of a national health data governance framework. This process will ensure that organizations are proactive in relation to the implementation of data standards and data governance, and that all staff with access to any health data are suitably trained and adhere to established ethical principles and codes of practice [113].

With the current adoption of person centered care, efforts are needed to address barriers to health literacy of the population, and to facilitate people's access to their own medical records and health information [113]. Evaluation of health IT is influenced by cultural bias and individual frames of reference based on cultural behaviors, prior knowledge, skills and experiences. Every stakeholder has a unique interest and perspective. It's been noted that "access to, and the processing of, personal health data can serve health-related public interests and bring significant benefits to individuals and society" [113].

Advances in genomics mean that treatment pathways are becoming tailored to individual patients. The increasing use of medical devices employing digital communication tools, digital innovations and big data are enabling people to manage their health more independently. These data are being used for clinical purposes. This means that improvements are required in health system governance.

### Genomics data and personalized medicine

Precision medicine requires greater use of any type of *omics data, consisting of highly granular and coded data. This is a growing area of research increasingly being used for personalized medicine. Genomic medicine needs high performance computing, data storage infrastructure and services. It also requires local and national networking infrastructure to enable access to, and data exchange with, large internationally federated databases. The Australian Genomics Health Alliance (AGHA) consists of more than 50 partners across Australia and internationally including the Global Alliance for genomics and Health and the Global Genomic Medicine Collaborative [114]. See also the USA based Electronic Medical Records and Genomics(eMerge) network [115] funded by the National Human Genome Research Institute that combines DNA biorepositories with electronic medical record (EMR) systems for large scale, high throughput genetic research in support of implementing genomic medicine. Electronic health record systems will need to be able to accommodate these types of data in the future to enable the provision of personalized medicine [116,117].

Another funded initiative was the 4 year p-medicine project [118] that concluded in 2015. This was about accelerating personalized medicine and overcoming the current problems in clinical research and paved the way for a more individualized therapy to better suit individuals [119]. This has formed the foundation for STaRC (Study Trial and Research Center) established by that project's core team. This group collaborates with ECRIN (European Clinical Research Infrastructure Network) [120,121], a distributed research infrastructure connecting multiple sites across Europe. This infrastructure creates added value through access to expertise and patients, increasing the reach, diversity, and the quality of clinical trial results. This work is enabling the merging of clinical data with biomolecular findings and imaging studies of individual patients. A similar research activity is in progress by researchers at the Victorian Comprehensive Cancer Centre using data from 10 organizations [122]. This type of data merging is proving to be difficult to achieve, given the many infrastructure limitations.

### Gap analysis and digital transformation

There is a need to identify the gap between an organization's vision and associated digital ambitions, and its current capabilities including the identification of associated roadblocks that need to be overcome in order to realize the vision. The roadmap for a digital transformation supporting an organization's capacity to care needs to be focused on overcoming all constraints and filling the identified gaps. The required skills, resources and partnerships needed for the organization's digital health ecosystem to function optimally must be considered. The previous chapters have shown many data relationships between input, process and output data variables. Information systems need to be able to make use of this knowledge and enable the significance of such relationships to be explored, thus enabling decision makers to assess impact.

This requires the availability of well connected "learning systems" based on healthcare facility ecosystems that ideally support vendor-neutral clinical, patient generated and administrative data storage, where datasets are well defined and standardized to represent key input, process and output variables. Such systems are then also able to support querying to obtain new content in readiness for App building and other useful purposes. Current examples are research databases or registries that reflect imperfect representations of a persons' lived experiences. This is due to significant data gaps [52]. Deeny et al. emphasize that "ongoing feedback of insights from data to patients, clinicians, managers, and policymakers can be a powerful motivator for change as well as provide an evidence base for action" [17].

Healthcare's vision needs to be about maximizing automation by focusing on developing methods that enable data entry or capture once only at the operational point of use and facilitate multiple uses of that data to suit numerous purposes. Any successfully integrated health system is the result of a sound understanding of its past state, and its desired future state as defined by their business case, organizational readiness for change, business processes and technology assessments. The results of such assessments formed the foundation for gap identification, determining integration needs and subsequent planning activities. It's important not to allow technical solutions to overshadow unique clinical and business needs. For example, frequently used clinical systems such as Computerized Physician Order Entry (CPOE) systems are used primarily for ordering medications. These usually include the provision of electronic alert warnings. A review found that the generation of such alerts varied widely between systems, were often confusing and were dependent on numerous factors including data entry methods used. Testers illustrated various workarounds that allowed them to enter numerous erroneous orders potentially compromising patient safety [123].

Information technology staff need to work closely with senior decision makers to identify current system capabilities, data structures, collection processes, storage, retrieval, transmission and information display processes. All data sources, including data warehouses, system components such as health information exchanges, applications, data and technology compliance standards currently met, need to be considered as part of this analysis. It's important to consider not only existing legacy systems but also all shadow systems in use as these requirements are best incorporated as part of the digital transformation agenda. Such an inventory then needs to be compared with organizational requirements, including legislative, regulatory, professional, technical and data standard compliance needs, so that system connectivity and associated interoperability schema can be designed and implemented to achieve a closer alignment with patient and user needs and fill the identified gaps.

A gap analysis and subsequent activities require careful consideration regarding the possible reusability of existing technologies in terms of limitations to meeting the desired functionality and future risks. Some technologies may need to be reconfigured to meet short term gains and some will need to be replaced at some point in time. These need to be included as part of the adopted investment strategy timeline. Developing a solution that meets all of the organization's interoperability requirements is the most difficult task for enterprise and solution architects [124]. It is important to concurrently identify required key data standards and develop a data governance implementation strategy as part of your digital transformation program [125–127]. This includes data ownership and data stewardship within and beyond the individual healthcare facility. Decisions regarding the integrated solution, user interfaces, reporting information needs and access control management, need to be made consistently, and should be effectively communicated, implemented, monitored and evaluated. This requires strong leadership, good governance and an appropriate workforce skill mix able to make well informed contributions.

EHR data may be used to link with clinical guidelines to provide Clinical Decision Support (CDS) systems. It is of concern that many CDS systems apply different terminological systems to code data. CDS studies differ regarding interventions referred to, populations, settings, and outcomes [107]. This diversity hampers the possibility of sharing and reasoning with data within different systems. One systematic review concluded that both commercially and locally developed CDSSs are effective at improving health care process measures across diverse settings, but evidence for clinical, economic, workload, and efficiency outcomes remains sparse [107]. The results of two other systematic reviews confirm the hypothesis that data standardization is a critical success factor for CDS systems [11]. These results also provided convincing evidence of guideline based CDS systems improving clinical performance and patient outcomes, although failure rates in the introduction of guideline based CDS systems remains high [128].

Many terminological resources, incl. terminologies, classifications and code systems, are in use to meet multiple high level reporting requirements. These secondary data are derived from processing detailed source data and are predominantly used for statistical analytical purposes. In the absence of the use of mandated standard data as collected at the source, these organizationally based terminological resources are the result of source data collected and stored by a variety of different systems. As a consequence most health service providers have many silos of information that are unable to be linked with any accuracy or consistency [129].

New skills are needed regarding shared information models and concept representation systems which are significantly more complex than those of the terminology classification systems in use.

In recognition of the complexities of managing health terminology in a manner that enables semantic interoperability, there is an identified need for highly skilled Professionals to work as "Health Data Scientists." There is a need to specify these new job roles, key skills requirements and curricular competencies for all personnel whose role relates to the delivery of vocabulary and controlled terminology services in healthcare organizations that deploy health information and communication technology (HICT) products. Only then can semantic interoperability in healthcare be achieved. New job roles identified include data mapping specialist, data conversion analyst, interface analyst, coding specialist, data modeler, terminologist, terminology standards developer/manager, content manager (including Clinical Documentation Improvement [CDI] specialist).

There is also a need for business standards that guide the evaluation of data map quality and terminology. There are many standard terminologies, including classification and coding systems in use. Each terminology is managed and maintained by its own standards development organization (SDO). The use of each terminology standard has its own set of requirements, models and precise linguistic technology. Semantic interoperability requires that the context of any term from any terminology is retained at every point of data transfer. This is best achieved by making use of "object" or "constraint models" such as used by the computer language "Java." Objects relate to "classes" in a reference model. This solution solves the following exemplar issues.

> *For example, our gastroenterology department wanted to understand whether we are utilizing endoscopic procedures for the right indications. When we tried to analyze the data, we found out that there are 84 different indications, which were not actually 84 different indications, just indications written different ways in unstructured formats. Many were synonyms for similar indications.*

Or

> *Within LOINC, you can choose different LOINC codes for a lab, depending on how granular you want to make it. We're working with nursing documentation in trying to decide the level at which you code it. And it matters depending on the use you have for that data. You might have multiple uses of that data that would require different levels of granularity for mapping. So it's not as simple as we use LOINC for lab or LOINC for nursing [130].*

These examples highlight the need for consistent representation of meaning in information models and terminology models as used by the entire family of HL7 standards, including FHIR models, and other standards in use, as components of interoperability schema developed to technically connect different systems. This requirement is a pre-cursor to the ability to enter data once at the source and use it many times. The continuing use of multiple terminologies needs to change. *We cannot treat terminology management in a departmental fashion; it needs to be an enterprise-wide strategy, because it really spreads across so many areas, and it is complex* [130]. Meanwhile significant resources are expended to map one data representation to another to enable data from a source system to be used by a receiving system.

A sound knowledge of the health organization's clinical needs, business needs, enterprise architecture and available technologies are also essential as clinical and technical agendas are sorted out [124]. This requires a solid foundation and an in-depth understanding of management and IT best practices, governance and organizational change management as many operational processes may need to be changed to improve work and data flows while optimizing automation. This is where the end users need to make their contributions as they are the most knowledgeable about their very specific information and workflow requirements. Nurses and midwives should be educated in health informatics and project management methods to prepare for, participate in and/or to lead clinical systems implementation [131]. The implementation lead needs to be able to deal with unexpected issues regarding data sharing and exchange, data source accuracy, adoption and use of data processing algorithms, clinical alerts and patient safety. These are governance responsibilities.

Every health service's ecosystem's connectivity requirements need to reflect the current status and future needs of its network. There is no universal set of systems, complete with connectivity standards, to suit every healthcare ecosystem. There are far too many confounding variables that need to be considered to maximize each ecosystem's functionality. What is important is a high-level reference architecture, an agreed open platform, and an agreed set of high level technical, identifier, data types and data standards that every healthcare facility's IT infrastructure needs to comply with. An inaccurate or incomplete gap analysis relevant to each type of user can have costly consequences [124].

## Conclusion

This chapter has focused on the digital transformation needs primarily from a healthcare facility's perspective. They have a need to be able to measure, monitor and manage their capacity to care, making use of patient data that determines workload demand. It was evident that the use of nursing or midwifery data has to date not been seriously considered. Every healthcare provider has a need to be able to liaise with additional internal and external product and service providers. It is therefore imperative that any facility's digital transformation considers this context in the interest of patient centered care to make any patient's health care journey easier to manage. This chapter has attempted to convey the critical importance of a number of high-level concepts associated with data accuracy, data consistency and reliable data use. Clinical data processing requires accuracy at every level of use. This is to a large extent dependent upon system architectures, maintenance and data governance.

A complete understanding and appreciation of these concepts requires in depth software engineering expertise, this chapter has focused on explaining the concepts and impact. Also highlighted was the need to identify key data standards, adopt a data-centric interoperability approach and to implement a data governance strategy. Some of these concepts will be further explored in the chapters to follow. Our focus in this chapter is on providing the necessary knowledge about the many digital concepts to be consider by stakeholders, and how to obtain user participation and acceptance of the digital transformation. Success requires sound change management, which is addressed in the next chapter.

## References

[1] WHO/ITU. National eHealth strategy toolkit. [cited 17 April 2019]. Available from: https://www.itu.int/pub/D-STR-E_HEALTH.05-2012; 2012.

[2] WHO. Draft Global Strategy on Digital Health. [cited 17 April 2019]. Available from: https://extranet.who.int/dataform/183439; 2019.

[3] Levin SA. Ecosystems and the biosphere as complex adaptive systems. Ecosystems 1998;1(5):431–6.

[4] Jacobides MG, Cennamo C, Gawer A. Towards a theory of ecosystems. Strateg Manag J 2018;39(8):2255–76.

[5] Serbanati LD, Ricci FL, Mercurio G, Vasilateanu A. Steps towards a digital health ecosystem. J Biomed Inform 2011;44(4):621–36.

[6] Wilkinson MD, Dumontier M, Aalbersberg IJ, Appleton G, Axton M, Baak A, et al. The FAIR Guiding Principles for scientific data management and stewardship. Sci Data 2016;3:160018.

[7] Romanelli M. Towards sustainable ecosystems. Syst Res Behav Sci 2018;35(4):417–26.

[8] Australian-Government. Digital service standard. Canberra: Commonwealth of Australia; 2019. [cited 9 May 2019]. Available from: https://guides.service.gov.au/digital-service-standard/.

[9] Averian A. A reference architecture for digital ecosystems. In: Internet of things—technology, applications and standardisation. IntechOpen; 2018.

[10] Beale T. The crisis in e-health standards III—solutions. [cited 22 May 2019]. Available from: https://wolandscat.net/2009/10/18/the-crisis-in-e-health-standards-iii-solutions/; 2009.

[11] Ahmadian L, van Engen-Verheul M, Bakhshi-Raiez F, Peek N, Cornet R, de Keizer NF. The role of standardized data and terminological systems in computerized clinical decision support systems: literature review and survey. Int J Med Inform 2011;80(2):81–93.

[12] Atalag K, Beale T, Chen R, Gornik T, Heard S, McNicoll I. openEHR-A semantically-enabled, vendor-independent health computing platform—white paper; 2017.

[13] Kropf S, Chalopin C, Lindner D, Denecke K. Domain modeling and application development of an archetype- and XML-based EHRS. Practical experiences and lessons learnt. Appl Clin Inform 2017;8(2):660–79.

[14] Marco-Ruiz L, Maldonado JA, Traver V, Karlsen R, Bellika JG, editors. Meta-architecture for the interoperability and knowledge management of archetype-based clinical decision support systems. IEEE-EMBS International Conference on Biomedical and Health Informatics (BHI), 1–4 June 2014; 2014.

[15] Marco-Ruiz L, Moner D, Maldonado JA, Kolstrup N, Bellika JG. Archetype-based data warehouse environment to enable the reuse of electronic health record data. Int J Med Inform 2015;84(9):702–14.

[16] openEHR-Foundation. openEHR reference model (RM)—latest. openEHR Foundation; 2017.

[17] Pereira C, Frade S, Brandão P, Correia R, Aguiar A, editors. Integrating data and network standards into an interoperable e-Health solution. 2014 IEEE 16th International Conference on e-Health Networking, Applications and Services (Healthcom), 15–18 October 2014; 2014.

[18] openEHR/CKM. Clinical knowledge manager. [cited 21 February 2019]. Available from: https://ckm.openehr.org/ckm/.

[19] Gornik T. Industry news—application-centric and data-centric interoperability. London: openEHR Foundation; 2019. [cited 3 February 2019]. Available from: https://www.openehr.org/news_events/industry_news.php?id=264.

[20] OpenEHR. openEHR industry partners. London: openEHR Foundation; 2019. cited 3 February 2019. Available from: https://www.openehr.org/industry_partners/.

[21] ISO/IEC. 10746-1:1998 (en) Information technology – open distributed processing – reference model Part 1–4.

[22] Mead CN. Data interchange standards in healthcare IT—computable semantic interoperability: now possible but still difficult, do we really need a better mousetrap? J Healthc Inform Manage 2006;20(1):71–8.

[23] HL7. Health Level Seven International. [cited 6 February 2019]. Available from: http://www.hl7.org/.

[24] IHE-International. Intergrating the Healthcare Enterprise (IHE). [cited 22 May 2019]. Available from: https://www.ihe.net/.

[25] openEHR. Clinical Modeling Wiki; 2019.

[26] ISO-13606-2. Health informatics—electronic health record communication—Part 2: archetype interchange specification. International Organsiation for Standardisation; 2008. [cited 25 May 2019]. Available from: https://www.iso.org/standard/50119.html.

[27] Blobel B, Ruotsalainen P, Lopez DM, Oemig F. Requirements and solutions for personalized health systems. Stud Health Technol Inform 2017;237:3–21.

[28] USA. Federal Enterprise Architecture Framework (FEAF) v.2. [cited 6 February 2019]. Available from: https://obamawhitehouse.archives.gov/sites/default/files/omb/assets/egov_docs/fea_v2.pdf; 2013.

[29] Australian Government. Architecture reference models version 3.0. Australian Government Information Management Office (AGIMO); 2011. [cited 6 February 2019]. Available from: https://www.finance.gov.au/sites/default/files/aga-ref-models.pdf.

[30] OpenGroup. Making standards work: the TOGAF standard—version 9.2: The Open Group; [cited 6 February 2019]. Available from: https://www.opengroup.org/.

[31] Zachman J. The concise definition of the Zachman framework. Zachman International; 2008. [cited 6 February 2019]. Available from: https://www.zachman.com/about-the-zachman-framework.

[32] HIMMS. Call to action: achieve nationwide, ubiquitous secure electronic exchange of health information. [cited 6 February 2019]. Available from: https://www.himss.org/library/himss-call-action-achieve-nationwide-ubiquitous-secure-electronic-exchange-health-information; 2017.

[33] Beale T. What is an open platform: Woland's cat. [cited 7 February 2019]. Available from: https://wolandscat.net/2014/05/07/what-is-an-open-platform/; 2014.

[34] Hersh W, Totten A, Eden K, Devine B, Gorman P, Kassakian S, et al. Health information exchange. Evid Rep Technol Assess 2015(220):1–465.

[35] JASON. Report task force final report. Health IT; 2014. [cited 6 February 2019]. Available from: https://www.healthit.gov/sites/default/files/facas/Joint_HIT_JTF%20Final%20Report%20v2_2014-10-15.pdf.

[36] ONC. Health information technology (health IT) certification criteria. Federal Register, Rules and Regulations. 159.

[37] ONC. Health IT certification program: enhanced oversight and accountability 81 FR 72404; 2016.

[38] ONC. Certification of health IT. [cited 5 February 2019]. Available from: https://www.healthit.gov/topic/certification-ehrs/about-onc-health-it-certification-program.

[39] ONC. Certified Health IT Product List (CHPL). [cited 5 February 2019]. Available from: https://chpl.healthit.gov/#/search.

[40] The Digital Health Society. European Connected Health Alliance. [cited 6 February 2019]. Available from: https://echalliance.com/page/Digitalhealthsociety.

[41] openEHR.An open domain-driven platform for developing flexible e-health systems openEHR Foundation; [cited 6 February 2019]. Available from: https://www.openehr.org/.

[42] INTEROPen. IHE, HL7 FHIR & openEHR—a smorgasboard of standards: an action group to accelerate the development of open standards for interoperability in the health and social care sector. [cited 13 March 2019]. Available from: http://www.interopen.org/2018/07/19/ihe-hl7-fhir-openehr-a-smorgasbord-of-standards/; 2019.

[43] DHCRC. Digital Health Cooperative Research Centre. [cited 23 May 2019]. Available from: https://www.digitalhealthcrc.com/.

[44] Reynolds CJ, Wyatt JC. Open source, open standards, and health care information systems. J Med Internet Res 2011;13(1):e24.

[45] Beresniak A, Schmidt A, Proeve J, Bolanos E, Patel N, Ammour N, et al. Cost-benefit assessment of using electronic health records data for clinical research versus current practices: contribution of the Electronic Health Records for Clinical Research (EHR4CR) European Project. Contemp Clin Trials 2016;46:85–91.

[46] Coorevits P, Sundgren M, Klein GO, Bahr A, Claerhout B, Daniel C, et al. Electronic health records: new opportunities for clinical research. J Intern Med 2013;274(6):547–60.

[47] De Moor G, Sundgren M, Kalra D, Schmidt A, Dugas M, Claerhout B, et al. Using electronic health records for clinical research: the case of the EHR4CR project. J Biomed Inform 2015;53:162–73.

[48] Balka E. Ghost charts and shadow records: implication for system design. In: Medinfo2010. CapeTown, South Africa: IOS Press; 2010.

[49] Lund-Jensen R, Azaria C, Permien FH, Sawari J, Bækgaard L. Feral information systems, shadow systems, and workarounds – a drift in IS terminology. Procedia Comput Sci 2016;100:1056–63.

[50] Huber M, Zimmermann S, Rentrop C, Felden C. The relation of shadow systems and ERP systems—insights from a multiple-case study. Systems 2016;4(1):11.

[51] Sweeney E. Shadow IT systems leave healthcare vulnerable to attacks. Hospitals & Healthsystems; 2017. [cited 5 February 2019]. Available from: https://www.fiercehealthcare.com/privacy-security/shadow-it-systems-leave-healthcare-vulnerable-to-attacks.

[52] Deeny SR, Steventon A. Making sense of the shadows: priorities for creating a learning healthcare system based on routinely collected data. BMJ Qual Saf 2015;24(8):505–15.

[53] ISO/TR. 12300:2014 Health informatics—principles of mapping between terminological systems. International Organisation of Standardisation.

[54] ISO/PRF. TS21564 Health informatics—terminology resource map quality measures (MapQual). International Organisation for Standardisation; 2019.

[55] Walker J, Pan E, Johnston D, Adler-Milstein J, Bates DW, Middleton B. The value of health care information exchange and interoperability. Health Aff 2005;24(Suppl 1). https://www.healthaffairs.org/doi/full/10.1377/hlthaff.W5.10 Web Exclusives, Project HOPE; 2005 [cited 21 May 2019]. Available from:.

[56] Aequus communis sententia: defining levels of interoperability. Elkin P, Froehling D, Bauer B, Wahner-Roedler D, Rosenbloom S, Bailey K, Brown S, editors. Medinfo2007. Brisbane: IOS Press; 2007.

[57] ONC. Technology OotNCfHI, editor. Proposed interoperability standards measurement framework; 2017. Washington, DC.

[58] DICOM. Digital Imaging and Communications in Medicine. [cited 23 May 2019]. Available from: https://www.dicomstandard.org.

[59] Grieve G. Levels of interoperability possible with the use of FHIR. FHIR Developer; 2018.

[60] HL7. FHIR overview. HL7 International; 2017. Available from: https://www.hl7.org/fhir/overview.html.

[61] Beale T. A FHIR experience: consistently inconsistent. [cited 8 May 2019]. Available from: https://wolandscat.net/2019/05/05/a-fhir-experience-consistently-inconsistent/; 2019.

[62] Levinson DR. Using health IT for care coordination: insights from six Medicare accountable care organisations. US Department of Health and Human Services; 2019. [cited 23 May 2019]. Available from: https://oig.hhs.gov/oei/reports/oei-01-16-00180.pdf.

[63] HL7. Argonaut project. [cited 6 February 2019]. Available from: http://docs.smarthealthit.org/argonaut/; 2019.

[64] Beale T. FHIR v openEHR—concreta. [cited 21 February 2019]. Available from: https://wolandscat.net/2018/10/10/fhir-v-openehr-concreta/#more-156; 2018.

[65] Beale TA. FHIR experience—the formalism. [cited 9 May 2019]. Available from: https://wolandscat.net/2019/05/08/a-fhir-experience-the-formalism/; 2019.

[66] Chute C. Current trends regarding the adoption of FHIR and terminology standards. 13 March 2019.

[67] openEHR clinical knowledge manager. [cited 26 September 2018]. Available from: https://www.openehr.org/ckm/.

[68] HL7. Resource AllergyIntolerance FHIR. [cited 21 February 2019]. Available from: https://www.hl7.org/fhir/allergyintolerance.html.

[69] 13606-1:2008 I. Health informatics—electronic health record communication—part 1: reference model. Geneva: International Organisation for Standardisation.

[70] ISO. ISO 13606-2:2008 Health informatics—electronic health record communication—part 2: archetype interchange specification. ISO; 2008.

[71] McNicholl I, Mehrkar A, Shannon T. INTEROPen FHIR and openEHR. [cited 4 April 2019]. Available from: https://www.interopen.org/wp-content/uploads/2019/03/INTEROPen-openEHR-and-FHIR_March2019. pdf; 2019.

[72] Aylward B, Wassall R. Defining an open platform. Apperta Foundation; 2017. [cited 10 February 2019]. Available from: https://apperta.org/openplatforms/.

[73] Commonwell-Health-Alliance. Cerner, McKesson, Allscripts, athenahealth, Greenway and RelayHealth announce ground breaking alliance to enable integrated health care. [cited 7 February 2019]. Available from: https://www.commonwellalliance.org/news-center/commonwell-news/cerner-mckesson-allscripts-athenahealth-greenway-relayhealth-announce-ground-breaking-alliance-enable-integrated-health-care/.

[74] HSPC. Healthcare Innovation—Healthcare Services Platform Consortium (HSPC). [cited 7 February 2019]. Available from: https://www.hcinnovationgroup.com/interoperability-hie/article/13025041/intermountain-cmio-on-building-interoperability-building-a-clinical-app-store.

[75] Mozzaquatro BA, Agostinho C, Goncalves D, Martins J, Jardim-Goncalves R. An ontology-based cybersecurity framework for the internet of things. Sensors (Basel, Switzerland) 2018;18(9).

[76] Opengroup. Open Platform 3.0™ white paper: the Open Platform 3.0 Forum. [cited 8 February 2019]. Available from: https://publications.opengroup.org/w147; 2014.

[77] Opengroup. Open Platform 3.0™ Snapshot. [cited 8 February 2019]. Available from: http://www.opengroup. org/openplatform3.0/op3-snapshot/.

[78] Furstenau D, Auschra C. Open digital platforms in health care: implementation and scaling strategies. In: ICIS2016: ISHealthcare. Association for Information Systems; 2016.

[79] Hovenga E, Garde S, Heard S. Nursing constraint models for electronic health records: a vision for domain knowledge governance. Int J Med Inform 2005;74(11 – 12):886–98.

[80] Ogeil RP, Heilbronn C, Lloyd B, Lubman DI. Prescription drug monitoring in Australia: capacity and coverage issues. Med J Aust 2016;204(4):148–148-e1.

[81] McMorrow D. A robust health data infrastructure. Agency for Healthcare Research and Quality; 2013.

[82] Safran C, Bloomrosen M, Hammond WE, Labkoff S, Markel-Fox S, Tang PC, et al. Toward a national framework for the secondary use of health data: an American Medical Informatics Association White Paper. J Am Med Inform Assoc 2007;14(1):1–9.

[83] Weiskopf NG, Weng C. Methods and dimensions of electronic health record data quality assessment: enabling reuse for clinical research. J Am Med Inform Assoc 2013;20(1):144–51.

[84] Flemming D, Hübner U. How to improve change of shift handovers and collaborative grounding and what role does the electronic patient record system play? Results of a systematic literature review. Int J Med Inform 2013;82(7):580–92.

[85] Nykänen P, Kaipio J, Kuusisto A. Evaluation of the national nursing model and four nursing documentation systems in Finland 2013; lessons learned and directions for the future. Int J Med Inform 2012;81(8):507–20.

[86] Urquhart C, Currell R, Grant MJ, Hardiker NR. Nursing record systems: effects on nursing practice and healthcare outcomes. Cochrane Database Syst Rev 2009;1:.

[87] Tresp V, Overhage JM, Bundschus M, Rabizadeh S, Fasching PA, Yu S. Going digital: a survey on digitalization and large-scale data analytics in healthcare. Proc IEEE 2016;104(11):2180–206.

[88] Meystre SM, Lovis C, Bürkle T, Tognola G, Budrionis A, Lehmann CU. Clinical data reuse or secondary use: current status and potential future Progress. Yearb Med Inform 2017;26(1):38–52.

[89] CDISC. Clinical Data Interchange Standards Consortium. cited 23 May 2019. Available from: https://www. cdisc.org/.

[90] REDCap. Research Electronic Data Capture secure web application. [cited 23 May 2019]. Available from: https://www.project-redcap.org/.

[91] Hyppönen H, Saranto K, Vuokko R, Mäkelä-Bengs P, Doupi P, Lindqvist M, et al. Impacts of structuring the electronic health record: a systematic review protocol and results of previous reviews. Int J Med Inform 2014;83(3):159–69.

[92] Vuokko R, Mäkelä-Bengs P, Hyppönen H, Lindqvist M, Doupi P. Impacts of structuring the electronic health record: results of a systematic literature review from the perspective of secondary use of patient data. Int J Med Inform 2017;97:293–303.

[93] WHO. International Classification of Diseases (ICD). World Health Organisation; 2018. [cited 18 August 2018]. Available from: http://www.who.int/classifications/icd/en/.

[94] Hripcsak G, Ryan PB, Duke JD, Shah NH, Park RW, Huser V, et al. Characterizing treatment pathways at scale using the OHDSI network. Proc Natl Acad Sci U S A 2016;113(27):7329–36.

[95] OHDSI. Observational Health Data Sciences and Informatics. [cited 11 February 2019]. Available from: https://www.ohdsi.org/.

[96] Hripcsak G, Duke JD, Shah NH, Reich CG, Huser V, Schuemie MJ, et al. Observational Health Data Sciences and Informatics (OHDSI): opportunities for observational researchers. Stud Health Technol Inform 2015;216:574–8.

[97] OHDSI. Observational Health Data Sciences and Informatics—data standardisation. [cited 18 March 2019]. Available from: https://www.ohdsi.org/data-standardization/.

[98] Jiang G, Kiefer RC, Sharma DK, Prud'hommeaux E, Solbrig HR. A consensus-based approach for harmonizing the OHDSI common data model with HL7 FHIR. Stud Health Technol Inform 2017;245:887–91.

[99] HL7. Health Level Seven International—vocabulary. [cited 11 February 2019]. Available from: http://www.hl7.org/Special/committees/Vocab/index.cfm.

[100] Marco-Ruiz L, Malm-Nicolaisen K, Pedersen R, Makhlysheva A, Bakkevoll P. Ontology-based terminologies for healthcare. [cited 11 February 2019]. Available from:https://ehealthresearch.no/files/documents/Prosjektrapporter/NSE-rapport_2017-08_Ontology-based-terminologies-for-healthcare.pdf.

[101] Gøeg KR, Birov S, Thiel R, Stroetmann V, Piesche K, Dewenter H, et al. Assessing SNOMED CT for Large Scale eHealth Deployments in the EU—D3.3 cost-benefit analysis and impact assessment—final report.

[102] Kalra D, Schulz S, Karlsson D, Stichele RV, Cornet R, Rosenbeck-Gøeg K, et al. Assessing SNOMED CT for Large Scale eHealth Deployments in the EU: ASSESS CT recommendations; 2016.

[103] LOINC. Logical Observation Identifiers Names and Codes: Regenstrief. [cited 25 February 2019]. Available from: https://loinc.org/.

[104] openEHR. openEHR deployed solutions. [cited 19 March 2019]. Available from: https://www.openehr.org/openehr_in_use/deployed_solutions/.

[105] openEHR. Task planning model specification. [cited 12 February 2019]. Available from: https://specifications.openehr.org/releases/PROC/latest/task_planning.html.

[106] Kannry J, McCullagh L, Kushniruk A, Mann D, Edonyabo D, McGinn T. A framework for usable and effective clinical decision support: experience from the iCPR Randomized Clinical Trial. eGEMs 2015;3(2):1–17.

[107] Bright TJ, Wong A, Dhurjati R, Bristow E, Bastian L, Coeytaux RR, et al. Effect of clinical decision-support systems: a systematic review. Ann Intern Med 2012;157(1):29–43.

[108] OpenEHR. Archetype Query Language (AQL). [cited 12 February 2019]. Available from: https://specifications.openehr.org/releases/QUERY/latest/AQL.html.

[109] OpenEHR aware multi agent system for inter-institutional health data integration. Vieira-Marques P, Patriarca-Almeida J, Frade S, Bacelar-Silva G, Robles S, Cruz-Correia R, editors. 2014 9th Iberian Conference on Information Systems and Technologies (CISTI), 18–21 June 2014; 2014.

[110] Data integration for clinical decision support. Yuchae J, Yong Ik Y, editors. 2016 Eighth International Conference on Ubiquitous and Future Networks (ICUFN), 5–8 July 2016. 2016.

[111] Kahn MG, Callahan TJ, Barnard J, Bauck AE, Brown J, Davidson BN, et al. A harmonized data quality assessment terminology and framework for the secondary use of electronic health record data. eGEMs 2016;4(1):1–21.

[112] Vayena E, Dzenowagis J, Langfeld M. Evolving health data ecosystem. WHO; 2016. [25 January 2017]. Available from: http://www.who.int/ehealth/resources/ecosystem/en/.

[113] OECD. The next generation of health reforms, In: Ministerial statement following OECD Health Ministerial Meeting 17 January 2017; 2017.

[114] Australian Genomics Health Alliance. [cited 11 February 2019]. Available from: https://www.australiangenomics.org.au/.

[115] eMerge network Electronic Medical Records and Genomics: Vanderbilt. [cited 2019 11 February]. Available from: https://emerge.mc.vanderbilt.edu; 2014.

[116] The promise of personalised medicine [press release]. Elsevier; 2016.

[117] Wilkinson MD. Genomics data resources: frameworks and standards. Methods Mol Biol (Clifton, NJ) 2012;856:489–511.

[118] p-medicine. Personalised medicine. [cited 11 February 2019]. Available from: http://p-medicine.eu/.

[119] Big data to speed up personalised medicine [press release]; 2016.

[120] ECRIN. European Clinical Research Infrastructure Network. [cited 11 February 2019]. Available from: https://www.ecrin.org/.

[121] Ohmann C, Canham S, Cornu C, Dreß J, Gueyffier F, Kuchinke W, et al. Revising the ECRIN standard requirements for information technology and data management in clinical trials. Trials 2013;14:97.

[122] VCC. Victorian Comprehensive Research Centre. [cited 11 February 2019]. Available from: https://www.ecrin.org/.

[123] Slight SP, Eguale T, Amato MG, Seger AC, Whitney DL, Bates DW, et al. The vulnerabilities of computerized physician order entry systems: a qualitative study. J Am Med Inform Assoc 2016;23:311–6.

[124] CGI. Health information integration: using gap analysis to develop relevant solutions—white paper. [cited 10 February 2019]. Available from: https://www.cgi.com/sites/default/files/white-papers/cgi-health-integration-gap-analysis-paper.pdf; 2014.

[125] Hovenga E, Lloyd S. Working with information and knowledge. In: Harris M, Associates, editors. Managing health services: concepts and practice. 2nd ed.. Australia: Elsevier; 2005.

[126] Hovenga EJS. National healthcare systems and the need for health information governance. In: Hovenga EJS, Grain H, editors. Health information governance in a digital environment. Amsterdam: IOS Press; 2013.

[127] Hovenga EJS. Impact of data governance on a nation's healthcare system building blocks. In: Hovenga EJS, Grain H, editors. Health information governance in a digital environment. Amsterdam: IOS Press; 2013.

[128] Kilsdonk E, Peute LW, Jaspers MWM. Factors influencing implementation success of guideline-based clinical decision support systems: a systematic review and gaps analysis. Int J Med Inform 2017;98:56–64.

[129] Hovenga E. Scoping review: Digital Health evidence. Unpublished report prepared for the Australian Digital Health Agency, Canberra; 2017.

[130] Bazzoli F. Roundtable: resolving terminology conflicts. United States: Faulkner & Gray; 2015. [cited 5 June 2019]. Available from: http://www.healthdatamanagement.com/news/roundtable-resolving-terminology-conflicts.

[131] Gruber D, Cummings GG, LeBlanc L, Smith Dl. Factors influencing outcomes of clinical information systems implementation: a systematic review. Comput Inform Nurs 2009;27(3):151–63.

# CHAPTER 10

# A digital transformation strategy enabling nursing data use

Digital transformation, in even the most minimal form, requires changes in operational processes that enables any healthcare facility to measure, monitor and manage their capacity to care. Effective use of nursing data is dependent upon successful change management in concert with the implementation of a suitable nursing resource management/acuity system. Successful change is dependent upon people behaviors throughout the change journey and is influenced by the associated organizational culture.

The previous chapter outlined the many aspects that need to be considered prior to the implementation of a digital transformation. This includes an examination of how best to make use of available technologies to meet short term and long term strategic objectives. The implementation process often requires major changes regarding any number of entrenched workflow and work method processes including how individuals communicate, how documents are shared, used and managed and how data/information are shared, exchanged and used. This requires a clear and effectively communicated vision. Implementing change can begin by starting small with a focus on working with people who may need to make changes to how they work [1]. Clinical leadership is essential when making workflow changes to meet system requirements. Without effective leadership changes can become incompatible with essential clinical workflow and decision making imperatives.

## System implementation and change management

Within any health care facility it is imperative that the implementation of any system is led by a small group of users/champions working collaboratively with staff from the IT department. It should not be led as any other IT project implementation. It needs to be well planned, adopt an holistic organizational perspective and have a clear vision regarding outcomes to be achieved [2]. Change needs to benefit every individual employee in order to be successful. Successful change management requires an intimate understanding of human behaviors, work and information flow processes. Automation should result in the availability of more timely and complete information, an easy to use system, a saving of staff time as well as improved patient flows and outcomes, providing value in return on investment [3]. Systems must be configured for healthcare professionals, by healthcare professionals, with early and

Measuring Capacity To Care Using Nursing Data. https://doi.org/10.1016/B978-0-12-816977-3.00010-1

continued engagement across all stages of development and implementation. Clinicians and administrators need to participate in co-creating and leading the change [4].

Part of the planning process must ensure that everyone has an opportunity to ask questions, and to alert implementers to possible issues requiring resolution. Everyone needs to fully understand what is being proposed. Leaders cannot assume that others have an equal appreciation of the need nor a desire for change as they do. The cultural landscape at every level and within every department plays an important role. This needs to be well understood as it influences what people value, their perceptions and their behaviors. Leaders need to make use of this knowledge to enable them to influence and drive the change process in the desired direction.

Changes may be initiated to accommodate new external regulatory or funding requirements or to meet new strategic directions associated with the implementation of new information systems. Healthcare organizations need to transform to be able to meet growing consumer demands, and to leverage digital technologies to lower operational costs. This requires all stakeholders, especially medical, nursing midwifery and allied health staff to be on board. Chief Information Officers (CIOs) need to take on new roles, adopt new skills and have access to all the necessary resources to lead a successful digital health transformation. Surveys have found that everyone wants full access to health information, and greater real-time control of health data, from any place, at any time. Services providing patient centred care also need to consider using remote technologies and mobile devices to minimize patient travel to where required services are physically located. A diverse set of agendas needs to be considered [5].

Change management requires strong, top down collective leadership and/or a bottom up approach. It needs to be led by clinicians and service managers supported by good program governance [6], in a manner that achieves workforce engagement, participation, effective teamwork and vision alignment by all concerned. Unexpected difficulties and issues need to be able to be resolved effectively throughout the change process. This is especially important for end users where real changes are acutely felt. This requires the adoption of a clear communication process managed by a system implementation lead for each user role type and required functionality. End-users need to acquire a sense of ownership regarding every work process change, for them to be comfortable and accepting. If not achieved this may lead to an inability to meet original project timelines. Such realities need to be accepted to avoid implementation failure and/or non-acceptance of change.

## Changing organizational digital health infrastructures

Organizational change is about the coordination of multiple personal change journeys. Some individuals require more time than others to process what is required and what the proposed change means for them. This requires honest, explicit and timely communication and staff support throughout the change process. End users need to feel that they are listened to and that their concerns are seriously addressed [7]. It may be helpful to appoint some end-users to local leadership positions and/or to take on a liaising role between end users and the change management and/or system implementation team. Some individuals may find it necessary to resign from their position when the proposed changes do not meet their specific needs. This is especially true if there isn't a willingness or ability to address issues identified by end users. A common issue is the insufficient time allocated to learn how to use any new system, this impedes productivity and leads to many staff frustrations and dissatisfaction. In some situations it may be necessary to retrench individuals who stand in the way of change, to facilitate progress and reinforce the organization's commitment to the proposed changes.

This chapter is focusing on changes required to be implemented as a result of changes to digital infrastructures which are likely to impact on workflow, work processes and information based decision making. Nurses and midwives may well be using sophisticated electronic health records (EHRs), and spend hours documenting care plans, interventions and observations. Unfortunately these data are rarely analyzed in a timely manner or in a form that enables them to intervene and address issues while the patient is still in the ward. As a rule they are unable to undertake their own data analytics at the same speed as an Internet search, or to pinpoint what is working well and why. This type of information should be available at any time, easy to read and be incorporated into daily practice.

An exploration of how big data can improve patient safety, care quality and nursing practice [8] identified the following databases that are ideally collected into a single data platform that can be related to the variables known to influence resource usage and in-patient costs presented in Table 6 in Chapter 3:

- EHR data
- Nurse-sensitive indicators
- Genomic data
- Past adverse and sentinel events
- Socioeconomic data
- Financial and claims data
- Social determinants of health
- Air quality data
- Staffing and workforce data
- Data from other care encounters
- Information from social media

It is particularly important for nurses and midwives at the policy level to make the case for the value of nursing and midwifery analytics. At the organizational level nurses and midwives need to identify and analyze frequently occurring delays or incomplete data availability at the time it is required, so that these deficits can be addressed as part of any local change process. Changes to any digital healthcare infrastructure ideally result in productivity and quality improvements along with operational cost savings.

Previous chapters have focused strongly on the patient and organizational data ecosystems, as such data constitute the core business of health service delivery and associated resource management. The patient and organizational data ecosystems, of which clinical information systems such as EHR/EMR systems are fundamental, represent service demand, which in turn determines human and other resource requirements. Other components of patient and organizational data ecosystems are data quality, data stewardships and governance. These latter concepts are dealt with separately in other chapters.

Here the focus is on clinical systems forming a critical component of the digital health infrastructure and the management of change to workflows and staff communication strategies. Successful clinical systems are, as a rule, designed with considerable input from clinicians enabling them to meaningfully accommodate clinical practice [7]. Training to use any proposed system is ideally undertaken immediately prior to implementation to avoid staff frustrations, adverse productivity and a non-acceptance of the implemented system and the associated changes to work processes.

Change implementation should lead to improvements to quality, safety and efficiency of patient care via a combination of digital interventions. However, the successes of these interventions are dependent on ensuring a rigorous implementation process [6]. The following digital health interventions represent key technologies to be considered, relative to ensuring patient safety and the ability to deliver quality health care for any organizational digital health infrastructure. Fundamental to all, is the patient master index or administration system, complete with unique identifiers for both patients and individual providers, to enable all recorded activities to be linked to individual patient records and to the team of service providers at the point of care. Key technologies to be considered are:

1. Clinical, including nursing and midwifery resource management system,
2. Electronic patient portals,
3. Electronic patient reminders and other mobile technologies,
4. Information sharing at discharge,
5. Computerized provider order entry (CPOE), including electronic prescribing,
6. Electronic health records, including clinical decision support.

Any number of niche applications, including 'shadow systems' in place may be associated with any one of the above key technologies.

## Common barriers

The literature has identified numerous common barriers to change when instituted to achieve a digital transformation. These tend to be the result of a lack of the following factors [9]:

- Awareness and meaning amongst many stakeholders which often leads to the non-acceptance of sharing data.
- Confidence and trust regarding privacy, confidentiality and security protection.
- Interoperability enabling free flow of data at the technical and semantic levels.
- A clear legal framework related to trust, regarding data access management and data use.
- Staff training to ensure correct data collection and use in day to day health professional practice. This includes medical and nursing students engaged in workplace practice.
- Meaningful and integrated solutions leading to behavioral changes.
- Innovation in funding models.

As part of any change management planning process one needs to consider all the above and include strategies to address these factors. In all instances it is important to consider all possible anticipated and possible unanticipated consequences of proposed changes, as well as the positive or negative severity of their impact when developing the plan for the change process. As a first step of the planning for change process it is important to document both the current and desired status of the following realities:

- technical infrastructure inventory, functionality and capacity,
- patient and organizational data ecosystem,
- workflows and patient journeys describing sequential steps of care provision associated with every clinical specialty, type of patient and/or type of health service for which standard care plans may or may not be available,
- the available workforce knowledge and skills capacity relative to need,
- organizational and departmental cultures and staff behaviors,
- leadership styles and management effectiveness.

One study noted that functional organizational structures often lack the capability to control the work flow across departments and thus the coordination of the care activities within a patient care trajectory [10]. Within such organizations, resources tend to be duplicated, causing waste. The autonomy in using the specialty's resources often prevails over accountability, in some cases reducing the effectiveness of treatments. Many hospitals are now changing to patient-centred models of care where the care delivery processes involved in delivering hospital care are structured and designed according to the needs of the patients. According to Fiorio, the core principle of the patient centred (PC) model consists of the delivery of the appropriate amount of cure and care to patients in the most suitable setting according to their health conditions. As the PC model requires integrated care, multi-professional and multi-specialty teams are strengthened and required to collaborate [10]. Information systems need to be able to support such organizational changes. This PC model fits well with the team nursing approach as described in Chapter 6.

Other features of the PC approach are that patients are no longer transferred across different units or departments; rather, physicians and technologies move to the patients' bed. This may require new managerial roles responsible for the appropriateness, timeliness, flow and integration of patient care delivery processes (e.g. the bed manager or case manager). Changes may require the adoption of new concepts in the physical environment to maximize resource pooling and patient grouping, based on each patient's clinical severity and on the complexity of the assistance required [10].

Successful change management is dependent upon effective collaboration, teamwork, leadership, innovation, workforce engagement, obtaining contributions from exceptional individuals able to think outside the box from within the organization, staff empowerment and management support. The first step towards driving any change is to use this knowledge to recruit the right people to occupy the various change management roles. In some circumstances this may require the inclusion of professional development activities to prepare the change management team members for their new roles.

Well performing organizational cultures are the result of effective leadership where agreed values and objectives are realistic, clear and shared by all. There are clear lines of accountability along with integrated teamwork, performance transparency, quality based on evidence, and a desire to learn from errors made. Innovation is essential to ensure new work processes can be developed, implemented and result in positive outcomes for all concerned.

A very extensive literature review of large-scale health system project planning, implementation and evaluation includes an outline of the challenges encountered when organizational/contextual, human/social, and technological dimensions are interrelated and co-exist. It was found that the exact nature of the relationship between these dimensions is less clear and requires more attention [11]. This review provides a useful foundational resource for organizations and evaluators implementing and evaluating health information systems. System implementation needs to be combined with the adoption of a productivity improvement or 'Lean' strategy. This may require considerable professional development as few healthcare professionals possess the necessary in-depth knowledge and skills required to implement a productivity improvement program.

## Using 'Lean' and 'Six Sigma techniques' to design new work processes

The Lean philosophy is about eliminating defects, increasing efficiency, streamlining work processes, and optimizing the use of resources. The Six Sigma concepts are about reducing defects and variations in processes. Lean Six Sigma combines these concepts. It involves adopting a problem solving approach to improve operational efficiencies (performance) and effectiveness (quality). Numerous tools have been developed and are in use by cross functional teams to facilitate continuous improvements using Lean management principles. One in common use is the 5S methodology: Sort, Straighten, Shine, Standardize, and Sustain [12]. There are numerous Lean methods and tools described in the literature [13].

Lean Six Sigma uses a five-step approach: Define, Measure, Analyze, Improve, Control (DMAIC), it represents a problem solving approach to improve work processes. Lean focuses on waste reduction using continuous improvements, workplace organization and visual controls. Unnecessary work can also be remembered using the acronym DOWNTIME: Defects, Overproduction, Waiting, Non-utilized talent, Transportation, Inventory, Motion, Extra-processing. Six Sigma emphasizes variation reduction. It's about the adoption of standard procedures and makes use of statistical data analysis, design of experiments and hypothesis tests [14]. Six Sigma methodologies or tools essentially apply extensively researched fundamental theoretical underpinnings that remain constant. These are well documented in traditional 'work study', 'industrial engineering', 'operational research' approaches and methodologies [15–17] applied by Toyota, resulting in their Lean approach. To apply this approach to any industry requires the identification of input, process and output factors that collectively determine productivity. These same principles were described in some detail in Chapter 4.

Processes can be studied at various levels of detail, as every process, operation, activity or procedure consists of any number of operations or steps including transports (any movement by a person, document, equipment, sample, information transfer), delays, inspection/decision making points and endpoints such as storage. An outline process chart can diagrammatically portray a sequence of key events to visualize any process relative to any type of work flow, including those associated with logistics, supply chain or service delivery. Such an outline chart or map provides the big picture and can be used to identify every other significant operation/activity that needs to be studied in some detail in order to make improvements by producing a new Lean process.

There are a series of questions to ask about each process/activity in terms of purpose, place, sequence, person, and means, to identify DOWNTIME [18]. The information gathered can then be used to eliminate, combine, re-arrange, or simplify an activity. These questions are:

- What is the purpose and what is achieved? (value proposition)
- Why is the activity necessary, what else could or should be done?
- Where is it being done, why in this location, where else might it be done, where should it be done?
- When is it done, why at that specific time, when could or should it be done?
- Who is doing it, why that specific person, who else could or should do it?
- How is it being done? Why is it being done in that specific way, how else could or should it be done?

Systematically posing these questions about every process/activity not only assists reflection, learning, and the adoption of evidence-based practice, it is essentially the first step towards problem-solving and can be applied to any industry. These questions can be posed by any manager/supervisor at any time to enable the person undertaking the activity to reflect. It's common for people to say, 'we've always done it this way.' New solutions should not lead to unintentional adverse consequences that may be the result of dysfunctional people behaviors, non-compliance with professional codes of conduct, unethical practices, a lack of knowledge or interpersonal skills.

These methods may now be referred to as 'Value Stream Mapping' (VSM), a Lean-management method for analyzing and documenting current operational processes and designing a Lean future process for any operational value stream. This methodology is no different to undertaking 'system', 'business' or 'clinical workflow' analysis activities. Flow charting is a well-known technique used by Health ICT professionals to visualize a work flow. Numerous charting and mapping techniques and methodologies are available, each with their own set of symbols used for diagrammatic purposes. The output factors are the result of both overall organizational performance and the quality of every service delivered for any individual patient.

All 'Lean', 'Six Sigma' or any of their equivalent methodologies require an evaluation of the current (prior to change) status, a proposed change and an evaluation of the implementation effectiveness post change. Successful change management can only be demonstrated by 'before' and 'after' studies. With any digital transformation it is important to evaluate outcomes. This requires us to consider Health IT evaluation methodologies. Health Informatics is a scientific discipline that relies on evidence relating to individual applications, systems, settings and clinical contexts. These vary significantly within any national health system. Another consideration is the need to apply a 'wider lens', that is, one cannot evaluate any one system in isolation due to the many interrelationships. System linkages in terms of types and functionality within any healthcare facility, determine overall benefits that can be achieved.

## Potential use of nursing data

The following resource management functions use data that are interrelated; all are candidates for change in order to improve any healthcare organization's capacity to care. Each of these functions need to be evaluated prior to the implementation of any change management project aiming to improve any healthcare facility's capacity to care. They may be used as a general functional evaluation guide. It is important to note that there are many data overlaps between these functions. It is therefore highly desirable to consider developing a fully integrated plan enabling most, if not all of these functions to be attained as a result of your change management strategy.

### Patient acuity/nurse dependency/nurse-patient ratios

This functionality has been described in some detail in previous chapters. Such systems make use of any number and type of nursing service demand indicator variables relevant to specific types of patients. Patient administration, Human Resource Management, Rostering, Bed management, Dietary and Payroll systems need to be connected to this system to realize its full functionality and prevent any-data re-entry. Data importing or integration and linkage is required to enable the production and use of timely dashboards, touch screens, various management reports.

### Work hours per patient day/visit/procedure/attendance/birth/occasion of service/operating minute etc.

To demonstrate productivity, specific units of work relevant to each service type as displayed in the heading above, are used as denominators to measure resource intensity, with hours worked being the numerator. By adding nursing/midwifery intensity measures, allied health intensity measures and other journey data (pharmacy, consumables, etc.) for individual patients, and linking it with the financial system the total cost incurred per patient and/or patient type treated per length of stay can be calculated. These measures can also be utilized within the financial system to cost patient groups coded by ICD or DRG.

### Workload management

Workload management involves matching demand with supply, maximizing the use of scarce resources and matching staff skills with patient needs throughout all hours of service delivery, including staff meal breaks. This should also include identifying budget cost overruns that can be due to: breaches of staff contracts, inappropriate use of available skill mix, inappropriate roster patterns, unbalanced leave allocations, misuse of time off in lieu of overtime, excessive use of casual/bank and agency staff and unplanned leave allocations leading to excessive accrued leave for individual staff.

Workload management should be closely linked with bed allocations, admission/discharge policies/practices, patterns of late and canceled discharges, patterns of planned and unplanned admissions, operating room and departmental service hours schedules. Workload data by department and shift enables the production of trend data useful for planning purposes. This includes the use of check lists for staff to indicate their perceptions relative to various workload indicators, such as no tea break, short meal break, work was exhausting or easy, quality of care sub-standard etc.

## Workforce planning

Details regarding workforce planning are provided in Chapter 8. This is about any facility's ability to project staffing requirements by skill mix and availability over time relative to service demand trends. This requires links with human resource management, staff development, education facilities and experiential workplace learning opportunities. Staff turnover rates by service type and skill mix including nurse specialists need to be monitored over selected periods to determine trends and changing needs.

## Care capacity management

This requires the use of real time and historic data to manage and project patient throughput, measure and manage clinical workloads, adjust available staffing to meet unplanned and unpredictable peaks in acuity and throughput, allocate patients to beds where there are appropriate resources to provide the care required and develop accurate roster profiles which match projected throughput.

Dashboard displays can be used to display real time bed capacity, pending discharges, late discharges, care hours required by acuity and hours available to provide care, levels of over and under resourcing, etc. Hospital at a Glance dashboards viewable in all clinical departments, bed management and resource (staffing) management departments, hospital operation control rooms and executive offices enable all levels of control to see a view of all clinical departments and their capacity to care. This real time view is generated by combining data from numerous live systems such as; patient administration system, e-Health record, patient/nurse dependency and workload management system, emergency management system, operating theater management system and patient flow system.

Retrospective analysis of an organization's capacity to care includes mapping daily actual activity by department/ward, variances between hours required for care and hours available to provide care, identifying reasons for overtime worked per weekday and weekends, casual/bank hours utilized per weekday or weekends, sick leave relief, staff shortages, clinical emergencies, operating theater time overrun, after hours emergencies resulting in unplanned admissions.

Productivity and efficiency reviews may include daily/monthly/year to date data relating to; staff hours worked, variances between clinical hours required and clinical hours available, casual/bank and agency hours used, sick leave hours incurred and staff redeployments within the facility. Common measures of staffing productivity and efficiency used in acute care hospitals include; hours per patient per day (HPPD), hours per operating minute (HPOM), hours per procedure (HPP), hours per case, (HPC), hours per attendance (HPA), hours per occasion of service (HPOS), hours per birth (HPB), etc. These measures together with rates of absenteeism, overtime, casual and agency utilization are compared with budget profiles, and trended over time.

Another aspect is to explore clinical pathway variances per clinical pathway and/or patient type and the reasons for extended lengths of stay, which may be due to a pre-existing cause such as patient comorbidities or any type of adverse event. Every clinician (medical, nursing midwifery, allied health, and clinical support staff) needs to be able to quickly view a consolidated list of patients by location for whom they have caring responsibilities at any one time. This view may be in the form of a patient journey board displaying; patient type, alerts/risks, length of stay, expected date of discharge, clinical dependencies, etc. for each patient. This information is enabled by linking the display board to the patient administration system, the electronic record, the patient flow system/bed management system and the patient acuity/dependency system. A pro-active view of potential discharges and transfers by specialty/ward can also be displayed on Journey boards and Hospital at a Glance dashboard displays. Dashboards are useful in providing key information at any instance when decisions need to be made.

### Pathways and care plans with outcome reporting

Clinical pathways and care plans directly relate to clinical/EHR/EMR systems linked with patient administration and any nursing/allied health/medical, diagnostic and therapeutic services workload systems. Outcome reporting should include; patient outcomes (complications, the patients response to treatments/care and the impact of co-morbidities/pre-existing conditions), clinical protocol variances and their impact, other system/organizational/social factors and their impact and variances to the pre-set length of stay. Other measures that should be reviewed together with patient outcomes include, nursing intensity by DRG relevant to the pathway being reviewed, incident reports related to adverse patient outcomes and financial reports related to activity based funding for specific episodes of care compared to actual costs related to all of the above.

### Nursing intensity measures

Nursing Hours Per Patient Day, Nursing Care Hours Required, Nursing Care Hours Available and Nursing Hours Worked are all measures of nursing intensity, used for different purposes and often interpreted by health care professionals inappropriately.

NHPPD — This measure is commonly derived by dividing nursing hours worked by occupied bed days (midnight census) during any number of 24 hr periods. A link with a payroll system will enable its use for costing nursing services. This measure is limited in measuring nursing intensity for individual patient care episodes or in accurately measuring demand. The risk associated with using this measure to determine demand for nursing hours is that it is likely to perpetuate the staffing as it currently exists, regardless of whether there is under or over supply.

HPPD — Measured hours per patient day is used extensively in health care organizations to identify the intensity of hours used to provide care (by a variable skill mix of staff) in nursing/ midwifery services per patient day, using a variety of bed day calculations as the denominator. These include; the midnight census patient count, fractional bed days, an average of two or three bed census counts across the 24 hr period and bed days according to bed utilization across three 8 hr periods of the day.

In Nursing and Midwifery services using patient nurse dependency/acuity systems on a continual basis to determine daily and shift by shift staffing, HPPD measures are used to measure care hours required per patient per day according to acuity, hours available to provide care per patient per day and total nursing/midwifery hours worked in the nursing/midwifery services per patient per day. By using the same units for measurement, variances between demand and supply can be measured, and FTE/WTE numbers can be easily calculated. Measuring required hours of care for every patient provides a gold standard for costing patient episodes of care/DRGs. Using patient dependency data to measure required HPPD will provide evidence of peaks and troughs in demand and patient type mix across shifts of the day and days of the week and months of the year, facilitating the development of accurate roster patterns that match peaks and troughs in workloads.

### Retrospective and proactive discharge analysis

Retrospective analysis of discharge data is undertaken by nursing services to identify the reasons associated with the occurrence of late or canceled discharges including late transfers to a discharge lounge. Late discharges can be very costly to acute care facilities as these patients continue to absorb nursing resources, on average, for an additional 5 to 6 hr in many large acute hospitals. In busy public hospitals late discharges are a major cause of bed blocking which restricts admissions from emergency and transfers out of critical care areas into ward beds.

Reasons for late discharges generally include late doctors rounds (most common reason), late completion of medical discharge documentation, pharmacy delays, late arrival of relatives, late arrival of transport home and late pathology results. Clinically appropriate late discharges have a legitimate clinical reason such as recovering from early morning surgery, or the need for a patient to stay until they had passed urine. Reasons for late discharges that are not clinically appropriate are generally due to process issues that can be reduced through change management. Unplanned discharges make up the greatest proportion of all late discharges. Fig. 1 displays a graph displaying late and unplanned discharges with reasons grouped to show clinically appropriate and not clinically appropriate late discharges as an example of analytical findings.

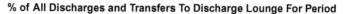

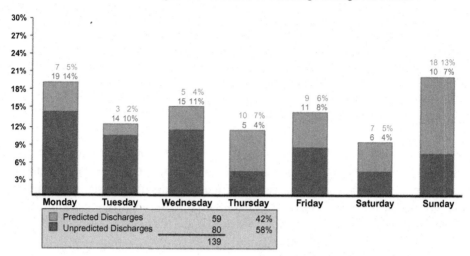

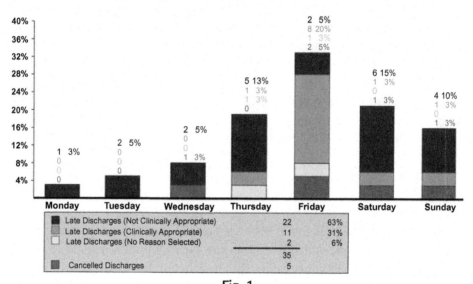

**Fig. 1**

Late and unplanned discharges by clinical appropriateness. *Produced from the TrendCare®
Software Version 3.6.*

Nursing workloads can be reduced by acting on these findings and redesigning work practices
of various services and therefore increasing capacity to care and improving patient flow.

### Diet ordering

This is another important function performed by nursing staff with the support of dieticians
for special and or complex diets. Compliance to clinical diet ordering is greatly enhanced when

it is integrated into a patient acuity system. This enables patient diets to be reviewed and updated when patient acuity predictions and actuals are completed. Ensuring clinical diet ordering and the management of food allergies is an important aspect of patient safety. Through interoperability, clinical diet requirements and known food allergies can be exported to menu systems limiting patient selections to only safe options.

### Rostering for clinical and non-clinical departments

For e-rostering solutions to be effective and accepted by users, the minimum interoperability requirements include; a daily import of staff details via an automated interface with the HRM system, an interface with the organizations work load management system, an automated interface with the organization's internal or external temporary staffing system (casual/bank and agency), an automated interface with a leave planner if one is not integrated with the roster system, an export to a mobile APP which enables staff to see their roster on their phone, and an export to payroll providing hours worked, leave and on-call.

Numerous commercial rostering systems are implemented by health care services to plan resource utilization for wards/departments. Rostering systems do not measure demand, they plan resource utilization and measure actual resources used. In clinical departments, workload management systems use patient acuity/dependency systems to measure demand. Electronic, evidence-based systems are now available to measure nursing hours required for care for a wide range of patient types by clinical specialty. By importing the roster into the workload management system, hours required for care can be compared to hours available for care, demonstrating the ward/services capacity to care.

### Clinical handovers

Clinical services require systems to have the capacity to produce comprehensive handover reports containing all the relevant information for each individual patient to enable a comprehensive handover to the next carer. This is critical to maintain patient safety and to continue the care provided as planned in acute, sub-acute or residential care. This includes; shift to shift handovers, service to service handovers, discharge summaries, referrals and the identification of patients for follow up. Through the process of interoperability, systems can include data imported from other systems as well as data that they specifically generate. For example, a patient acuity and workload management system can combine demographic data from the patient administration system and clinical data from the e-record to the data they generate such as, acuity indicator details, diet details, allied health details and staff allocation details. Providing the most comprehensive best practice handovers enhances the capability of clinicians from all disciplines to have the capacity to provide care at the highest level.

### Allied health intervention register and reporting

Allied health clinical staff provide a range of patient services/interventions that need to be scheduled in a timely manner within individual patient journeys that do not conflict with other interventions. They require a system that will measure the demand for their services and assist them to manage their workloads in a productive manner. The system needs to also provide reports for individual patients, for individual clinicians, by discipline and teams within a discipline. Allied health departments also need to track consumables used on each patient and equipment lent to patients at the time of discharge.

To be effective and user compliant, the Allied health system requires links with the patient administration system, HRM system, roster system, clinical/EHR/EMR system and the financial costing system. This level of interoperability is essential to minimize data entry, to maximize the reporting functionality and to enhance the resource planning capacity of the system.

To measure the demand for Allied health hours, the system must measure all of the following:

1. the actualized clinical hours used to complete all care for inpatient
2. the actualized clinical hours used to deliver outpatient care,
3. clinical hours spent doing clinical activities not related to inpatients or outpatients,
4. hours spent completing essential non-clinical activities, and
5. the additional hours that were required to complete all care that was not completed.

The use of this information together with the required uplift to cover leave, training, orientation, absenteeism, etc., enables allied heath managers to accurately calculate the FTE/WTE required for each discipline to meet service demand.

### Patient risk assessments with action plans

A significant component of nursing practice is patient risk assessment from which preventative measures are implemented. Such assessments are based on patient history, nutritional and general health status. Possible risks include falls, pressure injuries, infection, post surgery complications, skin tears, dehydration, medication errors etc. Every possible risk needs to have an associated set of minimization strategies such as the use of a special mattress to prevent pressure injuries, or a two hourly patient repositioning/turning regime.

Customized risk assessments with revision tracking can be developed and used across an organization or a group of organizations by making use of electronic assessment tools. Action plans/lists can be automatically generated as the assessment is completed. Real time audits can be run to identify any non-completed actions. Electronic assessments can be made available to be used at any stage of the patient journey, e.g., pre-admission, inpatient, and post discharge. Electronic assessments are generally part of the electronic record system, or if an electronic record has not been implemented may be part of an advanced patient acuity and workload management system. Patient assessment data should be linked with any incident reporting system so that risks, actions taken to mitigate risks and actual harm incidents can be analyzed.

### Human resource management registers and staff health profiles with reports

HRM systems will provide nursing services with the most useful accurate and effective information when they link with the payroll system, the rostering system and the workload management system. HR systems should include a comprehensive profile of every staff member including title, position, role, contract details and conditions of employment, contact numbers, address, next of kin, registrations, qualifications, skills, competencies, training, conference attendances, performance reviews, in-service attendance, languages spoken, work experience, work service awards and exit interview details.

Department managers require access to this HRM data for the staff they manage and comprehensive reporting from the system is required to identify training needs, compliance to mandatory training, competency achievements, registration status and much more. It is also important for HRM systems to provide a search function which enables managers at any time to locate staff by qualification, skill, experience, competency, or languages spoken.

### Staff health system

Health status records such as vaccination history, allergies, pre-existing disabilities, hospital acquired infections, staff injuries, staff work hazards, and return to work programs, all need to be monitored and reported on for nursing services. It is important that this information is available to managers so that when staff are caring for patients with infectious diseases staff safety can be managed effectively. This should not include the staff members medical record which may contain very sensitive and confidential information. Each individual staff member should have access to their health history held within the system.

### Efficiency measures/benchmarking all departments

Hospitals using the same patient acuity system and data standards are in the best position to aggregate their data by patient type and develop Hours Per Patient Day (HPPD) benchmarking standards. Where large volumes of data are collected from any individual service or department for any specific time period such as a year, annual trends and norms by month or season can be identified through data processing. By eliminating extreme out-liers, benchmarks can be developed against which current performance can be measured.

## Patient acuity and workload management system implementation project plan — A generic example using legacy systems

### Aim of the plan

To support hospitals in the ideal implementation, utilization and maintenance of a Patient Acuity and work load management system to facilitate the use of patient acuity data in the delivery of safe and efficient staffing plans and practices.

### Objectives of the plan are to provide

1. Information, timelines and guidance to enable hospitals to identify the key requirements and strategies to achieve system implementation in the most effective and efficient way.
2. An overall implementation plan to assist hospitals to develop their own customized system implementation plan.
3. A comprehensive "train the trainer" training plan which will expedite the training of all clinical staff using the system, interpreting the data and utilizing the data to generate short, medium and long-term safe staffing plans.

### Organizational benefits of implementing the system

1. Ability to measure patient acuity/nurse dependency for a wide range of patient types.
2. Efficient use of available clinical hours for the delivery of patient care.
3. Equitable distribution of workloads to nursing and midwifery staff.
4. Development of electronic rosters that match peaks and troughs in workloads.
5. Measurement of Safety CLUEs on each patient at the end of each shift.
6. Development of electronic patient care plans, encompassing the organization's own standards, with outcome measures and the ability to customize these plans of care to meet individual patient needs.
7. Development and maintenance of a dynamic electronic clinical pathway system with comprehensive variance reporting.
8. Development and maintenance of electronic patient assessments which autogenerate action plans and provide comprehensive patient risk analysis and reporting.
9. Measurement of clinical hours utilized in the provision of patient care for specific episodes of care for episode of care costing.
10. Fast and accurate clinical diet ordering and food allergy alerts system which links to menu systems.
11. Tracking and reporting of all allied health and specialist nurse/midwife interventions, including time spent completing clinical activities, reasons for incomplete activities, logging of consumables and loan equipment.
12. Development of effective and efficient nurse midwife, allied health and medical patient handovers.

13. Effective discharge analysis relating to the reasons for late and canceled discharges and compliance to discharge planning.
14. Access to real-time and predictive bed management information including predicted discharges, transfers, patients going on leave, empty beds and potentially empty beds.
15. Maintenance of staff HRM registers for registration renewals, mandatory and other training, conference attendance, competency achievements, annual performance reviews, exit interviews and numerous other HRM functions.
16. Management of staff immunization and other staff health details.

The schedule outlined in this document highlights the most efficient and practical approach to expedite the rollout of the system. This will ensure a rapid realization of benefits within the organization. These include improved patient outcomes, increased staff satisfaction and improved productivity and efficiency.

### Scope of project

1. Support Health Department/Ministry of Health, or hospital group IT personnel to identify and establish infrastructure and the appropriate IT environment to host the system software.
2. Implement the patient acuity and workload management system into all wards of the selected acute hospitals/sub-acute hospitals/and/or residential facilities.
3. Implement the integrated rostering solution into those hospitals that currently do not have a rostering solution.
4. Assist in the development of effective interfaces between systems, such as the patient management system, the rostering system, the HRM system and the patient menu system.
5. Support all hospitals in their use of the system in relation to training, IT support and clinical knowledge.

### Project priorities

- Finalization of a software license agreement that includes intellectual property protection clauses and a support and maintenance agreement that is fair and equitable for the organization and software developer.
- Finalization of the scope of the system implementation and training for each of the organizations hospitals/facilities. This relates to the scope of system features to be implemented (e.g. Roster, HRM, etc.) and the number and types of interfaces to be developed.
- Identify the most effective, manageable and cost-effective model for the system software installation centrally as a multi-campus system considering hardware, 3rd party licensing and deployment from the central servers to the relevant hospital/facility sites. The system should provide a range of options for the set up. These all need to be considered for production, training and user acceptance testing (UAT) environments.

- Development of an agreed project management plan that incorporates the organization's project management team and the system's project management team.
- Develop agreed roles and responsibilities for the organization's IT support and software developer IT support for the system's implementation project and system's support thereafter.
- Development of a Gantt chart that clearly outlines milestones and timeframes agreed to by the organization and the software developer for the implementation of the system.
- Configuration of the software utilizing the system's settings, and hospital and ward maintenance settings for each hospital/facility and all wards within each hospital/facility.
- Development of HL7 messaging standards' interfaces utilizing the systems' HL7 service to each of the hospitals/facilities, and the testing of these messages to ensure all patient transactions are accurately captured.
- Development of effective interfaces between existing rostering solutions and the patient acuity and workload management system for the group of hospitals/facilities being implemented.
- Development of effective interfaces between existing HRM solutions and the system for the group of hospitals/facilities being implemented.
- Development of effective interfaces between existing patient menu solutions and the system for the group of hospitals/facilities being implemented.
- Export data to other external systems (e.g. payroll, clinical costing, etc.), dashboards or spreadsheets.

*Project prerequisite*

- Adequate resourcing for IT project support at the start of the project and for the life of the project so that the setup of the servers and the maintenance of the servers is a priority.
- An agreed and signed software license which protects systems' intellectual property before the software is delivered to the organization.
- Appointment of an organizational project manager to manage the project.
- Each hospital needs to appoint an executive sponsor such as the Director of Nursing (DON) and have an effective steering committee that meets at least monthly to evaluate the progress of the project.
- Appointment of a full-time system co-ordinator/administrator as the administrator of the system in each hospital. This role is best occupied by an experienced RN with good computer skills.
- Ability to provide local IT support to assist with the rollout of the system and the development and/or testing of interfaces from the patient administration system (PAS), human resource management (HRM), and roster systems.

- An agreed project plan for the organization's project team and the software developer's project team. Activity schedules and timeline suggestions should be displayed in a *Gantt chart*.

The implementation plan outlined aims to efficiently expedite system rollout using a practical approach to ensure a quick realization of benefits to both staff and management within the organization. This requires the appointment of a suitably qualified Program Manager, an on-site co-ordinator and a defined number of project team members to suit the size of the organization and project outcome objectives. The appointed project coordinator needs to be present at all workshops undertaken by the software supplier and be able to devote significant time to this project implementation effort. This is a major task where proactive planning is of critical importance whilst working through the following implementation steps. A technical reference guide needs to be made available by the software vendor to the Information Technology Department.

### Software development project team

Project lead: Primary (lead) and secondary

A shared position where both persons will be involved in meetings and activities for the first 6 months is preferred. This arrangement ensures there are no gaps in the project knowledge and support when one member goes on leave or in the event that the project is extended in the future to include more hospitals. The recommended time dedication is 0 0.5 WTE. Both should be full-time permanent software developer employees.

Responsibilities:

- Co-ordinate the software developer project team throughout the implementation.
- Attend meetings with the organization as required to provide information relating to the negotiation of mutually acceptable dates for Train the Trainer workshops.
- Collaborate with the organization's project team to develop a realistic shared project plan which incorporates all required actions, persons responsible for the actions and an agreed date for the completion of each action.

IT support: Primary (lead) and secondary

The software developer needs to provide onsite and email support during the implementation phase. Two IT full-time permanent software developer staff should devote the equivalent of 1 WTE for the first year of the project. This can be shared across both dedicated IT staff so that there are no support gaps when one staff member is sick or on leave. These staff should have back up from other support staff with speciality skills such as organizational infrastructure and HL7 messaging.

Responsibilities:

- Provide support and guidance to the organization's IT team for the installation of the system software, the setting up of the HL7 service and the deployment of the software to the relevant clients.
- Respond promptly to all IT issues that arise in accordance with the signed service level agreement.
- Assist with the development and testing of the roster and HRM interfaces.
- Provide ongoing support and guidance to the organization's IT system support and hospital/facility based IT system support.

Clinical support

Clinical support needs to be managed by the nurse consultants available at the time of the call and should be provided via email or personally when on site. The time dedicated to the project needs to be determined based on the number and types of wards being implemented. All enquiries should be sent via the system co-ordinators from each site. The response time for all enquires needs to be in accordance with the time frames set out in the service level agreement.

Responsibilities:

- Assist with organizing and conducting system training workshops
- Provide clarification to system co-ordinators/administrators in relation to questions asked about acuity indicators, their related variables and how they are interpreted.
- Provide clarification and guidance to system co-ordinators/administrators on how to interpret and use the appropriate staffing areas when measuring workloads.
- Provide guidance on how to use all relevant features in the system to all hospital system co-ordinators/administrators as required.

*Governance structure*

It is recommended that each hospital sets up a governance board/group to oversee the system implementation. Membership should include but is not limited to:

- Project Sponsor — Director of Nursing
- Project Manager (organization wide)
- IT Project Lead (organization wide)
- System Co-ordinator/administrator
- Business Manager
- Assistant Director of Nursing
- Hospital based IT support for the project
- Ward Managers x 2 representatives

- Ward Champions x 2 representatives
- Union representative
- Organization wide senior nurse representative

N.B. Choosing the right people for this group is critical. They need to be motivated to keep the momentum of the project moving and to ensure that any barriers to meeting time targets are sent through to executive for action.

### Hospital project sponsor

Executive Sponsor is required for each hospital. This role requires a deep understanding of the organizational culture and an awareness of how the project will help the organization to achieve its goals. This position provides support to the organization's project manager, the system Co-ordinator/administrator and the IT lead, and reports progress and barriers to the executive for prompt action. In most acute hospitals, this is the Director of Nursing.

### Project manager

Appointed at organization wide level. This position provides a source of guidance for the project by providing policies, processes and methods documentation and metrics. The Project manager follows up on project activities and reports on risks, problems, barriers and delays to the hospital executive for action. This position needs to liaise between the functional and technical leads at each hospital and the system's Steering Committee at each hospital.

### System co-ordinator/administrator

The functional lead in each hospital needs to be a full time registered nurse with a strong conviction to see the organization be successful. Recruitment from within the organization is essential as a sound knowledge of the hospital's functional operations is required. The position requires leadership, management, strong communication skills, charisma, knowledge of software systems currently used by the nursing services, ability to write clear documentation and the ability to motivate staff.

### IT lead for the project *(organization wide)*

Appointed at organization level as the system may be installed centrally on servers or locally at one or more of the hospitals. This IT professional will need to liaise with the relevant organization's IT personnel to prepare servers, install the software and liaise with hospital-based IT in relation to hospital-based testing and rollout.

Technical lead *(hospital based)*

Ideally recruited from within each hospital. This position needs to be full-time dedicated to the project for a minimum of 3 months and then as required (average of 2 days per month) so that full interoperability is established and all user issues are reviewed and referred to the software developer support team if required. The technical lead liaises closely with the organization wide project manager, organization wide IT lead for the project and hospital-based IT system support team. This position requires an IT professional with good communication skills, ability to prioritize work with a sound knowledge of the hospitals' infrastructure.

## Executive lead for motivational strategy

The Director of Nursing at each hospital and/or an executive nurse from the organization is in the best position to announce the project. This needs to be followed up with a meeting of all nurse managers and IT staff to provide an overview of the product to be implemented and the anticipated organizational benefits to be realized by the continuing use of the software. This encourages co-operation and compliance regarding system use. The project plan, associated timelines and milestones needs to be communicated to all managers.

## Resource allocation and task allocation for system implementation

*– EXAMPLE ONLY*
Box 1 provides an overview of the seven project implementation phases. Suggested resource and task allocations per project implementation position are detailed in the tables to follow.

---

**BOX 1** **Project implementation phases**

**Phase 1** Project Initiation
**Phase 2** Data Gathering and Policy Development
**Phase 3** System Set Up, Configuration and testing
**Phase 4** Interface Building and Testing
**Phase 5** Training
**Phase 6** Go Live
**Phase 7** Post Go Live Support

---

Tables 1–6 detail project tasks associated with each phase for the project manager organisation wide, the system co-ordinator/administrator, the organisation wide technical support, nurse ward managers, the IT lead and the RN ward champions roles.

**Table 1  Project phases and related tasks for project manager organization wide (full time position for a minimum of 6 mths)**

---

**Phase 1 — Project Initiation**

- Set up an organization wide system implementation project steering group.
- Set up a meeting with the organization's Steering group and the project team to discuss the development of an organization wide project implementation plan with agreed objectives, strategies and timelines.
- Set up a system implementation project team in each hospital which includes an organization wide senior nurse representative (ex officio member), a registered nurse project lead (system co-ordinator/administrator) and an IT technical lead.
- Set up a system Steering Committee in each hospital with terms of reference. Members should include; Director of Nursing, organization wide project manager, system co-ordinator/administrator, IT support, nurse manager representatives, nurse champions, union representatives.
- Organize hospital executive nurse and/or, an organization wide senior nurse, to provide a motivational presentation, regarding the benefits of the system
- Ensure all project contracts/agreements are signed and necessary payments are made prior to software delivery.

**Phase 2 — Data gathering and policy development**

- Organize and attend a product demonstration with the vendor project team at each hospital for the hospital executive, the organization wide and hospital project teams and ward/department nurse managers to discuss requirements.
- Organize a meeting with the organization wide senior nurses aligned with the project, organization wide project manager, TrendCare consultants, organization wide IT Lead, system co-ordinators/administrators from each hospital and the IT support from each hospital to discuss their expectations in relation to the features that are going to be used in the first roll out, reporting requirements, the IT environment in each hospital, (PAS, HRM and roster systems used). This meeting will follow a product demonstration by the vendor consultants.
- Conduct a meeting with DONs, system co-ordinator/administrator and IT project managers to discuss the most appropriate system configuration, e.g. national multi-campus database, multi-campus database for hospital groups or individual hospital databases.
- Develop a project plan for the implementation of the system in liaison with all key stakeholders, including all system co-ordinators/administrators, IT project managers and the system project team.
- Organize a discussion session with the organization wide project team and the system project team to discuss the system's server requirements, set up and maintenance, outline roles and responsibilities, reporting lines and strategies to manage the set up and ongoing maintenance.
- Develop a risk management plan that includes risk mitigation in collaboration with the Vendor's project team.
- Organize a meeting with the vendor's consultants and the hospital system co-ordinators/administrators to discuss the system implementation forms detailing ward/department demographics required for system customization.
- Liaise with system co-ordinators/administrators to ensure that they have completed the system implementation forms for each ward/department, and contents for selection banks in the system.

**Table 1    Project phases and related tasks for project manager organization wide (full time position for a minimum of 6 mths)—Cont'd**

---

**Phase 3 — System set up, configuration and testing**

- Liaise with the organization wide IT support to ensure servers are acquired and prepared for the system installation as set out in the system's Technical Reference Guide.
- Liaise with organization wide IT support to ensure that a request to download the system software has been made to the vendor.
- Liaise with the organization wide IT support to ensure that the system software is installed in the UAT environment as set out in the system's Technical Reference Guide.
- Liaise with system co-ordinators/administrators to ensure that system properties, hospital and ward maintenance, and all relevant selection lists are set up in the system according to the systems Maintenance Guide.
- Check that the system co-ordinators/administrators have completed testing the UAT setup using the system test plan.
- Liaise with the organization wide IT support to ensure the HL7 service is installed, switched on and tested.
- Check that staff security using ADID has been set up and tested.

**Phase 4 — Interface Building and Testing**

- Liaise with organization wide IT support to ensure that the HRM interface is completed and tested.
- Liaise with organization wide IT support to check that the Roster interface is completed and tested.

**Phase 5 — Training**

- Check with all system co-ordinators/administrators to ensure that the Train the Trainer implementation workshop has been organized, an appropriate training room is booked and agendas distributed
- Check with IT to ensure that the software has been deployed to the training room computers and that the system co-ordinators/administrators have tested the program and set appropriate default wards on each of the training room computers.

**Phase 6 — Go live**

- Liaise with system co-ordinators/administrators and hospital-based IT support to check that a go live date has been identified for each hospital.

**Phase 7 — Post Go Live Support**

- Organize a meeting with all hospital executives, system co-ordinators/administrators and IT leads to review project outcomes.
- Develop a final report on the implementation of the system, what went well, problems/barriers, outstanding issues and lessons learned.

**Table 2  Project phases and related tasks for project lead (RN) system co-ordinator/administrator (Full time position for each hospital)**

### Phase 1 — Project Initiation

- Attend a meeting with organization wide senior nurses aligned with the project, organization wide project manager, vendor consultants, organization wide IT lead, system co-ordinators/administrators from each hospital and the IT support from each hospital to discuss their expectations in relation to the wards involved and the features that are going to be used in the first roll out, reporting requirements, the IT environment in each hospital, (PAS, HRM and roster systems used). This meeting will follow a product demonstration by the vendor consultants.
- Participate in a session with the organization wide project team and the vendor project team to discuss the system implementation forms, maintenance set up, roles and responsibilities, the project plan, project timelines, and strategies to manage the set up.
- Utilize the system Implementation Plan for system Co-ordinators/administrators as a checklist for the setup phase.

### Phase 2 — Data gathering and policy development

- Identify all wards to be involved in the initial training phase and forward the details to the vendor.
- Assist all nurse managers to complete the implementation forms for their wards/departments.
- Ensure all necessary clinical documentation provided for the system is available to the relevant users on the shared drive.
- Review the clinical documentation provided for users of the system and identify which training booklets will be required for the Train the Trainer workshops in consultation with the vendor nurse consultants.
- Develop policies and procedures, cheat sheets and reminder flags specific to the organization and/or ward/department to assist with the fast and effective roll out of the system. Examples of these should be provided by the vendor.

### Phase 3 — System set up and configuration

- After the system has been set up in the UAT environment complete the hospital and ward maintenance set up utilizing the system Maintenance Setup Guide under the guidance of the vendor project manager and clinical consultants.
- Test the system set up using the test plan provided by the vendor. Report any issues to the software supplier. Load relevant hospital selection banks with the agreed lists

### Phase 4 — Interface Building and Testing

- Ensure that the staff details imported from the staff master into the system are accurate. The nurse managers from each ward should be involved in checking these details.
- Assist with the testing of the HRM interface in collaboration with the organizational wide project manager, IT support (hospital-based) and the Systems project team.
- Test ADID sign on is working correctly
- Assist with the testing of the roster interface in collaboration with the organizational wide project manager, IT Support (hospital-based) and the project team.

**Table 2**   **Project phases and related tasks for project lead (RN) system co-ordinator/administrator (Full time position for each hospital)—Cont'd**

---

### Phase 5 — Training

- Attend a one-day product orientation conducted by the vendor for the project team which should include maintenance setup and system testing.
- Negotiate appropriate dates for the Train the Trainer workshop with the nursing executive team, considering the timelines for the setup of the live system with working interfaces. It is ideal to have the HL7 messages flowing into the system at least one week prior to the workshop.
- Ensure nurse managers have rostered themselves and at least one nurse champion from their ward to attend the Train the Trainer workshop days.
- Co-ordinate the set up for the four-day Train the Trainer workshop. Check that there are enough computers for the workshop. There should be a minimum of one computer per ward/unit being trained and no more than 3 participants per computer.
- Check that the system is working on each training computer. Tag a computer for each ward/unit and default each computer to the relevant ward/unit under the direction of the project team.
- Attend the four-day Train the Trainer workshop. Distribute and collect the workshop evaluation forms.
- Provide support and guidance to nurse managers and RN champions in the training of their staff. Once staff are trained, they must categorize all of their allocated patients for each shift thereafter.

### Phase 6 — Go Live

- Co-ordinate all wards in the first roll out. This involves all wards/units using the system in the production environment.

### Phase 7 — Post Go Live Support

- Review the progress in data collection across all wards.
- Co-ordinate and attend the IRR training workshop.
- Ensure 100% of RN's have been IRR tested in all wards.
- Utilizing actualization audits and the established benchmarks for HPPD per patient type, conduct regular reviews on all wards to monitor data integrity.
- Co-ordinate and attend the report interpretation workshop — 1 day.
- Co-ordinate and attend an advanced report interpretation and roster re-engineering workshop — 1 day.
- Conduct a user satisfaction survey and share results with the project director, the steering committee and the vendor project director.

---

**Table 3 Project phases and related tasks for technical support organizational wide (full time for 3 mths then 2 days/mth)**

<table>
<tr><td colspan="1" align="center">Phase 1 — Project Initiation</td></tr>
</table>

**Phase 1 — Project Initiation**

- Attend a meeting with the organization's senior nurses aligned with the project, the organization's project manager, vendor consultants, system co-ordinators/administrators from each hospital and the IT support from each hospital to discuss their expectations in relation to the wards involved and the features that are going to be used in the first roll out, reporting requirements and the IT environment in each hospital (PAS, HRM and roster systems used). This meeting will follow a product demonstration by the vendor consultants.
- Review all hardware and software requirements to install the system, and identify IT implementation requirements, e.g. servers, SQL and deployment software. Read relevant chapters of the system's Technical Reference Guide.

**Phase 2 — Data gathering and policy development**

- Attend a meeting with the organization's senior nurses, hospital DONs, the organizational wide project manager, system co-ordinators/administrators, hospital-based IT leads and the project team to discuss the most appropriate system configuration, e.g. national multi-campus database, multi-campus database for hospital groups or individual hospital databases.
- Assist with the development of a project plan for the implementation of the system into the organization's hospitals/facilities in liaison with all key stakeholders, including the organizations project manager, system co-ordinators/administrations, hospital-based IT leads and the project team.
- Organize a discussion session with the organization's project team and the vendor's project team to discuss the systems server requirements, set up and maintenance, roles and responsibilities, reporting lines and strategies to manage the set up and ongoing maintenance.
- Develop a risk management plan that includes risk mitigation in collaboration with the organization's project manager and the vendor's project team.
- Participate in a session with the organizational project manager and the hospital project teams to discuss the TrendCare maintenance set up, outline roles and responsibilities, reporting lines and strategies to manage the set up.

**Phase 3 — System set up and configuration**

- Set up the required servers to run the system software.
- Install the system software into the UAT, live and training environments and conduct user acceptance testing as outlined in the system's Technical Reference Guide. The vendor's IT support team should assist with this process.
- Complete the system setup maintenance and the hospital maintenance as outlined in the Maintenance Setup Guide.

**Table 3**  **Project phases and related tasks for technical support organizational wide (full time for 3 mths then 2 days/mth)—Cont'd**

---

### Phase 4 — Interface Building and Testing

- Install the HL7 service and organize compliance testing for events such as admissions, bookings, transfers, discharges, leave, patient updates, bed moves, etc. as outlined in the system's Technical Reference Guide.
- Work with vendor IT developers to develop and test the HRM interface.
- Work with vendor IT developers to develop and test the roster interface.
- Import staff details and import staff security profiles from the staff master into the system in collaboration with the vendor's IT support
- Organize the setup of the training computers for the four-day Train the Trainer workshops for end users.
- Roll out the system to all relevant workstations in the wards implementing the system

### Phase 5 — Training

- Attend the product orientation for the project teams.
- Review all relevant chapters in the system's Technical Reference Guide

### Phase 6 — Go Live

- Collaborate and co-operate with the system co-ordinators/administrators to ensure that the transition from the training environment to the live environment is successful.
- Report any system issues to the vendor.

### Phase 7 — Post Go Live Support

- Review backup and daily maintenance plan for the live system. Ensure that a server maintenance plan is maintained.
- Develop a recovery plan for the systems servers and SQL environments as outlined in the system's Technical Reference Guide.

**Table 4 Project phases and related tasks for nurse managers all wards (for each hospital) (7 days training, rest of time incorporated in daily practice)**

| |
|---|
| **Phase 1 — Project Initiation** |
| • Attend a presentation by the vendor's project team which will give an overview of the objectives and scope of the project |
| **Phase 2 — Data gathering and policy development** |
| • Complete IMP (Maintenance Setup) forms for relevant ward/department under the guidance of the system co-ordinator/administrator. |
| • Check data integrity of imported staff data for relevant ward/department and notify system co-ordinator/administrator of any errors. |
| **Phase 3 — nil** |
| **Phase 4 — nil** |
| **Phase 5 — Training** |
| • Attend the Train the Trainer workshop (4 days). If unable to attend send second in charge and at least one RN champion |
| • Immediately following the TrendCare Train the Trainer workshop implement a ward based system training program for all RNs and any other staff who will be required to use the system. |
| • While training, identify additional system champions to assist with system training. |
| • Utilize the training register on the shared drive to register all staff trained. |
| • Ensure 100% of registered nurses are trained in the system prior to the IRR workshop |
| • Attend the IRR workshop (1 day). |
| • Implement IRR testing for all staff who use the system. Utilize system champions to assist with the testing. |
| • Attend the Report Interpretation workshop (1 day). Identify the most useful reports and add them to the managers favorites list. |
| • Attend the Advanced Report Interpretation and roster re-engineering workshop (1 day). |
| **Phase 6 — Go Live** |
| • Ensure all patients are predicted every morning by 10 am. Ensure all patients are actualised for each 8 hr period of the day. Utilize the Ward Actualization Audit to measure compliance |
| **Phase 7 — Post Go Live Support** |
| • Maintain actualization rates at 100% |
| • Maintain the HPPD for each patient type within the international benchmark range |
| • Maintain 100% compliance of all staff being IRR tested annually |

**Table 5 Project phases and related tasks for IT lead (at each hospital) (1 to 2 mths full time then 2 days per mth)**

**Phase 1 — Project Initiation**
- Attend a meeting with the organization's senior nurses aligned with the project, organizational Project manager, vendor consultants, the organization's IT lead, system co-ordinators/administrators from each hospital and the IT support from each hospital to discuss their expectations in relation to the wards involved and the features that are going to be used in the first roll out, reporting requirements, the IT environment in each hospital, (PAS, HRM and roster systems used). This meeting will follow a product demonstration by the vendor's consultants.
- Participate in a session with the organizational project team and the project team to discuss the system IMP forms, maintenance set up, roles and responsibilities, the project plan, project timelines, and strategies to manage the set up.

**Phase 2 — Data gathering and policy development**
- Collaborate with system co-ordinator to deliver a list of workstations in each ward that will be used to access the system, including workstation name, location, description and closest phone extension

**Phase 3 — nil**
**Phase 4 — Interface Building and Testing**
- Work with organizational IT and vendor IT developers to develop and test the HRM interface.
- Work with organizational IT and vendor IT developers to develop and test the roster interface.
- Import staff details and staff security profiles from the staff master into the system in collaboration with the vendor IT support.
- Organize the Setup of the training computers for the four-day Train the Trainer workshops for end users.
- Assist with the roll out of the system to all relevant workstations in the wards implementing the system in the first round.

**Phase 5 — Training**
- Attend the product orientation for the system project teams.
- Review all relevant chapters in the system's Technical Reference Guide.
- Attend the user Train the Trainer workshop (4 days).
- Assist with training ward RNs in acuity categorization and the use of the staff work allocation screens. Attend the IRR workshop (1 day).

**Phase 6 — Go Live**
- Collaborate and co-operate with the system co-ordinator/administrator to ensure that the transition from the training environment to the live environment is successful.
- Report any system issues to the vendor.

**Table 6 Project phases and related tasks for RN champion all wards (2 days training)**

**Phase 1 — nil**

**Phase 2 — Data gathering and policy development**

- Contribute to IMP (Maintenance Setup) forms for relevant ward/department under the guidance of the Nurse Manager and system co-ordinator/administrator.

**Phase 3 — nil**

**Phase 4 — nil**

**Phase 5 — Training**

- Attend the Train the Trainer workshop (4 days).
- Immediately following the system Train the Trainer workshop implement a ward-based system training program for all RNs and any other staff who will be required to use the system.
- While training, identify additional system champions to assist with system training.
- Utilize the training register on the shared drive to register all staff trained.
- Ensure 100% of registered nurses are trained in the use of the system prior to attending the IRR workshop
- Attend the IRR workshop (1 day).
- Conduct IRR testing for staff who use the system.
- Attend the Report Interpretation workshop (1 day). Identify the most useful reports and add them to the managers favorites list.
- Attend the Advanced Report Interpretation and roster re-engineering workshop (1 day).

**Phase 6 — Go Live**

- Ensure all patients are predicted every morning by 10 am. Ensure all patients are actualised for each 8 hr period of the day

**Phase 7 — Post Go Live Support**

- Train new staff in the use of the system within 1 week of commencement
- Continuously lead good practise in system data entry
- Test IRR for staff, annually

## Risk assessment

A system implementation project has any number of potential risks resulting in time delays, budget blowouts or system errors. It is important to identify all potential barriers that may hamper the successful implementation and to assess both the likelihood and the significance of the consequence for each identified potential risk, based on the probably of occurrence. One risk assessment study [19] showed that the framework used was very robust and suitable for conducting a thorough risk analysis and that there are links between the quality of the risk assessment and the level of project outcomes. Their framework used the following risk dimensions: technological, human, usability, managerial and strategic/political, against which potential risk factors were identified. Their framework was based on a vast number of information system-related success factors or risk factors from the literature. Each identified risk factor needs to be critically examined and described in a risk management plan so that such risks can be mitigated.

### Risk rating matrix scale

A risk rating matrix should be used, such as the example of one assessment shown in Table 7, to assess the level of risk when developing a risk management plan consisting of mitigation strategies. Tables 8–11 details a number of potential risks known to be associated with the implementation of a patient acuity and workload management system that may be used as a guide. Every organization and system implementation needs to consider its own unique context when developing their risk management plan. Most risks tend to be technical, especially when systems need to rely on system connectivity and the use of interoperability schema. The highest priority rank for developing mitigating strategies must go to those items that have the greatest potential risk and consequence.

**Table 7  Risk rating matrix scale**

| Likelihood | Consequence | | | | |
|---|---|---|---|---|---|
| | Insignificant | Minor | Moderate | Major | Catastrophic |
| Certain | high | high | | | |
| Likely | medium | high | high | | |
| Possible | low | medium | high | | |
| Unlikely | low | low | medium | high | high |
| Rare | low | low | medium | medium | high |

**Table 8  Patient acuity and workload management system implementation identified potential risks and mitigation strategies — an example of technical risks**

| Risk description | Impact H/M/L | Probability H/M/L | Rank H/M/L | Mitigation strategies |
|---|---|---|---|---|
| Delay of the project due to time taken to set up the HL7 service and to test the required HL7 messaging. | H | M | H | The review and development of the required HL7 messaging for each site must be a high priority. Prioritize HL7 messages required for the project. This work should commence immediately. The software developer should assist with the testing of these HL7 messages as a high priority once the contract has been awarded and the software has been installed. |
| No HL7 messaging available to notify the system of new bookings and admissions, making the system time-consuming for nursing staff leading to incomplete data entry. | H | L | H | If HL7 is not available, the system should have options to import patient data via a flat file interface that is refreshed on a regular basis e.g., hourly. The text files contain bookings, admissions, and other patient movement data. |
| Delay in having all necessary interfaces operable in time to support each site going live. Without these interfaces, the system generates a more time-consuming task for nursing staff, that can lead to inaccurate data. | H | M | H | The review and development of the required HRM and roster interfaces must be a high priority. The software developer needs to have documented specifications for these interfaces in the system's Technical Reference Guide. |
| The HL7 feed available from the patient administration system has malformed messages that are unable to be successfully processed by the system as intended. | H | M | H | 1. Correct the deficiencies in the feed to bring it to be compliant. OR 2. Investigate the possibility of the software developer being able to manipulate the message content. |

**Table 8   Patient acuity and workload management system implementation identified potential risks and mitigation strategies — an example of technical risks—Cont'd**

| Risk description | Impact H/M/L | Probability H/M/L | Rank H/M/L | Mitigation strategies |
|---|---|---|---|---|
| Implementing a single central instance of a wider system may lead to an entity-wide outage should there be a server or intrastructure failure. This may occur during times of server maintenance, e.g. hardware and operating systems back up or database upgrade. This could also be due to database re-organization or a localized issue such as weather events which cause a power outage. This would make the system unavailable to all users nationally. | H | L | H | Install split fail over localities to mitigate this risk. OR Implement a distributed/network topology. The system should provide a management and reporting portal to support this type of environment. |
| Software faults leading to the failure of a function to be performed in the system. | H | L | H | The system design needs to have separate code modules for each function so that if one fails, others are usually not affected. In most cases, this allows support staff to target, correct and replace the code module without the need for a system wide downtime. |

**Table 9   Patient acuity and workload management system implementation identified potential risks and mitigation strategies — an example of managerial risks**

| Risk description | Impact H/M/L | Probability H/M/L | Rank H/M/L | Mitigation strategies |
|---|---|---|---|---|
| Insufficient infrastructure investment to cater for usage demand leading to a slow interface and/or database response times. | H | M | H | 1. Provide sufficient resources for the server infrastructure and ensure sufficient bandwidth and reliability of the network infrastructure for the client sessions. 2. Provide sufficient expansion capability for the infrastructure to cater for abnormal peaks in activity and future growth. |

**Table 10  Patient acuity and workload management system implementation identified potential risks and mitigation strategies — an example of human risks**

| Risk description | Impact H/M/L | Probability H/M/L | Rank H/M/L | Mitigation strategies |
|---|---|---|---|---|
| Delay in the progress of collecting reliable data due to the lack of commitment from nurse managers in the wards. | M | L | L | Ensure that the nurse managers are highly committed to the project and are well-informed of the benefits of the system and together with their champions receive comprehensive training in the system. |
| Breach of patient privacy by staff due to non-compliance to GDPR policy. | H | M | H | The software developer needs to have a clear policy on patient privacy and the management of personal details which complies with GDPR regulations. All its staff should have read and signed this policy. A breach of this policy should be considered as a serious breach of an employees' work contract. Software developers need to request hospitals to de-identify any patient and staff details in support emails. The system needs to de-identify all HL7 transaction logs prior to exporting them for analysis. |
| Inaccurate and incomplete data due to nurses' non-compliance to use the system correctly. | H | M | H | The software developer needs to provide comprehensive training to all nurses and make use of a formalized inter-rater reliability system to test the accuracy all nurses using the system. The provision of international benchmarks to measure the accuracy of HPPD per patient type and to identify under and over categorizing by nursing staff is a useful mitigation strategy. Actualization audits which measure the level of completeness of the data needs to be an integral component of the system. |

Table 11  Patient acuity and workload management system implementation identified potential risks and mitigation strategies — an example of strategic risks

| Risk description | Impact H/M/L | Probability H/M/L | Rank H/M/L | Mitigation strategies |
|---|---|---|---|---|
| Nurses have limited access to computers to input the acuity/dependency data, extending the time taken to get complete and accurate data. | M | L | L | Ensure there is adequate access for nursing staff to the ward computers. At least two ward-based computers should be available for the sole use of nursing staff during the hours of 9.30 am to 10.30 am (peak activity period of the day). |

## Desired outcome measures benefitting nurses and their patients

A nursing resource management system needs to be able to gather evidence of real care required by patients as well as all the actual work undertaken by nurses in any in-patient situation. This needs to be based on nursing work measurement as described in Chapter 4, including the measurement of nursing work done in addition to patient care in the ward area. This includes the work that is done to manage the ward and coordinate the shift, escort patients, wait in X-ray etc., and importantly, all the other invisible work nurses also do that is often not known and therefore not budgeted for or acknowledged. E.g. Care given to outpatients either on the ward or via phone. Nurses use their clinical judgment and identify their patient's status and needs when completing patient dependency data. These data can assist in decision making, to safely and effectively manage patients and staff. The information provided and optimum system use is about ensuring:

- **The right amount of staff with the right skill mix at the right time in the right place**. Nurse managers can make sure the available staffing resource match the hours required for care before the shift starts by predicting required nursing hours ahead of time. Over time these measurements assist in the identification of trends, identifying shifts that are frequently under or overstaffed so that rosters can be re-engineered to better match work peaks and troughs based on acuity actuals.
- **Fair and safe workload allocations**. Nurses can be assigned equitable workloads when nursing hours required for the upcoming shift are measured, thus ensuring workloads for individual nurses are fair and facilitate patient safety. Other required ward activities can be allocated based on nurses' ability to see who needs additional help and who can provide additional help. Communications about staffing on a shift can be noted and shared with ward managers, team leaders and duty managers to ensure all considerations are taken into account when making staffing decisions afterhours.
- **Automated acuity and patient details updates to enable the provision of customized handovers.** Such handovers need to be designed to suit each in-patient area.

This automation facilitates historical storage of shift by shift handovers enabling retrospective auditing. This may also be useful for medico-legal purposes.

- **Provision of evidence pertaining to departmental work completed by nursing staff, including patient care services provided.** This includes tracking of all nursing work such as escorts, being a team leader co-ordinating a shift, helping the manager do rosters, providing in-service training, conducting services for groups of patients, undertaking specific projects or quality activities and talking to outpatients by phone. Many wards do not know how much time is given to these common and essential activities and they are often not included in their budgets therefore the hours spent on them are taken out of the hours budgeted to give inpatient care. It is important to know how much time is needed for these activities so that informed decisions can be made around who should do what, when.

- **Provision of comprehensive records of all types of work undertaken by any staff member during their employment.** This requires tracking the number of hours worked on night shift, on call, acting up in a higher position, assisting managers, facilitating patient groups, working on projects, performing escorts, undertaking professional development activities and more. These are then added to a personal staff profile and summary. These data can be invaluable when going for a promotion, or used as a record of all the experience gained complementing an individual's work history and can be made available for any registration audits or when employment is terminated.

## Data collection methods

### Measuring patient acuity on a shift

*Just before heading to morning tea* each nurse, using their clinical judgment, selects acuity indicators for each of their patients, predicting the care their patients will require for the next 24 hr work cycle. (Evening period, Night period and tomorrow's Day period) *For an average nurse's workload (4–5 patients) this should take 2–5 min.*

*Towards the end of the shift* the nurse will review the indicators selected for each of their patients for this shift, upgrading predictions for that shift to actuals to reflect the actual care that was given. If there have been any major changes to treatment or the condition of the patients, the predictions for the evening or night shifts can be adjusted. *This should take between 1 and 3 min.*

*Nurses on evening and night shifts* do not need to predict patients for 24 hr unless the patient is a new admission on their shift as the patients have already been predicted by the day staff. Nurses on these shifts should review predictions for future shifts and may adjust predictions if a patient's condition changes and their care requirements change. Nearing the end of their shift, they will need to upgrade the predictions for their shift to actuals. *For an average nurse's workload this should take between 1 and 3 min.*

*The actualization of predictions is important, especially if you are busy,* as it is how the work that was actually needed by the patients this shift is recorded. Past timing studies undertaken by the authors indicate that a nurse's workload can increase by up to 12% from unpredicted work. This information can then be used by the Nurse Manager to ensure the staffing for those shifts matches the hours required by the patients and any other work that needs to be done other than patient care. The information is considered along with other factors such as the skill mix of the available staff, expected admissions, and transfers and is used to plan and adjust the staffing for the next 24 hours. Over time this data is then used as evidence to explain the actual intensity of work that is required in the department.

## Local nursing acuity data use

### Allocating staff to workloads

*At the commencement of a shift*, the Team leader can identify all staff rostered and make use of these data to allocate staff to the work they are rostered to do. E.g. who is rostered to do a half day training, to do their quality portfolio or to run patient groups. *The hours available to give care are identified for each staff member and can be allocated to teams or individual workloads of patients.* Importantly, hours worked but not available to provide care (such as Quality activities or Training etc.) are recorded but not measured against the patient workload. Extra staff or last minute changes can be easily added as required. Nurses can instantly see the patients, the hours of care they require and the staff allocated to that workload. This allows fair and even distribution of work amongst all staff. Meal breaks can be allocated and additional duties assigned. An electronically available *Workload Summary* is a great way for all staff to be aware of which team or nurse may need help.

For departments that do not measure patient acuity, *Work Area allocations* can be assigned. E.g. Theater 1, Theater 2, Clinic, Procedure room etc. Work allocations should be kept for historical purposes. *At the end of the shift* the hours are checked and adjusted by the Team Leader to reflect where all staff hours were actually worked. Team Leaders can communicate with assistant ward managers and/or ward managers using Shift notes. These notes can be left in advance by the manager to give instructions to assistant ward managers or Team leaders about staffing or used in retrospect to give explanation for decisions made or events that occurred that affected staffing.

### Handovers

Customizing handovers for each ward using the style of choice, should only include the data the next person needs to see under headings that make sense to the staff in the ward. Notes added to patients can be on-going or only allocated to a specific shift. Notes by medical or allied health staff can also be included. A hospital wide standardized format can be adopted for free text notes however labeled.

The data is about ward staff and ward patients, it is only useful if it is accurate and complete. A couple of minutes of work from each staff member on each shift can provide the manager with an incredible amount of evidence of ward patient needs and the activities required to support the ward's service. This data can be used to improve patient care and outcomes, and staff safety and satisfaction.

### Workforce planning

Workforce planning requires the use of any number of routinely generated reports to optimize nursing workforce management and for monthly reviews comparing actual to budgets. It is recommended that a roster re-engineering exercise is undertaken on a six monthly basis depending on the variability encountered in patient acuity, patient type and volume based on suitable reports. **Inter Rater Reliability** should be completed on all staff annually to ensure data collection consistency and accuracy throughout the year.

## Ward/unit manager/senior nurse daily routines to ensure data accuracy

08:30 am

- Review all predicted to actual work undertaken (actualizations) for previous day/s and update as required to ensure accurate nursing workload representation.
- Update current and next 24 hr roster with any changes to future shifts (sick leave, emergency leave, shift swaps). On Fridays review this for the weekend and following Monday.
- Check all in-patient types, locations, planned discharges, transfers & admissions.
- Review and adjust all staff work allocations.

10:00 am

- Check all 24 hr nursing work predictions are completed for all patients by the nurses responsible.
- Update patient details, treatment and care changes and diets where necessary.
- Review all variances for the next 3 shifts, arrange staffing changes where required, document rationale for adjustments made, identify need for extra staff or ability to deploy staff to another area.

14:00 pm

- Check all new admissions are included in work allocations for current and future shifts and flag new potential discharges.
- Allocate patients to afternoon work teams.
- Allocate staff meal breaks for evening shift.
- Update patient notes and handover details for evening staff if required.
- Ensure discharge analysis is completed.

15:00 pm

- Check that all day details were completed.
- Check diets and update if necessary for evening meals.
- Re-adjust staffing for the next 24 hr where necessary.
- Complete any shift notes required by after hours co-ordinator.

After hours and weekend co-ordinators need to undertake similar activities to continue to provide effective workload management at all times.

## Health IT evaluation methods

Previous chapters have identified numerous features associated with nursing workload management systems. These include the nursing workload or work measurement and/or acuity system used to differentiate care requirements between patients, the models of care adopted, the skill mix, rostering and resource management practices. Collectively these factors are interactive and influence outcomes in terms of patient safety, discharge status, incidence of adverse events, well-being and comfort, as well as overall organizational performance in terms of productivity and costs. A large variety of studies reported in the literature have focused on any number of different outcome combinations in order to provide evidence of the efficacy of the systems in use. Chapter 3 explored nursing workload measurement systems in use and concluded that:

> '......there is no common or agreed framework enabling comparative evaluations of these various methodologies to be made. There is no consistency regarding the variables considered, their definitions and/or measures used for any of these methods, even where the nursing care demand measuring methods appear to be the same. At best one can group the methodologies employed according to some common features'.

Now that we've explored the need for digital transformation, including the need for systems to be able to link and share data, we're in a better position to explore the possibility of adopting the possible use of health IT evaluation guidelines for nursing workload system evaluation purposes. It is important to obtain evidence from which systematic judgments can be made. This requires the adoption of a stringent scientific approach to the systematic reporting of comprehensive, relevant descriptions and measurements regarding technology use, the individuals using it and the environment within which it is used.

The EQUATOR (Enhancing the Quality and Transparency Of health Research) network [20] was established to improve the reliability and value of published health research literature by promoting transparent and accurate reporting and the wider use of robust reporting guidelines. Guidelines for the main types of studies are made available for this purpose, although nursing isn't one of its clinical areas for study, health informatics is included. A health IT evaluation study that is observational in nature can make use of the STROBE

(Strengthening the Reporting of Observational Studies in Epidemiology) reporting guideline [21]. Also available is the Statement on Reporting of Evaluation Studies in Health Informatics (STARE-HI) which is a guideline for writing and assessing evaluation reports in Health Informatics [22]. This guideline has been endorsed by the board of the European Federation of Medical Informatics (EFMI) and adopted as an official document by the International Medical Informatics Association (IMIA) [23]. Another useful evaluation methodology is the Guideline for good Evaluation Practice in Health Informatics (GEP-HI) [24]. This guideline was the result of identifying sixty issues that are of potential relevance for planning, implementation and execution of a health informatics evaluation study.

## A nursing workload management system and change management evaluation framework

Previous chapters have highlighted the many influencing service demand, operational and outcome factors that may need to be considered when designing a before and after evaluation study. The following framework has been developed based on the GEP-HI [24], STARE-HI [22] and STROBE [25] guidelines. The undertaking of a before and after evaluation is highly recommended as it ensures a critical assessment of all change management activities is undertaken. This in turn assists in identifying areas where additional changes may be beneficial. The planning phase consists of a consensus seeking process amongst all stakeholders resulting in a preliminary outline of the study to be undertaken. This ensures there is agreement and full collaboration from all concerned. This is followed by a more detailed study design phase including the development of the evaluation methods to be adopted and activities to be operationalized.

The guideline detailed in Table 12 provides a generic summary of what needs to be considered and documented for inclusion in the study design. Any exclusions need to be identified and documented with a rationale for such exclusion.

**Table 12 Before and after system implementation evaluation guideline**

| | |
|---|---|
| Title | Needs to be informative, use commonly used terms, give a clear indication of system or process evaluated, study objective and design. |
| Summary/Abstract | Provide an overview of every aspect associated with the study. This needs to be balanced in terms of what was done and found in terms of objectives, setting, participants, measures, study design, major findings and conclusion. |
| Keywords | Preferably MeSH [26] based to enable associated literature services. Keywords are used to provide a link with other reported studies to enhance retrievability of relevant associations. |
| Introduction | *Scientific background* — This relates to a literature review, knowledge gaps, applicable theories to provide any reader with context to make the rest of the report more meaningful and explain why this study meets an identified need. |

**Table 12**   **Before and after system implementation evaluation guideline—Cont'd**

| | |
|---|---|
| | ***Rationale for the study*** — Provide an explanation of what motivated this study, why this study was important and for whom, such as the nursing service, the nursing union and its members, organizational management, funders, governments, software developers or implementers, enterprise systems/network managers. |
| | ***Study objectives*** — Provide evidence to support a desired outcome, prove or disprove a pre-specified hypothesis or anticipated truth, identify efficacy of changes introduced, permissions obtained where relevant. In this case and in general big picture terms it is about measuring any aspect of an organization's capacity to care. This can include outcomes associated with any relationship between input and process factors, including any digital transformation strategy implementation as detailed in previous chapters. There should be a list of questions to be answered by the study or proposed system implementation. |
| Study context | ***Organizational setting*** — Identify locations, type of health and nursing services, departments included, types of patients/clients for whom services were provided, study dates, observer recruitment, training and preparation, organization climate or culture, potential data collection processes. |
| | ***System details and usage*** — Describe the availability of information systems, data formats, access and retrieval methods, operation systems by type of system, connectivity and interoperability schemas in use, result of any previous functionality gap analysis. This detail should enable the reader to understand how the system(s) work or is intended to work and relevant phase in the system's life cycle in terms of original design functionality vs current needs. |
| Methods | ***Study design & size*** — Provide details of the overall design and arguments for choosing it describing the overall approach adopted including the type of study such as observational, qualitative, quantitative, comparative, use of benchmarks or control group, work measurement, use of administrative data, triangulation, desk audit, performance monitoring, analytical etc. The description provides a frame of reference for the interpretation of the results and possible limitations and/or biases. |
| | ***Theoretical background*** — This is obtained as a result of a literature review and can be used as a foundation or modified for the study to guide study methods selection. This may relate to a specific model of care, information models, standards or practice guidelines. |
| | ***Participants, objects or concepts studied*** — Describe the primary focus of the study and perspective adopted, including skill mix relative to patient type, selection methods adopted, inclusion or exclusion criteria used, information flows, types of documentation, system users, systems, processes, patient journeys by type. |
| | ***Bias*** — describe how potential sources of bias were addressed. |
| | ***Study size & flow*** — List start and finish, study periods and interruptions, with clear descriptions of interventions throughout the study period. The timeframe in which the study is run can influence outcome variations. Study periods need to be representative of commonly occurring work processes and should not include holiday periods or unusual periods of activity. |
| | ***Outcome measures or evaluation criteria*** — Identify variables of interest included or excluded and confounding variables to be considered. Include definitions of key concepts studied. |

*Continued*

**Table 12**   **Before and after system implementation evaluation guideline—Cont'd**

| | |
|---|---|
| Results | ***Methods for data acquisition and measurement*** — List data sources, terminology standards in use and types of data collected for each variable of interest, and measurement/assessment methods adopted. Describe data comparability if the study covered several groups. This needs sufficient detail to enable others to assess appropriateness, representation, data accuracy, completeness and possible limitations to enable study replication. |
| | ***Methods for data analysis*** — Detail data processing methods and the statistical and analytical methods used. Describe how confounding variables and relationships, and cause and effect were identified and considered. Highlight any frame of reference used for before and after evaluative comparisons. |
| | ***Demographic and other study coverage data*** — This relates to casemix, users, workers, patterns of behavior, resources participants had access to, frequency of system use or other characteristics that the results relate to. This enables others to assess if the results apply to their situation. |
| | ***Unexpected events during the study*** — Describe anything unplanned that may have influenced the study results or outcomes, such as staff or process changes, political/management interference, change resistance. |
| | ***Study findings and outcome data*** — Display all details presented in any visual form relative to each study question, objective, evaluation criterion. |
| | ***Unexpected observations*** — Describe any unexpected positive or negative findings that were outside the scope of the study. |
| Discussion | ***Answers to study question*** — Present a critical analysis and interpretation of findings, lessons learned relative to the study itself or the methodology or study design adopted. A reasoning related to any unexpected findings. |
| | ***Strengths and weaknesses of the study*** — Provide a critical discussion of the study approach adopted and methods used based on reflection. |
| | ***Results in relation to other similar studies*** — Highlight the differences between this study and other previous studies, focusing on new learnings and knowledge gained. |
| | ***Meaning and generalizability of the study*** — Highlight the implications of the study findings for any one of the many possible stakeholders in the short and long term for both the healthcare facility and any possible wider implications for the profession or other healthcare facilities or software developers. |
| | ***Unanswered and new questions*** — Identify future research needs and opportunities to further progress knowledge discovery focusing on new lines of enquiry. |
| Conclusion | Provide an objective, concise and balanced summary of the main findings, including their impact and how these relate to the bigger picture. It is an important account of any evaluation study to assist readers to quickly assess the study's contribution. |
| Authors' contribution | This is relevant for individual study contributors. |
| Competing interests | Declare these where applicable, such as contributions by a system developer. |
| Acknowledgements | Acknowledge financial or in-kind support from external parties who deserve to be credited for their contributions |
| References | List references used to serve as evidence used to support arguments, and statements made. |
| Appendices | Attach supporting material such as detailed descriptions of various aspects that formed part of the study. |

The preliminary outline is about documenting the purpose of the study, including a rationale for why to undertake it, who will benefit and how the evaluation should be conducted. Every study needs to have clear objectives and a desired outcome. Project planning is about operationalizing the study, including the identification of who is responsible for each activity, complete with an associated timeline. This timeline needs to consider the environmental status quo as a starting point. Once the plan and study process details are finalized the study can be executed. This is followed by a report. When undertaken before any changes are made, this type of study enables the clear identification of delays, issues and gaps which need to addressed as part of the change management process. It therefore serves as a very useful precursor to any major change process. It also documents a baseline against which the outcomes of the change process can be measured and evaluated.

## References

[1] Hyland. Digital transformation: a traveler's guide for the journey. Hyland Software Inc; 2018. eBook available from: https://www.hyland.com/en-ZA/resources/trending-topics/digital-transformation/onbase-product-digital-transformation-ebook.

[2] AIIM. State of information management: are businesses digitally transforming or stuck in neutral? AIIM Industry Watch; 2017.

[3] PWC. Strategy and ten guiding principles of change management. [cited 14 February 2019]. Available from: https://www.strategyand.pwc.com/media/file/Strategyand_Ten-Guiding-Principles-of-Change-Management.pdf; 2004.

[4] Forsythe J, Carroll S, Norton C, Mackenroth E, Norton R, Strain R. The DNA of health IT change management. PWC Report, available from: https://www.pwc.com.au/industry/healthcare/assets/dna-health-it-nov12.pdf; 2012.

[5] Accenture. CIO perspectives on digital healthcare. Norway, Sweden and Finland: Accenture & Oxford Analytica; 2017. [cited 17 February 2019 ]. Available from: https://www.accenture.com/t20171220T105159Z__w__/no-en/_acnmedia/Accenture/Conversion-Assets/DotCom/Documents/Local/no-en/PDF/Accenture-Health-Nordics-CIO-Survey.pdf.

[6] Shaw T, Hines M, Kielly-Carroll. Impact of digital health on the safety and quality of health care. Australian Commission on Safety and Quality in Health Care (ACSQHC); 2018.

[7] Kelay T, Kesavan S, Collins RE, Kyaw-Tun J, Cox B, Bello F, et al. Techniques to aid the implementation of novel clinical information systems: a systematic review. Int J Surg 2013;11(9):783–91.

[8] Linnen D. The promise of big data: improving patient safety and nursing practice. Nursing 2016;46(5):28–34. quiz-5.

[9] ECHAlliance. The digital health society declaration. [cited 18 February 2019]. Available from: https://echalliance.com/page/DHSdeclaration; 2017.

[10] Fiorio CV, Gorli M, Verzillo S. Evaluating organizational change in health care: the patient-centered hospital model. BMC Health Serv Res 2018;18(1):95.

[11] Sligo J, Gauld R, Roberts V, Villa L. A literature review for large-scale health information system project planning, implementation and evaluation. Int J Med Inform 2017;97:86–97.

[12] Witt CM, Sandoe K, Dunlap JC. 5S your life: using an experiential approach to teaching lean philosophy. Decis Sci J Innov Educ 2018;16(4):264–80.

[13] Gao T, Gurd B. Organizational issues for the lean success in China: exploring a change strategy for lean success. BMC Health Serv Res 2019;19(1):66.

[14] ASQ, Lean six sigma in healthcare, [cited 27 January 2019]. Available from. http://asq.org/healthcaresixsigma/lean-six-sigma.html.

[15] ILO. Introduction to work study. Geneva International Labour Office; 1979.

[16] Barnes RM. Motion and time study: design and measurement of work. 7th ed. John Wiley & Son; 1980.

[17] Zandin KB, editor. Maynard's industrial engineering handbook. 5th ed. New York: McGraw-Hill; 2001.

[18] Hovenga E, Hazelton LM, Britnell S. Using six sigma lean and other tools for measuring quality. In: Saba V, McCormick K, editors. Essentials of nursing informatics. 7th ed. New York: McGraw-Hill; 2019.

[19] Sicotte C, Paré G, Moreault M-P, Paccioni A. A risk assessment of two interorganizational clinical information systems. J Am Med Inform Assoc 2006;13(5):557–66.

[20] EQUATOR. Enhancing the quality and transparency of health research. EQUATOR network; 2019. [cited 21 February 2019 ]. Available from: http://www.equator-network.org.

[21] Ammenwerth E, de Keizer NF. Publishing health IT evaluation studies. Stud Health Technol Inform 2016;222:304–11.

[22] Talmon J, Ammenwerth E, Brender J, de Keizer N, Nykanen P, Rigby M. STARE-HI—statement on reporting of evaluation studies in health informatics. Int J Med Inform 2009;78(1):1–9.

[23] IMIA, IMIA endorsed documents, International Medical Informatics Association, [cited 22 February 2019]. Available from. https://imia-medinfo.org/wp/imia-endorsed-documents/.

[24] Nykänen P, Brender J, Talmon J, de Keizer N, Rigby M, Beuscart-Zephir MC, et al. Guideline for good evaluation practice in health informatics (GEP-HI). Int J Med Inform 2011;80(12):815–27.

[25] STROBE. Statement—checklist of items for inclusion in observational studies. [cited 22 February 2019]. Available from: https://www.strobe-statement.org/fileadmin/Strobe/uploads/checklists/STROBE_checklist_v4_combined.pdf; 2007.

[26] MeSH, Medical subject headings browser, US National Library of Medicine, [cited 22 February 2019]. Available from. https://www.nlm.nih.gov/mesh/mbinfo.html.

# CHAPTER 11

# Measuring health service quality

## What is quality?

Should quality be measured in terms of overall health service performance? Braithwaite argues that the key measures of health system performance have not changed for decades. He notes that 60% of care is based on evidence or guidelines, that the system wastes 30% of all expenditure and that around 10% of patients experience adverse events [1]. Quality is a concept that can also be applied to any health care service or nursing activity. One could argue that quality is in the eye of the beholder; what is considered 'quality' by some is not considered to be so by others. There is a judgmental aspect to measuring quality. Quality is part of a holistic system which includes the patient as user of the processes involved in care and knowledge. Quality is necessary but not sufficient, as a high quality service may lead to low-value healthcare. Value is moving from bureaucracy based healthcare to the delivery of patient-centred and knowledge based health service programs.

How should quality nursing be described or measured? Is it about patient satisfaction regarding care received or about the degree of health care provider/supervisor satisfaction regarding nurses' collective performance? Is it about measuring the quality of key nursing activities or the degree of compliance to standard procedures? Should quality be measured based on actual performance by any individual or should quality measures also consider associated organizational aspects that are likely to influence that performance? Seven big nursing errors are mostly associated with patient falls, infections, medication errors, documentation errors and equipment injuries [2].

Quality in any nursing ecosystem is about the likelihood of delivering services that lead to patient comfort and wellbeing. An ecosystem essentially refers to the complex and interconnected organizational systems and services that collectively create the workplace 'atmosphere' or culture, that directly influences one's ability to be productive. It essentially refers to system wide dimensions of quality, including informatics, technologies, the use of entrepreneurial mindsets, professional practice, standards, education, culture and ethics. Professional integrity, accountability and transparency are fundamental to the delivery of quality healthcare services.

There are many confounding variables that independently and collectively influence the quality of healthcare services delivered. Quality should be assessed, evaluated, measured or

Measuring Capacity To Care Using Nursing Data. https://doi.org/10.1016/B978-0-12-816977-3.00011-3

perceived in terms of cause and effect or productivity. Productivity is the ratio of output to input. It applies to any business or industry or an economy as a whole. It's therefore essential to consider all resources used (input), processes or activities undertaken to create a change, and what was achieved (output). Quality is about measuring the return on investments made not just in dollar or quantitative terms but also in qualitative terms. In other words it's about optimizing the output to input ratio. Quality applies to any operational process at any level within an organization. Large and complex processes can, and should be, broken down into many smaller and simpler processes in an effort to better understand the bigger picture.

Leaders and managers are instrumental in raising productivity and quality as they are in positions to create a favorable organizational climate where collaboration and co-operation can flourish. Managers are responsible for obtaining the facts, planning, directing, co-ordinating, controlling and motivating staff in order to produce a high-quality health service. They need to ensure that all resources needed (input) to undertake all operational processes are made available at the right time and place. The health workforce needs to have and apply the right knowledge and skills to ensure that every process is undertaken correctly and that available resources are used appropriately. A quality health service may be defined in terms of the following six dimensions [3]:

*   Safety — avoid harm
*   Effectiveness — avoid overuse of inappropriate care and underuse of effective care
*   Person-centeredness — respectful of and responsive to individual preferences, needs and values that must guide clinical decisions.
*   Accessibility, timeliness and affordability — avoid harmful delays, reduce access barriers and financial risk for patients, families, communities.
*   Efficiency — avoid waste, good use of available resources
*   Equity — provide care at the same quality for all, irrespective of gender, ethnicity, race, geographic location, socio-economic status.

There are many workplace limitations influencing staff performance over which staff have little or any control. Quality measures can apply to any situation, activity, procedure or system, but these must be assessed in context. For example, nursing models of care operationalize the constructs of nursing practice [4] but this is dependent upon available resources including staff skill mix. Similarly nursing documentation is a high level activity that pertains to many subsequent nursing practices and patient outcomes. The quality of nursing documentation relating to patient care, its accessibility and use, influences the quality of nursing practice. Fig. 1 provides an overview of the many factors that need to be considered for managing quality as well as for interpreting any outcome measures.

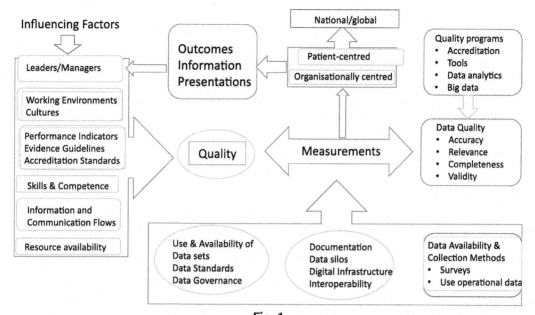

**Fig. 1**

An overview of quality influencing factors and management issues to be considered.

One review [5] found a lack of alignment between evidence based quality indicators of documentation and the nursing process. It did identify the importance of the use of standard terminologies, user friendly formats and systems as precursors for high-quality nursing documentation. Another review found some evidence that the adoption and use of electronic nursing documentation, promoted or improved the quality of care and patient safety in acute hospital settings [6]. These examples demonstrate that quality is about measuring outcomes, or the result of any activity or process undertaken by any individual as well as the workforce as a whole.

The American Society for Quality (ASQ) [7] promotes the use of Lean Six Sigma, as described in Chapter 10, to improve patient safety by eliminating life-threatening errors and addressing inefficiencies. ASQ offers a certification service and defines Lean Six Sigma as:

> '..a fact-based, data-driven philosophy of improvement that values defect prevention over defect detection. It drives customer satisfaction and bottom-line results by reducing variation, waste, and cycle time, while promoting the use of work standardization and flow, thereby creating a competitive advantage. It applies anywhere variation and waste exist, and every employee should be involved'.

## Quality programs

Health service quality programs have many features. Such a program needs to have a defined purpose and must consider costs, appropriateness of assessors, evaluation criteria to be adopted and data collection methods. Performance evaluation decisions need to be made against the stated values of the organization as well as the nursing discipline. Nursing documentation provides the evidence of nursing's contribution to patient outcomes [8].

The literature indicates that there are numerous quality measurement tools in use. It's a matter of developing a well suited quality program and choosing the most suitable set of tools to meet the organization's specific objectives. Some tools are more suited to measuring nursing's contribution to quality outcomes than others. One of these is the MISSCARE survey, a seminal tool used to explore missed care [9]. Missed nursing care is a major contributor to poor outcomes [10]. This is the result of both external and internal factors within the nursing ecosystem. The MISSCARE survey tool measures the most common influencing factors contributing to missed care [11–13].

## Nursing practice environments influencing quality

What are the characteristics of a nursing practice environment that facilitates the delivery of high quality care? This question was posed during the 1980's when researchers began to study these relationships. The American Academy of Nursing undertook studies in hospitals with low staff turnover rates due to their reputations of good places to work, to identify conducive work environmental factors. These factors became known as the 'forces of magnetism' and formed the foundation for the development of the Essentials of Magnetism (EOM II) unit level scores [14] adopted for further study by others [15, 16].

This instrument is based on Donabedian's Structure-Process-Outcome theory, not unlike the 'input-process-outcome' work study principles discussed in previous chapters. Kramer et al. [14] identified the following eight work processes/relationships as essential features making up a healthy work environment, as confirmed by nurses from numerous hospitals:

1.  Working with clinically competent peers
2.  Collegial/collaborative nurse-physician relationships
3.  Clinical autonomy
4.  Support for education
5.  Perception of adequate staffing
6.  Supportive nurse manager relationships
7.  Control of nursing and midwifery practice
8.  Transmission and adoption of patient-centred cultural values.

Empowering workplaces have been found to have positive effects on the quality of care such as fewer falls and risk minimization [17]. Both individual and organizational characteristics that create the nursing/midwifery work environment are known to play a significant role in patient safety, the quality of care provided and care outcomes [18, 19]. When nurses perceive their workplaces to be good, as measured by staff satisfaction surveys, their behavior has a positive influence on their patients' satisfaction rates [19]. It's about people and cultures that create or impede opportunities for the identification and implementation of Lean Six Sigma activities, that need to be supported by suitable systems and procedures. Leadership styles of managers at various levels within any healthcare organization, along with co-workers influence the 'vibes' that create the overall ecosystem in either positive, inspiring, engaging and supportive, or negative and non-collaborative organizational cultures. A collaborative workplace is paramount in being able to successfully implement and achieve the benefits of adopting Lean Six Sigma principles.

The foundation of a collaborative workplace includes: organizational structures, hierarchies, status symbolisms as identified by position titles, roles, boundaries of practice, mutual respect or its absence, collaboration versus competition, teamwork, including multidisciplinary teams, agreed values and the degree of compliance to these values by individuals. It's critical to align processes, workflows and organizational systems with the desired values and culture and to have a method that reinforces and sustains that, to ensure positive experiences for patients/clients as well as all workers/employees. The adoption of collaborative infrastructures, cloud computing, mobile applications and other digital technologies are key to facilitating work transformations, requiring new and innovative organizational workplace design concepts. A vibrant digital workplace ecosystem enables productivity improvements to be realized whilst enhancing overall patient/client and workforce satisfaction.

Nursing and midwifery practice environments are complex. Many factors influence the demand for nursing and midwifery services and their ability to meet these demands. Nurses and midwives need to work and deal with numerous stakeholders, care, treatment, management and administrative concepts that collectively make up their ecosystem. Nursing practice environments became a focal point relative to nursing shortages and patient safety during the early 2000s. This resulted in a major publication [20] guiding the design of nurses' and midwives' work environment to enable them to provide safer care.

This led to a research study evaluating the utility of several multi-dimensional instruments that found the Practice Environment Scale of the Nursing Work Index (PES-NWI) to be the most useful instrument [21]. This index includes factors that enhance or attenuate a nurse's ability to practice nursing competency and deliver high quality care. A follow up review of this instrument's use found that most users reported significant cause and effect findings between the nursing practice environment and outcomes. Although some modifications have been made, the instrument has remained primarily unchanged since its development [22]. These authors recommended that further research should shift toward identifying interventions that improve the environment in which nurses and midwives practice, determining if changing the environment results in care quality improvements.

Within healthcare, patients are, in essence, customers of services provided. As such they can play an important role in co-creating service value as they influence service provision in a variety of ways. This includes accessing and sharing information, interacting with the care and treatment processes and providing feedback. A practical value co-creation or innovation collaborative model for healthcare services, must consider the processes of providing, collecting and analyzing customer feedback. Such patient centred information may then be used to improve the quality of services provided [23].

## Collegial cultures

Collegial cultures refer to work environments where responsibility and authority is shared equally by one's colleagues. This determines the collective mindset that differentiates professional contributions to patient and organizational performance outcomes. It's about value systems, multidisciplinary teamwork, interdisciplinary teamwork, mutual respect of the relevant discipline's knowledge and contributions to satisfactory outcomes. Collegiality emphasizes trust, independent thinking and sharing between co-workers. It's an important aspect of effective teamwork which contributes significantly to the quality of care delivered.

According to Padgett [24], accountability is a crucial part of collegiality as nurses are answerable to each other for their collective practice. Collegiality was found to influence missed nursing care, along with other factors traditionally defined to make up the nursing practice environment [25]. It is therefore a major contributor to the quality of nursing services delivered and needs to be included as a performance measure. The nursing practice environment, including supportive peer relationships among nurses (collegiality), has been shown to influence both nursing and patient outcomes [21].

## Data quality

Patient data which reflects care delivered, treatment plans and nursing resource data need to be treated as assets. These collectively form the basis for providing meaningful information regarding a healthcare facility's capacity to care. These data reflect raw facts and figures until processed to produce new information, making such data more meaningful and valuable to users. New data or information may then be added to existing knowledge to create new knowledge.

Information and knowledge are used by many different decision makers at all levels within any healthcare facility. Ease of use is greatly facilitated where all relevant internal health information systems are able to connect, at all levels, in a manner that facilitates data sharing and linkage as presented in previous chapters. The importance of adopting data standards was also highlighted in previous chapters. Choices made regarding the local digital infrastructure and the adoption of data standards are influenced by what decision makers value most. Accurate, comprehensive and timely information improves decision making, the capacity to care and consequently health outcomes.

Governance principles relate to the information itself, as well as to health information systems and their relationships within any specific digital health ecosystem. Data collected for operational use by any healthcare provider, are also processed and used by higher level decision makers within the healthcare organization. In addition, systems need to be in place to facilitate reporting to external stakeholders such as other healthcare providers, researchers, funders, and Governments. It is therefore highly desirable to optimize data accuracy at the source. The quality of health services delivered by individual healthcare organizations within any country is of interest to both providers and consumers of care services. The adoption of globally agreed data standards, enables comparisons to be made and allows evidence of best practice to be shared and used for policy development. The following local and global recommendations on health data and governance, maximize clinical standards and quality outcomes within health services. On a macro scale, these outcomes then provide an international benchmark for quality assurance relevant to health services.

The Organization for Economic Co-operation and Development (OECD)'s recommendations regarding purposes of personal health data use associated with health reforms include:

- improving health care quality, safety and responsiveness,
- reducing public health risks,
- discovering and evaluating new diagnostic tools and treatments to improve health outcomes,
- managing health care resources efficiently,
- contributing to the progress of science and medicine,
- improving public policy planning and evaluation, and
- improving patients' participation in and experiences of health care.

It is recommended that governments establish and implement a national health data governance framework to encourage the availability and use of personal health data to serve the public interest [26].

The OECD Council on Health Data Governance has identified core elements to strengthen health data governance and thereby maximize the potential of using health data while protecting individuals' privacy. This fosters better care co-ordination across providers and across the health and social care sectors to provide a more complete picture of what happens to people throughout their health journeys, including individual care pathways [27]. Data relating to key areas of health care are currently not being linked to measure quality in most countries. The OECD identified the following list of key national personal health data available in any number of countries that are ideally shared, linked and analyzed to regularly monitor quality and health system performance, relative to outcomes of patient health journeys/care pathways. This does require the consistent use of unique identifiers as well as data standards.

- Hospital in-patient data
- Cancer registry data
- Mortality data

- Emergency healthcare data
- Mental hospital inpatient data
- Cardiovascular disease registry data
- Prescription medicines data
- Population census or population registry data
- Population health survey data
- Formal long-term care data
- Primary care data
- Patient reported health care outcomes data
- Diabetes registry data
- Patient experiences survey data.

All the above datasets include data collected and/or used by nurses. Nursing data most accurately reflects reality, as these are the data that directly relate to the people nurses care for. This is very different from data collected from patient records for the purpose of statistical reporting. Within appropriately designed digital health ecosystems, as discussed in chapter 9, data retrieval, data linkages, processing and reporting to suit these and many additional purposes can be automated.

### Health data uses and links to nursing data

Once a patient is admitted to a health service, nurses make use of the data that were used to determine that admission, to form the basis for the development of the patients' treatment and nursing care plan. Data and information are added to the record from other clinical service providers. This information is also used by nurses for their decision making. Clinical care documentation consists of five distinct data entry types [28],

1.  Observation — results of observations made, anything measured by a nurse, other clinicians, a laboratory, or reported by the patient as a symptom, event or concern.
2.  Instructions — regarding proposed interventions including medication and treatment orders.
3.  Action — documentation regarding actions taken (e.g. wound care), medication administration, completed treatment interventions, etc.
4.  Evaluation — opinions based on evaluations made such as risk potential, goals to be achieved, recommendations for follow up care.
5.  Admin-entry — administrative information including, appointments, consent etc.

Nurses make use of and contribute to this information, as they are the ones who may need to prepare patients for proposed interventions and/or actually complete the required interventions. They add their own observations, instructions and opinions for other clinicians to use, to ensure care continuity. Nurses are usually responsible for coordinating patient centric activities, including patient transport as required, as well as allocating additional resources or redistributing available resources to meet ad hoc unplanned new service demands at any point in time. Clinical care is very much a team effort requiring good collaboration and communication practices. Nurses often take on the responsibility of ensuring that continuing and timely clinical decision making is not compromised. Delays concerning information flow are costly and have the potential to lead to adverse clinical events due to missing information.

In a digital health ecosystem, such information is stored in a digital format and accessible from any device by those who need to make use of it and who have authorized access. Nurses are frequently required to access these data to develop nursing care strategies that facilitate continuity of care and minimize the risk of delayed care. An analysis and mapping of data and information flows tends to identify areas of shortcomings where changes can make great improvements to reduce costs and improve operational efficiencies. This process enables the identification of summary data required at various points of decision making throughout a patient's health journey. This knowledge forms a useful basis for designing information system dashboards, where data from multiple sources are presented on one screen to assist decision making at any level within the healthcare facility. Dashboards are ideally designed to suit decision making roles.

## Using data to support decision making

The ability to provide safe, high-quality care can be dependent upon any decision maker's ability to reason, think, and judge, which can be limited by a lack of experience [29]. These thinking and cognitive processing activities are supported by the data and information available and the ability to link these to prior knowledge. The expert performance of nurses is dependent upon informed decision making, continual learning and evaluation of performance. Critical thinking is fundamental to making effective decisions and directly relates to the need to match skill mix with service demands, as discussed in chapter 5.

Critical thinking is a core skill component enabling nurses, midwives and other clinicians who use data and information to prioritize and make key decisions to provide safe and quality care. In some situations these decisions have life or death consequences, in other situations decisions influence patient safety, the quality of services provided and outcomes. Nurses and midwives need to be able to identify problems, determine the best possible solution and choose the most effective method to address these [30]. This requires them to skillfully perceive, analyze, synthesize and evaluate collected information through observation and the application of knowledge gained from past experiences. The application of the nursing process, as described in chapter 6, is directly influenced by a nurse's ability to think critically.

## Data sets and data repositories

A foundational requirement enabling nursing data to be integrated into clinical data repositories for big data and science, is the implementation of standardized nursing terminologies, common data models, and standardized information structures within Electronic Health Records (EHRs) [31, 32]. An information model is a 'structured specification, expressed graphically and/or narratively, of the information requirements of a domain. An information model describes the classes of information required and the properties of those classes, including attributes, relationships, and states' [33].

Most of these data sets could be used to identify nursing service demands linked to nursing resource availability and usage data, relative to patient outcomes by patient type. This is subject to appropriate data linkage possibilities. Data use decisions should be taken by weighing societal benefits and risks within a data governance framework that maximizes benefits and minimizes risks [27].

### Data governance mechanisms

Eight data governance mechanisms, designed to maximize benefits to patients and to societies from the use of privacy protective health data, were identified by the OECD and described. These are listed below:

1. The health information system supports the monitoring and improvement of health care quality and system performance, as well as research innovations for better health care and outcomes.
2. The processing and the secondary use of data for public health, research and statistical purposes are permitted, subject to safeguards specified in the legislative framework for data protection.
3. The public are consulted upon and informed about the collection and processing of personal health data.
4. A certification/accreditation process for the processing of health data for research and statistics is implemented.
5. The project approval process is fair and transparent and decision making is supported by an independent, multidisciplinary project review body.
6. Best practices in data de-identification are applied to protect patient data privacy.
7. Best practices in data security and management are applied to reduce re-identification and breach risks.
8. Governance mechanisms are periodically reviewed at an international level, to maximize societal benefits and minimize societal risks as new data sources and new technologies are introduced.

Most of these governance mechanisms also apply to data collection and data linkages associated with the use of patient acuity data for the purpose of monitoring the capacity to care within individual healthcare facilities. Nurses depend on accurate and timely access to appropriate information to perform the great variety of professional activities associated with inpatient, outpatient, residential and community care. Nursing information integrates technical knowledge, quality control, and the clinical and administrative documentation of services provided. Nurses need information about available resources, science development and patient needs for decision making, program planning, delivery and supervision of clinical practice, management interventions and evaluating care outcomes [34]. The need for standardized terminologies to describe, compare and communicate nursing care activities across settings, population groups and countries was recognized during the early 1990s. By the early 2000s it became apparent that in addition to the use of standard terminologies, the development and adoption of clinical models or archetypes would improve this capacity even further [35–38].

Such computable definitions or specifications for a single, discrete clinical concept are inclusive of all data elements that make clinical sense about that concept and are designed for all imaginable clinical situations. These definitions are kept broad and constraints are minimal in order to maximize interoperability by being able to share and re-use the archetype across many types of healthcare, reflecting the broadest range of clinical scenarios. These specifications are expressed in Archetype Definition Language (ADL), which is an ISO standard, but able to be viewed and reviewed in 'clinician-friendly' formats, as structured definitions and mind maps. By design, they provide structure and specify content which means that archetypes can be both clinically meaningful and interpretable by EHR systems. Archetyped data will have the same meaning no matter what context it is used within the EHR and, similarly, no matter which EHR system it is used or what language is used. Their value and use for nursing documentation has been demonstrated [32, 37, 38].

We argue for a greater use of nursing service data as nurses and midwives represent significant care contributors influencing health outcomes. As discussed in previous chapters; data accuracy, integrity and completeness is another key requirement for meaningful use. Patient acuity data quality and governance needs to be considered as foundational within a big picture context.

## Standards, accreditation and governance

Previous chapters have referred to the use of numerous standards. Standards refer to [39]:

- Something considered by an authority or by general consent as a basis of comparison; an approved model.
- An object that is regarded as the usual or most common size or form of its kind.
- A rule or principle that is used as a basis for judgment.
- An average or normal requirement, quality, quantity, level, grade etc.

Standards serve as a basis of weight, measure, value, comparison, or judgment of recognized excellence or established authority. The health industry makes use of a variety of types of standards, including professional, industry and best practice standards as developed and accepted by expert communities. In addition, there are numerous formal standards as published by many standards development organizations (SDOs), that may be adopted voluntarily or adopted as legal or ethical requirements for professional and/or government (jurisdictional) initiatives. Standards need to not only be consensus based but they also need to be tested for efficacy. For any given context, information system standards address issues of order and compatibility in the design, development, implementation, and operation of information systems and information technology [34].

Data and terminology standards form the basis for effective communication and collectively they reflect the health language in use. Terminology standards include structured vocabularies, nomenclatures, code sets and classification systems. Health data can be configured in any language including any type of terminology standard, where data elements represent the atomic level from which meaning is derived. These types of standards are intricately linked to Health Informatics standards that apply to system interoperability as described in Chapter 9. The World Health Organization is one SDO responsible for a family of classification systems including the widely used International Classification of Diseases [40].

### Accreditation standards

A variety of international guidelines for quality and accreditation standards are available [41–44]. A legal framework for the provision of accreditation services across Europe became effective in 2008. There is the International Laboratory Accreditation Co-operation (ILAC) and the International Accreditation Forum (IAF) that collectively determine the nature of accreditation globally. Their role is to confirm integrity. Accreditation then provides the basis for the acceptance of products and assurance to its markets [45]. This is relevant for all surgical supplies and equipment used by health care workers in any healthcare organization. The International Society for Quality in Health Care (ISQua) is a member-based not-for-profit global community and organization with an extensive network of health care professionals across a large majority of countries [46]. Its mission is *to inspire and drive improvement in quality and safety of health care worldwide through education and knowledge sharing, external evaluation, supporting health systems and connecting people through global networks'*.

Most countries have established an authority to develop patient safety and quality standards that form the foundation for healthcare organizations' accreditation and certification services. Examples are the Joint Commission in the United States [47], the Australian Commission on Safety and Quality in Health Care [48, 49] and Accreditation Canada [50]. In the United Kingdom the Trent Accreditation Scheme (TAS) that was National Health Service (NHS) based, withdrew from all hospitals' accreditation related activities in 2010. In response a group of acknowledged clinicians and other experts actively working within the NHS and private medical practice, established the Quality Healthcare Advice, providing a surveying and accreditation service for hospitals and clinics located anywhere in the world [51].

Clinical pathology, imaging services and physiological accreditation services, are provided in the UK by a longstanding Government funded calibration and measurement service that continues to grow. This organization became the United Kingdom Accreditation Service (UKAS) and was appointed as the National Accreditation Body by the Accreditation Regulations 2009 in July 2016 [45].

The Australian Council on Healthcare Standards (ACHS) manages the EQuIP program that is part of its hospital and day surgery accreditation program and publishes clinical indicator reports. EQuIP has mandatory criteria used for performance measurement and includes four criteria for Information Management:

1. Health records management systems support the collection of information and meet the consumer/patient and organizational needs.
2. Corporate records management systems support the collection of information and meet the organization's needs.
3. Data and information are collected, stored and used for strategic, operational and service improvement purposes.
4. The organization has an integrated approach to the planning, use and management of information and communication technology [52].

### Types of standards

Industry standards are those developed, adopted and governed by an industry, for example: the healthcare industry. Such standards may or may not be adopted by all healthcare organizations depending on ownership and terms of use. Software vendors who have undertaken research and development to produce evidence based systems, incorporate industry standards and develop their own data standards, which are then used by those who purchase or acquire a user license.

Open standards are those developed, maintained and governed by communities. For example, the openEHR specifications that describe the management, storage, retrieval and exchange of health data in electronic health records, managed and governed by the openEHR Foundation [53], a not for profit foundation supporting the open research, development and EHR implementations. These specifications include information and service models for the EHR, demographics, clinical workflow and archetypes. They are designed to be the basis of a medico-legally sound, distributed and versioned EHR infrastructure. Key foci of these specifications are included in the ISO 13606 standard, part 1 and part 2 [54]. Open standards are technically managed and governed by its community and publicly available. Such standards are made available for commercial use via a Creative Common license arrangement.

Professional standards are those developed, maintained and governed by Professional organizations. Such standards detail ethical or legal duty of a professional to exercise the level of care, diligence and skill and/or reflect a demonstrated level of competence, promoting quality practice as prescribed in the code of practice of that profession. These standards reflect a desired level of performance. For example, the International Federation of Information Processing (IFIP) has published a global professional IT standard based on the Skills Framework for the Information Age (SFIA), used globally to certify IT professionals [55].

National legal and regulatory frameworks can include the requirement to comply with any number of these standards and/or include locally developed standards for the inclusion in any jurisdictional legislation. Countries may establish Government funded agencies/departments/institutes that may be allocated a regulatory role, responsible for the governance, maintenance and monitoring of the use of standards and/or legal requirements. Collectively such organizations have a responsibility to protect health information privacy and security, review patient consent practices, provide adequate protections to link datasets and facilitate multi-country statistical and/or research projects.

The external assessment of health service delivery methods is used globally as a means of regulating, improving and marketing health service providers, especially hospitals. This consists of peer review, accreditation, statutory inspection, ISO certification and evaluation against a specific framework describing clinical effectiveness, quality and safety expectations as defined by each of these types of authorities. Many sets of hospital accreditation standards are in use. Every nation has established its own data system of performance indicators for quality measurement and accreditation evaluation purposes. There are numerous reference centres around the globe providing clinical guidelines and health technology assessment methods, but these tend not to be effectively shared [44].

Quality standards, performance indicators and measurements can be grouped according to type as follows:

- Each of these will benefit from the use of nursing data.
- Population and community — public health.
- Consumers, care recipients, clients — patient expectations, satisfaction and experiences.
- Workforce wellbeing — personal, social and moral organizational cultures.
- Staff competency, skill mix — achieving, maintaining and applying knowledge and skills.
- Clinical practice — defining and testing effectiveness against scientific, practice based evidence including clinical indicators such as those used by Australia's Clinical Indicator Program [56].
- Service delivery — information and communication flows, patient journeys, organizational management, teamwork and scheduling.
- Risk, health and safety — promoting and maintaining safe working environments, minimizing risk of adverse events.
- Resource management — avoiding waste of skills, time, materials and money.
- Communications — internal, external information connectivity, record management and reporting.

Each of the above are applicable to every department, organization, region or nation in some form. The availability and possible uses of metadata standards were explored in Chapter4 where it was noted that the numerous confounding variables and data quality continue to be significant limiting factors for meaningful data use. It's a formidable challenge to set and define relevant standards, assessment and measurement criteria or protocols in a manner that enables not only one-off assessments but also ongoing monitoring. It is crucial that standards chosen to be used reflect the interests of patient safety, comfort and best practice, and do not conflict with desired individual/organizational provider performance criteria.

The International Society for Quality in Health Care (ISQua) [46], a NfP member organization, has developed guidelines for the design of external evaluation programs, including accreditation [41]. In January 2019, ISQua established its External Evaluation Association (IEEA) to deliver external evaluation services. Compliance with accreditation standards may be voluntary or mandatory. Many organizational enterprise systems incorporate data collection and processing in accordance with accreditation requirements. Many published standards lack readily observable data requirements, that can only be evaluated based on judgments made by the assessor relative to the published standard using a Likert rating scale [52].

### Standards governance

All standards need to be governed. An associated process is required when making decisions that define expectations, systems and its management, including a review timeline. Standards need to be used consistently in accordance with accepted norms, rules and actions throughout an organization, within any jurisdiction or by any profession. Governing bodies, boards of directors or governance committees are held accountable regarding governing systems in place for the publication, availability, transparency, use of and compliance with agreed standards specified for adoption. This is particularly important for the governance and management of key data standards as used within information systems. The World Health Organization (WHO) referred to the importance of data standards and noted:

> *"The use of eHealth and mHealth should be strategic, integrated and support national health goals. In order to capitalize on the potential of ICTs [information and communication technologies], it will be critical to agree on standards and to ensure interoperability of systems. Health Information Systems must comply with these standards at all levels, including systems used to capture patient data at the point of care. Common terminologies and minimum data sets should be agreed on so that information can be collected consistently, easily and not misrepresented. In addition, national policies on health-data sharing should ensure that data protection, privacy, and consent are managed consistently."* [57].

Governance guides the existence of a strategic policy framework applicable to a well-functioning health system, that is combined with effective oversight to promote and maintain population health. This is a political process that involves balancing competing influences and demands by collaborating with other sectors and many public and private stakeholders [58]. An organizational data governance program can shape its local directions for data acquisition, processing, management and use. The many uses of patient acuity data were described in previous chapters, this should guide the development of a strategic policy framework. Such a policy sets out high level principles and responsibilities associated with data capture and use. This extends through to the following elements [59]:

- *Effective oversight* which infers the existence of mechanisms to ensure that the policy is effectively enacted,
- *Coalition building* indicating that this is a co-operative and participatory behavior (rather than being autocratic or automatic) and that governance includes pro-actively encouraging, if not arranging or requiring this,
- *Providing appropriate regulations and incentives* so that a meaningful set of rules are clearly defined and there are explicit measures to encourage and facilitate compliance,
- *Attention to system design* inferring that governance has a role to play in ensuring that systems are well-aligned with policy and support policy enactment. (Here "system" refers to the holistic system including IT systems and business processes and related elements), and
- *Accountability* to ensure that there is clear responsibility for compliance and outcomes.

Data governance extends throughout the organization including at the point of care or data collection source, as this is where data context is created. Many people are directly involved in any number of different activities associated with data collection, management or use. Everyone needs to be familiar with the organizational data government policy or framework.

It is important to apply such governance principles to the acquisition, management and use of all patient acuity and associated data. The many potential data uses were described in previous chapters. Governance processes form a major component of data quality management. A data governance process includes identifying what data is required and how feasible it is to obtain and use such data. Selected data elements or concepts to suit any specified purpose need to be adopted as data standards and governed accordingly, to ensure uniform data collection, processing and reporting to improve the provision of quality information.

## Reliability and quality measures associated with patient acuity data

Chapter 3 detailed various nursing service demand factors, followed by nursing workload and nursing work measurement methods. This included reference to the many possible data sources and numerous standard datasets, that may be used to describe the many variables associated with measurements in use to determine any organization's capacity to care. It was noted that there is no common or agreed framework to enable comparative evaluations of these various methodologies to be made, no consistency regarding the variables to be considered and their definitions and/or measures used apart from some common features.

For example the literature revealed that even the frequently used Nursing Hours Per Patient Day (NHPPD) measure did not make use of standard definitions for 'direct patient care responsibilities' (i.e. data selection to represent nursing hours), 'patient day' (i.e. per occupied beds or midnight census), 'in-patient unit'(i.e. applicability for types of patients) or 'productive hours' (i.e. data selected for processing). There may be consistency regarding data selection and use within any health service to note trends, but such results tend not to be comparable with the results obtained from other similar studies. There are no agreed metadata standards either. The complexity of measuring nursing service demand and the difficulties associated with defining the multitude of variables according to agreed data standards was highlighted.

Chapter 4 detailed various nursing work measurement methods including the data sets required to undertake such studies. Such studies form the foundation for patient acuity data. It was noted that such studies can be undertaken from a number of different perspectives or study objectives, and that a number of different methods may be used to analyze what aspects of nursing work can or needs to be measured. The time study and work sampling work measurement methodologies were described in some detail including the conversion of work measurement data to nursing workload measures, relative to types of patients.

As a consequence of the many confounding variables influencing any organization's capacity to care, it is essential to consistently make use of a standard nursing work measurement tool across multiple healthcare facilities. This enables health care facilities to reliably aggregate such data resulting in a very large data sample, from which patient acuity benchmarks by patient type can be determined. Large data samples provide new opportunities for nursing research to identify best practice. Benchmarks can provide reliable measures which can be used to predict nursing care demand for each defined patient type. These benchmarks can also be used to identify inefficient work and operational practices that would benefit from re-engineering and change management. Benchmarks, developed from accurate acuity data which have been validated through inter-rater reliability testing, will provide the best available nursing service demand measures, with the smallest possible range of likely variation when applied to individual patients. This also provides the ability to identify reasons associated with such variations.

Reliable use is dependent upon consistent interpretation and selection of indicators/ratings/criteria used to define categories for each patient type, that are associated with a standard time measure for three 8 hr periods of the day. Useful outcomes associated with the use of a patient acuity system are dependent upon data collection accuracy and completeness. Patient acuity data collected at each point of care form the foundation for all subsequent resource management decision making. Data collection accuracy (Inter Rater Reliability) and completeness need to be audited at least annually to avoid gradual bias creep and/or to identify those who are having difficulty in the correct interpretation of data collection indicators/criteria. Acuity data integrity can be compromised due to staff turnover, when new staff have not undertaken system orientation prior to their involvement in patient acuity data collection. Patient acuity system implementation processes, including data collection and use, were described in some detail in the previous chapter.

## Clinical data management issues

Data management should not be considered in isolation but within an environment where other services such as data storage and communication are provided. Common data management services are those required to define, store, retrieve, update, maintain, backup, restore and communicate applications and dictionary data. These also apply to services for distributed database management. A data management standard defines services provided at an interface. Data management services are particularly important for the management of persistent data, that is, data retained in an information system for more than one data management session, as is frequently the case for health data.

Current database management systems include large volumes of unstructured health and medical data, including text, images, bio-signals, video and audio. These data are unstructured and tend to be ignored beyond immediate use, making it difficult to use for secondary purposes [60]. This also means that medical artificial intelligence is working with incomplete data. This is expected to change once these types of data can be quality controlled, anonymised, visualized, extracted and managed using an agreed standard for the purpose of big data analysis. The Digital Imaging and Communications in Medicine (DICOM) standard, complied with by most organ imaging software, contains the patient ID within the file making it impossible to separate the image from this identifying information. This standard uses three different Data Element encoding schemes, although the same basic format is used for all applications, including network and file usage [61–64].

Another issue is the need for the ability to migrate data from legacy systems to a new platform without loss of meaning [65]. This is dependent on the quality and format of the legacy data. It is most desirable to transfer such data in a manner that retains the original data quality, without adversely affecting the legacy system's performance. This requires a thorough understanding of the original data storage structure and careful planning of the data migration process as discussed in Chapter 9. The size of the data stored, plus transmission speed, determines the time required for the migration, limiting continuing use during the migration period.

Some clinical functions may be dependent upon proprietary data, those functions are likely to be lost following data migration. This is a major issue for many devices with embedded software that cannot run on platforms other than those related to the original system design and development. In some situations, old systems need to continue to be operational for some time, despite the adoption of new systems. Various approaches are in use to transpose data into a common format to facilitate collaborative research, large-scale data analytics and to share various tools and methodologies. Issues associated with data linkage, using application centric versus data centric approaches were discussed in Chapter 9. The biggest issue associated with application-based data exchange is the need to map data from one system to another to suit each individual use case. This provides opportunities for error and may compromise patient safety in some instances.

## Outcomes research and big data

With fully integrated EHRs containing computable source data, outcomes research can be significantly enhanced, including any type of 'best practice' assessment. It is extremely beneficial to have agreed sets of outcome indicators that can be monitored in a fully automated and timely manner. The current state of play as described below, reflects the continuing use of distributed legacy information systems, containing data silos with associated data exchange issues as discussed in Chapter 9. Global adoption of data-centric solutions has the potential to generate huge databases in real time, to support the availability of practice based evidence to meet very specific patient centred needs. This will require significant transformation of the many existing outcomes research networks.

> *Outcome quality benefits that would bring evidence to the system use* vs. *quality loop are unlikely to be obtained in a near future since they require integration with population-based outcome measures including mortality, morbidity, and quality of life that may not be easily available'* [66].

There have been a number of studies that identified adverse events, representing unintended, and at times, harmful occurrences associated with the use of medicines, equipment, or various health service practices employed, including associations with staffing levels and skill mix [67–71]. ICHOM, the International Consortium for Health Outcomes Measurement Inc. (US) [72], is developing a new paradigm focused on health outcomes in a form that matters most to patients. This represents a future view of value-based health care. Their published standard sets are standardized outcomes, measurement tools, time points and risk adjustment factors for given medical conditions, based on patient priorities.

The International Society for Pharmacoeconomics and Outcomes Research (ISPOR), a USA incorporated not-for-profit (NfP) organization, have members around the globe including an Australian Chapter [73], ISPOR describes itself as '*the leading global scientific and educational organization for health economics and outcomes research and their use in decision making to improve health*'. There is little evidence of any health informatics input, yet one of their strategic objectives is to '*encourage access, appropriate use, and an understanding of opportunities and limitations of health care data to inform health care decisions'*. ISPOR has published a taxonomy of patient registries that provides a working definition, describes distinguishing characteristics of patient registries and defines terms used in the analysis, reporting and publishing of registry data.

Another USA based NfP organization is the National Quality Forum whose members include representatives from the Agency for Healthcare Research and Quality, the Centres for Disease Control and Prevention, the Centres for Medicare & Medicaid Services, and the Health Resources and Services Administration. This organization works to catalyze improvements in healthcare via the development of consensus-based quality standards and guidelines, including the advancement of electronic performance measurement. Their health IT initiatives are designed to support that move. They have produced a large number of useful reports [74].

Patient registries refer to organized systems that make use of observational study methods. These study methods collect uniform clinical and other data to evaluate specified outcomes for a population defined by a particular disease, condition, or exposure. They serve predetermined scientific, clinical or policy purposes [75]. This author noted that numerous patient registries exist, including patient generated registries, although there is a lack of data collection standardization and potential competition for registered patients across registries, fracturing sets of patient data and compromising data accuracy. Many Clinical specialities develop their own set of performance indicators requiring secondary use of data.

A framework for Australian clinical quality registries was endorsed by Health Ministers in 2014. An assessment for cost effectiveness of five of these registries found that the return on investment varied, with benefit-to-cost ratios ranging from 2:1 to 7:1. Problems identified included: low coverage, inadequate reporting and inadequate collection of information about patient outcomes. This limits the effect of some clinical quality registries, and their value to the health system [76].

The World Health Organization (WHO) has published a Global Reference List of 100 Core Health Indicators to provide concise information on the health situation and trends. The aim was to streamline and reduce the reporting burden on individual countries [77]. In addition, WHO has published a handbook of indicators and their measurement strategies for monitoring six building blocks that make up any nation's health system as whole. These consist of health service delivery, health workforce, health information systems, access to essential medicines, health system financing and leadership/governance [78]. This describes any nation's healthcare business processes [79].

WHO makes use of a number of databases, some of which are maintained by a range of other organizations including: the United Nations International Telecommunication Union (ITU), the United Nations Department of Economics and Social Affairs (UNDESA), the United Nations Educational, Scientific and Cultural Organization (UNESCO), the United Nations Children's Fund (UNICEF) and the World Bank. WHO's Global Health Observatory (GHO) provides access to data and data analysis for its monitoring of the global health situation [80]. Health related data is compiled for its 194 member states although only 34 of its members are able to provide reliable quality health data. The accuracy and amount of information collected and processed varies considerably between nations. Few are able to provide data that directly relate to services provided by nurses and midwives.

Historically data sets and collections were largely fiscal or statistical and collected administratively. Such data collections had varying levels of governance. The emergence of the electronic health record and clinical decision support systems, require a consistent structure of the data as well as consistent representation of each individual thing in the record (each concept). These structures and codes must be meaningful at the time of collection in the clinical environment. They must also retain their original meaning in computer systems over time to support extraction and aggregation of data to meet the long-standing reporting requirements of fiscal, planning and statistical collections. The WHO's health system framework [81] lists its values as improved health, responsiveness, social and financial risk protection and improved efficiency. A nation's desired outcomes should determine what data is required to be collected for the purpose of evaluating its health system.

Information collected, provides an indication of the effectiveness of the health system within any nation and provides a key foundation for the development of health policy initiatives. With the increasing use of ever changing technologies, including the introduction of electronic health records, there is a changing landscape from which to collect quality health data. This data should be used effectively by many, at all levels, within any nation's health industry, including various types of decision support systems. Nations have adopted a multitude of combinations from the many and varied health industry components to make up their nation's health care sector. Any national health care service industry or system has its own complexity and may be described from any number of different perspectives. The most common aspect of any national health system, guiding data collection and reporting requirements, are the many and varied funding arrangements, including health insurance protocols plus the need to meet national and international statistical data collection requirements. Outcome measures may also be referred to as performance indicators.

### Performance indicators and health system frameworks

The Organization for Economic Co-operation and Development (OECD) aims to measure and compare the quality of health service provision in different countries via its Health Care Quality Indicators project. It has developed a Health Care Quality Framework for Health System Performance Measurement. The OECD notes that although information on health care spending is expanding, the information on the value that health services create is still limited [82]. The OECD 's 2018 data collection process was based on 59 indicators covering multiple themes from 35 countries.

The OECD use these data to assist governments by identifying drivers of high-quality care as the cornerstone of quality improvement. However, many countries, even those with advanced data systems, have difficulty linking practice performance to outcomes because of limitations in data availability and poor capabilities to link data [83]. A review of performance indicator frameworks from eight countries revealed great variations, although each typically included reference to monitoring and improving quality and efficiency of the healthcare system. There are substantial definitional inconsistencies regarding individual performance indicators. Countries tend to make annual changes to indicators and/or their definitions, limiting the usefulness of these data collections. Braithwaite et al. [83] noted that:

> *Although there is a substantial literature dealing with the design, properties and scientific soundness of individual indicators, there is considerably less attention given to how indicators are used in practice and the impact they may have on the behavior of health professionals, or on the quality of care.*

Performance indicators in use in Australia regarding its health system performance relate to; accessibility, continuity of care, effectiveness, efficiency, sustainability, responsiveness and safety. Australia has defined these metadata standards as follows [84]:

- Accessibility — People can obtain healthcare at the right place and time irrespective of income, physical location and cultural background.
- Continuity of care — Ability to provide uninterrupted, coordinated care or service across programs, practitioners, organizations and levels over time.
- Effectiveness — Care/intervention/action provided is relevant to the client's needs, is based on established standards. Care, intervention or action achieves desired outcomes.
- Efficiency & Sustainability — Achieving desired results with the cost effective use of resources. Capacity of system to sustain workforce and infrastructure, to innovate and respond to emerging needs.
- Responsiveness — Service is client orientated. Clients are treated with dignity, confidentiality, and encouraged to participate in choices related to their care.
- Safety — The avoidance or reduction to acceptable limits of actual or potential harm from health care management and the environment in which health care is delivered.

The actual performance indicators and data collection requirements are frequently changed to reflect current government policy and funding initiatives. Some of these requirements influence local operational processes to ensure their compliance can be demonstrated. For example, one performance indicator included in the National Healthcare Agreement (PI 21b) states: Waiting times for emergency hospital care: proportion of patients whose length of emergency department stay is less than or equal to 4 hr (2018). This has resulted in the introduction of short stay spaces located separately from but close to the emergency departments in hospitals, so that anyone potentially staying longer than 4 hr can be transferred to this location to receive continuity of care. Delays are usually the result of non-bed availability for new admissions, suggesting that the bed management processes are not optimum or that demand exceeds service capacity.

Similarly the New South Wales Government's performance indicators are aligned to its key strategic programs listed as Premier's and State Priorities, Election Commitments, Safety and Quality Framework, Better Value Care, mental Health Reform and Financial Management Transformations [85]. For example, under the safety category it has numerous performance indicators or outcome measures which are not necessarily synchronized with national data collection requirements. Their information bulletin lists many performance indicators under each strategic KPI such as 'provide world-class clinical care where patient safety is first'. This document again demonstrates the numerous changes made to their annual data collections.

The Australian Commission on Safety and Quality in Health Care (ACSQHC) has developed a number of performance indicators, data sets, clinical standards and guidelines for routine monitoring use in accordance with the National Health Reform Act (2011) [48]. The Commission works in collaboration with jurisdictions, the private hospital and primary care sectors, the Australian Digital Health Agency, the National Health Chief Information Officers (CIO) Forum, Health Outcomes Australia and other national bodies to promote the safety and quality agenda within national *E*-Health programs.

ACSQHC has a national list of hospital acquired complications, many of which could be used as nursing sensitive indicators. These can be used as a measure to demonstrate the value of providing highly skilled nursing services and may also measure other outcome indicators. Clinical Care Standards have been developed in line with best practice [49], and are used to identify and define the care that patients should be offered for a specific clinical condition. These standards need to be considered in conjunction with clinical guidelines and are intended to be used to guide the delivery of appropriate care and to reduce unwarranted variation. Potentially, these need to be incorporated in decision support systems for use at the point of care when care and treatment options are being discussed with the individual concerned. There is no indication that any of this knowledge is being stored in computable formats, limiting the usability of these data.

The National Health Service in the United Kingdom (UK) identified around 50 outcome indicators pertaining to five areas of quality [86, 87].

1. Preventing people dying prematurely,
2. Enhancing quality of life for those with long term conditions,
3. Helping people to recover from episodes of injury or ill health,
4. Ensuring people have a positive experience of care,
5. Treating and caring for people in a safe environment and protecting them from avoidable harm.

These were designed to be of use to the Secretary of State for Health and Social Care to hold the NHS to account. The NHS Scotland based their quality strategy around the following priorities [88]:

• Caring and compassionate staff and services,
• Clear communication and explanation about conditions and treatment,
• Effective collaboration between clinicians, patients and others,
• A clean and safe care environment,
• Continuity of care,
• Clinical excellence.

The USA based Patient-Centred Outcomes Research Institute (PCORI) [89] focuses on funding projects designed to improve care and outcomes for patients living with high-burden health conditions. This relates to helping patients and those who care for them, to make better informed decisions about healthcare choices. It's critical for the nursing profession to contribute and gain new knowledge from these initiatives by ensuring we have the capacity to contribute nursing data [90]. The Nursing profession needs to ensure that it can collect such data and make the nursing contributions to human health visible. These transformational activities have major implications for nurse leaders [91].

### Measuring caring as an outcome measure

This book has a focus on the capacity to care. When measuring quality, we need to be able to determine how well the caring functions associated with healthcare service delivery are being achieved. Caring is a highly valued human function most applicable for anyone receiving a healthcare service. Patient satisfaction is directly influenced by the caring received throughout their interactions with health service providers. Good caring experiences are beneficial to a patient's overall sense of wellbeing and comfort. Caring is person centred and involves interpersonal behaviors of all healthcare team members, responsible for supporting meaningful communication with the patient/client, their family and/or their significant others. These behaviors reflect the 'ethics of care' underpinning the caring process. These behaviors may be referred to as personalization, participation and responsiveness, as applied when meeting a person's health and care needs whilst making them feel 'cared for' [92]. These three concepts were defined following extensive research by Strachan as follows:

Personalization is the degree to which the healthcare team gets to know the person. This includes those interpersonal behaviors that demonstrate; connecting, knowing and empathizing.

• Connecting is showing genuine interest in the person and making them feel at ease.
• Knowing is getting to know the person as an individual and understanding how their health affects their life.

- Empathizing is showing concern and understanding of a person's experiences and feelings.

Participation is the degree to which the healthcare team respects the involvement of the person, and those close to them, in their healthcare. This includes those interpersonal behaviors that demonstrate; involving, goal setting and sharing decisions.

- Involving is giving the person, and where relevant their family and those close to them, understandable information and encouragement to contribute to their healthcare.
- Goal setting is exploring the person's expectations and possibilities for their health and wellbeing.
- Sharing decisions is exploring options for the person's health & care needs and agreeing a plan of care together to meet their health goals.

Responsiveness is the degree to which the healthcare team monitors and responds to the person's health and care needs. This includes those interpersonal behaviors that demonstrate; being attentive, anticipating and reciprocity.

- Being attentive is being with, or available for the person, to help with their health needs and goals.
- Anticipating is planning ahead and working together to co-ordinate the person's healthcare.
- Reciprocity is monitoring and encouraging the person's progress and, if necessary, adapting their plan of care accordingly.

These definitions provide useful criteria for the identification or development of suitable patient satisfaction measurement instruments. Effective workforce management at every level, including education, workforce planning, and working conditions, is the precursor to staff being able to exhibit these caring behaviors.

### Impact of funding arrangements on the selection of performance indicators

Every funded health service program makes reference to a specific data set used for reporting purposes. This is unique to every jurisdiction or funding entity. Individual programs may change with budget cycles and/or change of Government, making it difficult to monitor trends. What gets measured, gets paid for. Funding strategies play a crucial role. Funding models can be used to transform health systems by providing incentives for the responsible use of resources to benefit patients and taxpayers. Funding models relate to performance and outcome measures, they include:

- Fee for service — requires a catalogue of services complete with an associated pricing structure.
- Capitation based in primary care — this is usually population based.
- Activity based funding — requires a catalogue of activities or outcome measures. A commonly used method requires all discharges to be coded using the ICD-10/11

classification system along with codes for all procedures. There usually is an associated grouping strategy based on resource usage by type of service. Payments are usually made irrespective of clinical outcomes reducing a focus on quality measurement.

• National budgets for secondary care.

Health care outcomes should be regarded as a partnership between providers and service recipients. Many of the latter do not necessarily follow all instructions given to maintain the best possible health status over time. Funding models need to ensure that healthcare providers, both individuals and organizations, are not solely held accountable for outcomes achieved. Chosen performance indicators and/or outcome and quality measures used for funding purposes, need to realistically represent resource requirements to meet service demands. According to PWC [93] the right funding model can:

• Reward providers for securing good outcomes for patients, by providing incentives to explore and deploy better connected and integrated models of care, that align providers and patients around shared goals.
• Increase market competition and depth to drive innovation and new investments.
• Encourage funding to be efficiently pooled around individuals (patient journeys) and issues, and therefore avoiding the spread of funding multiple agencies.
• Support patient choice and participation in shaping their healthcare experience.
• Lead providers to optimize resource allocation and reduce waste.
• Shift the focus from the 'here and now' to longer term 'invest to save' thinking.
• Reduce health inequalities between social groups.

Australia has a long way to go to bring about change to a patient centred fully integrated health service. Government funding for the public sector is based on a three-year data plan, reviewed annually. This is the main feature of the current Australian National Reform Agreements between the Federal, State and Territory Governments. This plan contains details regarding file specifications, calculation policy, business rules (data matching and activity-based funding), data compliance policy, data privacy and the secrecy and security policy. The Administrator will make all non-identifiable aggregated and patient level data collected, under the Data Plan, available to jurisdictions based on the patient's place of residence, where such release is legally permitted [94]. In addition, each Australian jurisdiction has its own data requirements. None have a patient journey focus. The private sector has its own arrangements with health insurers. Many patients move between public and private sectors throughout their health journeys.

The Australian Government has adopted Activity Based Funding(ABF) principles to provide funding for all public hospitals managed by the States and Territories [95], who in turn have adopted their own ABF methodologies to fund and monitor public hospital performance. For example, the Queensland Government includes all details in one publication [96], whereas the Victorian Government lists the following reporting requirements, most of which reflect service demand, none make use of nursing data:

- Victorian Admitted Episodes Dataset (VAED) — a minimum dataset collected from public and private hospitals, rehabilitation centres, extended care facilities and day procedures centres.
- Victorian Emergency Minimum Dataset (VEMD) — de-identified demographic, administrative and clinical data detailing presentations at Victorian public hospital emergency departments.
- Elective Surgery Information Systems (ESIS) — patient level collection of elective surgery waiting lists data from approved Victorian public healthcare services.
- Agency Information Management System (AIMS) — summarized financial and statistical data from Victorian hospitals.
- Victorian Integrated Non-Admitted Health Data set (VINAH) — data from various funded programs including Family Choice, Enteral Nutrition, Hospital Admission Risk, Hospital based palliative care consultancy team, Medi-Hotel, Specialist clinics, Palliative care, Post-Acute care etc.
- F1 financial reporting data collections.
- Victorian Cost Data Collection (VCDC) — dataset reflecting the cost and mix of resources used to deliver patient care.

## Big data management and governance

An important requirement prior to undertaking any data analysis is to ascertain data accuracy, completeness and appropriate representation of the concepts being evaluated. There are many opportunities for data errors to occur during data entry, archiving, data transfer or linkage between systems. Data may need to be pre-processed and cleaned prior to the application of any data analytics. Data cleansing relates to the identification of and correcting erroneous inputs and removing or making missing data consistent. There is a need to access source data prior to applying any mathematical algorithm used for data processing as is required to generate new information and knowledge. The application of quality measurement tools may not require the same rigor as is required for various research studies, but care must be taken to ensure the results produced have meaning and consistency to enable trend analysis.

Secondary use of data includes the use of 'big data' in healthcare, a growth area with great potential to provide new useful insights into various health conditions, healthcare delivery and health outcome domains. Secondary data use requires an infrastructure that enables accurate data processing to produce valuable evidence. The ability to make good use of health data for secondary purposes requires sound management of data volume, variety, velocity (frequency of production), veracity (data uncertainty) and value by ensuring accuracy, data integrity, and semantic interpretation.

Herman and Williams have created a categorization of big data use in healthcare and published their systematic review used to develop these [97]. They found four high level categories; administration and delivery, clinical decision support (with a sub-category of clinical information), consumer behavior and support services. This provides a suitable baseline from which to assess the proliferation of the use of big data in healthcare.

Secondary data use includes the use of what is often referred to as 'big data'. There are numerous research entities concerned with measuring various health related issues. For example the Institute for Health Metrics and Evaluation (IHME), an independent population health research centre for health trends and forecasts, provides rigorous and comparable measurement of the world's most important health problems and evaluates the strategies used to address them [98]. A systematic literature review of this topic concluded that the reliable use of 'big data' within any healthcare ecosystem is yet to be established [97]. This study provided a picture of the disparate and diverse uses of big data in healthcare, indicating that its use is not systematic but opportunistic. In addition, it identified the need to question the errors in such data and data sources, as well as the metrics with which the use of big data can be measured. Access to big data must be adopted as an aspect of healthcare, health informatics and data science.

Current quality big data issues associated with secondary use of data were identified during an interview with the Victorian Comprehensive Cancer Centre's Head of Research Development [99]. One Australian example concerned a study that made use of data from 10 Victorian hospitals concerning 40 million admitted episodes which were organized based on ICD codes. Analytics were biased, as many cases are treated exclusively in outpatients, or a mix of inpatient and outpatient care. Admitted episodes only code the primary reason for admission, such as cataract extraction, however many such patients have any number of co-morbidities such as diabetes that aren't coded. Data coding and/or data abstraction refers to a process of taking away or removing characteristics from something (a concept) in order to reduce it to a set of essential characteristics. Such new data presentations loose their power of use due to such simplification. This is especially of concern when applied to clinical data as such abstraction may result in a loss of meaning.

Codes set up for payment purposes are not well suited for use in clinical records. This example highlights the need for quality data collection at the source, to enable meaningful use of big data. Other issues identified included:

• limitations associated with the ability to link data from the Victorian State-wide registry for cancer, due to difficulties encountered when there is a need to match patient identity, as Australia does not allow anyone other than the Australian Government to know the unique identifiers, established for the MyHealthRecord initiative. Every healthcare provider has their own system for identification requiring demographic data from multiple providers for one patient to be mapped, to consolidate their health/medical record.

- The potential value of 'big data' can only be realized if analytics are able to match data required to answer specific research questions around specified co-horts. This is difficult to achieve in the absence of data standards compliance.
- Clinical trials on the other hand make use of small samples, but need to undertake very granular (deep) analytics. This impacts the degree of outcome accuracy if and when applied to the general real-world patient population.
- Current practice means that only funded research can pay for data collection which can be very costly in the current environment. Adoption of data-centric systems significantly reduces such data collection and data cleaning costs.
- Clinical research requires a good match between multidisciplinary skill sets and research questions.
- Clinicians are time poor and will only invest their time if there is a return in value for them. Reaching agreement on outcome measures to be monitored within and between specialties/disciplines at one or more organizational levels can be very time consuming.

These comments were supported by a report from the newly established Digital Health Cooperative Research Centre. It noted that in spite of the abundance of digital data held in Australia, health and medical researchers often spend several months and even years assembling data required for their research, severely impacting scientific advances.
This is due to a myriad of legislative, ethical and other barriers resulting in long delays to access and the inefficient use of data stored in various databases. Issues identified were; health service delivery fragmentation across primary, secondary, in-patient and allied healthcare settings, complex funding and ethical approval processes and ad hoc policies and data government strategies that differ between and across State, Territory and Federal data custodians. The existing processes lack consistency and transparency.

Better access to health data for researchers could save Australia $3 billion and improve the health of its population over the next 15 years [100]. This requires transparency and clarity regarding all data policies and processes. This is vital to building a trusted environment and a nationally consistent, streamlined approach to data governance at every point of data collection and storage. Australia is well behind other developed nations, such as the United Kingdom and the United States of America, who have recognized the opportunities afforded by coherent data policies.

> *Only through embracing the expanded role of data in the health and medical research land-scape and the support of a national mandate at the highest level can we achieve our shared goals of supporting the health system, delivering world class healthcare and ensuring eco-nomic sustainability and success [100].*

## Health quality measurement issues

This chapter has explored a broad scope of different perspectives associated with quality, the measurement of health service quality and health system performance. This included a focus of data quality, standards and governance. Individual countries, professional disciplines, unique health programs and numerous consumer groups have developed their own outcome/performance indicators and measurement frameworks to identify, define, measure and monitor health outcomes. These ultimately determine data collection requirements. There are few if any standards that cross these domain boundaries, yet there tends to be one conceptual high level voice regarding desired outcomes, as reflected in some of the metadata, but no commonly accepted standard definitions, nor consistency regarding data collections from one year to the next. This is limiting our ability to learn from each other's experiences or to acquire large, timely, reliable and big data bases for more sophisticated data analytics. The next two chapters summarize what has been covered in our quest to document how leaders in health care, both locally and globally can ensure that every health system and health service organization has the capacity to care.

The final chapter concludes with a future vision on how health services can maximize their capacity to care within a digital health ecosystem.

## *References*

[1] Braithwaite J. Changing how we think about healthcare improvement. BMJ 2018;361:k2014.
[2] Delamont A. How to avoid the top seven nursing errors. Nursing made Incredibly Easy 2013; 11(2):8–10.
[3] IOM. Crossing the global quality chasm: improving health care worldwide. Washington DC: Institute of Medicine, The National Academies press (US); 2018.
[4] Bender M, Spiva L, Su W, Hites L. Organising nursing practice into care models that catalyse quality: a clinical nurse leader case study. J Nurs Manag 2018;26(6):653–62.
[5] De Groot K, Triemstra M, Paans W, Francke AL. Quality criteria, instruments, and requirements for nursing documentation: a systematic review of systematic reviews. J Adv Nurs 2018;75(7):1379–93.
[6] McCarthy B, Fitzgerald S, O'Shea M, Condon C, Hartnett-Collins G, Clancy M, et al. Electronic nursing documentation interventions to promote or improve patient safety and quality care: a systematic review. J Nurs Manag 2018;27(3):491–501.
[7] ASQ. Lean six sigma in healthcare n.d. [cited 2019 27 January]. Available from: http://asq.org/healthcaresixsigma/lean-six-sigma.html.
[8] Mykkanen M, Miettinen M, Saranto K. Standardized nursing documentation supports evidence-based nursing management. Stud Health Technol Inform 2016;225:466–70.
[9] Marven AC. Missed nursing care—a nurse's perspective. an exploratory study into the who, what and why of missed care. Melbourne: The University of Melbourne; 2016.
[10] Kalisch BJ, Xie B, Dabney BW. Patient-reported missed nursing care correlated with adverse events. Am J Med Qual 2013;29(5):415–22.
[11] Kalisch BJ, Landstrom G, Williams RA. Missed nursing care: errors of omission. Nurs Outlook 2009;57(1):3–9.
[12] Kalisch BJ, Landstrom GL, Hinshaw AS. Missed nursing care: a concept analysis. J Adv Nurs 2009;65(7):1509–17.

[13] Kalisch BJ, Lee H, Rochman M. Nursing staff teamwork and job satisfaction. J Nurs Manag 2010;18(8): 938–47.

[14] Kramer M, Maguire P, Brewer BB. Clinical nurses in magnet hospitals confirm productive, healthy unit work environments. J Nurs Manag 2011;19(1):5–17.

[15] de Brouwer BJM, Kaljouw MJ, Kramer M, Schmalenberg C, van Achterberg T. Measuring the nursing work environment: translation and psychometric evaluation of the essentials of magnetism. Int Nurs Rev 2014;61(1):99–108.

[16] Oshodi TO, Crockett R, Bruneau B, West E. The nursing work environment and quality of care: a cross-sectional study using the essentials of magnetism II scale in England. J Clin Nurs 2017;26(17–18):2721–34.

[17] Purdy NM. Effects of work environments on nursing and patient outcomes [Doctoral Thesis]. Ontario, Canada: The University of Western Ontario; 2011.

[18] Lee SE, Vincent C, Dahinten VS, Scott LD, Park CG, Dunn LK. Effects of individual nurse and hospital characteristics on patient adverse events and quality of care: a multilevel analysis. J Nurs Scholarsh 2018;50(4):432–40.

[19] Copanitsanou P, Fotos N, Brokalaki H. Effects of work environment on patient and nurse outcomes. Br J Nurs 2017;26(3):172–6.

[20] IOM. Keeping patients safe: transforming the work environment of nurses. Washington (DC): Institute of Medicine, National Academies Press (US); 2004.

[21] Lake ET. The nursing practice environment: measurement and evidence. Med Care Res Rev 2007;64(2 Suppl):22s–104s.

[22] Swiger PA, Patrician PA, Miltner RSS, Raju D, Breckenridge-Sproat S, Loan LA. The practice environment scale of the nursing work index: an updated review and recommendations for use. Int J Nurs Stud 2017;74:76–84.

[23] Zhang L, Tong H, Demirel HO, Duffy VG, Yih Y, Bidassie B. A practical model of value co-creation in healthcare service. Procedia Manuf 2015;3:200–7.

[24] Padgett SM. Professional collegiality and peer monitoring among nursing staff: an ethnographic study. Int J Nurs Stud 2013;50(10):1407–15.

[25] Menard KI. Collegiality, the nursing practice environment, and missed nursing care. Doctoral, University of Wisconsin-Milwaukee; 2014.

[26] OECD. The next generation of health reforms—Ministerial Statement following OECD Health Ministerial Meeting 17 January. 2017.

[27] OECD. Health Data Governance: Privacy, Monitoring and Research—Policy Brief. 2015.

[28] openEHR. n.d. openEHR ENTRY Types FAQs [cited 2018 13 September]. Available from: https://openehr.atlassian.net/wiki/spaces/resources/pages/4554768/openEHR+ENTRY+Types+FAQs.

[29] Benner P, Hughes RG, Sutphen M. Clinical reasoning, decisionmaking, and action: thinking critically and clinically. In: Hughes R, editor. Patient safety and quality: an evidence-based handbook for nurses. Rockville: Agency for Healthcare Research and Quality (US); 2008.

[30] Papathanasiou IV, Kleisiaris CF, Fradelos EC, Kakou K, Kourkouta L. Critical thinking: the development of an essential skill for nursing students. Acta Inform Med 2014;22(4):283–6.

[31] Westra BL, Latimer GE, Matney SA, Park JI, Sensmeier J, Simpson RL, et al. A national action plan for sharable and comparable nursing data to support practice and translational research for transforming health care. J Am Med Inform Assoc 2015;22(3):600–7.

[32] Min YH, Park H-A, Chung E, Lee H. Implementation of a next-generation electronic nursing records system based on detailed clinical models and integration of clinical practice guidelines. Healthc Inform Res 2013;19(4):301–6.

[33] JIGSH. n.d. Joint Initiative for Global Standards Harmonization (JIGSH) Health Informatics Document Registry and Glossary, Standards Knowledge Management Tool (SKMT) [web page]. Available from: http://www.skmtglossary.org/.

[34] PAHO, editor. Building standard-based nursing information systems. Washington DC: Pan American Health Organisation (PAHO); 2001.

[35] Min YH, Park H-A. Applicability of the ISO reference terminology model for nursing to the detailed clinical models of perinatal care nursing assessments. Healthc Inform Res 2011;17(4):199–204.

[36] Hovenga E, Garde S, Heard S. Nursing constraint models for electronic health records: a vision for domain knowledge governance. Int J Med Inform 2005;74(11–12):886–98.

[37] Park H-A, Min YH, Jeon E, Chung E. Integration of evidence into a detailed clinical model-based electronic nursing record system. Healthc Inform Res 2012;18(2):136–44.

[38] Park H-A, Min YH, Kim Y, Lee MK, Lee Y. Development of detailed clinical models for nursing assessments and nursing interventions. Healthc Inform Res 2011;17(4):244–52.

[39] Dictionary. n.d. [Web page]. Available from: http://www.dictionary.com/browse/digital.

[40] WHO. n.d. Family of International Classifications Geneva [cited 2018 8 September]. Available from: http://www.who.int/classifications/en/.

[41] Fortune T, O'Connor E, Donaldson B. Guidance on designing healthcare external evaluation programmes including accreditation. International Society for Quality in Health Care (ISQua); 2015.

[42] Kluge E-HW. Health information professionals in a global ehealth world: ethical and legal arguments for the international certification and accreditation of health information professionals. Int J Med Inform 2016;97:261–5.

[43] O'Connor S, Hanlon P, O'Donnell CA, Garcia S, Glanville J, Mair FS. Understanding factors affecting patient and public engagement and recruitment to digital health interventions: a systematic review of qualitative studies. BMC Med Inform Decis Mak 2016;16:1–15.

[44] WHO. Quality and accreditation in health care services: a Global review. Geneva: World Health Organisation; 2003.

[45] UKAS. United kingdom accreditation service london: United Kingdom Accreditation Service; n.d. [cited 2019 1 June].Available from: https://www.ukas.com/about/about-accreditation/.

[46] ISQua. n.d. International Society for Quality in Health Care Ireland [cited 2019 1 June]. Available from: https://www.isqua.org/.

[47] JCR. The Joint Commission, n.d. Standards [cited 2019 1 June ]. Available from: https://www.jointcommission.org/standards_information/standards.aspx.

[48] ACSQHC. National safety and quality health service standards. Australian Commission on Safety and Quality in Health Care; 2017.

[49] ACSQHC. Overview of the Clinical Care Standards 2019 [cited 2019 20 March]. Available from: https://www.safetyandquality.gov.au/our-work/clinical-care-standards/overview-of-the-clinical-care-standards/.

[50] Accrediation-Canada. n.d. Health and Social Services Standards [cited 2016 1 June]. Available from: https://accreditation.ca/intl-en/standards/.

[51] QHA. n.d. QHA Trent Accreditation United Kingdom [cited 2019 1 June]. Available from: https://www.qha-trent.co.uk/.

[52] ACHS. EQuIP national table—standards. Australian Council on Healthcare Standards (ACHS); 2017.

[53] openEHR. n.d. An open domain-driven platform for developing flexible e-health systems openEHR Foundation; [cited 2019 6 February]. Available from: https://www.openehr.org/.

[54] ISO. 13606-1:health informatics—electronic health record communication—part 1: reference model. ISO.

[55] IFIP-IP3. Professional IT standards: international federation of information processing (IFIP)—International Professional Practice Paetnership (IP3) [cited 2019 5 March ]. Available from: https://www.ipthree.org/gain-ip3-accreditation/ip3-accreditation-program/it-professional-standards/.

[56] ACHS. Clinical indicator program: The Australian Council on Healthcare Standards (ACHS); n.d. [cited 2019 23 March]. Available from: https://www.achs.org.au/programs-services/clinical-indicator-program/.

[57] WHO. Forum on health data standardization and interoperability. Geneva: World Health Organisation; 2012.

[58] WHO. Health Systems: World Health Organisation; n.d. [cited 2019 13 March ]. Available from: https://www.who.int/healthsystems/topics/stewardship/en/.

[59] Andronis K, Moysey K. Data governance for health care providers. In: Hovenga E, Grain H, editors. Health information governance in a digital environment. vol. 193. Amsterdam: IOS Press; 2013.

[60] Kong H-J. Managing unstructured big data in healthcare system. Healthc Inform Res 2019;25(1):1–2.

[61] DICOM. Digital Imaging and Communications in Medicine n.d. [cited 2019 23 May]. Available from: https://www.dicomstandard.org.

[62] ISO. ISO 12052:2006—health informatics—digital imaging and communication in medicine (DICOM) including workflow and data management. ISO; 2006.

[63] Piliouras TC, Suss RJ, Yu PL, editors. Digital imaging & electronic health record systems: implementation and regulatory challenges faced by healthcare providers. 2015 long island systems, applications and technology, 2015; 2015. p. 1.

[64] Wang KC, Kohli M, Carrino JA. Technology standards in imaging: a practical overview. J Am Coll Radiol 2014;11(12 Pt. B):1251–9.

[65] Healthcare-IT-News. Carestream white paper- medical data migration, [cited 14 February 2019]. Available fromIn: Rochester. Carestream Health Inc; 2018. https://pages.healthcareitnews.com/rs/922-ZLW-292/images/Carestream_whitepaper_medical_data_migrations_for_IT_Managers.pdf.

[66] Degoulet P. The virtuous circles of clinical information systems: a modern Utopia. Yearb Med Inform 2016;1:256–63.

[67] Butler M, Collins R, Drennan J, Halligan P, O'Mathúna DP, Schultz TJ, et al. Hospital nurse staffing models and patient and staff-related outcomes. Cochrane Database Syst Rev 2011;7.

[68] Kohn LT, Corrigan JM, Donaldson MS. To Err is human: building a safer health system. Washington DC: Institute of Medicine, National Academy Press; 2007.

[69] Spence Laschinger HK, Leiter MP. The impact of nursing work environments on patient safety outcomes: the mediating role of burnout engagement. J Nurs Adm 2006;36(5):259–67.

[70] Van den Heede K, Clarke SP, Sermeus W, Vleugels A, Aiken LH. International experts' perspectives on the state of the nurse staffing and patient outcomes literature. J Nurs Scholarsh 2007;39(4):290–7.

[71] Olley R, Edwards I, Avery M, Cooper H. Systematic review of the evidence related to mandated nurse staffing ratios in acute hospitals. Australian Health Review; 2018.

[72] ICHOM. International Consortium for Health Outcomes Measurement, Inc.(US) n.d. [cited 2018 31 October]. Available from: https://www.ichom.org/.

[73] ISPOR. The professional society for health economics and outcomes research n.d. [cited 2019 19 March ]. Available from: https://www.ispor.org/.

[74] NQF. National Quality Forum n.d. [cited 2019 19 March]. Available from: http://www.qualityforum.org/what_we_do.aspx.

[75] Workman T. Engaging patients in information sharing and data collection: the role of patient-powered registries and research networks. Rockville (MD): Agency for Healthcare Research and Quality; 2013.

[76] ACSQHC. The australian commission on safety and quality in health care. economic evaluation of clinical quality registries: final report. Sydney: ACSQHC; 2016.

[77] WHO. Global reference list of 100 core health indicators. Geneva: World Health Organisation; 2015.

[78] WHO. Monitoring the building blocks of health systems: a handbook of indicators and their measurement strategies, [cited 22 March 2019]. Available from:Geneva: WHO; 2010. http://www.who.int/healthinfo/systems/WHO_MBHSS_2010_full_web.pdf.

[79] Hovenga EJS. Impact of data Governance on a Nation's Healthcare System Building Blocks. In: Hovenga EJS, Grain H, editors. Health information governance in a digital environment. Amsterdam: IOS Press; 2013.

[80] WHO. Global Health Observatory (GHO) n.d. Data [cited 2019 22 March]. Available from: https://www.who.int/gho/en/.

[81] WHO. The WHO Health Systems Framework n.d. Available from: http://www.wpro.who.int/health_services/health_systems_framework/en/.

[82] OECD. Data for Measuring Health Care Quality and Outcomes 2019 [cited 2019 28 March]. Available from: http://www.oecd.org/els/health-systems/health-care-quality-indicators.htm.

[83] Braithwaite J, Hibbert P, Blakely B, Plumb J, Hannaford N, Long JC, et al. Health system frameworks and performance indicators in eight countries: a comparative international analysis. SAGE Open Med 2017;5:2050312116686516.

[84] AIHW. National health performance framework canberra: Australian Institute of Health and Welfare; n.d. [cited 2019 28 March]. Available from: https://meteor.aihw.gov.au/content/index.phtml/itemId/392569.

[85] NSWHealth. Information bulletin—KPI and improvement measure data supplement. NSW Government; 2018–19. [cited 2019 28 March]. Available from: https://www1.health.nsw.gov.au/pds/ActivePDSDocuments/IB2018_048.pdf.

[86] NHS-Digital. Information and technology for better care. Health and Social Care Information Centre Strategy; 2015. 2015–2020. [cited 2019 3 May]. Available from: https://digital.nhs.uk/article/249/Our-Strategy & www.hscic.gov.uk.

[87] Digital NHS. Data, insights and statistics n.d. [cited 2018 31 October]. Available from: https://digital.nhs.uk/data-and-information/data-insights-and-statistics.

[88] NHS-Scotland. The healthcare quality stratgey for NHS Scotland Edinburgh. Scottish Government; 2010. [cited 2019 27 March]. Available from: https://www.gov.scot/binaries/content/documents/govscot/publications/report/2010/05/healthcare-quality-strategy-nhsscotland/documents/0098354-pdf/0098354-pdf/govscot%3Adocument.

[89] PCORI. Patient-centred outcomes research institute (PCORI);2018. [cited 2018 8 September]. Available from: https://www.pcori.org/about-us.

[90] Brennan PF, Bakken S. Nursing needs big data and big data needs nursing. J Nurs Scholarsh 2015;47 (5):477–84.

[91] Westra BL, Clancy TR, Sensmeier J, Warren JJ, Weaver C, Delaney CW. Nursing Knowledge: big data science—implications for nurse leaders. Nurs Adm Q 2015;39(4):304–10.

[92] Strachan H. Person-centred caring: its conceptualisation and measurement through three instruments (Personalisation, participation and responsiveness). Glasgow, Scotland: Glasgow Caledonian University; 2016.

[93] PWC. Towards a value based approach 2018 [cited 2019 22 March]. Available from: https://www.pwc.com.au/publications/pdf/funding-thought-leadership-18apr18.pdf.

[94] Administrator. National health funding pool canberra. Australian Government; 2019. [cited 2019 22 March]. Available from: https://www.publichospitalfunding.gov.au/Media/Administrators%20Data%20Plan%202018-19%20to%202020-21.pdf.

[95] IHPA. Independent Hospital Pricing Authority Canberra n.d. [cited 2019 26 March]. Available from: https://www.ihpa.gov.au/.

[96] Queensland-Government. Health funding principles and guidelines. Brisbane: Queesland Government; 2018–2019. [cited 2019 22 March]. Available from: https://publications.qld.gov.au/dataset/service-agreements-for-hhs-hospital-and-health-services-supporting-documents/resource/88daaf6b-883e-4424-a883-f5a63717c4b5.

[97] Hermon R, Williams PAH, editors. Big data in healthcare: what is it used for?2014. Perth, WA: Edith Cowan University; 2014.

[98] IHME. n.d. Measuring what matters University of Washington: Institute for Health Metrics and Evaluation; [cited 2019 22 March ]. Available from: http://www.healthdata.org/projects.

[99] Layton M. Melbourne: Dr. Meredith Layton| Head-Research Development, Victorian Comprehensive Cancer Centre. 2017.

[100] Srinivasan U, Ramachandran D, Quilty C, Rao S, Nolan M, Jonas D. Flying blind: Australian researchers and digital health. Health data series, vol. 2. Sydney: CMCRC and Digital Health Cooperative Research Centre; 2018.

# CHAPTER 12

# Residential and community care management

## Introduction

Our focus for the previous chapters has largely been on acute in-patients. Many patients treated in the acute care sector are from an older age group; they are known to make far greater use of healthcare services than do younger groups. The demand for services for older adults will rise substantially in the coming decades putting increasing pressure on the capacity of the health workforce to deliver those services [1]. Most people now expect to live well beyond their 70s which has consequences for their ability to continue to undertake the many activities of daily living, including lifting, carrying, walking, daily hygiene activities, shopping, cooking, using technology, taking medications, managing finances, and driving. There are also serious consequences for the health industry as a whole. The size of the community and long-term residential care sector is growing in response to a rapidly accelerating aging of the world's population. Between 2005 and 2030 the population of older adults is expected to almost double [1].

Managing nursing and midwifery resources is most complex and demanding in the acute sector. The complexity of the acute healthcare sector is primarily due to the random nature of service demand as well as the many consequential rapid changes regarding caring and treatment priorities. Managing nursing resources in the community and long term residential care sector differ significantly from the care provided in the acute care sector. This poses unique challenges in different ways.

Most older people will eventually experience multiple health problems that need to be managed by national health systems, the health workforce and budget allocations [2]. Therefore it is important to include a chapter that addresses the capacity to care for this sector. All theoretical underpinnings of topics covered in previous chapters can be applied to residential and community care as well as any other unique health services where the capacity to care matters. The healthcare workforce in general receives very little geriatric training and is not well prepared to deliver the best care to older citizens [1]. This is slowly being addressed.

Meeting residential caring service needs must focus on providing support for those living with continuing health and functional impairments with the aim of maximizing functional ability,

Measuring Capacity To Care Using Nursing Data. https://doi.org/10.1016/B978-0-12-816977-3.00012-5

comfort and general wellbeing as opposed to focusing on curing, treating and providing relief from acute health related episodes. Changes in care requirements occurring for residents in long term residential care, are mostly due to a population dealing with multiple co-morbidities, geriatric syndromes and a decline in any one of multiple functional abilities due to the aging process. Such changes influence workforce needs. These changes occur at a much slower pace than is experienced in the acute sector. Residential care services consist of a variety of daily routines although models of care, funding arrangements and data collections differ significantly between types of facilities and countries.

The use of patient/resident acuity is also applicable to the residential care sector, however its use is most commonly related to funding which in turn determines its workforce capacity to care within the context of physical, social, cultural, environmental and financial constraints. Such systems are frequently the result of pragmatic activity identification often not based on sound professional knowledge or research. Most of the funding for this sector relates to Aged care services.

We have shown that achieving the capacity to care for acute care health services and the effective management of its nursing and midwifery workforce, can be achieved using patient acuity data. The same principles apply for the management of healthcare and functional ability issues in the long term residential sector. However as these facilities are homes for those residents, additional consideration needs to given to also provide a caring service to meet their lifestyle, social, spiritual and entertainment needs. This chapter aims to explore the differences between the acute and residential long-term care sectors and evaluate the implications for the capacity to care.

## Residential care environments

Residential long-term care environments are relevant for an aging population who are no longer able to manage independent living, as well as for those who are born with, or have an acquired disability in similar circumstances. Physical and culturally appropriate residential environments for these types of individuals requiring long term care are many and varied. Australian statistics indicate that 86% of those in permanent residential care on 30 June 2018 had at least one diagnosed mental health condition (most commonly depression), or behavioral problem and 52% had dementia [3]. This is in addition to declining functional capacity amongst an aging population and any number of health issues due to one or more chronic co-morbidities. These factors define the complexity associated with the provision of residential care. Prevalent variations in the knowledge and skill sets of care staff, inequity in the level of healthcare support able to be provided, the challenges associated with building good relationships with multiple stakeholders and existing funding mechanisms define the challenges to be addressed [4].

Meeting caring needs for an aging population is compounded by the presence of intellectual disabilities, mental health needs or dementia. This is especially difficult for those who are living in single person households. Personality and individual lifestyle preferences also need to be considered when evaluating the capacity to care by providing community-based support services. This means that those eligible for residential care tend to be over 75 years of age, have some form of cognitive impairment, require extensive assistance with activities of daily living and have complex needs [5]. Many countries have experienced significant increases in care demands for their long-term residents. In many instances, workforce capacity, in terms of numbers, knowledge and skill, have not been able to meet these growing demands.

The aim of caring is to prevent adverse consequences, accommodate adversity, manage disease and disability whilst minimizing adverse impacts on desired lifestyles. It is imperative that personal, interpersonal and societal factors associated with aging are also able to be addressed to ensure individuals can maintain a sense of overall comfort and wellbeing irrespective of individual ill health, disability or adversity. This usually requires teams of people, each providing a unique type of service. Team members need to be able to communicate and collaborate not only with each other but also with the resident and their significant others within the context of the resident's own social/cultural network. Caring activities provided for such individuals in long term care remain constant for much longer periods of time when compared with the acute sector.

The philosophy of care is about providing for and supporting daily living activities in accordance with desired individual lifestyles, whilst maintaining general wellbeing and comfort in culturally appropriate ways. The latest trend in service provision for this population is to provide person centred community based support services in the home and their community, to enable residents to live independently wherever and for as long as possible [5, 6]. Ideally physical and social environments available for the provision of residential care, can support desired lifestyles for most of its residents, whilst responding appropriately to the biological and physical changes associated with the aging process of people. Staff need to have access to supporting occupational health and safety devices as their work requires a considerable amount of manual handling of people which is associated with a high incidence of musculoskeletal disorders for staff. Improving resident mobility capacity also minimizes these risks for staff. Workplace environments, including staff capacity and care models, need to be constructed to benefit both residents and staff [7].

Many supportive digital devices and assistive technologies are now available to support a wellness model of care, empower individuals to prevent illness, minimize risk of adverse events and provide residents with caring solutions that are focused on health and wellbeing. Those in residential environments require others to organize a suitable device that will be of benefit to them. This includes providing advice to design or to select the most suitable device. Health care workers who have a good appreciation of functional abilities and who work closely with the residents concerned should be involved at the design or selection stage to ensure usability. Supportive work environments where health care workers work together as a collaborative team are known to have positive impacts on residents.

## *Measuring care service demand and funding mechanisms*

Our focus here is the measurement of care service demand for the aged care residential/
long term care sector as this is currently one of the largest and fastest growing type of health
service. A Cochrane systematic review found that despite an overwhelming volume of
publications suggesting that there is a distinct relationship between staffing levels in
residential aged care and outcomes of improved well-being, no evidence was found to
substantiate these claims [8]. Another literature review found no evidence of the use of
nurse-patient ratios in residential aged care [9]. These authors identified that there is a need to
further explore the impact of context, patient/resident care needs, staff experience,
education and skill mix. Nay et al. [9] concluded that the determination of staffing should take
account of resident mix, environmental design, staff expertise, model of care and other
contextual factors that influence care. It was noted that leadership, participatory
management and staff development are clearly linked to quality outcomes and staff
recruitment and retention.

The measurement of care service demand in residential facilities is based on various assessment
tools which determine a category status for each resident. These tend to relate directly to
residential status eligibility and funding options. Assessing care needs of the elderly in a
consistent and equitable manner is a priority for any Government. It is desirable to have a
national assessment framework able to refer individuals to the most appropriate funded
care program in the first instance.

Sansoni et al. identified seven types of assessments in use for aged care and evaluated a number
of assessment domains and tools in terms of their ongoing need identification properties [10].
This 2012 review identified numerous issues that need to be considered when developing an
assessment framework from which to determine eligibility for residential and community care
services, as well as for the identification of specific service and caring needs [10]. These include
depth and breadth rules regarding 'can do' vs 'does do' when determining functional capacity,
rules for periodic assessment, prioritizing service needs and care planning. Other assessment
tools are available for specific conditions such as diabetes, pain management and
palliative care.

The ability to undertake health and caring needs assessments is not only dependent on using the
right assessment tools, it also requires skill and time. Undertaking a comprehensive
assessment is a time consuming activity. Assessments should be completed prior to or
following admission, at regular intervals for follow up and adjustments, or following any
changing event such as a change in health status such as post CVA (Cardio-Vascular Accident).
Effective assessments require managerial support, a knowledgeable, suitably educated and
skilled workforce, access to the right equipment, positive staff attitudes and a cooperative
resident. Admission to a long term/residential care facility is due to an inability to meet
home care service needs, increasing frailty, chronic ill-health, dementia, cognitive

impairments or a high risk of adverse events such as delirium, falls and disability. All have complex health and care needs [11].

Assessments serve many different purposes, such as eligibility for care support or admission. There may need to be a follow up assessment to provide a more comprehensive resident profile on which to base further care requirements or a more comprehensive assessment to form the basis for a detailed care plan. The latter may require the use of multiple assessment tools such as one of many pain scales [12], the Braden scale to assess pressure sore risk [13] and/or the Barthel Index to assess self care and mobility activities of daily living [14].

Assessment tool examples include the National Framework for Documenting Care in Residential Aged Care services (NATFRAME) [15], Psychogeriatric Assessment Scales [16] plus a suite of Resident Assessment Instruments (interRAI) [17]. The interRAI assessment items are tailored to each particular care setting or purpose of use. Some of its instruments in use are very extensive, containing hundreds of assessment items, making the assessment very time consuming to use [10]. A primary United States example is the mandated Resident Assessment Instrument Minimum Data Set (RAI-MDS), that serves administrative, quality, remuneration and research purposes, developed and in continuing use by the Centres for Medicare and Medicaid Services [18]. This has since been adapted for use in other countries.

Canada uses the Resident Assessment Instrument-Minimum Data Set 2.0 (RAI-MDS 2.0) and the new interRAI Long Term Care Facilities (interRAI LTCF©) to develop their Resident Utilization Groups (RUG-III Plus) casemix system [19]. This consists of 7 categories ordered in a hierarchy from lowest to highest clinical complexity. Each category has numerous groups totalling 44 groups for all seven categories as follows:

1. Reduced physical functions — (10 groups)
2. Impaired cognition — (4 groups)
3. Behavior problems — (4 groups)
4. Clinically complex — (6 groups)
5. Special care — (3 groups)
6. Extensive services — (3 groups)
7. Special rehabilitation — (this has 5 subcategories and 14 groups)

There is an overall acceptance that resident mix by type, as identified by any one of many different classification or grouping systems, is a useful approach to determining staffing/funding requirements. Such acuity or dependency systems (case-mix) require the identification of factors applicable to individual residents as identified via each system's associated assessment and funding instrument. The Resource Utilization Groups (RUG-IV), a case-mix system originally developed in 1994 [20], has been in use by the USA's Centres for Medicare & Medicaid Services for many years [21]. This grouping methodology has seen many different versions over the years. It is based on clinical and estimated resource utilization similarities of the individuals assessed. It has been examined, and as a consequence either validated or rejected [6, 22–28]. A recent overview of international staff time measurement validation studies of the RUG-III casemix system, found it difficult to assess the accuracy of the cost measures used to evaluate the predictive validity of the algorithm [29]. Canada has developed its own version, RUG-III Plus [19].

A synthesis of international approaches to aged care funding and an evaluation of a set of options for changes to the model used in the Australian residential care sector [23] has resulted in a further major study into needs, costs and classification of residential aged care known as the Resource Utilization and Classification Study (RUCS) [30]. As a result a national classification and funding model for residential aged care, the AN-ACC has been developed [31]. This consists of 13 classes providing a mechanism to consider care required by different types of residents as follows:

| | |
|---|---|
| Class 1 | Admit for palliative care |
| Class 2 | Independent without confounding factors (CF) |
| Class 3 | Independent with CF |
| Class 4 | Assisted mobility, high cognition, without CF |
| Class 5 | Assisted mobility, high cognition, with CF |
| Class 6 | Assisted mobility, medium cognition, without CF |
| Class 7 | Assisted mobility, medium cognition, with CF |
| Class 8 | Assisted mobility, low cognition |
| Class 9 | Not mobile, higher function, without CF |
| Class 10 | Not mobile, higher function, with CF |
| Class 11 | Not mobile, lower function, lower pressure sore risk |
| Class 12 | Not mobile, lower function, higher pressure sore risk, without CF |
| Class 13 | Not mobile, lower function, higher pressure sore risk, with CF |

It is anticipated that once in use it will be possible to define staffing requirements for each of the above classes as well as best practice models of care. This tool is also intended to be used as an information tool to measure and fund outputs and to meaningfully measure resident outcomes. The AN-ACC assessment system separates assessment for funding from assessment for care planning [31]. In addition the Australian resource utilization and classification study (RUCS)'s proposed funding model has adopted the following key design principles [32]:

1. A fixed payment per day for the costs of care management and care that is shared equally by all residents.
2. A variable payment per day for the costs of individualized care for each resident based on their AN-ACC casemix class.
3. A one-off adjustment payment for each new resident that recognizes additional, but time-limited, resource requirements when someone initially enters residential care.

Additional new proposed concepts are; the use of the AN-ACC classification system, six different fixed payment per day rates to account for structural cost differences between facility types and the adoption of assigned National Weighted Activity Units (NWAUs), based on a single national care price of an NWAU of 1.00, to the adjustment payment, each AN-ACC class and each base care tariff based on an annual costing study [32]. The funding model is based on a costing study [33] that led to an understanding of differences in cost drivers between different types of facilities including size and location as well as differences resulting from seasonal effects. It is said to be clinically meaningful, easily explained, conceptually sophisticated and simple administratively.

The methodologies used to classify patients/residents into similar groups for funding purposes, as presented in the previous section, were derived in a variety of ways. Work measurement methods are rarely used for these purposes. The focus of funding instruments as detailed above, is on assessing service needs [10], not on measuring the time or knowledge/skills required to provide those services. As a consequence of adopting these methodologies, funding strategies have often been found to be inadequate and unable to ensure suitable staffing levels. Those managing these resources are often frustrated regarding their inability to provide a suitable skill mix. This has resulted in reported harm and poor outcomes for residents.

New quality regulatory standards became applicable in Australia in July 2019 as one effort to address these issues [34]. The Australian Aged Care Funding Instrument (ACFI) is no longer considered to be applicable [15]. As a consequence, the Australian government received more than 5000 submissions from aged care consumers, families, carers, aged care workers, health professionals and providers regarding adverse consequences of the current system resulting in the establishment of a Royal Commission of Enquiry into Aged Care Quality and Safety [35]. The results of the work undertaken over recent years as described in the previous section, will now form part of what needs to be considered. An interim report was provided by 31 October 2019 (https://agedcare.royalcommission.gov.au) and a final report is due no later than 30 April 2020.

The New Zealand's Ministry of Health has mandated the use of the interRAI assessment tools for all older people requiring long term disability support at home and within residential care facilities. Its Long Term Care Facility assessment tool (interRAI LTCF) is applied six monthly and used to support care planning, to report quality measures and indicators as well as for grouping people with similar needs in accordance with the Resource Utilization Groups (RUG). The latter is used as a funding tool [24]. RUG-III was comprehensively tested using Bupa New Zealand data across all its care home facilities during 2017.

Patient classification principles described in Chapter 4 equally apply to the use of RUG and residential care services generally. This type of classification is often referred to and known as 'case-mix', a popular method in use by many countries to classify patients/residents on the basis of resource usage to support health service funding mechanisms. A good casemix system should provide meaningful clinical descriptions of individuals within each group. Such profiles can then be used to identify specific service needs and treatments from which average costs can be calculated based on time, knowledge and skills required to meet these needs. Data inconsistencies across various funded programs need to be addressed to enable optimal data use and data analytics.

## Residential service work measurement methods and outcomes

Long-term residential service work measurement studies, based on substantial sample sizes within the long term residential sector, have primarily focused on the need to identify specific resident characteristics that determine human resource and funding requirements. Chosen key characteristics and their definitions, as these apply to the funding method in use, continue to be reviewed and modified. Such modifications, with a few exceptions, tend not to be based on actual work measurement studies as described in Chapter 4.

Staff time measurement data were collected by the Canadian interRAI group and the University of Waterloo via the Canadian Staff Time Resource Intensity Verification (CAN-STRIVE) study. This study's results were used to derive the new RUG-III Plus casemix index (CMI) values [36]. The methodology used for this purpose has not been published. Insufficient sample sizes to complete data for all RUG groupings were identified as a limitation of this study by another researcher and provider of long term care in Ontario who had consulted with Dr. John Hirdes, the CAN-STRIVE project leader [37].

One major Australian study made use of six typical resident profiles [38]. For each of these profiles time assessments were undertaken to arrive at a total resident nursing and personal care time per day by combining the direct and indirect nursing and personal care per intervention per resident multiplied by the frequency per shift. The lowest care needs profile time was 2.5 hours per day ranging to the highest of 5 hours per day/24 hours. These six profiles with their associated staff time allocation were then presented to seven national focus groups across the country who determined the descriptive validity of the interventions and timings.

These focus group results were complemented by a MISSCARE survey based on 922 respondents [39], and a Delphi survey that was undertaken with 102 invited experts (residential site managers), to determine resident profile changes over time, the associated staffing and skill mix changes required. Desktop modeling was then used for 200 residents representing a typical mix. This resulted in an average of 4.30 hours per resident day and a skill mix requirement of registered nurse (RN) 30%, enrolled nurse (EN) 20% and personal care workers (PCWs) 50%. This study's findings confirmed that existing staffing levels and skill mix in residential care requires improvement.

Little has changed over the years in terms of the time required to care for specified types of residents as measured by acuity. What has changed is the mix of resident types. A far greater proportion of residents in residential facilities now require complex care needs (high care intensity), and as a consequence, the average hours required for care (acuity) has increased significantly. The work sampling methodology applied to the patient assessment information system (PAIS) patient classification system explained in chapter 4 was also used in extended (residential) care during the late 1980s. These studies consisted of 47,365 sampling observations covering 2200 resident days and resulted in time values for five categories of dependency (resident profiles) with time values ranging from 2.86 hours per resident day to 5.37 hours per resident day (24 hours) for each of the six dependency categories (resident profiles). These time values exceed the more recent time values arrived at in the 2016 study. The TrendCare acuity/dependency and workload management system currently used in some Queensland residential facilities, has defined 8 resident types and 12 category timings for each resident type. Measures of hours required for care over the past ten years clearly identifies a change in the mix of resident types, more complex resident care and an increase in demand for care hours.

Another work sampling study undertaken in Victoria in the late 1980s reviewed the actual average nursing and personal care hours used per resident, relative to the expected staff requirements. This study used the Care Aggregated Module (CAM) funding model in use at that time, which was based on a Resident Classification Instrument (RCI), an additive classification model using weights derived from a zero based mean overall service need score and weighted by regression values. The hours associated with each resident category included one hour of administration time per resident per week. The funded skill mix was RN 32.5%, other nursing/caring staff 59.5% and therapy staff 8%.

A total of 540 residents were included in this study covering 3399 resident days across publicly funded (46.3%), not-for-profit (27.8%) and private (25.9%) nursing homes. A total of 59,158 work sampling observations were recorded. The results demonstrated that the RCI classification method did not consistently discriminate on the basis of nursing/caring resource usage between its categories. There were also significant differences between the three sectors. This study concluded that the RCI did not provide a reliable basis for predicting the level of individual residents' service needs.

The highest RCI funded dependency was greater than actual resource usage providing an incentive to create greater dependence. The lowest RCI funded dependence was found to be more than 42% less than actual resource usage. This funding model has had numerous changes since then. These results demonstrate that actual residential aged care staffing requirements have changed due to the change in the type of residents, with a greater number of higher dependent residents, and very few low dependent residents. Over the years it has also been recognized that it is far more cost effective to provide consumer/person centred community and home care services for as long as possible. This is resulting in a changing mix of types of residents in residential care over time. This in turn impacts on the acuity mix and the human resource skill mix needs required to provide the necessary services.

## Identifying skill mix needs

Little evidence has been found regarding the precise skill mix required to improve residential aged care and outcomes of improved wellbeing [8]. The principles presented in Fig. 1 in Chapter 5 apply equally to the aged care sector. There are specialties within this sector as well, such as palliative care, brain injury, dementia care, psychogeriatric, intellectually disabled and respite care. Carers may need to spend additional time with some of these residents to establish a good rapport, prior to them being able to successfully undertake various caring tasks, avoid abusive or disruptive resident behavior, and the use of restraints. There also needs to be a focus on skills required to support life-style, culture, hobbies and interests such as music, gardening, craft work, cooking and reading. Countries such as Australia, who continue to experience additions to its population via immigration, need to consider the cultural and language diversity of its older citizens and its aged care workforce.

Residential healthcare staff, care for those living with any number of chronic diseases and co-morbidities. The focus of the care they provide is on maintaining comfort, wellbeing and a quality of life irrespective of any number of physical, physiological, mental or behavioral limitations. This requires carers to engage with residents and to become familiar with not only their personal limitations and how to manage them, but also with their individual personalities and preferences regarding lifestyle, language and culture. This type of carer and resident engagement results in more settled and co-operative residents, improved carer job satisfaction and stability and a continuing learning environment for carers.

This model of care demonstrates that cultural awareness and carer personality, skill and competence matches are far more important than boundaries of practice between aged care workers. A creative use of skill mix that best matches the assessed care needs in this sector is highly desirable. There needs to be balance of resident centred clinical and social care. This requires a skill mix that includes not only registered staff but also all other staff who collectively provide services to benefit residents. Available community resources such as volunteers and relatives should also be considered as contributors to the variety of skills required.

## Organizational and nursing models of care

Given the trend of 'ageing in place', along with the provision of greater home and community based support services, there is a need to also consider providing day-care facilities for respite care. This enables home based carers to have daily respite from the stress of family members who, for example, exhibit aggressive, socially unacceptable or demanding behaviors or whose care is a heavy burden on carers.

Twenty four hour respite facilities are also required to share the caring burden and reduce stress on home based carers allowing them to have a significant break for a period of weeks.

The availability of such facilities does reduce the demand on long term residential care and are cost effective. This service may also be regarded as an intermediary step for those waiting for a place in a residential care facility, providing suitable support services during the transitional phases from independent living to long term residential care. The provision of different types of services does require changes in traditional staffing patterns and provider roles based on local, cultural, social and environmental needs.

The quality of care provided for residential long term care is influenced not only by staffing numbers and skill mix but also by the model of care adopted. The model of care defines how all caring activities are organized to be undertaken for each resident. Workforce stability, an adequate number of staff with a sound knowledge of individual resident care needs, ensures care continuity can be provided. Such consistency enables carers to organize their workload relative to individual resident care needs and preferences, which in turn reduces unanticipated interruptions resulting in less time spent walking from one room to another. This model of care provides carers with more time to engage and talk to the resident while completing their work.

It is desirable to group residents by type to enable them to support each other as well as to generate a sense of community with shared needs and interests. This is difficult to achieve given the numerous cognitive and behavioral problems amongst these populations. Generational differences are particularly difficult to support due to significant variations in value systems, music, hobbies and entertainment tastes and interests. There are many incidences of young people with disabilities needing to be cared for in aged care facilities which is detrimental to their ability to learn how best to maximize their independence or engage in suitable external social participation. This is where different models of care need to be employed to address such challenges.

The ability for staff to engage with residents is especially important for those residents who exhibit behavioral problems due to psychogeriatric conditions, addiction or dementia or a combination of such conditions. The ability to create calm and nurturing environments enabling appropriate responses to challenging behaviors by all concerned, requires intensive levels of support to result in more stable behaviors [5]. This is thus a pre-cursor to not having to resort to the use of any kind of restraint. In addition, one needs to consider how service provision is best organized. Bowers [40] found that nurses who organized themselves by tasks, spent less time interacting with individual residents and more time traveling between tasks. The provision of care continuity by one carer or caring team to suit one resident, was found to reduce the time necessary to complete the various service activities.

The establishment and maintenance of a comprehensive caring relationship with the resident and family, and a care plan based on a caring need assessment based on the resident's frame of reference, is highly recommended. Team nursing within long term residential facilities is where the team ideally includes a registered nurse (RN) in the actual delivery of services as described in some detail in Chapter 6. In such instances the RN can fulfill the role of assessment and care coordinator. Continuity of care can be difficult to achieve in situations where individual residents are difficult to manage. An alternative model is where lesser skilled staff categories provide all the care with RNs primarily focusing on medication administration, wound care and other technical activities. This task based approached is not as cost effective, nor does this generate the opportunity for establishing the required rapport with individual residents, a homely culture with greater staff satisfaction and a better overall quality of care.

It is important to consider innovative care models to best suit local conditions, caring and support needs. We concur with the Ontario long term care association that this requires six key ingredients [41].

1. Leadership — Leaders who believe in change and are able to implement and support it.
2. Teamwork — A team-based approach that goes beyond traditional care (e.g., supporting activities of daily living) by employing other personal support workers or making use of volunteers with a variety of different skills supporting residents' quality of life.
3. Culture — Residential facilities that demonstrate success in innovation have an embedded culture of change by empowering staff to make decisions in the best interest of individual residents.
4. Programmatic approach — New models of care are resident-centred and provide unique programming to a specific population.
5. Additional funding — Additional investments beyond provincial funding to support the new model of care. This may include the need for space re-design.
6. Quality of life and care — Where traditional long-term care has been structured and funded to primarily focus on health care and physical care, innovative models integrate and value both quality of care and quality of life outcomes. It's about alleviating boredom and loneliness. This needs to be designed to suit specific groups or types of residents.

The principles of care guiding the development of new organizational and nursing models of care are [1]:

• The health needs of the older population need to be addressed comprehensively.
• Services need to be provided efficiently.
• Older persons need to be active partners in their own care.

A literature review of various models of care concluded that features of new models showing positive results were [1]:

• Interdisciplinary team care — Providers from different disciplines collaboratively manage the care of a resident, they communicate regularly about those they care for.
• Care management — The provision of a combination of health assessment, planning, education, behavioral counseling and coordination involving the resident and their families/significant others.

- Chronic disease self-management programs — Designed to provide health information and empower patients/residents to assume an active role in managing their chronic conditions. The program may be structured and include time-limited interventions and may be led by health professionals or trained laypersons.
- Pharmaceutical management — Pharmacists provide advice about medications to patients/residents either directly or through the actions of interdisciplinary teams to encourage safe, effective use of both prescribed and over the counter medications.
- Preventive home visits — Regular visits by nurses or others to monitor health and functional status, encourage self-care and appropriate use of health services and who communicate with their patients/residents' primary care providers.
- Proactive rehabilitation — Rehabilitation therapists provide outpatient assessments and interventions to assist their patients/residents to maximize their functional autonomy, home safety and quality of life as a supplement to primary care services.
- Caregiver education and support — Designed to assist informal carers of older persons with chronic conditions by providing various combinations of information, counseling, emotional support or training.
- Transitional care — A nurse or advanced practice nurse works with the patient and where relevant formal and informal carers, to manage any short term recovery period at home or in long term residential care.

Another emerging trend is the development of purpose built facilities that enable the adoption of the small household living concept especially relevant for those living with dementia. This includes offering home cooked meals, providing the privacy of real home spaces, personal laundry and the provision of 'help on hand' when residents need it. Such homes are connected via a village square with shops, where residents can walk safely, share a coffee or enjoy time out with friends and family. Such models of care require a workforce with varied skill sets and specific rostering patterns.

## Staffing resource management

Chapter 7 has covered this topic in some detail. It includes a section that specifically applies to geriatric, disability residential services and long term rehabilitation. All principles covered apply equally to the residential and community aged care settings. The primary concern is the ability to match service demand with the number of staff and skills required at any point in time, in accordance with routines and the probability of likely unexpected scenarios. It is particularly important for staff to be able to comply with the caring needs of residents, their family and significant others as described in Chapter 11. Staffing resource management within this sector is largely driven by the funding mechanisms in use in terms of staffing numbers, rostering requirements and the ability to employ a given skill mix as well as regulatory and accreditation requirements. This sector also needs to consider the management of unpaid volunteers and informal carers.

For example the new Australian Aged Care Quality Standards [34] requires organizations to be able to demonstrate:

- the workforce is planned to enable, and the number and mix of members of the workforce deployed enables, the delivery and management of safe and quality care and services,
- workforce interactions with consumers are kind, caring and respectful of each consumer's identity, culture and diversity,
- the workforce is competent and the members of the workforce have the qualifications and knowledge to effectively perform their roles,
- the workforce is recruited, trained, equipped and supported to deliver the outcomes required by these standards,
- regular assessment, monitoring and review of the performance of each member of the workforce.

What isn't clear is how compliance is measured or what performance indicators should be used to measure outcomes. In other words there isn't an agreed set of metadata standards that could be used to measure outcomes in a digital environment. Performance data is currently used to meet reporting requirements relating directly to the funding mechanism in place as described previously.

## *Aged care workforce planning*

The nursing workforce planning principles outlined in Chapter 8 apply equally to this sector. It needs to be recognized that this sector is the second largest employer of nursing staff after the acute care hospital sector in most developed countries. Reliable workforce data collections pertinent to this sector are few and far between.

In addition, agreed data standards and collection mechanisms from the diverse range of aged care industries and occupations are absent [42]. As demand is growing at a rapid pace there is a need to plan for a rapid growth in the size and mix of the aged care workforce. Demand growth is not only determined by the increasing number of residents but also by the increasing complexity of care requirements and a growing number of single-person households, resulting in a decline in the ratio of informal carers. Our ability to meet future demand also requires an increase in overall productivity to be considered. This may be achieved via digital transformation, changes to workforce educational strategies, boundaries of practice and models of care [43]. It's important to ensure that the future workforce is well prepared for a technology driven world. Innovative and emerging approaches to care have the potential to increase staff engagement which improves care and service delivery as discussed previously.

A 2018 aged care taskforce worked with thousands of aged care stakeholders to develop numerous strategic actions, based on significant opportunities for transformational change, for an Aged care workforce skills and qualifications framework. This requires investment to improve workforce planning, implement better career pathways, build leadership across the industry at all levels, complete with strategic succession plans, attract and retain the right people with the right fit and to keep valued skills and talented people [44]. This workforce should be prepared in a manner that reflects future trends. Age care workers must meet the needs of those in residential care, as well as meeting the needs of an increasingly larger numbers of people requiring supportive home care, a precursor for residential care. Workforce planning modeling activities need to be able to make use of agreed standards applicable to both home care and residential care settings based on holistic care plans for a range of different categories of staff.

## Use of informatics, digital transformation

Chapter 9 covered numerous issues associated with digital transformation. Residential and community care is part of any nation's digital ecosystem. Data collected when individuals are assessed to determine care needs and/or eligibility for entry into a residential facility should fit with an agreed high level metadata set to better enable data sharing, linking and processing. These data determine care needs and funding possibilities but can also be used to manage workforce resource needs and allocations. The applicable regulatory and funding data collections should be automated, preferably directly from internal operational systems.

These digital transformation principles are no different from other health services. It's important for all types of patient centred health services to be connected to enable information about any individual to be accessible as and when required. Such information needs to be able to follow anyone as they move through the health system from home/ community care, to acute and residential care. Long term residents also have short term acute care needs at times. There is an increased risk of resident's information being miscommunicated or lost at every transfer point due to their complex and chronic healthcare needs [45].

There is a lot of scope for the use of any number of digital solutions, assistive technologies or supportive digital devices, such as remote monitoring services, movement sensors, alert buttons, wearables, smart home integration systems and many others. These devices are applicable to support home care and long term residential care to benefit residents and carers alike. Such devices, systems and technologies also need to be interoperable where applicable to maximize their benefits. Virtual and gaming technologies may be used for carer educational purposes such as to gain a better appreciation of what it is like to live with dementia [46]. The adoption of electronic medication management systems is especially beneficial. Ideally these are digitally connected with the local pharmacy and prescribing medical officers' systems.

Research undertaken to enhance our understanding of the role that such technological disruption and innovation are playing now and in the future has resulted in a technology roadmap for this sector [47]. This roadmap includes numerous issues to be addressed most of which are largely covered in Chapter 9. Issues such as: under-developed technological readiness within the sector, amongst the workforce and carers, and limited videoconferencing infrastructure, are particularly important and in urgent need of being addressed to enable this sector to benefit from a digital transformation. Technology in aged care involves more than assistive care, it may be viewed as an additional supportive partner.

The roadmap details six fundamental technology requirements:

1. Choice and control as this is applicable to the use of technology designed to enhance older people's quality of life.
2. Sufficient flexibility to support varying consumer needs and preferences.
3. Personalized technology integration into private homes.
4. Involvement of end-users for co-design of technology development for application in providing care and support.
5. Co-evaluation with end users of the effectiveness of technologies developed to support quality care and individual wellbeing.
6. Technology integration into aged care policy and processes.

The aged care sector must transform to technology enabled environments by applying agreed standards and protocols that define the interfaces between the different components, to achieve ecosystem wide connectivity and interoperability. Care needs to be taken regarding patient privacy and data use as device manufacturers may be collecting data, which may be viewed as digital specimens, to their 'cloud' from devices in use. Such organizations are not necessarily bound by ethical research principles or professional codes of conduct or relevant legislation. The potential for data use that is not in the residents' best interest will vary by country and their health funding mechanisms. For example the 'for profit' health sector is more likely to benefit commercially from such data use as explained by Perakslis and Coravos [48]. These authors propose a three-pronged strategy for avoiding harm and protecting the privacy of digital specimens:

1. Properly categorize digital specimens by data type or format, by level of permission such as consented or unconsented or informed but not consented, and by level of risk to the data donor.
2. Enabled by this categorization, new and more practically usable methods or consumer notification that is currently in place.
3. Consumer protections must be put in place to inform and protect the public but also to enable adequate penalties for privacy violations.

There needs to be secure business to government (funders) data exchange opportunities as a means of automating all reporting activities. Essentially there needs to be choice at all application levels, data need to be governed to ensure consistency and a reference model should be used, such as openEHR, that ensures technical consistency for all data modeling and long term sustainability at minimal cost.

An important requirement to support those receiving home care, or in long term residential care, is to connect with the acute care services, community care services and social support services using videoconferencing and telehealth services. A high level benchmark for such services is the International Code of Practice for Telehealth Services [49] which aligns with the relevant ISO standard [50]. Telehealth is an important support service for chronic disease management.

## Documentation, reporting and change management

Community and residential care providers should have access to documentation policies, developed by each organization which incorporate regulator, governing or accreditation bodies that provide facility specific guidelines. These should include the documentation of medication and incident management. Data/information collection processes used in community based settings, are currently far less extensive than in residential care settings, although they are of equal importance. Documentation represents a significant component of nursing/residential carer work. It needs to include risk assessments such as the potential for falls and an advanced care plan, detailing resident preferences for end of life care, to support risk management and decision making. An advanced care plan represents an important indicator for decisions to be made regarding a potential transfer to acute care. Staff need to have ready access to the relevant information prior to taking definitive action to ensure the resident's preferences are adhered to [51].

Documentation requirements within the community and long term residential sector are extensive at the time of and immediately following admission. Documentation requirements are much less, on a daily basis, when the 'exception reporting' methodology is used. It is important that any variation in a resident's condition is recorded by carers, including any action taken, such as, who an issue/incident was reported to. More detailed documentation is usually associated with the timing of reporting requirements for funding and accreditation purposes. In addition, a standard set of information must be accessible to every carer at all times. This ensures they can be well informed and able to communicate care needs effectively at handover times to staff unfamiliar with the resident or following any incident, to ensure appropriate care continuity. The fundamental principles as presented in Chapter 10 are equally valid to meet these requirements in this sector.

Assessments undertaken to determine service eligibility and/or for care planning purposes, require extensive documentation, as do the use of funding mechanisms and associated reporting. An Australian study found that close to 70% of staff reported the need to transfer information from paper to a computer system taking a median of 30 min per shift. There was a large amount of faxing and telephone communication between facility staff and General Practitioners and community pharmacies. These high volume information exchange activities and inefficient procedures, contribute to the evidence demonstrating the need for well connected interoperable information systems [52] as discussed in Chapter 9.

One review found significant duplication of data collected across the different data collection frameworks in use [53]. A variety of aged care minimum data sets are in use for reporting purposes relative to various funding programs. In Australia these include: the Aged Care Assessment Program (ACAP), the Commonwealth Home Support Programme (CHSP), the National Respite for Carers Program (NRCP), the Veteran's Community Nursing Services Program used in addition to the primary funding instrument used as previously discussed. Every country has its own set of data collection requirements. As a consequence, significant data capture variations exist across commercially available Aged Care client documentation systems. These authors noted that the results of their review provides an argument for, at a minimum, a national approach to information management in aged care to address multiple stakeholder information needs and more effectively support the provision of care [53].

Data capture requirements within one long term residential care facility becomes complex and time consuming when funded by a variety of programs, due to data capture inconsistencies between such programs. Data capture processes at points of care are largely designed to ensure care continuity, optimum client service provision and quality outcomes rather than to ensure optimum communication between individual care providers. Agreed metadata sets as discussed in Chapter 3 may overcome these data inconsistencies, time consuming data capture and information sharing issues.

The use of whatever technologies are in place, does need to be evaluated to ensure these technologies provide the desired value and benefits. One example of such an evaluation is regarding the use of electronic health records in residential aged care facilities that found substantial perceived benefits for staff and residents had been the result of their use [54]. The Lean Six Sigma and change management principles as discussed in Chapter 10 also apply to the aged care sector, although the barriers may be greater when dealing with older staff and entrenched routines. Nursing or Care acuity data again provides valuable information for numerous secondary data use supporting decision making at various management levels. It's beneficial to be able to make use of clinical and administrative process data for reporting purposes rather than having to capture required data separately to suit a specific purpose. This requires a suitable digital infrastructure.

## Measuring service quality

Attributes associated with quality measurement and management as discussed in Chapter 11 apply equally to the community and long term residential sectors, although the selection of performance and outcome indicators may differ to account for specific community based and long term residential care risk factors that need to be monitored to enable effective management.

One quality issue of particular concern is elder abuse by other residents, staff, family or significant others. A review of staff to resident and resident to resident abuse [55] identified a wide range of abusive behaviors. It found a lack of common understanding of what constitutes elder abuse, making it difficult to develop effective measures, interventions, institutional structural or other changes to address this quality issue. Long term residents represent a particularly vulnerable community. Elder abuse refers to any act or lack of appropriate action, that causes harm or distress to a long term aged resident. Quality measures need to take account of these types of incidents. The most common forms of abuse identified were physical, care giving neglect, psychological, verbal, sexual, financial, medication and material. Other types of abuse were referred to as loss of dignity, emotional abuse and disrespect. Abuse may refer to acts of commission as well as omission and may be confused with methods adopted to ensure resident safety, such as the use of any type of restraints, or methods adopted to feed those who are unwilling to cooperate.

There is a strong relationship between residents and their family's satisfaction rates and staff satisfaction [56]. Using these types of well designed quality measures may be sufficient to infer that elder abuse is minimal or non-existent. These two parameters are strongly associated with workforce issues such as staff numbers and skill mix. A review of aged care workforce issues identified as relevant to the quality of care delivered, highlighted the importance of improving staff to resident ratios as well as staff training, especially for non-registered carers [57]. These were considered essential to underpin a positive staff culture and supportive relationships between stakeholders, pre-cursors to staff satisfaction. Having the capacity to care ensures that necessary unscheduled tasks, such as responding to call bells and assisting residents with their toileting needs are met in a timely manner. Such activities are frequently delayed due to staff busyness and/or shortages.

Studies [58–60] have shown that not-for-profit aged care residential facilities deliver higher quality care than for-profit facilities as they have distinctly different goals. There is a significant link between high organizational and personal support for carers, the delivery of quality care, the minimization of missed care and reductions in staff turnover [60, 61]. It is therefore highly desirable to include quality indicators regarding aged care provider characteristics in accreditation standards.

Another major quality issue is the incidence of pressure injuries as older residents have aging skin, and may have numerous other predispositions such as reduced mobility, altered sensory perception, incontinence and poor nutrition. Reliance on carers to move and reposition their bodies exposes them to friction, skin tears and a variety of pressures. Prevention requires careful, slow, skilful and frequent maneuvering of residents unable to do so themselves, to reduce such occurrences. The cost of time and resources required should be considered against the cost of treating pressure ulcers, which is considerable [62]. It is highly desirable to align accreditation standard reporting measures with these types of quality measures.

## *Qualify of life-future vision*

Around the globe we are witnessing a desire to improve the quality of care for a rapidly aging population. This is a major social priority. The WHO has suggested that the adoption of the 'life course' conceptual framework will assist with the development of efficient and equitable responses to an aging population and the increasing longevity challenge [63]. This approach recognizes that all stages of a person's life are intertwined with each other, with the lives of others born in the same period, and with the lives of past and future generations. There is a global realization that our capacity to provide the many needed and desirable supportive services is in crisis. We must make significant workforce and service delivery changes as well as consider and resolve various upcoming ethical dilemmas to enable us to meet these challenges.

Not only is it necessary to consider future growth in this sector, there is also a need to consider the desires and preferences of newer generations of elderly people who are expected to demand, not only improved standard care but also a greater consideration for their quality of life, which is a difficult to measure concept [59]. The move to a much needed 'person centred' care model, along with an increasing adoption of the 'life course' approach is adding to these challenges. There is a concern that residents with multiple co-morbidities tend not to receive whole-person care of sufficient quality. Such individuals are often overtreated medically resulting in depersonalized and fragmented experiences [64]. This is where the adoption a 'life course', person centred, multidisciplinary team model of care approach is most beneficial.

Quality of life perceptions relate to the physical environment such as private rooms and geographic location, access to the outdoors and opportunities to be actively engaged in various social and purposeful activities including cooking, gardening or handywork. Other facility measures are their for-profit, not for profit or government funded status, staff engagement and/or stability as well as their capacity to not only provide quality care but also to respond to requests for assistance in a timely manner. Innovative resident programs for the alleviation of resident boredom are also measures of a quality service. The latter is very much related to individual resident characteristics' likes and dislikes as well as culture and socio-demographic factors. The mix of resident population characteristics as well as levels of acuity are expected to change over time and will require changes to be made in the models of care adopted and/or physical facilities, for example the types of meals and choices offered. Quality of life issues are multidimensional.

One study found that the ability to provide personal attention was the most amenable to improvement using facility level characteristics. This suggests more funding and increased mandatory hours for staff could improve engagement and relationships between residents and staff. This study also found that severe (high acuity) casemix lead to lower quality of life, suggesting that programs are needed to better accommodate sicker and frailer residents [65]. Longitudinal population changes need to be carefully monitored to ensure that steps are taken to enable a continuing match between workforce and service demands. This does require the adoption of agreed metadata standards.

## *References*

[1] IOM. Retooling for an ageing America: building the health care workforce. Washington DC: National Academies Press; 2008.

[2] WHO. World report on ageing and health. Geneva: World Health Organisation; 2015.

[3] AIHW. People's care needs in aged care. Canberra: Australian Institute of Health and Welfare; 2018. [cited 2019 8 May]. Available from: https://www.gen-agedcaredata.gov.au/Topics/Care-needs-in-aged-care.

[4] Dudman J, Meyer J, Holman C, Moyle W. Recognition of the complexity facing residential care homes: a practitioner inquiry. Prim Health Care Res Dev 2018;19(6):584–90.

[5] OLTCA. This is long-term care 2019. Ontario: Ontario Long Term Care Association; 2019. [cited 2019 22 April]. Available from: https://www.oltca.com/OLTCA/Documents/Reports/TILTC2019web.pdf.

[6] Parsons M, Rouse P, Sajtos L, Harrison J, Parsons J, Gestro L. Developing and utilising a new funding model for home-care services in New Zealand. Health Soc Care Community 2018;26(3):345–55.

[7] Coman RL, Caponecchia C, McIntosh AS. Manual handling in aged care: Impact of environment-related interventions on mobility. Saf Health Work 2018;9(4):372–80.

[8] Hodgkinson B, Haesler E, Nay R, O'Donnell M, McAuliffe L. Effectiveness of staffing models in residential, subacute, extended aged care settings on patient and staff outcomes. Cochrane Database of Systematic Reviews 2011. Art.No.CD006563(6).

[9] Nay R, Fetherstonhaugh D, Garratt S. Innovative workforce responses to a changing aged care environment. Melbourne: State of Victoria, Department of Health; 2011. [cited 2019 20 April]. Available from: http://www.health.vic.gov.au/agedcare/publications/pubs.htm.

[10] Sansoni J, Samsa P, Owen A, Eagar K, Grootemaat P. An assessment framework for aged care. Centre for Health Service Development, University of Wollongong; 2012.

[11] Bauer M, Fetherstonhaugh D, Winbolt M. Perceived barriers and enablers to conducting nursing assessments in residential aged care facilities in Victoria, Australia. Australian Journal of Advanced Nursing 2018;36 (2):14–22.

[12] Pain-Doctor. 15 Pain Scales -and how to find the best pain scale for you. [cited 2019 21 April]. Available from: https://paindoctor.com/pain-scales/.

[13] Braden BJ, Bergstrom N, Ball J. Braden Scale for predicting Pressure Sore Risk: Prevention Plus; [cited 2019 21 April]. Available from: http://www.bradenscale.com/.

[14] RACGP. Medical care of older persons in residential aged care facilities—The Silver Book Barthel Index. Melbourne: Royal Australian College of General Practitioners; [cited 2019 21 April]. Available from: https://www.racgp.org.au/clinical-resources/clinical-guidelines/key-racgp-guidelines/view-all-racgp-guidelines/silver-book/tools/barthel-index.

[15] Australian-Government. Suggested assessment tools for aged care funding instrument (ACFI). Canberra: Depatment of Health-Ageig and Aged Care; 2018. [cited 2019 19 April] Available from: https://agedcare.health.gov.au/aged-care-funding/residential-care-subsidy/basic-subsidy-amount-aged-care-funding-instrument/suggested-assessment-tools-for-aged-care-funding-instrument-acfi.

[16] Jorm A, Mackinnon A. Psychogeriatric assessment scales user's guide. 4th ed. Canberra: Commonwealth Department fo Health; 2016. [cited 2019 21 April]. Available from: https://agedcare.health.gov.au/sites/default/files/documents/12_2016/pas_user_guide_-_4th_edition_may_2016.pdf.

[17] interRAI. Instruments: an overview of the interRAI suite. USA: International collaborative network; 2018. [cited 2019 21 April] Available from: http://www.interrai.org/instruments/.

[18] CMS.gov. Resident assessment instruments (RAI) minimum data sets (MDS 3.0). Baltimore: USA Government; 2019. [cited 2019 10 May] Available from: https://www.cms.gov/Medicare/Quality-Initiatives-Patient-Assessment-Instruments/NursingHomeQualityInits/MDS30RAIManual.html.

[19] CIHI. Resource utilization groups version III plus: Canadian Institute for Health Information; [cited 2019 22 April]. Available from: https://www.cihi.ca/en/resource-utilization-groups-version-iii-plus.

[20] Fries BE, Schneider DP, Foley WJ, Gavazzi M, Burke R, Cornelius E. Refining a case-mix measure for nursing homes: Resource utilization groups (RUG-III). Med Care 1994;32(7):668–85.

[21] CMS. Skilled nursing facility (SNF) prospective payment system—legislative history. Centers for Medicare & Medicaid Services; 2018. [cited 2019 21 April]. Available from: https://www.cms.gov/Medicare/Medicare-Fee-for-Service-Payment/SNFPPS/Downloads/Legislative_History_2018-10-01.pdf.

[22] CIHI. Information sheet: RUG-III grouping methodology. Canadian Institute for Health Information; 2016. [cited 2019 21 April]. Available from: https://www.cihi.ca/sites/default/files/document/rug-iii_infosheet2016_2017_en.pdf.

[23] McNamee J, Poulos C, Seraji H, Kobel C, Duncan C, Westera A, et al. Alternative aged care assessment, classification system and funding models final report. Centre for Health Service Development, Australian health services research institute, University of Wollongong; 2017.

[24] Parsons J, Parsons M, Rouse P. The development of a case-mix system for aged residential care—public report. Auckland: The University of Auckland; 2017. [cited 2019 21 April]. Available from: https://media.bupa.com.au/download/544969/casemixpublicreportmay2018.pdf.

[25] Bjorkgren MA, Hakkinen U, Finne-Soveri UH, Fries BE. Validity and reliability of resource utilization groups (RUG-III) in Finnish long-term care facilities. Scand J Public Health 1999;27(3):228–34.

[26] Dellefield ME. Using the resource utilization groups (RUG-III) system as a staffing tool in nursing homes. Geriatric Nursing (New York, NY) 2006;27(3):160–5.

[27] Mueller C. The RUG-III case mix classification system for long-term care nursing facilities: Is it adequate for nurse staffing? J Nurs Adm 2000;30(11):535–43.

[28] Poss JW, Hirdes JP, Fries BE, McKillop I, Chase M. Validation of resource utilization groups version III for home care (RUG-III/HC): evidence from a Canadian home care jurisdiction. Med Care 2008;46(4):380–7.

[29] Turcotte LA, Poss J, Fries B, Hirdes JP. An overview of international staff time measurement validation studies of the RUG-III case-mix system. Health Services Insights 2019;12. 1178632919827926.

[30] Eagar K, McNamee J, Gordon R, Snoek M, Duncan C, Samsa P, et al. The Australian national aged care classification (AN-ACC)-the resource utilisation and classification study: Report 1. Australian Health Services Research Institute, University of Wollongong; 2019. [cited 2019 8 April]. Available from: https://agedcare.health.gov.au/reform/resource-utilisation-and-classification-study.

[31] Eagar K, McNamee J, Gordon R, Snoek M, Kobel C, Westera A, et al. AN-ACC: a national classification and funding model for residential aged care: synthesis and consolidated recommendations: The Resource Unitilisation and Classification Study-Report 6. Australian Health Services Research Institute, University of Wollongong; 2019. [cited 2019 8 April]. Available from: https://agedcare.health.gov.au/reform/resource-utilisation-and-classification-study.

[32] McNamee J, Snoek M, Kobel C, Loggie C, Rankin N, Eagar K. AN-ACC: a funding model for the residential aged care sector—the resource utilisation and classification study: Report 5. Australian Health Services Research Institute (AHSRI)—University of Wollongong; 2019. [cited 2019 8 April]. Available from: https://agedcare.health.gov.au/reform/resource-utilisation-and-classification-study.

[33] Westera A, Snoek M, Duncan C, Quinsey K, Samsa P, McNamee J, et al. The AN-ACC assessment model: the resource utilisation and classification study: Report 2. Australian Health Services Research Institute (AHSRI), University of Wollongong; 2019. [cited 2019 8 April]. Available from: https://agedcare.health.gov.au/reform/resource-utilisation-and-classification-study.

[34] ACQSC. Aged care quality and safety commission's standards. Australian Government Department of Health; 2018. [cited 2019 7 May]. Available from: https://www.agedcarequality.gov.au/providers/standards/standard-7.

[35] Australian-Government. Royal commission into aged care quality and safety. Canberra: Department of Health; 2019. [cited 2019 22 April]. Available from: https://agedcare.health.gov.au/royal-commission-into-aged-care-quality-and-safety.

[36] CIHI. RUG-III plus decision support guide. Ottowa, Ontario: Canadian Institute for Health Information; 2018. [cited 2019 22 April]. Available from: https://www.cihi.ca/sites/default/files/document/rug-iiiplus-decision-support-guide-draft-1.3-en_0.pdf.

[37] Akinsulie S. Healthcare management forum. [cited 2019]. Available from: https://healthcaremanagementforum.wordpress.com/2016/10/21/developing-a-framework-for-case-costing-in-long-term-care-in-ontario/; 2016.

[38] Willis E, Price K, Bonner R, Henderson J, Terri G, Hurley J, et al. Meeting residents' care needs: a study of the requirement for nursing and personal care staff. Melbourne: Australian Nursing and Midwifery Federation (ANMF); 2016. [cited 2019 20 April]. Available from: http://www.anmf.org.au/documents/reports/National_Aged_Care_Staffing_Skills_Mix_Project_Report_2016.pdf.

[39] Henderson J, Willis E, Xiao L, Blackman I. Missed care in residential aged care in Australia: an exploratory study. Collegian 2017;24(5):411–6.

[40] Bowers BJ, Lauring C, Jacobson N. How nurses manage time and work in long-term care. J Adv Nurs 2001;33 (4):484–91.

[41] Wilkinson A, Haroun V, Cooper N, Chartier C. Long-term care plus: realizing innovative models of care for the future. Ontairao Long Term Care Association (OLTCA); 2018.

[42] Commonwealth-of-Australia. Future of Australia's aged care sector workforce. Canberra: Community Affairs References Committee; 2017. [cited 2019 7 May]. Available from: https://www.aph.gov.au/Parliamentary_Business/Committees/Senate/Community_Affairs/AgedCareWorkforce45/Report.

[43] Raguz A, Martin D. Inquiry into the future of Australia's aged care sector workforce. HammondCare Submission; 2016.

[44] Commonwealth-of-Australia. A matter of care, Australia's aged care workforce strategy. Canberra: Aged Care Workforce Stategy Taskforce; 2018. [cited 2019 8 May]. Available from: https://agedcare.health.gov.au/sites/default/files/documents/09_2018/aged_care_workforce_strategy_report.pdf.

[45] Manias E, Bucknall T, Hutchinson A, Botti M, Allen J. Improving documentation at transitions of care for complex patients. Australian Commission on Safety and Quality in Health Care (ACSQH); 2017. [cited 2019 11 May]. Available from: https://www.safetyandquality.gov.au/wp-content/uploads/2017/06/Rapid-review-Improving-documentation-at-transitions-of-care-for-complex-patients.pdf.

[46] Dementia-Australia. Virtual reality and computer games to bring a positive change in dementia care. [cited 2019 9 May]. Available from: https://www.dementia.org.au/videos/dementia-australia-virtual-reality-and-computer-games-to-bring-a-positive-change-in-dementia-care; 2019.

[47] ACIITC. A technology roadmap for the Australian aged care sector. Adelaide: Medical Device Research Institute, Flinders University for the Aged care Industry IT Council (ACIITC); 2017. [cited 2019 8 May] Available from: http://aciitc.com.au/wp-content/uploads/2017/06/ACIITC_TechnologyRoadmap_2017.pdf.

[48] Perakslis E, Coravos A. Is health-care data the new blood? The Lancet-Digital Health 2019;1(1). e8.

[49] Telehealth-Quality-Group. International Code of Practice for Telehealth Services—A quality benchmark that will transform digital health and care. 2017 [cited 2019 9 May]. Available from: http://www.telehealth.global/download/2017-V2-INTERNATIONAL-TELEHEALTH-CODE-OF-PRACTICE-MASTER.pdf.

[50] ISO/TS13131. Health Informaics: Telehealth Services-Quality planning guidelines, 2014 [cited 2019 9 May]. Available from: https://www.iso.org/standard/53052.html.

[51] Leong LJP, Crawford GB. Residential aged care residents and components of end of life care in an Australian hospital. BMC Palliat Care 2018;17(1):84.

[52] Gaskin S, Georgiou A, Barton D, Westbrook J. Examining the role of information exchange in residential aged care work practices—a survey of residential aged care facilities. BMC Geriatrics 2012;12:40.

[53] Davis J, Morgans A, Burgess S. Information management in the Australian aged care setting: An integrative review. Health Information Management Journal 2016;46(1):3–14.

[54] Zhang Y, Yu P, Shen J. The benefits of introducing electronic health records in residential aged care facilities: a multiple case study. Int J Med Inform 2012;81(10):690–704.

[55] Radermacher H, Toh YL, Western D, Coles J, Goeman D, Lowthian J. Staff conceptualisations of elder abuse in residential aged care: A rapid review. Australas J Ageing 2018;37(4):254–67.

[56] Plaku-Alakbarova B, Punnett L, Gore RJ. Nursing home employee and resident satisfaction and resident care outcomes. Saf Health Work 2018;9(4):408–15.

[57] Wells Y, Brooke E, Solly KN. Quality and safety in aged care virtual issue: What Australian research published in the Australasian journal on ageing tells us. Australas J Ageing 2019;38(1):E1–6.

[58] Baldwin R, Chenoweth L, dela Rama M. Residential aged care policy in Australia—Are we learning from evidence? Aust J Public Adm 2015;74(2):128–41.

[59] Xu D, Kane RL, Shamliyan TA. Effect of nursing home characteristics on residents' quality of life: a systematic review. Arch Gerontol Geriatr 2013;57(2):127–42.

[60] Henderson J, Blackman I, Willis E, Gibson T, Price K, Toffoli L, et al. The impact of facility ownership on nurses' and care workers' perceptions of missed care in Australian residential aged care. Aust J Soc Issues 2018;53(4):355–71.

[61] Xerri M, Brunetto Y, Farr-Wharton B. Support for aged care workers and quality care in Australia: A case of contract failure? Aust J Public Adm 2019;1–16. https://doi.org/10.1111/1467-8500.12379.

[62] Wilson L, Kapp S, Santamaria N. The direct cost of pressure injuries in an Australian residential aged care setting. Int Wound J 2019;16(1):64–70.

[63] WHO. The implications for training of embracing a life course approach to health. Geneva: World Health Organisation; 2000. [cited 2019 12 May]. Available from: https://www.who.int/ageing/publications/lifecourse/alc_lifecourse_training_en.pdf.

[64] Shippee ND, Shippee TP, Mobley PD, Fernstrom KM, Britt HR. Effect of a whole-person model of care on patient experience in patients with complex chronic illness in late life. American Journal of Hospice and Palliative Medicine 2017;35(1):104–9.

[65] Shippee TP, Hong H, Henning-Smith C, Kane RL. Longitudinal changes in Nursing home resident–reported quality of life: The role of facility characteristics. Res Aging 2014;37(6):555–80.

# CHAPTER 13

# Current and future vision

## Global health and capacity to care

Leaders of national health systems have a responsibility to plan, govern and facilitate the creation of resources, both financial and human. The provision of education and training services for potential and incumbent health care workers at all levels is critical to meeting these responsibilities. These collective resources are then used to deliver health services, with the aim to maintain population health at the highest level, in a manner that is equitable, safe, quality assured and sustainable in the long term to enhance the country's productivity. This is about the capacity to care.

Health systems represent complex ecosystems, consisting of a great variety of stakeholders each with their own perspective in relation to health care priorities. Today's trend is for all stakeholders to focus on effective access and use (coverage) of health services, maintaining public health, preventing ill-health and injury, and providing patient centered care at every point of care delivery. Ensuring healthy lives and promoting well-being for all, at all ages is the third goal within the United Nations' sustainable development goals that are about transforming our world. Their 2030 agenda is a plan of action [1]. Every Government's health policies should reflect this transformational global vision.

The digital transformation now in progress is enabling healthcare organisations to better measure, monitor and manage health care delivery performance at every level within the health system. The measurement of capacity to deliver health care is dependent upon the adoption of agreed sets of data and technology standards as described in Chapter 9. Throughout this text we have identified the many current short comings that are inhibiting the accuracy of performance measurements now in use. Incomplete and/or inaccurate information leads to inadequate monitoring and limits the creation of new knowledge which is required to better manage and improve performance outcomes.

Financial sustainability is a key factor for sustaining health systems as global health care spending continues to increase dramatically due to an increasing annual growth rate [2]. Factors impacting health care costs include increased life expectancy and associated co-morbidities, newly emerging infectious diseases, pressure to minimize existing communicable diseases, a growth in diseases such as cancer, heart disease and diabetes, a surge in mental health

Measuring Capacity To Care Using Nursing Data. https://doi.org/10.1016/B978-0-12-816977-3.00013-7

conditions such as dementia, substance abuse and other debilitating conditions. Deloitte's global health care outlook suggests the following likely stakeholder responses:

- Use of innovative technologies and personalized programs to engage with consumers and improve patient experiences.
- A move to the forefront of data interoperability, security, and ownership.
- Radical changes to healthcare delivery sites, including who provides care, the type of care given, and how health services are delivered, with an aim of transforming to value and outcomes-based care.

There is a mal distribution of health workers globally. Nurses and midwives constitute more than 50% of the health workforce in many countries, working across almost all service delivery settings. Around 35 million nurses and midwives represent the largest group of healthcare professionals worldwide. They make a substantial contribution to health service delivery systems in primary care, acute care and community settings, yet they rarely take part in high level decision making and policy development [3, 4]. A 2001 World Health Assembly resolution (WHA54.12) established the imperative for all Member States to give urgent attention to improve nursing and midwifery in their perspective countries [5]. Global health is based on networks and partnerships that extend beyond national boundaries.

A 2016 white paper on health systems and their sustainability provides a good overview of growing issues that need to be addressed, such as an aging, technologically advanced world, creating the need to reconceptualize and redesign healthcare systems for the future [6]. Their summary of the health care paradigm shifts needed for a sustainable future reflects a vision of:

- Personalized medicine.
- Integrated two-way patient information flows.
- A person centric focus as opposed to provider and diagnosis centric.
- Community-centric as opposed to centralized hospital centric.
- Less invasive and image-based treatment options.
- Episode-based, outcome-based, predicated on the long-term condition.
- Preventing illness by promoting wellness.

Precision medicine is another growing trend. This is about guiding healthcare decisions towards facilitating the most effective evidence-based treatment and care, for any individual patient, or patient type based on genetic and biological make-up, their environment and lifestyle. This includes prevention and wellness measures through the ability to identify predisposing conditions or high-risk patients for specific conditions or diseases. This requires ready access to real world evidence. Given that the nursing and midwifery workforce is three times as large globally than any other health profession, it is imperative that their contributions to person-centered care are visible, measured and evaluated to generate scientific evidence of their contributions and workloads.

Half the world's population is online and most access the Internet daily. Personal data and cybersecurity management are major issues. The Internet of Things (IoT) usage has been growing exponentially over the last few years. The scale of IoT adoption and device interoperability has created an insecure environment that is more vulnerable to personal data leaks. Clients/patients use of external monitoring devices or Apps to connect with healthcare providers for multiple purposes is increasing. Nurses are at the forefront, educating these clients/patients regarding device use and data security measures to be taken.

Innovations are driven by the need to curtail an unsustainable rise of healthcare delivery costs, a rapid evolution of science and medicine and the digitalization of health and healthcare. The nursing and midwifery professions need to be prepared for, and need to fully participate in the digital healthcare transformation now in progress.

> *The combined advances in discovery and clinical sciences, data science and technology and their convergence through the Fourth Industrial Revolution, are paving the way for unprecedented changes, which will profoundly transform health and healthcare to become much more connected, efficient, pre-emptive, precise, democratized and affordable.* [7]

## *Nurses and midwives' unique contributions to global health*

Nurses and midwives, are highly valued members of multidisciplinary healthcare teams, and need to be able to deal with diplomatic, humanitarian, political, governmental and non-governmental issues, contribute their ideas and perspectives [8] and to add value to every person's health journey. Governments must see jobs for nurses and midwives not as a cost, but as an investment in the development of effective, sustainable healthcare and wellness [4].

A United Kingdom's government all-party Parliamentary Group on Global Health concluded that universal health coverage cannot possibly be achieved without strengthening nursing globally. This was not only about numbers of nurses and midwives, but crucially about making sure that nurses and midwives' contributions are properly understood, and about enabling them to work to their full potential [9]. This group noted that 'the nursing contribution is unique, because of its scale and the range of roles nurses (and midwives) play'. Their unique contributions were defined as:

- Intimate hands-on care,
- Professional knowledge,
- Person-centered and humanitarian values.

Nurses and midwives can play a far greater role in primary care, improving care for women in pregnancy, labor and the post-partum period, managing non-communicable diseases (chronic disease management), by working with their patients and the wider community to promote health and health literacy and in the prevention and early detection of disease. Recommendations made were to:

1. Raise the profile and status of nursing and midwifery and make it central to health policy.
2. Support plans to increase the number of nurses being educated and employed globally.
3. Develop nurse leaders and nurse leadership.
4. Enable nurses to work to their full potential.
5. Collect and disseminate evidence of the impact of nursing on access, quality and costs, and ensure that this evidence is incorporated in policy and acted upon in practice.
6. Develop nursing to have 'triple impact' on health, gender equality and economies.
7. Promote partnership and mutual learning between the UK and other countries.

Fig. 1 visually demonstrates the critical roles occupied by nurses and midwives within any national health system, the many influencing factors that enable them to work to their full capacity to care and to contribute to the desired outcomes.

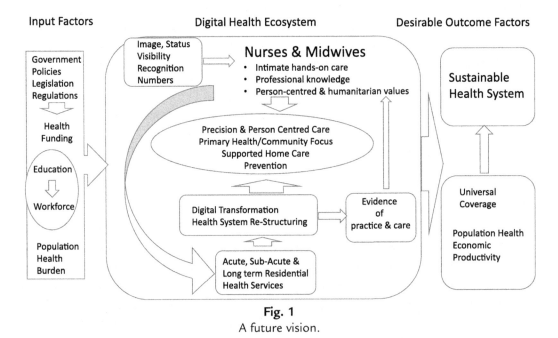

**Fig. 1**
A future vision.

In 2018 the World Health Organization and the International Council of Nurses launched the 'Nursing Now' campaign, a program managed by the Burdett Trust for Nursing, in an effort to raise the profile of nurses and midwives [10]. Its vision reflects this campaign's goal.

> *"The changing needs of the 21st Century mean nurses have an even greater role to play in the future. New and innovative types of services are needed – more community and home-based, more holistic and people-centered, with increased focus on prevention and making better use of technology. These are all areas where nurses can play a leading role. However, maximising nurses' contributions will require that they are properly deployed, valued and included in policy and decision-making."*

The Nursing Now campaign focuses on five core areas:

- Ensuring that nurses and midwives have a more prominent voice in health policymaking.
- Encouraging greater investment in the nursing workforce.
- Recruiting more nurses into leadership positions.
- Conducting research that helps determine where nurses can have the greatest impact.
- Sharing of best nursing practice.

This text contributes to some of the above recommendations through its focus on a better use of nursing data to measure any health service's capacity to care. It has explained the need to adopt value driven healthcare operations, and to accurately measure nursing and midwifery service demands. This requires an improved capacity to collect accurate data describing all resource input factors including the health status of the population at large and more specifically of individuals seeking a healthcare service.

Many hospital information systems collect clinical data at the bedside and make use of the information to improve, for example, fall prevention care. Choi and Choi [11] noted that most systems make use of administrative data not clinical nursing data. They developed a web-based Nursing Practice and Research Information Management System to process clinical nursing data that enabled them to measure nurses' delivery of fall prevention care and its impact on patient outcomes. Their pilot study led to the development of a computer algorithm based on a fall's prevention protocol. This is another example of why there needs to be greater use of nursing data.

Although the greatest global development opportunities are in primary care and public health, our focus has been on the acute sector as this represents a 24/7 all year round service and has the greatest complexity and cost in terms of nursing and midwifery human resource management. This text explains how to improve the visibility of nurses and midwives' contributions within the acute, sub-acute and long term residential (aged and disability) care settings.

Improved acute care management, a digital transformation, new technologies and innovations are expected to result in a greater transfer of patients from acute care to community and home based care. The realization of these types of changes will require new enabling legislation, effective regulation, accessible, affordable and high quality nursing education programs covering all specialties, knowledge and speciality skills to suit new roles and changed boundaries of practice. Employers need to be ready and committed to employ nurses and midwives in advanced and specialty roles.

Nursing work measurement requires a sound understanding of what nursing work consists of, as well as understanding practice theories and caring philosophies, that is, the science of nursing and midwifery that underpins nursing and midwifery practice. The ability to meet service demand requires the ability to match these needs with the right number of human resources, possessing the best possible mix of knowledge and skills and adopting the most appropriate model of care. There is a global agreement amongst the Nursing and Midwifery professions that their workload is underestimated and that workload data fails to capture the complexity of their roles. Previous chapters collectively provide a lot of information and guidance on how best to overcome this issue and enhance the visibility of nursing and midwifery contributions to any nation's health care services.

All data, information and knowledge nurses and midwives use and collect must be at the center to drive individual patient care, all associated resource management, as well as research, performance monitoring and reporting. This includes all documentation regarding patients' health status at various points in time along their health journey, treatment, care plans, all interventions, findings and observations regarding patient responses and surrounding circumstances. To maximize the use of these data, there needs to be a widespread adoption of an agreed set of standards. Investing in the nursing and midwifery workforce has a substantial positive impact on any nation's overall economic productivity.

## Our digital health ecosystem

To date the scope of digital health tends to be relative to specific international, regional, national or professional initiatives [12–29]. There is an internationally agreed vision for person-centered healthcare supported by fully integrated health systems. New strategies have been developed by most developing countries. These aim to transform infrastructures and digital health implementation approaches and to correct the way digital health has been approached by many healthcare organisations to date. There is international agreement that the primary hurdle to gaining maximum benefits of digital health is effective and semantic information exchange as explained in Chapter 9. This needs to be resolved to enable the provision of person-centric care, as the use of integrated EHRs and sound data analytics have great potential for cost savings. This has been demonstrated by Kaiser Permanente, replicated in Spain [30], and by the Canadian Health Outcomes for Better Information and Care (C-HOBIC) project that introduced systemic use of standardized clinical nursing terminology for patient assessments [31–33].

There is a requirement for national and preferably global compliance to a key set of standards. At the very least there needs to be a national health knowledge management strategy. Facilitating data use for clinical trial purposes requires international agreements and governance of key sets of data standards. There is a need for information system design and implementation pathways to be developed to reach cohesive solutions which are safer, more cost effective and technology agnostic than is currently the case. Morrow proposed the need for a new software architecture for the USA in 2013. It was noted that the resolution of interoperability issues to enable data integration across patients '*can be resolved only by establishing a comprehensive, transparent, and overarching software architecture for health information*' [34].

Countries at the forefront in the use of eHealth such as Denmark and Norway have created national bodies in order to ensure coordination and information standardization. These include the Danish Health Data Authority/Sundhedsdatastyrelsen, and the Norwegian Directorate of eHealth/Norska Direktoratet för e-helse (NDE). The Norwegian Directorate for *E*-health Research was created on the 1st January 2016, a sub-ordinate institution of the Ministry of Health and Care Services. It is responsible for the implementation of Norway's national policy on eHealth, establishing the requisite national standards and administrative use of eHealth methodology [35]. Their research and assessment center collects, produces and disseminates knowledge required by the authorities to develop a knowledge-based policy on e-health [36]. It's policy objective is *'one citizen – one record'*. By 2012 almost 100% of healthcare facilities were using EHRs, many make use of the openEHR platform discussed in Chapter 9.

Another positive European example is from the epSOS (Smart Open Services for European Patients) project that achieved technical interoperability between 17 countries. Others include the Australian Northern Territory's comprehensive shared electronic health record (EHR) based on Ocean's openEHR platform and New Zealand's national program that shares a common operating environment, a national infrastructure, secure connectivity, ICT capability and health identity information systems supported by telehealth initiatives and a set of nationally endorsed IT standards [13, 37–40].

Another example comes from Sweden, where all Swedish Counties invested in, and coordinated existing ICT projects to make them more effective and to increase their scope [41]. Joint development initiatives primarily focused on establishing a common infrastructure as a first stage. The Swedish National Information Structure developed their National Interdisciplinary Terminology. Their National Board of Health and Welfare laid the foundation for more structured, unambiguous and user-friendly documentation in both healthcare and social services. The prerequisites were created for a new Patient Data Act that provides for secure electronic information transfer between different caregivers, combined with strong protection of citizen integrity and greater scope for transparency and influence.

The national focus was on the use of eHealth to both streamline and renew Sweden's health and social care services. In some situations this required the development and deployment of entirely new eHealth services required to enable health and social care to be provided and operated in new ways. From a secondary data use perspective this strategy also included a project that has increased the availability of comparable data for quality performance reporting. Data analytics used for online reporting of patient and service providers' perceptions, facilitates free choices regarding available health care services. The long term aim is to provide quality registers with data directly from the healthcare services' health records system. The 2016 Final Report of the StandIN Project [19] provides a description of the business/operational, semantic and technical common starting points based on international standards for future healthcare information systems.

An analysis of the findings from an extensive scoping review on digital health evidence, undertaken in 2017 [42], revealed that there is an urgent need to invest in, and to prepare a global infrastructure that facilitates and supports a new digital health ecosystem [43] to;

> 'accelerate the mandate for health-data repositories, to integrate seamlessly with one another and to create a single data-exchange protocol' [17, 22, 44, 45].

A 2017 World Health Economic Forum white paper on digital transformation for healthcare notes that:

> The two expected big shifts will be disruptions to the location of care (for instance, moving care out of the hospital and closer to home) and disruptions to the type of care ('diagnose and treat' to 'prevent and manage') [17].

This means homecare will emerge as 'an important new location of care', and 'virtual care will broaden access to healthcare in rural areas, especially in emerging economies'. Such reforms will require significant legislative and policy changes, to facilitate changes in scope of practice for nurses and other health professionals, and to enable practice across jurisdictions. It is evident that progress is accelerated via Legislation, Government funded initiatives, mandatory compliance requirements, multidisciplinary cross sectional collaboration [27, 46–48] and research led by health informatics experts with a strong clinical focus to develop new solutions.

The analysis of scoping review findings concluded that:

- Successful eHealth strategies require commitment from political decision makers, healthcare leaders and stakeholders who all share the same vision and objectives.
- The effective IT adoption, use of EHRs, data use and data stewardship are best achieved when driven by Clinicians.
- Legal frameworks for data transfer between entities are essential.
- Adoption of a national communication infrastructure for healthcare permits safe, controlled health data transfer through a central hub.
- Nursing data that are compliant with nationally agreed system architectural principles, key data standards, technical and functional requirements, ensures support for and effective use of EHRs by nurses. All contract requirements for health information system development must also comply with these principles, standards and requirements.

Adverse findings were indicators of what hasn't worked well, these were:

- Integration between mobile devices and EHRs.
- Innovation take up is substantially slower in sectors dependent upon public funding.
- Information exchange between systems using different platforms.
- Incomplete data collection and transfers.
- Use of idiosyncratic clinical codes unique to a specific healthcare delivery system requiring mapping and/or translation.

- Use of non-standardized data for Clinical Decision Support systems limiting the possibility for sharing or comparing.
- Variable drug alert systems.
- Voluntary adoption of standards slowing down uptake.
- Heavy reliance on volunteers for standards development increased the time required to complete projects and has resulted in insufficient input from relevant experts.
- Continued use of legacy systems that need to connect, such as legacy patient administration systems (PAS) to new systems limiting functionality.
- Certification requirements and funding incentives resulting in some vendors only meeting specified requirements to receive payments rather than optimizing their products.
- Data inconsistencies in HL7 standards.
- Fragmentation of responsibilities between health agencies and projects.
- No data or information system standards included in Australian health agency accreditation requirements.

Some current practices may need to be abandoned, some may need to further evolve over time, others need to be accepted as lessons learned and enable new solutions to be developed. A key issue for many eHealth systems is the governance and management of health data, information, knowledge and their interaction with computing and communication technologies (information standards), connectivity and interoperability. Renewed and newly established infrastructure components need to be suitably prepared to support new healthcare delivery strategies made possible by the establishment of a global digital healthcare ecosystem. This requires a stronger focus on the use of specialized modular health software to enable integration and deployment; functions impossible to do well with the legacy systems in current use.

Storing information in a computable format in vendor neutral repositories enabling direct access to source data, have been shown to provide a source of new knowledge discovery. Such availability reduces the time required to discover both positive and negative health care treatment patterns, and to institute changed practices amongst clinicians. This work is complex and needs to be given the highest priority to achieve significant benefits. There is an urgent need not only to upskill health data scientists and overcome the dearth of expertise in this area, but also for strong collective national multidisciplinary clinical leadership, knowledge, governance and financial support for standards development and stewardship.

Kalra et al. indicated this with the following statement. This again highlights the invisibility of the nursing and midwifery professions. We need to ensure that every elaborate communication strategy includes three times as many nurses and midwives as any other health professional:

> *'An elaborate communication strategy with the scientific associations of health care providers (medical sub-disciplines, primary care physicians, allied health personnel) is needed to inform, educate and convince with regard to the necessity of semantic interoperability, well structured electronic health records, performant end user terminologies, suitable international reference and aggregation terminologies, and clinical documentation skills.'* [49]

## *Measuring health system effectiveness*

Health system performance assessments are guided by various World Health Organization publications and resolutions detailing a globally recognized direction towards universal access, patient-centered healthcare, and effective coverage [50–53]. Intermediate goals such as access, effectiveness, efficiency, acceptability and continuity may also be applied to assess any national health system's performance. Effective coverage is about use, and the quality of healthcare interventions formally defined as the fraction of potential health gain that can be delivered through a health care service such as prenatal and antenatal care and delivery, managing chronic conditions, vaccinations and child health. Current health system measurements are based on data that is routinely collected and available. As this data is often incomplete it can tend to inadequately indicate health gain. There is a need to design new performance assessment frameworks, approaches, methods and tools which focus on periodical, small size assessment programs rather than national wide evaluations [54].

Previous chapters have highlighted the dearth of useful data and reliable tools to effectively evaluate overall organizational and health system performance. There is considerable variation between healthcare organizations and information systems in use, regarding what data are collected and documentation structures generally. This limits our ability to make optimal use of data collected, limits our ability to identify priorities for action and leads to insufficient measurements of key health system functions and outputs. The Agency for Healthcare Research and Quality noted:

> *'Improved health care and lowered health care costs will be realized only if health related data can be explored and exploited in the public interest, for both clinical practice and bio-medical research 'the potential to enable a dramatic transformation in the delivery of health care, making it safer, more effective, and more efficient'* [34]

Data collections are fundamental to our ability to measure health system effectiveness. There is a need for an important expanded role for nurse leaders to be included in the strategic planning and implementation of digital health. There is a need to enable the collection of data once only at the source and to enable these data to be used many times for multiple purposes. There are essentially six types of data collections [55]:

1. Electronic health records (lifelong and episodic), including past and present clinical, pharmaceutical, nursing, patient behavior and sentiment data
2. Administrative data, including utilization and workforce data
3. Claims activity data, including all financial data
4. Disease registries
5. Health surveys
6. Clinical trials data

Better and more effective patient care requires a major digital transformation and large-scale data analytics at the point of care and for clinical trial use.

> *'... optimizing clinical trial design and execution with the EHR4CR platform would generate substantial added value for the pharmaceutical industry, as main sponsors of clinical trials in Europe, and beyond'* [24].

Another major issue identified by the 2015 HIMSS CNO-CNIO vendor roundtable [55] was the need to consider implications of health IT system version upgrades. Interdependencies and aggressive timelines create challenges for organizations, health IT suppliers and many industry partners involved with implementing systems. Collaboration and co-operation between all parties, negotiated realistic timeframes for upgrade releases and a commitment from organizations to apply upgrades within a limited time frame is essential for the effective management of system alignment within a digital ecosystem. Four key principles related to system upgrades were identified as; privacy and security of health information, data standards, interoperability and health data record immutability requiring the use of identifiers and timestamps. There were three recommendations; promote standards and interoperability, advance electronic quality measures and leverage the nursing informatics expertise.

## *Hospital performance statistics and costs*

Activity based costing methods using various casemix measures were described in Chapter 7. All data collections associated with these methodologies are used not only for financial purposes but also for performance statistical purposes. These national data collections began during the early 1990s. Australia established an Independent Hospitals Pricing Authority (IHPA) as part of its National Health Reform agenda in 2011 [56]. This Authority's key purpose is to promote improved efficiency in, and access to, public hospital service through the setting of the National Efficient Price (NEP) and National Efficient Cost (NEC) for public hospital services. This involves collecting hospital cost data nationally from hospitals using nationally consistent methods of costing hospital activity. These datasets are used for benchmarking, funding and the planning of hospital services. This dataset forms the foundation for the development of the NEP.

Hospital expenditure is split into five streams, admitted acute, emergency department, non-admitted, subacute and other products based on number of separations (discharges), presentations and service events [57]. Prices are adjusted based on numerous variables including location, indigenous status and pediatrics. Various data collection improvements have been adopted over the years providing increased confidence in the data collected and used for national reporting.

The output measures against which costs are identified are National Weighted Activity Units (NWAU) where the average public hospital service equals 1 NWAU. Each measure reflects the relative hospital resources used to account for the differences in the complexity of patients' conditions or procedures and a selection of patient characteristics. For example; a total knee replacement equals 4 NWAUs. A large component of these resource costs is nurse staffing. Nursing budgets constitute the greatest percentage of the overall resource budget in the least complex facilities. Property, plant and equipment costs are excluded from these cost measures which reflect operational running costs. Each hospital's NWAU can then be compared with other similar types of hospitals. The NWAU hospital cost between each group of similar hospitals varies considerably from, for example $3500 to $6300 across large metropolitan public hospitals in2014–15 [57]. Every hospital needs to be able to drill down and carefully examine why these differences occur.

Similar activity based costing methods are in use in other countries although it's not clear if similar benchmarking comparisons are available. Many European countries have adopted the use of Diagnosis Related Groups (DRGs) based payment systems. A 2013 review found considerable variation within and between systems [58].

A systematic review of high cost patient characteristics in the literature, that excluded the grey literature, found 55 studies however the authors were unable to assess the quality of these studies due to their methodological diversity [59]. It was found that no more than 30% of high cost patients were in their last year of life. It is evident that many hospital financial directors are under pressure to examine how best to reduce costs. From a finance perspective it is typical to work with readily available financial information containing line item expense categories such as personnel, space, equipment, supplies etc. Each of these are attractive targets as a spending reduction is expected to give immediate results.

Kaplan and Haas [60] make the point that, making such reductions without first considering the best mix of resources needed to deliver excellent patient outcomes, is counter productive. They detailed the following five common cost-cutting mistakes:

1. Cutting back on support staff — this can lower clinician productivity and raise treatment costs as professional staff will complete tasks that could be done by unqualified staff.
2. Underinvesting in space and equipment — for example if there is an inadequate number of operating rooms and infusion pumps, this can limit throughput, causing treatment delays and an increase in costs.
3. Focusing narrowly on procurement prices — not seeing the big picture of costs and underestimating human resource costs.
4. Maximizing patient throughput — if done without considering additional resources then clinicians will spend less time with individual patients and are more likely to have increased patient harm rates, increasing the overall cost.
5. Failing to benchmark and standardize clinical practice to reflect evidence based practice — can result in lower productivity, higher cost and lost opportunity for improved patient, staff and organizational outcomes.

These authors concluded that 'the only sustainable way to reduce costs is to start with an in-depth analysis of the current processes used to treat each type of patient'. This refers to the adoption of the Lean Six Sigma approach discussed in Chapter 10.

## Safe patient care vs costs

An OECD report [61] focusing on the economics associated with patient safety found that approximately 15% of total hospital activity and expenditure is a direct result of adverse events. The most burdensome adverse event types include venous thromboembolism, pressure ulcers and infections. These are nurse sensitive outcome measures and are preventable for most instances. General drivers of adverse events were identified as a lack of communication, lack of skills/knowledge, inadequate organizational culture, misaligned incentives. We have addressed these issues in previous chapters that focused on the link between patient safety and nurse and midwifery staffing levels and skill mix. The cost of prevention is typically much lower than the cost of harm in every instance [61, 62].

There has been an increasing public awareness over recent years about the high incidence of adverse events as well as the absence of necessary healthcare (care left undone) with negative impacts. This transparency has created a strong desire to ensure patient safety for all who have an encounter with the health system. Any occurrence of patient harm adds to the cost of healthcare as well as a personal cost to the patient and a potential reduction in productivity and associated personnel costs. It is important for health service executives to balance cost with patient safety in a sustainable manner and to accept that nurses and midwives spend far more time with their patients than any other type of health professional.

Nurses and midwives play critical roles in ensuring patient safety. This is achieved through monitoring patients for clinical deterioration, detecting errors and near misses, understanding care processes and local systems and performing numerous other tasks [63]. For registered nurses and midwives to achieve this requires having the right staff, with the right skills available at the right time and having adequate support staff, so that nurses and midwives can utilize their time more productively and spend more time with patients/women and babies.

## Benefits from using nursing data

Nursing data is generally very comprehensive and provides transparency. Nursing notes will indicate the need for risk management strategies to be implemented and the cause of any adverse event. Nurses are key to quality and safety and to ensuring that the best patient outcomes are achieved [64]. Nurses need access to data and they need to know how to make use of that data. Achieving Nursing Informatics competencies is highly desirable for nurses, as they need to be able to access and accurately interpret the information provided on a screen, navigate whatever EHR system is in place, appreciate the importance of data entry accuracy and value the ability of technology to support their decision making and practice. The Joint Commission expects all daily nursing work to include patient engagement, and for healthcare organizations to have developed and implemented patient-centered systems that aim to improve the quality of care and patient safety [65].

The research, development and use associated with patient acuity/nurse dependency systems over many years has resulted in new knowledge gained regarding the value of using nursing/ midwifery data. Patient acuity systems should not be viewed by health service executives as 'just a nursing or midwifery system'. They need to be a core requirement and a significant component of any healthcare provider's organizational digital transformation strategy. In summary the benefits achieved from adopting the use of electronic evidence based patient acuity/dependency and workload management system are extensive. The use of this nursing and midwifery data, for all stakeholders, is likely to result in improved staffing and significant cost savings across the organization, due to transparency and the ability to make decisions based on a single source of truth. System implementation is likely to reveal numerous previously unknown measures which can be used to improve safe staffing and increase efficiency and productivity. These include;

- The type of work completed by the nursing and midwifery teams.
- The demand for nursing and midwifery care hours by measured patient acuity.
- Nursing and midwifery work which was previously "invisible" and hence not accounted for in budgets (time spent out of the department waiting in X-ray, completing escorts, collecting blood from pathology, providing advice to other wards/services, finding equipment, etc.).
- The total number of patients cared for during the day, evening and night periods.
- Measured peaks and troughs in nursing/midwifery work for each shift on each day of the week.
- Inefficiencies due to roster patterns that do not match demand.
- Inappropriate allocation of one on one care/specialling.
- Agency and casual/bank utilization during peak and trough work periods.
- Occurrence rates and patterns for late and canceled discharges and their reasons.
- Overtime and sick leave rates and patterns for shifts of the day and days of the month.
- Inequity in staffing across wards/services when the same valid measurement tool is used for all services.

Nurses and midwives must be able to accurately interpret their data and to present the data as evidence when negotiating resources for their service. Nurses and midwives need to have the capacity to make use of evidence provided in their data and contribute to change management and re-engineering activities. Organizations should also have the capacity to measure workloads for other healthcare professionals such as allied health and medical services, and to measure other nursing and midwifery services workloads such as antenatal services, post-natal follow up care, sub-acute care, residential care, community care and outpatients services. This enables the calculation of resource usage relative to any patient's health journey. As discussed in Chapter 6 patient acuity data needs to be aligned with care and treatment plan data as documented in patient records. These types of information systems need to be accessible and incorporated in all health professional curricula.

The ready access to nursing, midwifery, allied health and medical data enables healthcare facilities to achieve improvements in productivity and efficiency by using new information and knowledge to:

- Project patient care requirements and proactively manage staffing to meet peaks and troughs in patient care demand.
- Measure, trend and reduce variances between the demand and supply of nursing and midwifery resources, in order to maintain safe and cost effective patient care. This is possible as a result of greater transparency of patient types, demand measured by acuity (hours required for care), and nursing and midwifery hours available to provide care.
- Allocate fair and equitable workloads to nursing and midwifery staff in each ward or department and across the organization. This is done according to acuity and includes the provision of a more reliable match between caregiver skills/knowledge and patient needs.
- Monitor and trend the actual staff skill mix in each nursing and midwifery department to ensure that care requirements are met and that the $ cost per hour is retained within budget.
- Reduce and expand roster profiles in accordance with increases and decreases in patient acuity and bed utilization to maximize patient safety, minimize waste and reduce nursing/ midwifery fatigue.
- Measure patient turnover (churn) on each shift for every day of the week and implement staggered shifts to manage peaks in workloads.
- Accurately re-engineer roster patterns and skill mix in line with patient needs, minimizing waste and maximizing quality patient outcomes.
- Track, report, analyze and manage all productive and nonproductive hours utilized daily and/or for any selected period, to identify service requirements and performance against budget.
- Track, trend and manage all labor hours including agency, casual, overtime, leave and absenteeism to retain a $ cost per hour that is within budgetary parameters.
- Identify and trend reasons for overtime and leave so that systems can be improved to enable overtime hours to be reduced and for leave to be better managed.
- Develop Rosters for all services in the organization with effective leave planning, and the ability to generate a large suite of reports related to "On Call", individual staff rosters, leave history, daily and weekly staffing and public holidays.
- Measure, trend and benchmark nursing/midwifery, and support worker hours utilized to provide one on one care to patients in all wards across the organization.
- Identify reasons for late and canceled patient discharges in medical and surgical wards/and implement improved systems so that patient turnover is efficient and nursing workloads are reduced.

- Identify reasons for canceled patient admissions so that strategies can be implemented to improve access and efficiency.
- Track length of stay and related variances to pre-set patient outcome goals so that processes can be implemented to reduce lengths of stay and variances to patient outcomes.
- Provide evidence of complication rates by patient, clinical pathway, speciality and doctor so that improvements can be made to reduce complication rates, improve patient outcomes and health journeys.
- Identify patient outcomes within clinical pathways and care plans, which will enable the organization to capture national clinical indicator data, and nurse/midwife sensitive indicator data, and measure organizational performance in relation to these indicators.
- Utilize computerized assessments in order to effectively identify and proactively manage all patient risks to maintain patient safety and minimize litigation.
- Maintain dynamic and comprehensive Clinical Handovers for all disciplines which can be retrospectively interrogated.
- Provide dynamic nursing, midwifery and allied health input of requests for medical actions into a comprehensive computerized Medical Handover for afterhours and weekends, based on the international standard of Situation, Background, Assessment, Recommendation (SBAR). SBAR is a reliable and validated communication standard to reduce the risk of clinical errors.
- Identify nursing/midwifery intensity per DRG/HRG, and develop realistic nursing and midwifery cost weights. Develop accurate ward budgets based on actual patient acuity data and actual nursing/midwifery resource usage.
- Record and track staff competencies, training and in-service (including mandatory requirements) and measure staff/department/organizational compliance to attendance.
- Record and manage staff registration, clinical privileges, indemnity and work clearances for all professional staff.
- Utilize a dynamic multi-disciplinary diet ordering system with extensive reporting and the ability to link to menu systems minimizing the risk of patient diet errors.
- Provide a multi-disciplinary Allied Health Intervention Register which reports on all interventions performed and interventions not completed for all patients in each ward, per hospital, discipline, patient, clinician and DRG/HRG in order to measure allied health workloads, maximize efficiency and assist with the accurate costing of Allied Health services.
- Provide a Specialist Nurse/Midwife Intervention Register which reports on all interventions per ward, hospital, discipline, patient, clinician and DRG/HRG in order to maximize efficiency and assist with the accurate costing of nursing and midwifery services.

Making nursing and midwifery data, information and knowledge readily available and accessible is not enough. It is imperative that every decision maker knows how to effectively analyze and interpret the information provided, apply their own prior knowledge and make the best possible decisions based on the evidence provided. Management is accountable for decisions they make, and has a responsibility for ensuring that the available workforce has both the necessary capability and capacity to provide quality care at all times. This includes making provision for tea breaks and meal breaks for all staff during their shift. Nurses and midwives will lose the incentive to undertake accurate data collection if they don't see any of the tangible benefits.

Our continuing research and development activities over many years, across numerous hospitals in several different countries, is now providing the best possible evidence of nurse staffing requirements relative to well defined patient types. When the evidence indicates that the required nursing or midwifery resources are not available then it is management's responsibility to take action. Fixed budgets may be considered a good reason not to take action by some, but we know that inaction has major adverse consequences. Clinical and managerial leaders need to foster a culture of professionalism and responsiveness. The adoption of a multi-disciplinary approach to problem solving enables all staff to provide safe and compassionate care.

## Optimizing our capacity to care in a sustainable health system

Our capacity to care is dependent upon many factors including:

- national health systems and how these are organized and funded,
- location of healthcare services,
- healthcare environments and cultures associated with organizational and individual providers,
- demographics and size of the demand relative to funded resources,
- available resources in terms of numbers as well as knowledge and skills,
- availability of support services, equipment, supplies and technologies,
- transport opportunities for clients/patients.

The manner by which health services are funded provides the primary foundation for the selection of data to be captured, processed and reported. Funding methodologies can also influence how organizations determine workforce planning, service demand measurements, day to day staffing and workload allocation. Variations between service types are particularly noticeable when comparing acute care funding relative to long term residential funding as explained in Chapter 12.

System conditions that are relatable to healthcare system sustainability include the following guiding principles [66] that should be adhered to by policy makers, administrators and managers.

1. Healthcare service delivery should occur in an 'open system' relating to many external stakeholders, such as; equipment and supplies manufacturers, regulators, education providers, populations etc., within its overall ecosystem as described in Chapter 9.

2. Resources that contribute to any health system's capacity to care should not be lost to the system. For example; Experienced and skillful health professionals leaving the industry prematurely, and a continuing reduction in funding levels without making other changes are both unsustainable losses.

3. Avoid accumulating costs to the health system such as increasing numbers of aged and chronically ill people endlessly trapped within acute care facilities. This will eventually clog the system and result in unsustainable delays to service access. An accumulation of excessive management reporting procedures in a system will eventually swamp the resources that should be available for other purposes.

4. Health systems need to be sufficiently adaptive so that they can respond in a timely manner to any external or internal changes without compounding possible adverse impacts. Adaptive health systems are able to respond and adapt processes in a timely manner to accommodate any change likely to influence either inputs or outputs.

5. Care for people as they are a key resource for the health system and society as a whole. For example, if the quality of healthcare is poor, then the ability of the community to actively participate in the workforce is ultimately diminished, thus reducing the size of a crucial input to the health system. Poor healthcare impacts on workforce and its tax contributions funding the health system as well as that system's workforce.

### Close the loop between resource flows into and out of the system

Part of the solution to health system sustainability is closing the loop between the flows into and out of the system and to continuously match resource availability with service demand. This requires consistent monitoring to identify trends, effective use of accurate data and the use of information and knowledge by every decision maker as well as a strong link with workforce planning activities. There must be a capacity to explore innovative designs for service delivery based on evidence, with the aim of maintaining a stable health system at all times.

This text has focused on the many and varied concepts associated with the work and services provided by nurses and midwives and their required capacity to care within the context of dynamic health care environments. Context was provided in the first chapter. A beginning exploration of issues associated with nursing workloads, identified the need to study nursing work at the unit/ward level. When measuring operational activity and efficiency there is a need to identify input criteria such as all resources and associated costs, plus the characteristics of care recipients, the many and varied healthcare service delivery processes and the end products in terms of the number of people treated/cared for and their outcome/health status following discharge.

Chapter 2 identified that data collections to measure operational activity and efficiency need to be presented in ways that enables effective use. Continuous improvements are best facilitated within learning health systems where the use of volume based statistics is being replaced by a focus on value based healthcare. This requires a comprehensive understanding of the key components facilitated by codifying best practice and identifying obstacles preventing health systems from delivering better outcomes that matter to patients and are cost effective. A major challenge is to bridge existing e-health data/information silos and facilitate the conduct of useful operational research studies that can explore relationships between demand, input, process and outcome factors. Our focus has been on the measurement of demand, input, process and outcome factors associated with the delivery of nursing and midwifery services, as this workforce is the largest group, with the most direct patient contact of the entire health workforce. Nurses and midwives must have the capacity to care.

## Nursing workload analysis

An analysis of what determines nursing workloads led to an exploration of the many and varied nurse staffing methods and nursing care demand measures in use. Common methods in use were identified as Nursing Hours Per Patient Day, Nurse Staffing Ratios and Patient Acuity/Dependency systems measuring demand and supply in Hours Per Patient Day by bed utilization. An analysis of each determined numerous inconsistencies regarding data collected, degree of accuracy, reliability and data collection methods.

Evaluating each method in a meaningful way requires a sound understanding of the many confounding variables that may need to be considered, their definitions, metadata and the many relationships between metadata items. Some variables indirectly influence a patient's/client's admission to any health service, other variables indirectly influence the length of stay or period of service and associated costs. The remaining variables are known to directly influence nursing workloads, patient outcomes and costs. Other process considerations are the efficacy of information and communication flows.

Digital environments are now technically able to better collect and process all data associated with these confounding variables and enable research to be undertaken to improve our understanding of these variables and the relationships between them. This requires the use of computable, standardized health data or 'artifacts' able to provide standardized meaning of all electronic communication undertaken within a given domain, such as within one organization, a network of healthcare organizations, nationally or globally. This, along with many other benefits, facilitates a more accurate calculation of nursing and midwifery staffing requirements by measuring actual service demand in real-time.

An examination of available nursing data standards revealed numerous nursing terminology systems and minimum data sets in use, but no agreed global or professional standard. Numerous health concept representation methods are in use and some high-level standards exist. The diversity of language, including within one language, is well recognized. It is therefore desirable to adopt an information reference model, with an associated set of data types, to represent a defined health domain, in order to maximize the benefits gained from using digital technologies. Reaching agreement is greatly assisted by the development of a set of metadata standards.

Metadata represent the 'containers' for more detailed data elements, with suitably described characteristics to reflect meaning in accordance with the FAIR (findable, accessible, interoperable, reusable) guiding principles. Metadata refers to structured information that describes, explains, locates and facilities retrieval of data for the use and/or management of an information source. Its use enables the content to be understood by both people and machines. Chapter 3 provides a suggested draft nursing domain metadata structure that reflects input, process and output metadata, based on the previously identified confounding variables. Some external nursing workload and workforce availability influencing variables, such as political, professional, managerial and industrial activities known to influence the nursing and midwifery capacity to care, are difficult to measure. Our own experiences and the literature overall has identified these variables as major obstacles frequently limiting nursing and midwiferie's capacity to care.

### Nursing and midwifery work characteristics and measurements

Many studies resulting in nursing and midwifery staffing and nursing/midwifery workload methodologies, identified in previous chapters, were based on actual nursing and midwifery work measurements. Again we identified few common features between such studies concerning the study purpose, level of detail, time investments, existing data availability, accuracy, completeness and reliability, boundaries of work measured or study scope, periods of study time, continuing use and updates over time, study costs incurred and use of work measurement data. Adopting one standard work measurement methodology in multiple wards and hospitals enables the results to be compared and aggregated, creating smaller degrees of error and facilitating the development of benchmarks. Prior to any study there is a need to analyze nursing/midwifery work in terms of what is included in that work domain.

Multiple datasets are required for any nursing or midwifery work measurement study, representing the many confounding variables. There also needs to be a recognition that all productive work consists of essential work content plus many inefficiencies that are the result of multiple factors, many of which not under the control of the worker. It is important to also consider possible ineffective outcome measures that are the result of poor practice, or insufficient resource allocation during a work measurement study period. Chapter 4 provides all details to be considered and examples of work sampling and timing studies. New technologies, whilst retaining the fundamental work measurement principles, enable more efficient and cost effective data collection and processing methods than studies reported in the literature. Methods adopted to convert work measurement data to valid, user friendly, intuitive and usable nursing and midwifery patient acuity/dependency, workload management and workforce planning systems are detailed in Chapter 4.

Scientific foundations of the nursing and midwifery disciplines guiding nursing practice, service delivery methods, and to some extent staff allocations, are presented in Chapter 6. Numerous factors influence models of care adopted as a means of distributing the workload amongst the various categories of staff. The formal visibility (or lack thereof), professional recognition and how nursing and midwifery services are valued within society, highly influence how clinical work is distributed and organized, and also impacts on the working environment culture. If nursing and midwifery care is to be accurately completed and completed on time, clinical pathways and care plans must include reference to all other planned clinical activities. Multidisciplinary plans of care provide for more collaborative decision making and co-ordinated care. We have identified through experience that collaborative models of care and small team nursing are the most effective nursing and midwifery models of care in terms of the quality of care delivered and cost effectiveness.

### The nursing and midwifery workforce

Along with technology changes and medical advances there is a growing requirement for special knowledge and skills amongst nurses and midwives. The increasing number of nursing and midwifery specialties has not been well recognized within the health system. The many and varied unique areas of knowledge and skills amongst the nursing and midwifery workforce, is not consistently acknowledged as well as it is in other health disciplines. Specialist nurses and midwives must collect their own evidence of work intensity, for work completed and for work which was not able to be done. Without this evidence the measurement of skill mix and its matching to service demand is challenging. It is important for the health workforce to be appropriately prepared to work effectively in their allocated area of practice. Suitably qualified staff in numbers that match service demand, reduce the potential of error, yet clinical speciality knowledge, skill measures and variances between supply and demand do not appear to be routinely monitored in many organizations.

We were unable to find meaningful detailed studies of this type in the literature other than studies that differentiated between registered nurses and lesser qualified nursing staff. Patient-centered and holistic caring strategies require flexibility in terms of adherence to boundaries of practice. There is a need to develop a standard set of metadata to consistently identify skill mix at the desired level of detail within any health staffing establishment. This may then be used to develop suitable educational and professional development programs, supporting succession planning and minimizing the likelihood of staff shortages in any clinical specialty.

It is highly desirable for every health professional to work at their full potential and not spend excessive amounts of time, undertaking tasks, that can be provided by staff with different sets of knowledge and skills. We have provided strategies on how to maximize the use of support staff in in-patient ward areas whilst ensuring the quality of care is maintained. This is one strategy that can be used to address existing nursing and midwifery shortages. It is imperative that nursing and midwifery support staff acquire the necessary caring knowledge and skills to assist patients with activities of daily living, and communication skills, which are fundamental for all staff interacting with people. Job evaluation processes and systems do not appear to be consistently in routine use in the health industry as is the case in many other industries.

Every registered professional discipline has its own set of criteria consisting of a required set of minimum skills and competencies that also guide educational programs. Maintaining such stringently defined scope and boundaries of practice is limiting workforce flexibility. A standard set of metadata or skills and competency framework covering clinical services for example, could be used for the development of a skill mix measurement methodology to suit nursing and midwifery as well as other health professions.

Once a skills and competency framework is available, various digital tools supporting its use can then be developed to improve overall human resource management and to ensure, that the most appropriately qualified people are matched with service demands. Such a framework enables individuals to map their skills and competencies to enable professional bodies to develop desired new career pathways, for healthcare services to identify organizational skill mix needs and for accredited training services to use as a foundation for the development of educational programs. The well established, internationally developed Skills Framework for the Information Age (SFIA) described in Chapter 5 could be used as an excellent generic starting point. Everyone working in the health industry needs to be better prepared to maximize the benefits to be gained from the digital transformation currently underway.

We have identified a variety of nurse staffing methods used nationally, that improve staffing at unit/department level and improve client/patient outcomes. These improvements are possible by utilizing best practice to develop staff establishment allocations, rostering methods and daily staffing. These practices enable future staffing demands to be predicted and allows for efficient planning and recording of staffing requirements and effective resource use. Data analysis is required to predict future demand for planning purposes, to record actual demand and resource usage and to monitor trends relative to every type of service delivered. Possible overlaps between services should also be considered when analyzing service data. A list of service demand and staffing data variables and definitions in use, as presented in Chapter 7, provides a good overview of the complexity associated with staffing resource management activities.

A number of methods in use for calculating and budgeting ward/departmental staffing need, based on various nursing workload measures in use were described. This demonstrated the variable outcomes and cost implications. The results of a very extensive international patient type hours per patient day (HPPD) benchmarking research study [67], based on data representing over five million patient days, demonstrated the HPPD range associated with 106 well defined acute care in-patient types. Service demand peaks and troughs must be identified for effective staff rostering to be achieved and to meet the 24/7 year-round service needs. Numerous rostering patterns and methodologies were described in Chapter 7 demonstrating how best to maintain continuity of care, provide quality care and to be cost effective. Effective nursing and midwifery resource management and associated financial management requires; the adoption of a data centric approach at the operational level, optimum connectivity between systems within an enterprise, and the ability to undertake data analytics.

Information about the many variable needs at operational levels within the health industry, in terms of service demand by specialty, workforce numbers, skills and competency mixes required should be collected in a standard manner nationally, to support national workforce planning activities. Turnover rates by speciality and service location vary. The lead in time to educationally prepare the workforce for specific areas of practice also vary, as do actual workforce participation rates. Health workforce planning activities to date have been inadequate due to incomplete data, data availability and a continuing simplistic view of the nursing and midwifery professions and the significance of contributions made by them to the health of any nation. There needs to be a coordinated approach to workforce planning between education providers, governments, employers and the profession. A suggested list of Metadata requirements to be considered for workforce modeling purposes is provided in Chapter 8. With the increasing clinical specializations it is imperative that workforce planning places a greater emphasis on the need to meet changing skill demands relative to changing roles by health service type.

## Digital transformation needs

Every chapter in this book has identified numerous data issues that need to be resolved if we are to maximize the benefits to be attained through the digital transformation now in progress. Health service delivery is a global endeavor. Global GPS, Internet services and various electronic social and professional networks have changed the way we work or need to work. Chapter 9 documented current practices and explored the many factors that need to be considered to produce an effective digital health ecosystem, able to effectively support the health workforce and meet service demands in a sustainable manner. There is an urgent need to adopt an agreed set of standards to facilitate system connectivity and semantic interoperability for maximum clinical data use. Current deficiencies were highlighted in every chapter.

Any digital transformation will require changes to be made to operational and related information and communication processes. The overall aim of digital transformation is to improve information and communication flows and to enhance patient journeys. Making use of Lean Six Sigma methodologies, as described in Chapter 10 is highly recommended along with a list of resource management functions that are candidates for change. Implementing these actions will support the identification of inefficiencies so that these can be addressed. This requires the evaluation of health IT currently in use and the adoption of comprehensive change management protocols. Recommendations are likely to require changes to work processes but also to the existing IT systems in use. These recommendations may form the foundation for system specifications for new systems to be developed or acquired. Improved system connectivity, along with the adoption of data standards, is expected to enable any senior member of the clinical workforce to undertake any type of data analytics, enabling them to intervene and address issues while the patient remains in their care.

Patient acuity data, the adoption of unique identifiers, data standards and governance, along with digital transformations are required to better monitor and manage the quality of services delivered, relative to patient outcomes and resource usage. Electronic nursing documentation is fundamental to finding the evidence of nursing's contribution's to patient outcomes and overall organizational performance. It is highly desirable to automate the monitoring of compliance to accreditation standards and to have systems in place that enable data to be captured once, at the source, and used for multiple purposes. Caring is an activity that every member of the health workforce needs to be able to exhibit. When measuring the capacity to care it is important to include caring outcome measures in accordance with the caring definition provided in some detail in Chapter 11. Making use of big data requires consistent use of agreed data standards and the ability to access source data to maximize the value to be obtained from undertaking data analytics.

## A future vision

We have demonstrated that nurses and midwives are making very significant contributions to health services delivered around the globe however these contributions remain largely invisible. We have shown how the nursing and midwifery professions can take steps to bring about change by embracing the current digital transformation and global policy directions towards patient-centered care, precision medicine, and the development of strategies that will enable a transfer from acute in-patient care to home and community care for many patients.

Nurses and midwives have an intimate understanding of the many operational aspects of health service delivery in any setting, as well as the many variables that determine health service demand, including contextual socioeconomic determinants of health. Nurses and midwives need to be equal participants in greater numbers at all levels of governance. Their contributions need to be supported by evidence readily available via the inclusion and use of nursing data within every health care organization. Every healthcare organization needs to embrace the many and varied solutions and innovations made possible through digital transformation. A well designed digital transformation includes the adoption and integration of nursing documentation, patient acuity and other related data into their enterprise systems.

The World Health Organization's (WHO) is providing a strong lead supporting this vision with their global strategy on digital health [68, 69], support for the Nursing Now program [10] and their commitment to strengthening the voice of nurses and midwives [5]. Digital health includes electronic health (eHealth), medical, clinical, nursing and health informatics, telemedicine/telehealth, mobile health (mHealth), the Internet of Things (IoT) and Artificial Intelligence

(AI). All of these technologies need to be able to integrate technically with each other in a manner that facilitates optimum data availability, information transfer without any loss of meaning, meaningful data linkages to explore conceptual relationships, data aggregation to produce 'big data' to suit data analytics for the provision of evidence. All of these functions are dependent upon agreed data standards, data stewardship and governance infrastructures as well as the adoption of other interoperability standards and associated legislative changes.

There needs to be greater recognition of the value associated with the availability of health data, and the many benefits that can be realized by individual healthcare organizations and the population at large, in the form of a highly functional sustainable health system. Every nation needs to consider the establishment of an 'Office of the National Data Custodian' as part of its legislative and regulatory arrangements pertaining to personal health data, data sharing, release and use, as specified in an agreed national health data framework [70]. There needs to be a differentiation between identifiable data sharing between trusted users, de-identified data and open access data. Such an office would be responsible for managing the implementation, stewardship, associated governance [71, 72] and education activities supported by accredited release authorities, research centers and educational institutions. Nursing data needs to be a primary component of any large health data set. Once collected using predefined data standards, nursing data will be able to provide significant insight in terms of evidence of cause and impact on numerous health service outcomes measures and can be used to support precision medicine.

Meanwhile the adoption and use of patient acuity data can, not only ensure equitable nursing and midwifery work distribution, but can also provide accurate nursing workload measures and provide numerous additional benefits towards achieving quality outcomes within cost efficient environments. Using nursing data for multiple purposes provides greater formal visibility of nursing and midwifery contributions. It also provides new opportunities for health executives and policy makers to gain a greater understanding and appreciation of the complexities associated with nursing and midwifery resource management. Nursing and midwifery data will provide the evidence needed by the nursing and midwifery professions to justify their staffing claims. It will also facilitate the undertaking of research by nurses and midwives to identify and value further benefits, resulting from their significant patient-centered contributions to global health.

# References

[1] UN. Transforming our world: the 2030 agenda for sustainable development, New York: United Nations; 2015. [cited 30 May 2019]. Available from: https://sustainabledevelopment.un.org/post2015/transformingourworld.

[2] Deloitte. Global health care outlook: shaping the future, [cited 30 May 2019]. Available from: https://www2.deloitte.com/global/en/pages/life-sciences-and-healthcare/articles/global-health-care-sector-outlook.html; 2019.

[3] WHO, Girardet E. Global standards for the initial education of professional nurses and midwives, World health Organisation; 2009. [cited 30 May 2019]. Available from: https://www.who.int/hrh/nursing_midwifery/hrh_global_standards_education.pdf.

[4] Crisp N, Brownie S, Refsum C. Nursing and midwifery: the key to the rapid and cost-effective expansion of high-quality universal health coverage. Doha, Qatar: World Innovation Summit for Health; 2018 [A report of the WISH Nursing and UHC forum 2018].

[5] WHO. Global strategic directions for strengthening nursing and midwifery 2016–2020, Geneva; 2016.

[6] ISQua. Health systems and their sustaionability: dealing with the impending pressures of ageing, chronic and complex conditions, technology and resource constraints. Ireland: The International Society for Quality in Health Care; 2016 [cited 1 June 2019].

[7] Wolrd-Economic-Forum. Health and healthcare in the fourth industrial revolution: global future council on the future of health and helathcare 2016–2018, [cited 2 June 2019]. Available from: http://www3.weforum.org/docs/WEF__Shaping_the_Future_of_Health_Council_Report.pdf; 2019.

[8] Preto VA, Batista JMF, Ventura CAA, Mendes IAC. Reflecting on nursing contributions to global health. Rev Gaucha Enferm 2015;36:267–70.

[9] APPG. Triple impact: how developing nursing will imporve health, promote gender equality and support economic growth. London: All-Party Parliamentary Group on Global Health; 2016 [A Report by the All-Party Parliamentary Group on Global Health].

[10] WHO. Health workforce: nursing now campaign, [cited 31 May 2019]. Available from: https://www.who.int/hrh/news/2018/nursing_now_campaign/en/; 2018.

[11] Choi J, Choi JE. Enhancing patient safety using clinical nursing data: a pilot study. Stud Health Technol Inform 2016;225:103–7.

[12] Infoway. Empowering Canadians: renewing pan-Canadian collaboration to deliver the next wave of digital health innovation—summary corporate plan, Canada; 2016.

[13] Government N-Z. Digital health 2020—overview. Health, New Zealand: Ministry of Health; 2016.

[14] Government Q. eHealth investment strategy. Queensland: State of Queensland (Qld Health); 2015. p. 64.

[15] Government A. Australia's cyber security strategy: enabling innovation, growth & prosperity. In: Cabinet DotPMa. Canberra: Commonwealth of Australia; 2016.

[16] World-Economic-Forum. Advancing cyber resilience—principles and tools for boards. Geneva: World Economic Forum; 2017.

[17] World-Economic-Forum. Digital transformation of industries healthcare industry—white paper, Geneva; 2017.

[18] Government V. Digitising health how information and communications technology will enable person-centred health and wellbeing within Victoria. Victorian Government; 2016. p. 25.

[19] Skjöldebrand AL, Nordgren H, Sandell R, Johannesson R, Gillespie C, Ehn G, et al. Common framework of standards for interoperability and change management final report—StandIN project, Phase 1; 2016.

[20] OECD. The next generation of health reforms—Ministerial Statement following OECD Health Ministerial Meeting 17 January; 2017.

[21] Sauermann S, Forjan M, Herzog J, Urbauer P, Frohner M, Pohn B, et al. eHealth strategies-scientific review, Wien: University of Applied Sciences; 2016. [cited 15 June 2015]. Available from: https://healthy-interoperability.at/fileadmin/downloads/D_eHealthresearchreport_201605_V01.00.pdf.

[22] World-Economic-Forum. Shaping the future implications of digital media for society valuing personal data and rebuilding trust: end-user perspectives on digital media survey: summary report—white paper, Geneva; 2017.

[23] ACSQHC. The Australian commission on safety and quality in health care. Economic evaluation of clinical quality registries: final report. ACSQHC: Sydney; 2016.

[24] Beresniak A, Schmidt A, Proeve J, Bolanos E, Patel N, Ammour N, et al. Cost-benefit assessment of using electronic health records data for clinical research versus current practices: Contribution of the Electronic Health Records for Clinical Research (EHR4CR) European project. Contemp Clin Trials 2016;46:85–91.

[25] De Moor G, Sundgren M, Kalra D, Schmidt A, Dugas M, Claerhout B, et al. Using electronic health records for clinical research: the case of the EHR4CR project. J Biomed Inform 2015;53:162–73.

[26] Dupont D, Beresniak A, Sundgren M, Schmidt A, Ainsworth J, Coorevits P, et al. Business analysis for a sustainable, multi-stakeholder ecosystem for leveraging the Electronic Health Records for Clinical Research (EHR4CR) platform in Europe. Int J Med Inform 2017;97:341–52.

[27] ONC. A shared nationwide interoperability roadmap v1.0. Office of the National Coordinator for Health Information; 2014.

[28] ONC. A 10-year vision to achieve an interoperable health IT infrastructure. Office of the National Coordinator for Health Information; 2014.

[29] NHS. Five year forward view. NHS England; 2014.

[30] González González AI, Miquel Gómez AM, Morales DR, Sierra VB, Romero AL, Rodilla JM. The development of a successful integrated care model/El desarrollo de un exitoso modelo de atención integrada. Int J Integr Care 2015;15(8):158–60.

[31] C-HOBIC. Canadian health outcomes for better information and care project, Canada, [cited 22 August 2018]. Available from: https://c-hobic.cna-aiic.ca/about/default_e.aspx; 2018.

[32] CIHI. Inclusion of nursing-related patient outcomes in electronic health records, Canadian Institute for Health Information; 2018. Available from: https://www.cihi.ca/en/c-hobic-infosheet_en.pdf.

[33] White P, Nagle L, Hannah K. Adopting national nursing data standards in Canada. Can Nurse 2017;113(3):18–22.

[34] McMorrow D. A robust health data infrastructure. Agency for Healthcare Research and Quality; 2013.

[35] NDE. Norwegian Directorate of eHealth (NDE), a sub-ordinate institution of the Ministry of Health and care Services. Oslo: Norwegian Government; [cited 1 June 2019]. Available from: https://ehelse.no/english.

[36] Norwegian Centre for e-Health Research. [cited 1 June 2019]. Available from: https://ehealthresearch.no/en/; 2016.

[37] Atalag K, Yang HY, Tempero E, Warren JR. Evaluation of software maintainability with a comparison of architectures. Int J Med Inform 2014;83(11):849–59.

[38] Park Y-T, Atalag K. Current national approach to healthcare ICT standardization: focus on progress in New Zealand. Healthc Inform Res 2015;21(3):144–51.

[39] Reid C, Osborne G. Strategic assessment: establishing the electronic health record. Ministry of Health: New Zealand; 2016.

[40] Bowden T, Coiera E. Comparing New Zealand's 'Middle Out' health information technology strategy with other OECD nations. Int J Med Inform 2013;82(5):e87–95.

[41] Swedish-Government. In:Affairs MoHaS, editor. National eHealth–the strategy for accessible and secure information in health and social care. Sweden; 2010.

[42] Hovenga E. Scoping review: digital health evidence, unpublished report prepared for the Australian digital health agency. EJSH Consulting Services: Melbourne; 2017.

[43] United-Nations. Government information systems: a guide to effective use of information technology in the public sector of developing countries, New York; 1995.

[44] WHO. Forum on health data standardization and interoperability. Geneva: World Health Organisation; 2012.

[45] World-Economic-Forum. Digital transformation of industries: digital enterprise—economic forum white paper in collaboration with accenture; 2016.

[46] ONC. Report to congress on health IT progress. USA: Office of the National Coordinator for Health Information Technology (ONC); 2016. p. 2016.

[47] ONC. Health IT enabled quality improvement. Office of the National Coordinator for Health Information Technology; 2014.

[48] ONC. Health IT certification program: enhanced oversight and accountability 81 FR 72404; 2016.

[49] Kalra D, Schulz S, Karlsson D, Stichele RV, Cornet R, Rosenbeck-Gøeg K, et al. Assessing SNOMED CT for large scale eHealth deployments in the EU: ASSESS CT recommendations; 2016.

[50] WHO. Global diffusion of eHealth: making universal health coverage achievable. Report of the third global survey on eHealth. Geneva: WHO; 2016.

[51] WHO. The WHO health systems framework, Available from: http://www.wpro.who.int/health_services/health_systems_framework/en/.

[52] WHO. Global strategy on people-centred and integrated health services, interim report, World Health Organisation; 2015. [cited 24 November 2018]. Available from: https://apps.who.int/iris/bitstream/handle/10665/155002/WHO_HIS_SDS_2015.6_eng.pdf.

[53] WHO. Background paper for the technical consultation on effective coverage of health systems, Geneva: World Health Organsaition; 2001. [cited 25 November 2018]. Available from: http://citeseerx.ist.psu.edu/viewdoc/download?rep=rep1&type=pdf&doi=10.1.1.111.1239.

[54] Jannati A, Sadeghi V, Imani A, Saadati M. Effective coverage as a new approach to health system performance assessment: a scoping review. BMC Health Serv Res 2018;18(1):886.

[55] HIMSS. Guiding principles for big data in nursing, using big data to imprpve the quality of care and outcomes, Healthcare Information and Management Systems Society; 2015. [cited 1 June 2019]. Available from: https://www.himss.org/sites/himssorg/files/FileDownloads/HIMSS_Nursing_Big_Data_Group_Principles.pdf.

[56] IHPA. Independent hospital pricing authority Canberra, [cited 2019 26 March]. Available from: https://www.ihpa.gov.au/; 2011.

[57] AIHW. Costs of acute admitted patients in public hospitals from 2012–13 to 2014–15 Canberra, Australian Governmnet, Australian Institute of Health and Welfare; 2018. [cited 31 May 2019]. Available from: https://www.myhospitals.gov.au/our-reports/cost-of-acute-admitted-patients/november-2018/pdf-files/if_costsofacuteadmittedpatients_2014-15.pdf.

[58] Busse R, Geissler A, Aaviksoo A, Cots F, Häkkinen U, Kobel C, et al. Diagnosis related groups in Europe: moving towards transparency, efficiency, and quality in hospitals? Br Med J 2013;f3197:346.

[59] Wammes JJG, van der Wees PJ, Tanke MAC, Westert GP, Jeurissen PPT. Systematic review of high-cost patients' characteristics and healthcare utilisation. BMJ Open 2018;8(9) e023113.

[60] Kaplan RS, Haas DA. How not to cut Helath Care costs. Harvard business review; 2014.

[61] Slawomirski L, Auraaen A, Klazinga N. The economics of patient safety: strengthening a value-based approach to reducing patient harm at national level, OECD; 2017. [cited 31 May 2019]. Available from: https://www.oecd.org/els/health-systems/The-economics-of-patient-safety-March-2017.pdf.

[62] Duckett S, Jorm C, Moran G, Parsonage H. Safer care saves money: how to improve patient care and save public money at the same time. Melbourne: Grattan Institute; 2018.

[63] AHRQ. Nursing and patient safety: patient safety network, Agency for Healthcare Research and Quality; 2019. [cited 2 June 2019]. Available from: https://psnet.ahrq.gov/primers/primer/22/Nursing-and-Patient-Safety.

[64] Glassman KS. Using data in nursing practice. Am Nurse Today 2017;12(11).

[65] JCR. The joint commission, standards, [cited 1 June 2019]. Available from: https://www.jointcommission.org/standards_information/standards.aspx.

[66] Coiera E, Hovenga EJ. Building a sustainable health system. Yearb Med Inform 2007;11–8.

[67] Lowe C. An international patient type hours per patient day (HPPD) benchmarking research study unpublished. TrendCare Systems Pty Ltd.: Brisbane; 2014.

[68] WHO. Draft global strategy on digital health, [cited 17 April 2019]. Available from: https://extranet.who.int/dataform/183439; 2019.

[69] WHO. Guideline: recommendations on digital interventions for health system strengthening, Geneva: World Health Organisation; 2019.Licence: CCBY-NC-SA3.0IGO: [cited 18 April 2019]. Available from: https://www.who.int/reproductivehealth/publications/digital-interventions-health-system-strengthening/en/.

[70] Productivity-Commission. Data availability and use, overvieew and recommendations Report No. 82, Canberra, [cited 5 June 2019]. Available from: https://www.pc.gov.au/inquiries/completed/data-access/report; 2017.

[71] Hovenga EJS. National healthcare systems and the need for health information governance. In: Hovenga EJS, Grain H, editors. Health information governance in a digital environment. Amsterdam: IOS Press; 2013.

[72] Hovenga EJS. Impact of data governance on a nation's healthcare system building blocks. In: Hovenga EJS, Grain H, editors. Health information governance in a digital environment. Amsterdam: IOS Press; 2013.

# Case study 1 — Patient Assessment and Information System (PAIS): Work measurement research and workload measurement methodology

## Study purpose

In 1982, the Victorian Government's Health Commission Victoria (HCV) had a legislative responsibility to determine the nursing staff numbers by skill mix, including associated budget allocations for all 164 public hospitals. Its Health Management Services group was given the responsibility for establishing a uniform and practical approach to meet this requirement. Three acute health service hospitals volunteered to be part of this original study. The resulting staffing methodology was tested in a fourth hospital [1]. The outcome provided information for the development of a nursing workload measurement methodology. The work sampling data were later used to measure non-nursing workloads.

## Original sample

Nine acute medical/surgical wards in three different volunteer acute hospitals plus a further 20 wards in a fourth hospital were used to validate the original findings. It was estimated that to achieve the desired accuracy and a representative sample, observations needed to cover 7 days in each hospital. In order to arrive at an average time per activity at the 95% confidence level, for an activity that had received 5% of all observations recorded, a total of 30,400 observations were required. The sample from four hospitals covered a total of 2371 patient days during which time 39,774 observations were recorded relative to 8748 allocated staff hours. This equals an average of 3.69 staff hours/patient day (excl. night shift).

## Methods

The literature review results provided essential attributes of a nurse staffing methodology [2,3] and a preliminary list of highly likely significant patient related indicators of staff resource usage. These were used for data collection form design which was guided by a need for ease of data collection and to enable subsequent computer processing.

### Research objectives

- Measure the workload generated by all in-patients in several wards in three hospitals.
- Identify how ward staff distribute their available time between patients and ward activities.
- Identify relationships between patient population characteristics and ward activities.

### Original study design

Data collection consisted of five different datasets plus contextual information regarding organizational factors including nursing models of care in use.

1. Patient characteristics, treatment and care plan-input by bed location — input.
2. Staff hours allocated to work each observed shift in hours by skill mix — input.
3. Observable nursing activities — consisting of work sampling observations recording the activity each staff member was engaged in at the instant of being identified by the observer, and where direct patient/nurse interaction was observed for which patient — process.
4. Sampling observations to determine quality of care delivered based on procedural compliance with procedure guidelines relative to nursing activities undertaken. This was further supported by the use of patient satisfaction questionnaires — process.
5. Occupied bed hours by patient during the observation period — output.

Patient data were collected by a registered nurse using a standard format. Data sources were medical records, bedside documentation and any other available paper based data sources. These were verified by the observer's visual assessment of the patient. Documentation was frequently found to be incomplete or not up to date. Staff rosters were matched with actual staff on duty. The bed number and patient last name provided the link between the patient data set and the work sampling data set. Admission and discharge times for individual patients were collected to enable occupied bed hours per patient to be calculated for any given time frame such as a shift or patient day.

Staff categories included in that data set were the nurse in charge, registered nurses, enrolled nurses and nursing students who were then part of the workforce in one hospital studied. Some wards included nursing attendants and/or ward clerks who undertook work that would be performed by nursing staff in their absence. As the primary objective was to arrive at a ward staffing methodology applicable to any Victorian hospital there was a need to include these staff categories where they were included in the allocated ward staff.

An analysis of nursing practice resulted in the development of a nursing taxonomy consisting of the data elements to be included in the work sampling data set. Data elements were selected not only on their likely frequency of occurrence but also on the basis of associated nursing skill required based on known boundaries of practice. Each data element representing a "group of mutually exclusive observable work" was defined. A total of 96 activities were included. These were organized in a hierarchical structure consisting of 28 groups within three major categories. All were listed on the work sampling data collection form.

An analysis of hospital procedure manuals formed the quality of care data set. Only major, relatively frequently occurring activities that were related to activities being measured were included for observation. Observation criteria consisted of a short list of yes/no questions to enable a qualified observer to note whether the nurse performed the various components of the activity in a compliant manner. Observations were undertaken randomly by senior nursing staff nominated by the hospital. Frequently occurring activities were audited more frequently than others due to the random sampling. Patient perception of adequacy of care was also studied via questionnaires. Questions were confined to criteria considered important to the patients' welfare.

A roster of observers enabled two observers to be allocated to collect all data as required for each of the five data sets throughout the study period. Each observer worked no more than 4 h straight. Only morning and afternoon shifts were observed to minimize study costs. Observer reliability was established. Work sampling observations consisted of five to six randomly selected start times for each observation ward round where the observer recorded the activity being performed at the instant of observing each staff member. The minimum interval between rounds was the minimum time required to complete an observation round and locate all staff on duty. Different sets of random times were used for each observation period. Where two activities were being performed concurrently, such as communication, the dominant activity was recorded. Walking was considered to be a component of an associated nursing activity that was then identified by that activity and where possible associated with a patient, alternatively it was associated with an indirect activity. Thus the universe of nursing work during the observation period was recorded.

Organizational data consisted of ward size, type, nursing philosophy, nursing delivery patterns, methods and available support services. A questionnaire was used for this purpose. All stakeholders were well prepared prior to the study and information leaflets were made available to inform patients and visitors during the study.

### Data analysis

All observations were totaled vertically and horizontally on each work sampling observation form. Staff hours were calculated for each observation period. A TRS 80 microprocessor, for which a special program was written in BASIC, was used for the remaining data analysis.

Sampling data were analyzed according to percentage of occurrence per nursing activity. Patient population characteristics were first analyzed and compared by hospital.

Data were examined to assess how the relationship between patient characteristics and resource usage could best be expressed. It became evident that some method of grouping patients according to these indicators could provide a reliable method for ascertaining nurse staffing requirements. The hypothesis that an increasing number of perceived significant indicators identified per patient would result in an increase in resource usage was formulated. Patients were then grouped accordingly. Each indicator was assigned one point with the exception of hygiene care, where total care was allocated two points. Patients were then classified into one of four groups. The highest common score was 12. Very few patients exceeded this number. The minimum dependency group (A) consisted of patients who scored between 0 and 3, group B patients scored 4–6, etc.

Direct sampling data by patient were then identified according to each patient's dependency classification. The computer was then used to identify the distribution of work sampling observations between all direct patient interactive nursing activities per patient category. Further calculations resulted in establishing the average direct care hours per patient day (excl. night shift) for each patient category. This confirmed the hypothesis to be true. Identified variations between hospitals were further examined in the light of organizational data collected to identify likely influencing factors to explain any variations identified. This patient dependency classification was tested in a fourth hospital where the work sampling study was replicated. All data from four hospitals were then combined prior to further analysis.

Relative values were calculated using total direct care values for each category. It was assumed that all indirect activities would increase at the same rate as the direct care hours distribution per dependency category. The indirect care hours were then distributed to each dependency category on the basis of these relative values. Standard time values for each patient category were then calculated to enable their use as benchmarks. The results of this analysis formed the foundation for the development of the Patient Assessment and Information System (PAIS) nursing workload monitoring system.

### Findings

The percentage of occurrence for the number of observations relating to each patient category ranged from 4.3% to 12.4% indicating that the resulting total direct time values for categories A, B & C have a degree of accuracy ($\pm$ 3%) that is greater than the 95% confidence level. Category D resulted in a $\pm$5% degree of accuracy. Nursing time for each significant direct activity measured increased with each increasing dependency category as shown in Table 1. This finding strongly supported the hypothesis and this patient classification's validity. Critical dependency indicators were found to be mobility, elimination, hygiene, nutrition, emotional support/teaching, and all technical activities. Collectively this only covered 36.13% of the total nursing workload.

Table 1 Average minutes per patient day (excl. night shift) by activity and dependency category

| | Average minutes per patient day (a.m.+p.m. shifts — 14.5h) | | | |
|---|---|---|---|---|
| Dependency category | A | B | C | D |
| N = Number of patients | N = 969 | N = 693 | N = 579 | N = 130 |
| Assist mobility | 4.16 | 7.82 | 11.58 | 17.33 |
| Elimination | 1.97 | 4.36 | 6.76 | 8.81 |
| Hygiene care | 4.30 | 10.64 | 17.18 | 22.57 |
| Nutrition | 3.44 | 5.95 | 12.82 | 10.46 |
| Intravenous therapy | 1.16 | 1.45 | 6.91 | 20.14 |
| Communication | 9.87 | 10.21 | 10.45 | 15.49 |
| Wound dressings | 3.86 | 5.15 | 6.10 | 12.10 |
| Oral drug administration | 1.78 | 1.98 | 2.72 | 2.52 |
| Parenteral drug administration | 1.04 | 1.27 | 3.91 | 10.36 |
| Miscellaneous procedures/ interventions | 1.39 | 1.22 | 3.02 | 9.78 |
| Vital signs & observations | 7.06 | 8.76 | 12.08 | 18.69 |
| **Total direct hours/patient day** | **0.90** | **1.21** | **1.87** | **2.88** |
| **Indirect hours/patient day = 1.66** | | | | |

Indirect ward activities accounted for 44.92% of all staff time allocated. The remaining 18.95% accounted for meal breaks and personal time. Daily patient and nursing activity variations by ward were found to be equally as different as the differences between wards. There were a number of organizational differences between the wards studied. The number of observations relevant to each ward were insufficient to determine if these variations were significant influencers of final nursing times per patient category.

The hospital found to have the smallest amount of average direct care hours per patient category, especially for the highest dependency category, was also found to have a lower performance compliance rate. This suggests that these wards were not as well staffed as other hospital wards studied and that the care for the most dependent group of patients was compromised. Another contributing factor was that the primary model of care was the use of task allocations as opposed to the adoption of team based patient allocation methods in other hospital wards.

## Results

This initial study undertaken in four hospitals resulted in the development and validation of a patient classification model with associated standard nursing staff time values. It was hypothesized that these preliminary standard time values could be used as a benchmark against which other hospital patient populations' nursing staff needs could be tested.

## Staffing methodology development

The indirect time was distributed evenly and added to each direct category time value plus a 15% addition to cover personal time and tea breaks. Meal breaks vary and are not part of the paid time. The resultant time standards per dependency category enabled their use as benchmarks in any medical/surgical ward in acute hospitals. This formed part of a valid nurse staffing methodology that became known as the Patient Assessment and Information System (PAIS).

## PAIS implementation and use

Patient classification principles were clearly defined to enable consistent use. The model was easy to use and sensitive to demand changes over time as additional activities identified for individual patients could easily be accommodated by extending the number of classification categories. New time values for higher dependency could be arrived at by extrapolated estimation in the first instance for later review following further work measurement studies [4]. One of the original study hospitals first trialed PAIS in two wards and gradually introduced it to all ward areas in that hospital. Two years later it had successfully been implemented in at least 26 Victorian hospitals.

PAIS was marketed through a Health Department Victoria publication [5] that explained the nurse staffing methodology. Training workshops were held on request. This included a list of the many positive features, options and detailed instructions for use, a sample patient assessment form and well defined criteria used for patient assessment to determine level of patient dependency. This form was replicated on a white board by some hospitals as shown in Fig. 1.

**Fig. 1**
White board used for patient dependency data collection — Prince Henry's Hospital 1984.

The HCV document included detailed instructions regarding many staffing management applications including the use of a productivity index comparing predicted with actual staff hours and rostering. PAIS was used to resolve a number of industrial disputes over nursing workloads, predictively and retrospectively to support nurse staffing management, and to estimate additional staffing needs for new services. Some hospitals were using it on a per shift basis and tied it to their admissions and discharge policies as a means of controlling and managing nursing workload in accordance with staff resource availability. A PAIS user group was established in Victoria.

One workload evaluation study in two major hospitals demonstrated a change in patient mix in 1 year that represented an 11% increase in nursing workload for an average 100 patients treated by those hospitals on the basis of increasing patient dependency resulting from reductions in average of length of stay. Efforts were made to reduce the number of low dependency in-patients by improving discharge planning and elective admission processing. Three years following the PAIS development, the dependency categories had increased from four to six to accommodate the greater number of high dependency patients.

An evaluation of administrative systems found that as a general rule nightshift staffing requirements ranged between 18% and 25% of the am plus pm requirements. Most data processing was undertaken manually. A stand alone microcomputer system (PANDA) was designed by a nurse working for Health Computing Services, a not for profit company that had provided computing services to public health care institutions since 1967. Numerous additional work measurement studies were subsequently undertaken in several Victorian hospitals using this work sampling methodology covering acute psychiatric inpatient care [6], midwifery [7,8], oncology [9] and aged care [10]. Nursing documentation improved as nurses had to be able to provide evidence to support the inclusion of significant activity indicators. This formed part of the continuing reliability testing. PAIS was used for many nursing services productivity reviews during the early 1990s.

## Use of PAIS in New South Wales, Queensland and Western Australia

Over the next decade, this methodology was adopted by more than 100 hospitals Australia wide, and computerized in various forms by several software providers. PAIS adoption expanded to New South Wales (NSW) where it was validated in a major teaching hospital in 1990 [11]. It was in use by a further 20 or so hospitals supported by the NSW PAIS interest group. The PAIS staffing methodology was also validated in the private sector in Western Australia [12].

A continuous self-recording approach was used to again measure nursing time per PAIS category in three large teaching hospitals from which new standard time values were calculated [13]. These were used to calculate nursing hours per Australian Diagnosis Related Group

(AN-DRG) and average nursing hours per day of care per separation for a total of 46,924 cases across eight teaching hospitals. The results of this major 2 year study were used to develop nursing intensity measures (nursing service weights) per AN-DRG. Twenty standards for nursing practice and patient outcomes were developed based on PAIS [14].

PAIS was adopted as the corporate patient dependency system in Queensland in 1993. Further work measurement validation studies were undertaken in several hospitals during 1995 using the original work sampling methodology [15]. At that time PAIS was also compared with the use of the TrendCare staffing methodology for Queensland Health [8,15,16]. The generation of demand for nursing resource estimations per patient was found to be appropriate from either system with a strong consistency between systems. The PAIS standard time values were updated as a result of this latest work sampling study as shown in Table 2. Staffing the night shift requires between 18% and 25% of 24h staffing depending on ward size. The required hours need to be rounded up or down according to the closest whole number of staff and the minimum number required to provide safe care irrespective of actual workload.

Table 2 PAIS Standard Time Values

| PAIS dep. cat. | Standard time values medical/surgical (excl. night shift) (%) | Standard time values maternity (excl. night shift) (%) |
|---|---|---|
| A | 2.91 ± 4 | 2.81 ± 4 |
| B | 3.56 ± 3 | 3.44 ± 3 |
| C | 4.51 ± 3 | 4.45 ± 4 |
| D | 5.51 ± 4 | 6.51 ± 6 |
| E | 7.51 ± 5 | 7.54 ± 6 |
| F | 9.05 ± 6 | 12.55 ± 9 |

## Discussion

The extensive continuing use of PAIS in large teaching hospitals as well as regional and country hospitals demonstrated that the PAIS nursing workload measurement methodology applied to any clinical specialty. For ease of use it was found to be sufficient to focus on the number of significant workload influencing activities relevant for individual patients, the greater the number, the greater the level of nurse dependency. It was important to retain the same classification rules as had applied during the work sampling studies. Differences between clinical specialities were found to be the mix of such activities not the number of activities identified by category. Similarly the work sampling studies had shown that the time distribution between activities measured varied by clinical specialty and patient characteristic including age. Significant variations of dependency mix between shifts in the acute sector, had also been identified during the work sampling studies. This is why one needs to make use of historical data to determine workload patterns and configure rosters accordingly. The time values used as benchmarks provided a good indication of nursing workload in any clinical specialty. Large hospitals made use of their data to more evenly distribute their nurse staffing establishment across all wards based on workload trends.

## Issues encountered

Despite the significant investment and widespread adoption by nurses and health care organizations for well over a decade, PAIS is no longer in use. This final outcome is likely to be due to a number of factors, such as commitment, governance, automation, political and industrial events, education/knowledge gaps, poor use and/or sharing of information. They are presented here as lessons learned to serve as a reminder for those investing in future nursing workload methodology development innovations.

Adoption and use of any nursing workload measurement system requires a lot of commitment from many to ensure consistency in the interpretation of the rules associated with its use and reliability of new information generated. Despite the provision of educational material, workshops and access to a user group, there were many nurses who continued to question the system's ability to fully account for all nursing work. The original intent of PAIS was ease of use. The PAIS categorization rules concentrated on identifying significant workload influencing activity. The time values were the result of measuring ALL nursing work, a concept that continued to be poorly understood. Nurses were continually debating the need to include additional activities in the list to determine level of dependency. A variation from the classification rules in use during the work measurement phase from which the benchmark time values were derived, is likely to inflate the subsequent nursing workload measure.

Another issue was the lack of feedback to nursing staff regarding the use of information provided by them and/or a lack of use. A constant disregard by management in terms of making staffing adjustments based on workload information generated, leads to overall dissatisfaction, a lack of commitment to provide reliable information and ultimately industrial unrest. Maintaining the relevance of any nursing workload estimation method, requires a governance structure and commitment to maintain the standard. Standards do need to be reviewed and updated to reflect ongoing changes in treatment and care protocols as well as in patient mix. This requires ongoing investment, which health service executives were reluctant to make. They failed to appreciate the many significant cost savings, improved outcomes and productivity improvements that were made by nurses using PAIS data.

Automation during the 1980s was very poor. System maintenance and use required a dedicated project officer. Self-contained computer systems required a lot of data re-entry. Systems developed by large software vendors who did develop modules within their system to automate PAIS data, were poorly designed. Despite these issues and associated political and industrial turmoil in Victoria that had a direct impact on PAIS, it is amazing that PAIS was accepted and used by many for more than a decade. It's a tribute to the hundreds of nurses who were truly committed to make it work.

## *Political interference*

Victorian hospitals each had a project officer who managed the data. There was a significant nursing shortage as a result of the implementation of the 38 h week and a gradual transfer of hospital based nursing students was underway to University based courses [17]. Student nurses were part of the workforce while undertaking their hospital-based programs.

PAIS data gave nurses information that enabled them to contribute to their workload management and ensure safe care. This was achieved by notifying ambulances, and/or canceling elective surgery when required. These strategies were in use when there were no bed vacancies and were now used to indicate insufficient nursing staff to provide safe care. This is in line with the 2018 ICN's position statement [18] that calls on healthcare employers to:

> *Ensure systems are in place to alter or stop patient flows and admissions to match the available nursing supply. Nurse leadership must have the authority to stop admissions when unsafe staffing situations arise and to authorise, at short notice, additional staff when patient safety is at risk.*

PAIS was officially owned by the State Government's Health Commission of Victoria (HCV), governed by a quasi-autonomous Board of Commissioners. In 1985 legislative changes meant it no longer had responsibility of establishing and funding hospital nurse staffing levels [17]. This responsibility was transferred to and divided among eight new regional directors who chose not to make use of senior nursing advice. The HCV became a State Government Department where the Chief General Manager had a direct reporting relationship with the Health Minister (David White 1985 and 1989), who refused to support this nursing workload management system. The Minister had adopted a policy to reduce surgical waiting lists and saw the use of PAIS as detrimental to his policy initiative. The use of PAIS was officially banned but nurses continued to make use of it. They were unable to continue once the Minister made all project officer positions redundant [19]. Senior nurses within the HCV were demoted as a source of advice on policy and operational matters in nursing. Considerable industrial unrest followed, including a 50 day nursing and midwives strike in 1986.

These events had major repercussions. Nurses lost the power of using workload data to manage safe patient care. A new government in 1992 made sweeping fiscal changes to Victoria's health system. Many nurses resigned from the public workforce, unable to come to terms with their professional and legal responsibilities to provide safe care in the face of unyielding workloads [20]. Nurses faced mass redundancies resulting in critical staff shortages over the next 8 years. The union again took action with a campaign titled "Nursing the system back to health." This resulted in the first mandated minimum nurse-to-patient ratios in 2000. As a result there was a need to fill 1300 vacancies [21,22].

## References

[1] Hovenga E. Casemix, hospital nursing resource usage and costs (PhD Thesis). Sydney, Australia: University of New South Wales (UNSW); 1995.

[2] Aydelotte M. Nurse staffing methodology: a review and critique of selected literature. National Institutes of Health; 1973 Contract No.: USDHEW Pub. No. (NIH) 73-433.

[3] Murphy LN, Dunlap MS, Williams MA, McAthie M. Methods for studying nurse staffing in a patient unit. DHEW Pub.No. HRA 78-3. 1978.

[4] Hovenga E. A patient classification model (PAIS) and nursing management information system. Melbourne: Health Commission Victoria; 1983.

[5] Hovenga E. Patient Assessment Information System (PAIS): a patient classification model based on patient/ nurse dependency and a nursing management information system. Health Deartment Victoria; 1985.

[6] Crowther E, Hiep A. Protocol PAIS psychiatric module. Royal Park Hospital; 1986.

[7] Hovenga E. The Patient Assessment and Information System (PAIS) for midwifery patients. Northcote, Victoria: EJSH Consulting; 1990.

[8] Hovenga EJS, Hindmarsh C, Nursing Area HIB. Queensland PAIS validation study. Brisbane: Queensland Health; 1996.

[9] Hovenga E. Work sampling at Peter MacCallum Cancer Institute and review of PAIS use. Northcote, Victoria: EJSH Consulting; 1988.

[10] Hovenga E. The Patient Assessment and Information System (PAIS) in extended care. Northcote, Victoria: EJSH Consulting; 1990.

[11] Goodwin M, Hawkins A. PAIS dependency system: a validation. Aust J Adv Nurs 1990;7(3):24–7.

[12] Coleman A. PAIS validation study 2nd report incorporating Dec 1996–Feb 1997 work sampling. Western Australia: St John of God Health Care Subiaco; 1999.

[13] Hathaway V. PAIS timing study. Sydney: Concord Hospital; 1992.

[14] Picone D, Ferguson L, Hathaway V. NSW nursing costing study. Sydney Metropolitan Teaching Hospitals Nursing Consortium; 1993.

[15] Hovenga E, Hindmarsh C, editors. Queensland Health—PAIS validation study: results and issues for nursing cost capture. Eighth Casemix Conference in Australia. Canberra: Commonwealth Department of Human Services and Health; 1996.

[16] Hindmarsh C. In: Nursing Area HIB, editor. Report on cross-referencing study between PAIS and TrebdCare. Queensland Health; 1995. p. 1–66.

[17] Fox C. Industrial relations in nursing—Victoria 1982–1985. The University of New South Wales, School of Health Administration; 1989. p. 208.

[18] ICN. Evidence-based safe nurse staffing. Position statement. Geneva: International Council of Nurses; 2018. p.1–7.

[19] Johnson H. A memorable decade: the history of the Nursing Projects Officer Group 1982–1993 and the PAIS Users Group 1987–1993, Melbourne [unpublished] 2004.

[20] Gerdtz MF, Nelson S. 5–20: a model of minimum nurse-to-patient ratios in Victoria, Australia. J Nurs Manag 2007;15(1):64–71.

[21] ANMF. Australian Nurses and Midwives Federation (Vic Branch) history. [cited 4 October 2018]. Available from: http://www.anmfvic.asn.au/about-us/history; 2018.

[22] ANMF VB. Nurse/midwife patient ratios—it's a matter of saving lives. Melbourne: Australian Nurses and Midwives Federation, Victorian Branch; 2015.

# Case Study 2 — Design, development and use of the TrendCare system

## Study purpose

To develop an electronic tool that could accurately measure the demand for nursing services so that nursing resources could be supplied to meet peaks and troughs in workloads and provide fair and equitable workloads to nurses.

## Original sample

Initial studies were undertaken in nine medical and surgical wards across two Mercy Acute care Hospitals in Queensland, Australia between 1985 and 1989 [1]. Once the prototype system was developed and partly computerized it was trialed and used by seven Mercy hospitals across Queensland. The specialities of maternity, pediatrics, and critical care were researched and developed during this time (1990–92) and were frequently updated to incorporate ongoing research findings. The success of the system led to requests for a trial from public and private hospitals outside of the Mercy group. Consequently a trial of the system, which was partly paper based and partly electronic, was conducted in a further 11 private and public Queensland hospitals during 1992–94 with results being presented at the world Nursing informatics conference in 1994 in San-Antonio [1]. More than 2500 nurses rated patients across 806,250 nursing shifts to test the reliability of the clinical indicators incorporated in the patient nurse dependency system. The validity of the nursing hours allocated to each patient category and reliability of acuity indicator definition interpretation was the focus of this trial.

## Research objective

- To identify the independent patient indicators and their related variables that consistently make nursing more time consuming, and to measure the effect they have on the nursing hours required to provide quality patient care, identified as the dependent variable.
- To develop an evidence-based patient nurse dependency system which is user friendly and able to accurately measure nursing hours required for care, for a wide scope of acute and non-acute patient types.

- To develop a valid and reliable inter-rater reliability system to support the ongoing reliability of the TrendCare patient nurse dependency system.
- To develop a workload measurement system which would proactively and retrospectively measure the total workload of a nursing service, including clinical and non-clinical work, and provide variance reporting relating to demand and supply (required hours vs available hours).
- To develop a workload allocation system which utilizes patient acuity measures to allocate fair and equitable workloads to nurses and/or nursing teams.
- To develop national and international benchmarks for clinical and non-clinical work in hours per patient per day (HPPD) utilizing TrendCare data from hundreds of nursing services across multiple countries.

## *Methods*

The first research model utilized was an action research approach based on timing specific patient care nursing interventions, developing an average time for each intervention and trialing a patient categorization system utilizing the sum of all interventions completed by nurses. This methodology did not provide accurate nurse/patient contact times and proved to be a very time consuming method for the nurse when categorizing patients, as all interventions had to be accounted for.

Time study methods were modified for the next phase to better meet the research objectives. A new timing study approach was designed and adopted which included timing all nurse patient contact times and identifying patient interventions without individually timing each intervention. This methodology was more easily implemented. It gave a more realistic nurse contact time, accounted for nurses conducting interventions simultaneously, and also clearly demonstrated that nursing intervention time was increased according to the patients varying levels of physical, emotional, behavioral and technical dependence.

Independent acuity indicators and their related variables were recognized as being the stimulus for change in the dependent variable. These were first identified and timed in medical/ surgical patient care areas in two Queensland Mercy hospitals. The most critical patient acuity indicators and their related variables were identified from a literature review. These were adopted as the basic criteria for developing the medical/surgical acuity indicators and related variables and were on the initial acuity categorization form.

The TrendCare timing studies involved all nurses and nursing support staff (RN, EN, AIN, HCA, etc.) who provided any direct or indirect care, related to the patient whose contact time was being recorded. All staff in the ward were aware of which patients were involved in the study and all carers wore a timing device in case they needed to time and record a contact. Formal training was conducted for all ward staff. All objectives, timing study processes and answers to frequently asked questions were clearly documented in the TrendCare Timing Study Guide which was accessible to all staff on the ward.

Patient contact timings were conducted across three 8 hr periods of the day with the day period commencing at 0700, the evening period starting at 1500 and the night period starting at 2300. It was important to note that this was a study of patient dependence across an 8 hr period and should not be confused with measuring nursing work across a nursing shift. The dependency system was designed to be utilized for any mix of nursing shift lengths and start times.

Further time and motion studies identified five levels of nursing intensity required for each 8 hr period of the day to care for a broad cross section of medical and surgical patient types. All relevant acuity indicators were then identified and patient category hours developed and tested for a range of specialties in the acute health care setting. This research and development was conducted over a 9 year period (1985–93). These included: Medical, Surgical, High-Dependency medical/surgical, Intensive Care, Coronary Care, Pediatrics, Pediatric ICU, Maternity Postnatal, Labor, Antenatal and Special Care Nursery. The results of these time studies formed the foundation for the development of a much wider range of patient types and future software design. These developments have been enhanced over time by continuing follow up research and development activities.

## Study design

An action research approach using time studies was adopted for the initial study. The number of acuity indicators and their relevant variables were kept to a minimum and defined broadly, so that all factors which significantly increased the time required to care for an individual patient were able to be acknowledged.

As the research data from the timing studies were reviewed the number of acuity indicators were extended and additional variables added to account for additional comorbidities, such as mental health conditions, obesity, chronic diseases etc. for the medical/surgical patient and subsequently for all patient types. Indicators such as thought processes, behavior, respiratory assistance and 1 on 1 care were added to accommodate these findings. Indicator variables were also extended for some indicators to account for a more defined level of dependence.

Time studies undertaken resulted in the ability to establish an average time value per patient type for three 8 hr periods of the day. Time studies focused on timing the total nursing process. Data analysis of dependency clusters resulted in five categories for medical surgical patients with varied times for each of the three periods of the day giving a total of 15 medical/surgical category times. Additional categories were also added for each patient type to accommodate varying levels and periods of one on one care (specialling). The system was constantly in use to manage numerous hospital's nursing resources while research activities continued. Feedback from user sites, in particular results from inter-rater reliability testing was used to inform ongoing research activities. Nurse satisfaction were all monitored throughout the trial with the use of the system, the quality of patient care and patient satisfaction.

The system was used to provide hospital wide efficiency statistics in hours per patient day for actual hours worked and evaluated against hospital targets and efficiency standards.

The system's reliability and validity was tested during a state-wide trial, involving 11 hospitals and undertaken over several months. The results of user surveys evaluating user satisfaction were used to improve the design of the electronic system. New computer concepts were employed resulting in the first release of the electronic TrendCare patient acuity and nursing workload measurement system in 1995. The research timeline has continued to be ongoing so that the system remains valid, reliable and relevant over time.

### Data analysis

The data collected consisted of time studies, first by activity and later by patient type. The timings included direct and indirect care activities for three 8 hr periods of a patients' 24 hr day (day, evening and night). This occurred throughout the initial and ongoing study period, enabling the system's development to progress and to be continuously refined, using spreadsheets for data analysis purposes.

Timing studies continue to be undertaken today to ensure the time values for all patient type categories and the acuity definitions remain valid, and that the system continues to measure nursing hours required accurately. A purpose built software program is currently used to collate, analysis and report on research data.

### Findings

The first timing studies undertaken demonstrated that timing specific nursing interventions was not a valid method for categorizing patients. Using a stop watch timings commenced from the beginning of the set up time through the intervention, to the completion of the clean-up process. Timing were conducted on a wide range of nursing interventions and showed that the time taken to do the same procedure on different, or even the same patient, varied greatly according to specific patient variables.

The impact of variances relating to doctors' treatment preferences and individual patient differences were found to have a significant impact on nursing interventions on particular shifts for the same diagnosis/medical procedure. This finding meant that the categorization of patients by disease and/or medical procedure was not useful.

The literature had indicated that mobility, hygiene, and nutrition were critical independent variables. These were adopted as basic criteria for the patient categorization model. Level of orientation, frequency of observations, continence, medications, treatments, teaching/counseling, emotional support, language/communication difficulties and whether the patient requires isolation or internal transfers were also included. The research indicated that specialty units required specific additional criteria.

The trial undertaken across 11 hospitals in 1993–94 identified that some clinical indicators required clearer definitions. The existing category hours were validated and additional categories were developed for some very highly dependent patients. Six months into the trial, the inter-rater reliability of the categorizing system was extremely high (92%). User surveys identified that nurses found the system to be extremely user-friendly and over 80% of nursing staff rating the system as "giving a fair workload."

Subsequent large samples of data from a large cross-section of healthcare services across three countries have clearly demonstrated that the TrendCare acuity system is sensitive enough to accommodate the nursing services provided in large tertiary health services, large regional health services as well as small rural health services. TrendCare International benchmarking studies have also demonstrated that the TrendCare acuity tool is sensitive enough to identify different staffing requirements for different models of care [1].

## Results

The results of the initial time studies were used to develop a factorial patient acuity classification system for medical and surgical patients using clearly defined indicators of dependency, with five levels of dependence and 15 acuity categories each with a specific allocation of time. This proto type research methodology and system design was used to develop acuity classification tools for other specialities such as Maternity, Pediatric and Critical Care.

The validation of the system was based on an evaluation of outcomes as measured via a range of quality assurance activities. This included the use of clinical and operational key performance indicators and nurse user satisfaction surveys. The trial across 11 hospitals resulted in the ability to generate a benchmark range of hours required for care, per patient, per day for each patient type.

In 1995 a new version of the TrendCare System was released. This version had made use of the results from timing studies, the trial feedback and the application of new software development technology, to become a fully electronic patient nurse dependency system. The enhanced computerization of the system reduced user contact time and improved reliability through automation and interoperability. The selected patient characteristics and care elements (indicator variables) were able to be combined using automated algorithms which placed patients into acuity categories and decreased the element of user subjectivity.

The number of patient types now identified and able to be differentiated based on average nursing dependency (acuity) have continued to increase. These continue to be updated based on the evidence gained through the analysis of large samples of data from large cross-sections of acute healthcare (tertiary, secondary, primary), sub-acute and residential services across six countries. These include: Australia, New Zealand, Singapore, Thailand, United Kingdom and Ireland.

## Staffing methodology development

The first standard time values developed within the TrendCare acuity categories acknowledged the dynamic nature of nursing practice and the unpredictable elements of patient care by factoring in additional time for unpredicted work. Additional time has also been factored in for paid tea breaks. Results of time studies and the analysis of data from user sites, identified the need to pre-set minimum staffing levels when patient numbers are low in small wards/units and/ or acuity levels are very low. The variance between required hours and available hours when calculated in TrendCare ensures that minimum staffing levels are considered.

The TrendCare workload management system tracks, monitors and reports on all hours worked. Planned work hours are imported from the organization's roster/scheduling system and are moved to staffing areas that clearly define retrospectively, how and where the hours were utilized. All hours available to provide inpatient care in the ward/unit are compared to hours required for care (determined by the integrated patient acuity/nurse dependency system) and variances displayed. Other clinical hours (shift co-ordination, time waiting in X-ray, external escorts, etc.), non-clinical hours (ordering supplies, clerical activities, cleaning equipment, environmental tasks, etc.), training and development and absenteeism hours are collected and reported in order to identify opportunities to improve productivity and efficiency.

To meet the goal of providing fair and equitable workloads for nurses, a workload allocation screen was developed in the TrendCare system. This enables patients to be allocated to nurses according to acuity and provides transparency in relation to patient allocation, workload allocation to each nurse/team, meal break times and other allocated activities.

To ensure that user sites of the TrendCare patient nurse dependency system maintained reliable and complete data, a formal inter-rater reliability system was developed based on the theories researched in an extensive literature review. Score sheets are able to be printed directly from the system. Audits were also built into the system so that incomplete data can be identified and rectified.

Nurses from a wide variety of specialties were involved in developing the criteria for the TrendCare patient classification component, the design of reports for prospective (predicted) and retrospective (actual) variance analysis, the process design of data input screens at ward level, the identification of the wide scope of staffing areas in the system to accommodate the scope of nursing work, and the development of educational packages for users.

## TrendCare implementation and use

TrendCare has continued to grow, develop and mature since 1995. The number of user sites has grown substantially. Ongoing time and motion studies are regularly conducted both nationally and internationally to ensure the time values for all patient type categories and the acuity definitions remain valid. The system has, over the past 20 plus years continued to measure nursing and midwifery hours required accurately. The average patient acuity in most hospital wards has continued to increase due to increased co-morbidities, shorter lengths of stay and changes in treatment methodologies. These increases in acuity have been measured, using the TrendCare system, and trended over time. This evidence was then used to obtain a corresponding increase in nursing and midwifery resources within health care services using the system. Some surgical services have shown evidence of lower acuity due to the use of laparoscopic surgery, however, with shorter lengths of stay and a fast turnover of patients, the TrendCare workload measurement system has shown an overall increase in nursing workloads for these ward types.

The reliability of the TrendCare acuity and workload management system has continuously improved over the past 20 plus years by using the results of a multitude of timing studies spread across four countries including Australia, New Zealand, Singapore and England. These ongoing studies use a non-experimental action research methodology and are a modified version of the traditional timing study method. This adapted methodology, using total patient contact timings, is unique to the TrendCare acuity system and provides a process that enables the reliable prediction and actualization of nursing and midwifery workloads.

TrendCare acuity studies are conducted across a large sample size and a wide cross-section of nursing and midwifery services with variations in models of care. A wider range of acuity indicators can be developed from these studies to provide a reasonable average for any group of patients within the same patient type. All TrendCare sites are invited to participate in timing studies and benchmarking. It is through these ongoing activities that the TrendCare acuity system is evaluated and updated so that it remains valid and reliable for all patient types.

In 2020 the TrendCare acuity system includes timings for a total of 226 uniquely defined patient types. These patient types were defined based on significant variations in nursing resource requirements and include: 34 medical, 37 surgical, 42 pediatric, 15 maternity, 6 neonatal, 24 mental health, 15 critical care, 11 rehabilitation, 6 sub-acute, 8 residential care, 10 emergency, 7 mass casualty, 6 community and 7 other specialty patient types. Detailed definitions, provided in help screens, clearly define all indicators and related variables for each patient type maximizing objectivity in relation to categorization (user selection of indicator variables). Individual patient characteristics are accounted for by having specific acuity indicators that collectively contribute to higher weightings for complex patients.

## *Discussion*

Staffing methodologies provide an estimate for nursing resource requirements, some more accurately than others. It is evident from this case study that achieving the best possible degree of accuracy required many years of research and development to cover the numerous clinical specialties, identify all relevant staffing areas and to account for the vast number of confounding variables influencing nursing service demands. Due to the many ongoing clinical and technical changes occurring over time, it was also necessary to continue this research and development effort to ensure continuing relevance, validity and reliability.

It became apparent that it was necessary to collect data detailing nursing and midwifery hours required for care, relative to patient types, within specific wards for each eight hour period of the day, for at least a 12 month period, to be able to identify seasonal trends. TrendCare data is extremely valuable as it facilitates management to consider various options to improve efficiency in terms of effective nursing and midwifery workforce management to match anticipated demand variations.

The goal of a patient dependency system is to identify the predicted staffing hours required within which complete care is expected to be delivered safely and effectively. For appropriate staffing to be available on the day, it is necessary to come as close as possible to estimating the future care requirements for each individual patient for the next 24 hr period. TrendCare was designed to provide the best possible estimate to minimize significant variations that compromise patient care and safety or add unnecessary cost. This has been achieved in the TrendCare methodology by differentiating patient types and their relevant care resource requirements for three eight hour periods of the day through large sample timing studies.

The TrendCare Patient Acuity and Workload Management System is now a mature, validated solution extensively utilized by nursing and midwifery services in Australia, New Zealand, England, Ireland, Thailand and Singapore hospitals. The system is used to determine patient acuity measures and nursing and midwifery hours required for care. Hours required for care are predicted for each period 24 hr in advance and actualized at the end of each 8 hr period. Many of the research activities undertaken have been conducted in collaboration with universities, unions, colleges and nursing advisory groups nationally and internationally to encourage external reviews and feedback on the integrity and reliability of the system in the healthcare environment.

The software solution provides a scientific, evidence-based approach for establishing the required level of staffing for a nursing service. The acuity measurement tool within the system is a form of automated reasoning that is underpinned by scientifically tested algorithms and weighted clinical indicators that have been identified as having an impact on patient/nurse contact time and nursing workloads. Making consistent use of data, process standards and strict version control has enabled continuous system improvements and updates to be made available to all users.

### Validation and endorsement

A four year comprehensive 2005 comparative review of the TrendCare patient dependency/ acuity system with the use of the nurse/patient ratio staffing methodology used in Victoria, Australia [2] concluded that "TrendCare can provide fairer and more equitable workloads and at a lower cost than the Victorian mandated nurse/patient ratios."

In 2007 TrendCare was selected as the national product to measure patient acuity, manage nursing and midwifery workloads and assist with nursing workforce planning in Singapore. The system was implemented into all public acute hospitals and subsequently into all sub-acute hospitals. TrendCare was awarded an AustCham Business Award [3] for this project.

In 2008 TrendCare won both the Australian national and state Microsoft eHealth iAwards for innovation in IT development and the National ICT exporter of the Year Award for its unique clinical management system [4,5].

TrendCare is regularly validated as new timing studies are undertaken for various patient types on an almost continuous basis, the results of which are used for continuous improvement. The research process adopted is explained in some detail in a publication describing the 2012, 2013 and 2014 validation studies initiated in three countries to validate the acuity tools used for women in labor, antenatal women and postnatal mothers and babies. The results of this study demonstrated the importance of regularly validating the TrendCare category timings with actual timings of the care hours provided. It was evident from those findings that variances to models of care and a decreased length of stay in maternity units have increased midwifery workloads particularly during the night period [6].

In 2015, the Safe Staffing and Healthy Workplace Unit of the New Zealand Ministry of Health commissioned a significant review of their safe staffing methodology, which included a review of the TrendCare acuity tool used to provide the acuity data for their care capacity management workforce calculations. This acuity tool was described in the final report as "currently the only validated patient acuity tool available" [7]. In 2018 the Health Minister announced a safe staffing accord that provides a commitment to the Care Capacity Demand Management (CCDM) program that relies on TrendCare data which includes the ability to calculate safe staffing levels for nursing and midwifery services. As a consequence every public hospital in New Zealand will be using the TrendCare system [8].

In 2015, The Salford Royal NHS Foundation Trust, one of the most highly regarded trusts in the United Kingdom conducted an international search for a validated, evidence based, patient nurse dependency and workload management system for nursing services. As a result of this search, TrendCare was selected, trialed and subsequently implemented into the Salford Royal Hospital Trust.

In 2018 Salford Royal NHS Foundation Trust were able to demonstrate improved staffing levels, a decrease in patient harm incidents, significant ongoing cost savings and achieved an award of high commendation from the Care Quality Council for the management of human resources. TrendCare is now embedded as the workload management and workforce planning system for nursing services and Allied Health across all acute and sub-acute services within the Salford Royal Foundation Trust [9].

In 2016, TrendCare was selected by the Ministry of Healthcare Executives in Ireland, as the most appropriate system to measure the nursing hours per patient, per day, to assist with their review of nursing workforce requirements for the total country [10]. A major research project across 3 public hospitals was conducted in 2016/2017. The final Irish taskforce report [11] on staffing and skill mix for nursing, provides external comprehensive evidence of the relevance and scientific value of the TrendCare staffing methodology. The report also confirmed the capability of the system in relation to providing intelligent data to predict and manage nursing workloads and to proactively plan and cost nursing resources accurately and efficiently. Nurse user feedback relating to the Irish project was extremely positive in relation to the ease of use and value of the data generated [12]. In 2020 TrendCare was selected as the national Safe Staffing and Skill Mix system for the Republic of Ireland.

In 2019 The United Kingdom's National Institute for Health and Care Excellence endorsed Trendcare as a nursing and midwifery solution for safe staffing.

## Lessons learned

A sustainable nursing workload and workforce management system development, implementation and ongoing management and governance requires:

- Strong leadership, commitment and persistence.
- Continuing high level Government, Hospital Executive Management, Nursing and Midwifery services collaborative support over many years.
- Extensive professional and industrial support and collaboration.
- Transparency in relation to data collection processes, reporting, performance monitoring and use.
- Seamless interoperability with patient management systems, e-Rostering, HRM systems, temporary staffing management systems, operational dashboards and costing solutions.
- The adoption of an integrated information system supporting ease of data collection.
- Governance of adopted standards and support for regular system use reviews and updates.
- Continuing workforce education and annual inter-rater reliability assessments.
- Ongoing research and development for validation and reliability.
- Continued awareness of international changes in nursing and midwifery practices and other health factors impacting on nursing and midwifery workloads.

## References

[1] Lowe C. Prospective and retrospective patient nurse dependency system: developed, computerised and trialled in Australia. In: Grobe S, Pluyter-Wenting E, editors. Nursing Informatics'94, San Antonio. Amsterdam: Elsevier; 1994, p. 842.
[2] Plummer V. An analysis of patient dependency data utilizing the TrendCare system [Doctoral thesis]. Melbourne: Monash University; 2005.

[3] Australian High Commission, Singapore. Medrel AustCham business awards 07, Singapore: Commonwealth of Australia; 2007. [cited 15 May 2019]. Available from: https://singapore.embassy.gov.au/sing/MEDIARELAUSBUS07.html.

[4] iTWire. TrendCare wins at the National AIIA iAwards 2008. [cited 15 May 2019]. Available from: https://www.itwire.com/freelancer-sp-720/search.html?searchword=trendcare.

[5] Dembo D. Microsoft Health Partners Win AiiA Awards: microsoft. [cited 15 May 2019]. Available from: https://blogs.msdn.microsoft.com/nickmayhew/2008/05/28/microsoft-health-partners-win-aiia-awards/; 2008.

[6] Lowe C. Validation of an acuity measurement tool for maternity services. Int J Med Health Sci 2015;9(5):1417–24.

[7] Hendry C, Aileone L, Kyle M. An evaluation of the implementation, outcomes and opportunities of the Care Capacity Demand Management (CCDM) Programme—final report. Christchurch, NZ: NZ Institute of Community Health Care; 2015.

[8] Broughton C. Safe staffing concerns remain despite new pay deal for nurses. Stuff 2018. August 9.

[9] Roddy R. E-Rostering, acuity and dependency, a presentation to NHS Group; 2018.

[10] Drennan J, Duffield C, Scott AP, Ball J, Brady NM, Murphy A, et al. A protocol to measure the impact of intentional changes to nurse staffing and skill-mix in medical and surgical wards. J Adv Nurs 2018;74 (12):2912–21.

[11] O'Halloran S. Framework for safe nurse staffing and skill mix in general and specialist medical and surgical care settings in adult hospitals in Ireland: final report and recommendations. Ireland: Department of Health; 2018.

[12] Hefferman M. Data doesn't lie. Focus 2018;26(9).

# Index

Note: Page numbers followed by *f* indicate figures, *t* indicate tables, and *b* indicate boxes.

Printed in the United States
By Bookmasters